Sourdough Microbiota and Starter Cultures for Industry

Elaine Berger Ceresino
Grazina Juodeikiene
Susanne Miescher Schwenninger
João Miguel Ferreira da Rocha
Editors

Sourdough Microbiota and Starter Cultures for Industry

Springer

Editors
Elaine Berger Ceresino
RISE
Gothenburg, Sweden

Susanne Miescher Schwenninger
Research Group for Food Biotechnology
Zurich University of Applied Sciences
Wädenswil, Switzerland

Grazina Juodeikiene
Food Science and Technology
Kaunas University of Technology
Kaunas, Lithuania

João Miguel Ferreira da Rocha
Chemistry and Biochemistry
University of Porto
Porto, Portugal

ISBN 978-3-031-48603-6 ISBN 978-3-031-48604-3 (eBook)
https://doi.org/10.1007/978-3-031-48604-3

This Springer imprint is published by the registered company Springer Nature Switzerland AG
The registered company address is: Gewerbestrasse 11, 6330 Cham, Switzerland

Paper in this product is recyclable.

Preface

Sourdough has been baked for thousands of years and increasingly consumed in our societies. In our time, the sourdough process is a meeting point with the old civilizations, when people have learned to cultivate and mill cereals and mix it with water forming a dough that spontaneously ferments and is baked into bread. The development of sourdough breadmaking occurred alongside the development of languages, brewing of beer, and when the baking oven was dug into the ground.

Lactic acid bacteria are the dominating microflora in sourdough, but naturally occurring yeasts have a similar important role. The first sourdough in the ancient times was probably a result of spontaneously started processes caused by the microflora found naturally in cereals. Nowadays, even starter cultures are sophisticatedly used to uplift flavor, aroma, and texture in bread consumed worldwide.

"A special feature of *Fructilactobacillus sanfranciscensis* is that it uses an efficient and low-energy way to break down maltose." Sourdough process is a complex system constructed upon various factors which are detailed explained in this book for baking enthusiasts, R&D professionals, milling industries, and scientists who wish a deeper understanding of the entangled details of sourdough that results in unique and high-quality bread.

The book was carefully written by experienced members of the network SOURDOMICS (COST Action 18101- Sourdough biotechnology network toward novel, healthier, and sustainable food and bioprocesses) across the globe. We thank all contributors of this book for your patience and professionalism. We hope that the readers will be satisfied with the content found in this book and that may even contact us with any comments, questions, or suggestions of improvements regarding its content.

Gothenburg, Sweden	Elaine Berger Ceresino
Kaunas, Lithuania	Grazina Juodeikiene
Wädenswil, Switzerland	Susanne Miescher Schwenninger
Porto, Portugal	João Miguel Ferreira da Rocha

Contents

1 **Overview of Sourdough Microbiota** 1
Elaine Berger Ceresino, Călina Ciont (Nagy), and Oana Lelia Pop

2 **Sourdough Microbiota Diversity in Southern Europe** 21
João Miguel Ferreira da Rocha, Pasquale Russo, Vittorio Capozzi, Mariagiovanna Fragasso, and Francisco Xavier Malcata

3 **Sourdough Microbiota Diversity in Central Europe** 61
Katarzyna Dybka-Stępień, Anna Otlewska, and Anna Diowksz

4 **Traditional and Novel Applications of Sourdough Microbes in Northern Europe and the Baltic Countries** 93
Clarissa Schwab and Maria Florencia Bambace

5 **Microbiological and Biomolecular Methods Applicable to the Isolation, Characterization and Identification of Microbial Sourdough Strains** 131
Antonio A. Câmara Jr, Larissa P. Margalho, Wilson J. F. Lemos Jr, Vanessa S. de Oliveira, André F. Guerra, and Anderson S. Sant'Ana

6 **Undesirable Microorganisms in Cereal Flours and Their Impact on the Stability of Sourdough Microbiome** 169
Anna Otlewska, Katarzyna Dybka-Stępień, and Katarzyna Rajkowska

7 **Metagenomics Analysis of Sourdough Microbiota** 211
Wilson J. F. Lemos Jr, Antonio A. Câmara Jr, Larissa P. Margalho, Vanessa S. de Oliveira, André F. Guerra, and Anderson S. Sant'Ana

8 **Metabolic Engineering of Lactic Acid Bacteria and Yeasts for the Production of Compounds with Industrial Applications** 223
Birsen Yılmaz, Hande Mortaş, Semra Navruz Varlı, and Duygu Ağagündüz

9 **Immobilization Techniques Applicable to Sourdough** 257
Zeynep Agirbasli, Mensure Elvan, and Sebnem Harsa

10 **Sourdough-Based Starter Cultures for Fermentation in Agri-food Industry** 281
Gabriel Albagli, Lillyan Nathalia Brito Silva, Nathália Moura Nunes, Daniel Perrone Moreira, Priscilla Filomena Fonseca Amaral, and Priscilla Vanessa Finotelli

11 **Sourdough Lactic Acid Bacteria as Functional Microorganisms for Technological Purposes** 309
Zühal Alkay and Enes Dertli

12 **Yeast Strains from Sourdough as Potential Clean-Label Starters for Fermentation Processes** 343
Oana Emilia Constantin, Bogdan Păcularu-Burada, and Gabriela-Elena Bahrim

13 **Health Promoting Features of Sourdough Lactic Acid Bacteria** 379
S. Paramithiotis, L. Bosnea, and M. Mataragas

14 **Rope and Mould Spoilage in Breadmaking – Risk Analysis** 395
M. Piotrowska, A. Nowak, and A. Koziróg

15 **Impact of Sourdough Microbiota on FODMAPs and ATI Content in Bakery Products** 425
Sabrina Geisslitz and Katharina Anne Scherf

16 **Recent Advances in the Use of Sourdough Fermentation to Improve the Quality of Gluten-Free Bakery Products** 461
Anıl Alper and Aylin Altan

Index 493

Chapter 1
Overview of Sourdough Microbiota

Elaine Berger Ceresino, Călina Ciont (Nagy), and Oana Lelia Pop

1.1 Overview of Sourdough and Sourdough Microbiota

Sourdough bread is now a worldwide phenomenon. Once sourdough bread is still widely manufactured at the farm level across Europe, it is highly appreciated by consumers for its distinct flavor, texture, healthy attributes, and natural origin (Garnweidner-Holme et al. 2022). Traditional breadmaking, which involves a long non-controlled fermentation process with unrefined whole-grain flour made of a variety of locally available grains and root grains, has boomed in the recent years. Opposing to mass production of bread from refined flour, which is added of commercial yeast starters and additives, sourdough revival has created a demand for healthy and aromatic bread uplifting our standard for bread quality (Marklinder 2015; Wang et al. 2023). Bread is one of the most ancient and yet relevant types of food that are fundamental for our diet, with an intake of 100–200 g/day per capita in the countries consuming Western-like diets (Wang et al. 2023). Sourdough bread is also one of the oldest types of bread in the world with its origins stemming from North Africa, reportedly dating back to 1550 B.C. (Samuel 1996). In ancient times, the science of obtaining leavened sourdough bread was carefully crafted throughout the years by the Egyptians using mainly wheat traditional fermentation (Canella-Rawls 2012). Despite of a rich history, only recently sourdough had its microbial diversity studied and associated with the superior quality of traditional bread (Novotni et al. 2021).

E. B. Ceresino (✉)
RISE, Gothenburg, Sweden

C. Ciont (Nagy) · O. L. Pop
Department of Food Science, University of Agricultural Sciences and Veterinary Medicine, Cluj-Napoca, Romania

Molecular Nutrition and Proteomics Laboratory, Institute of Life Sciences, University of Agricultural Sciences and Veterinary Medicine, Cluj-Napoca, Romania

E. B. Ceresino et al. (eds.), *Sourdough Microbiota and Starter Cultures for Industry*, https://doi.org/10.1007/978-3-031-48604-3_1

Sourdough, sourdough starter, levain, chef, lievito madre, massa mãe, mum, are some of the names given to sourdough (Jakobsson 2021). The mixture of flours and water fermented mainly by lactic acid bacteria (LAB) and yeasts is internationally called sourdough (De Vuyst and Neysens 2005), and this term will be utilized in this book. The traditional sourdough leads, through a spontaneous fermentation process, to the natural selections of microorganisms. The optimal and complete sourdough fermentation is obtained through repeated turnovers where the previous dough with its unique microbiome starts a fermentation process in a new mixture of water and flour. This process is called "backslopping" (Marklinder 2015). Flour, water, and ingredients from specific locations are the main sources of the sourdough microbial diversity originating distinguishable breads in a variety of quality parameters, such as aroma and taste profile (Suo et al. 2021). The positive attributes of sourdough, such as superior texture, flavor and nutritional properties are largely credited to the plant-associated microbiota (Novotni et al. 2021).

During the flour fermentation, in the "fermentome", the apparent volcanic fermenting mixture hides a rather stable microbial ecosystem. Such ecosystem is a result of metabolic changes and adaptations performed by the microorganisms due to the influence of the surroundings (Khlestkin et al. 2022). The metabolic activity of these microorganisms produces metabolites that are appealing to foods, such as organic acids, aroma compounds and exopolysaccharides, enhancing the flavor and texture of bread (Hansen and Schieberle 2005; Pételet al. 2017). Initial grain- and flour-derived *Entereobacteriaceae* including *Enterobacter* and *Cronobacter* are replaced by more acid tolerant bacteria *Leuconostoc* and *Weissella* spp. until *Lactiplantibacillus plantarum* and *Levilactobacillus brevis* or *Limolactobacillus fermentum* establish themselves in the mature dough (Novotni et al. 2021). Processing factors, such as temperature, backslopping method, and pH seem to also greatly influence the microbiome of the dough (Martín-Garcia et al. 2023) and are well described in other chapters of this book.

The first published studies on sourdough were performed by Wolfin in 1884 who was the first scientist to associate the dough acidification with the presence of lactic acid bacteria and yeasts (Röcken and Voysey 1995). Further studies on sourdough microbiota were only continued in 1970 utilizing sourdough from the San Francisco Bay area (Sugihara et al. 1970). Sugihara and colleagues had a pivotal role in showing that the dominant yeast in sourdoughs from that region was the non-spore forming *Kazachstania exiguus (*sin. *Saccharomyces exiguus,* anamorph *Candida holmii, prev. Torulopsis holmii)* and the LAB *Fructilactobacillus sanfranciscensis* was the dominant bacteria (Sugihara et al. 1971).

Yeast and bacteria exist in a ratio of 1:100, with the first mainly serving as leavening agent by producing carbon dioxide as a by-product of their metabolism. The bacteria play a unique role in lowering the dough pH by producing organic acids as by-products of their metabolism, besides volatile organic compounds, among other bread quality-enhancing substances (Calvert et al. 2021).

1.2 Bacteria and Yeasts in Sourdoughs

Sourdough hosts a rather modest number of 100 species of microorganisms (Fraberger et al. 2020). However, once the fermentome is stabilized, the mature sourdough microbiota is composed of only a few species of yeast and LAB. Yeasts are unicellular, acid-tolerant, facultative anaerobic fungi that undergo vegetative growth by budding or fission, occurring in a higher number in the sourdoughs. They are fermentative, facultative fermentative, or respirative (De Vuyst et al. 2017). Lactic acid bacteria (LAB) are a group of Gram-positive microorganisms naturally occurring in fermented food products and used as probiotics. They have as main characteristic the ability of synthetizing lactic acid as the primary product of the carbohydrate fermentation. Other characteristics comprise non-mobility, being anaerobic or microaerophilic. LAB function well in various environments, being able to survive and replicate in condition of low pH (4.5–7.0) and can grow in a temperature spectrum of 10–45 °C, with salinity up to 6.5% (Garbacz 2022).

The most common methods for performing the microbiological characterization of sourdoughs are culture-dependent methods, followed by culture-independent methods, a combination of both, and genetic sequencing (Arora et al. 2021; Calabrese et al. 2022; Landis et al. 2021). Using 16S and ITS rRNA amplicon sequencing, Landis et al. (2021) revealed that many mature communities were dominated by a single yeast and few bacterial species in a proportion of three bacteria to one yeast. However, this proportion may vary as in sourdough communities the interactions between yeast and bacteria have an enormous impact on the formation of a microbial consortia (Banik et al. 2019; Calvert et al. 2021).

1.2.1 Process-Associated Microorganisms in Sourdough

Lactic acid bacteria (LAB) of the *Lactobacillus* genus, now belonging to the family *Lactobacillaceae* include homofermentative and heterofermentative bacteria ubiquitous in sourdoughs. They have a safe tradition of food use and are awarded qualified presumption of safety (QPS) or generally regarded as safe (GRAS) by the European Food Safety Authority and US Food and Drug Administration, respectively (Qiao et al. 2022). It is worthy to mention that LAB nomenclature has been recently revised implicating in the change of many consolidated species names. The taxonomic reorganization of the lactic acid bacteria reclassified over 300 species in 7 genera and 2 families into one family, the *Lactobacillaceae*. This family comprises 31 genera including *Lactobacillus*, *Paralactobacillus*, *Pediococcus*, *Weissella*, *Fructobacillus*, *Convivina*, *Oenococcus*, *Leuconostoc* and other 23 new genera of the previously *Lactobacillus* species (Qiao et al. 2022; Zheng et al. 2020).

The genus *Lactobacillus* is the most abundant in sourdoughs, with 82 species being detectable. Within the family *Lactobacillaceae*, species like *Lactiplantibacillus plantarum*, *Leviactobacillus brevis*, *F. sanfranciscensis* and *Limosilactobacillus fermentum* are commonly found (Arora et al. 2021).

Sourdoughs are generally classified in 4 different types according to the technological process employed, which directly impacts the predominant microbiota in the doughs. Spontaneous sourdoughs that are inoculated with flour and water have plant-associated microbiota as fermentation starters. Before the 1900s, sourdough was produced using a large variety of ingredients, and not only cereals. In this type of sourdough, also called sourdough type 0, the spontaneous fermentation process is assured by LAB and yeast strains from the ingredient's wild microbiota. In sourdough type zero, also known for sponge doughs or pre-doughs, *Lactiplantibacillus plantarum* and *Levilactobacillus brevis* growth is favored, yielding in acidic environment and good aroma (Marklinder 2015). Ripari et al. (2016) have evaluated whether the microbiota of the ingredients had a major role in sourdoughs type 0. The authors have reported that the role of the inocula is minimum as the doughs mature to harbor a stable microbiota, except for *Acetobacter* spp. which has been identified in sourdoughs started with *Malus domestica* (apple) flowers.

Because the sourdough obtained in uncontrolled conditions could be sticky, sour, and bitter, Ernst Böcker, through his pioneering work, developed a new concept on sourdough fermentation process in 1908 in Minden, Germany. His first culture of sourdough starter became the cornerstone for many other patents, such as pumpable sourdough pastes or gluten free oat bread. Today's most common bakery's sourdough fermentation is initiated by *Fructilactobacillus sanfranciscensis* using Böcker's method, consisting of adding the starter culture to the mixture of flour and water. An effective availability of maltose, dough pH-sensitivity and a temperature of around 28 °C favors the propagation of *F. sanfranciscensis* in the sourdough process. This starter culture can be obtained commercially with the name Böcker *Reinzucht-Sauerteig*, which represents over 80% of the market in Germany (Marklinder 2015).

Sourdough type I, which is defined as a dough which ferments totally without the addition of baker's yeast, is often carried out in three-step process over 24 h. More specifically, type 1 refers to those processes that are initiated spontaneously and suffers a repeated reinoculation of 5 to 20% (m/v), using a new batch of flour and water. The material that has been obtained from a preceding fermentation is called mother dough. Mother doughs that have been passed through generations results in various unique baked goods, such as San Francisco sourdough bread, traditional Italian sourdough breads and the German three-stage rye sourdough bread (De Vuyst et al. 2017, 2021; Marklinder 2015). In this type of sourdough prevails the bacteria *F. sanfranciscensis*, *Limosilactobacillus pontis*, *Levilactobacillus brevis*, *Limosilactobacillus fermentum* and *Fructilactobacillus fructvorans*. Even *Companilactobacillus mindensis* and *Levilactobacillus spicheri* might be present. The yeasts *Kazachstania humilis*, *K. exigua* and *Saccharomyces cerevisae* are also present in this type of sourdough (Marklinder 2015). The most common yeast species detected in sourdoughs belong to *Saccharomyces* spp., *Candida* spp., *Kazachstania* spp., *Torulopsis* spp., *Yarrowia* spp., *Wickerhamomyces* spp., *Kluyveromyces* spp., and *Pichia* spp. (De Vuyst et al. 2017; Meroth et al. 2003).

Various sourdough ecosystems where single yeast was dominant had *Saccharomyces cerevisiae* accounting for over 50% of fungal ITS reads in 77% of the evaluated sourdoughs (Landis et al. 2021). The *S. cerevisiae* strains' ubiquitous presence is probably due to the existence of wild strains in the flour or, because of the widespread utilization of baker's yeast utilization during sourdough breadmaking (Arora et al. 2021). There is evidence that the yeast strains of *S. cerevisiae* have been often present in spontaneous sourdoughs because they can rapidly metabolize simple carbohydrates such as maltose and glucose for their multiplication and alcoholic fermentation (Lahue et al. 2020).

In addition to *Saccharomyces cerevisiae,* other yeasts were abundant in world widely selected sourdoughs, such as *Kazachstania humilis, K. servazzii, K. unispora Wickerhamomyces anomalus*, *Pichia membranifaciens*, *Naumovozyma castellii, Candida sake* and *Saccharomyces bayanus*. *Kazachstania barnettii, K. exigua, K. naganishii, Candida quercitrusa, Nakaseomyces* spp., *Pichia mandshurica, Pichia kluyveri, Vanderwaltozyma yarrowii, Candida solani, Pichia norvegensis, Hanseniaspora uvarum, Candida santamariae, Candida tropicalis, Cyberlindnera fabianii, Torulaspora delbrueckii, Candida parapsilosis, Meyerozyma guilliermo, Trichomonascus ciferrii, Issatchenkia orientalis,* and *Dipodascaceae* spp. were less abundant among the samples (Landis et al. 2021), but still relevant candidates sourdough starters. The yeast species which are the most relevant starters cultures are *C. humilis, C. krusei, C. milleri, K. marxianus, M. guilliermondii, P. kudriavzevii, W. anomalus, K. exigua and of course, S. cerevisiae* (Arora et al. 2021).

Sourdough fermentation is a complex system where yeasts play a major role, and this is explained in detail in the Chap. 12 of this book. One important functionality of yeasts is to release fructose from the dietary fiber fructans, which is abundant in rye flour. When fructose becomes available to lactic acid bacteria, acetic acid is formed, which gives a balanced and desirable acidity in the bread. If the yeasts had not been there, alcohol would have been formed instead (Brandt 2007).

In type-II sourdoughs, the most acid- and heat-resistant bacteria of the flour-water mixture are selected. These are often single-dough processes where doughs are fermented for 15 to 20 hours. The main purpose of type-II sourdough is pH lowering and aroma formation, thus sourdough fermentation results from the addition of a starter culture. A selected starter culture that fits well in sourdough tanks should be carefully chosen, because it should withstand low pH values for up to five days and temperatures around 30–40 °C. *Limolactobacillus pontis*, *Limolactobacillus fermentum*, *Limolactobacillus panis* and *Limolactobacillus reuteri* are found in this environment (Marklinder 2015). *Limolactobacillus* received this name because of the property of most strains of this genus to produce exopolysaccharides from sucrose, therefore they are considered a slimy lactobacillus. *Limolactobacillus* found in sourdough are characterized by facultative heterofermentative behavior, thermotolerance and acid resistance (Zhang et al. 2020). The spontaneous yeast flora in sourdough is unable to remain active in the higher temperature and lower pH value, and therefore extra baker's yeast is often added at the final stage of the fermentation process (Marklinder 2015). The objective of producing this type of sourdough is to have them commercialized under the form of semi-liquid, heat-treated for dried products (De Vuyst et al. 2017).

Type III sourdoughs are dominated by freeze-dried starter cultures mainly intended for the bakery industry. Both backslopped starter culture-initiated or mixed sourdough fermentation processes can be used. In type III sourdoughs, the process is initiated with a LAB starter culture, which is then followed by traditional back-slopping. Even with the addition of a starter culture, competition between microorganism may occur, but generally, the microorganisms of the starter culture prevail. *Pediococcus pentosaceus*, *Lactiplantibacillus plantarum* and *Levilactobacillus brevis* are relevant here. Even in this case where often there is only one-step process, baker's yeast is often added during dough preparation (De Vuyst et al. 2017, 2021).

1.2.2 Human Manipulation

Food safety concerns have increased the need to establish systems that provide more comprehensive coverage and monitoring to maintain food products. Hazard Analysis and Critical Control Point (HACCP) implementation varies by country and is concerned with the allocation of resources and legislation-mandated guidelines. Cross-contamination of foods from many sources, such as a food handler's hands, cooking utensils, equipment, workplace, etc., causes the majority of food infections (Josephs-Spaulding et al. 2016). The biology of the individual who first manufactures the starter is a possible contributor to sourdough microbiota. This is certainly relevant to the bacteria in the starter, including many species connected with the human microbiota (Modi et al. 2021). To better understand the impact of human manipulation on sourdough microbiota, the possibility of flour and hands action as contamination sources for the starting community was examined. Two species, *L. plantarum,* and *L. spicheri,* were found on the bakers' hands. Sourdoughs have been shown to have the same bacteria on the operator's hands. Handwashing procedures might help to avoid food contamination to a large extent (Reese et al. 2020).

To investigate the biopreservation impact of sourdough, survival curves of hygiene potential pathogens that produce mouldiness and tackiness were examined at traditional Lavash bread following intentional flour contamination with 4 log CFU/g *B. subtilis* and 3 log CFU/g *Neurospora sitophila.* The finding showed that increasing the time and temperature of sourdough fermentation reduced bacteria's survival, but the crust's roast aroma did not have the same intensity (Seyed and Alireza 2011). A pH of $\leq$4.2 inhibits harmful bacteria from growing, and baking inactivates any vegetative microbial form (Josephs-Spaulding et al. 2016).

The relevance of microbiota variety on bodies and in environments related to sourdough is highlighted by this sequential movement of bacteria between starters and operators (Seyed and Alireza 2011).

1.3 Microbial Succession

The variety of microorganisms of mature sourdoughs was previously examined, revealing over 60 LAB species from the genus like *Leuconostoc*, *Pediococcus*, *Enterococcus*, *Weissella*, and, in particular, *Lactobacillus* (Gobbetti et al. 2016; Minervini et al. 2014). For yeast microbiota, six yeast species (*Saccharomyces cerevisiae*, *Kazachstania exigua*, *Candida humilis*, *Pichia kudriavzevii*, *Torulaspora delbrueckii*, and *Wickerhamomyces anomalus*) are often found in sourdoughs (Boudaoud et al. 2021; Landis et al. 2021). Several of the challenges were detailed, such as the unregulated succession of lactic acid bacteria owing to natural fermentation, and the technological limits in measuring them precisely and promptly (Oshiro et al. 2020). However, the microbiota and metabolite formation variations while sourdough fermentation is uncertain.

According to previous research, in the maturation of sourdough, the action of one species progressively gives way to another and appears to go through three stages (Baek et al. 2021; Oshiro et al. 2019, 2020; Taccari et al. 2016). These three phases are controlled by the genus (i) *Enterococcus*, *Lactococcus*, and *Leuconostoc*; (ii) *Lactobacillus*, *Pediococcus*, and *Weissella*; and particular (iii) LAB species such *Lactobacillus brevis*, *Lactobacillus fermentum*, and *Lactobacillus plantarum* (Baek et al. 2021; Oshiro et al. 2019). As shown in Fig. 1.1, in the initial fermentation stage, there was a significant increase in LAB and yeast counts and an increase in amino acids, glucose, lactic acid, and acetic acid (Zhang et al. 2019a). For example, in type I sourdough, according to Baek et al. (2021), the dominant bacteria was *Fructilactobacillus sanfranciscensis* (<5 log CFU/g sourdough), *Lactobacillus curvatus* (9.5 log CFU/g sourdough), and *Lactobacillus brevis* (6.5 log CFU/g sourdough). As data analysis revealed, the diverse *F. sanfranciscensis* strains considerably impact the growth kinetics and the proportion of bacterial/yeast cells, whereas *K. humilis* adapts to the bacterial strain (Altilia et al. 2021). Then, the sourdough bacteria detected on the first day (i.e., *Lactococcus lactis*, *Leuconostoc*

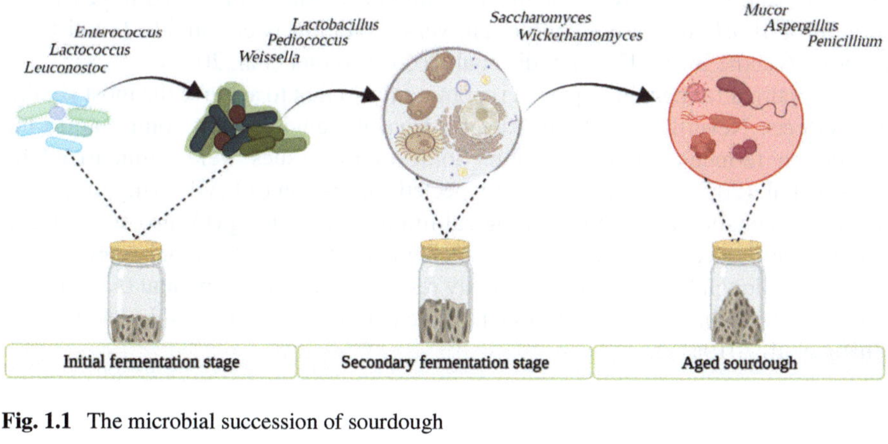

Fig. 1.1 The microbial succession of sourdough

holzapfelii/citreum, *Candida silvae,* and *Wickerhamomyces anomalus*) were gradually substituted by species more adapted to the sourdough biota (*Lactobacillus brevis, Lactobacillus alimentarius/paralimentarius*, *Saccharomyces cerevisiae*) (Taccari et al. 2016). To explain the final stage of the LAB community succession process, the influence of pH was examined. Under lower pH circumstances, the succession from *Weissella* to *Lactobacillus* was expedited. However, in the wheat sourdough simulation medium subcultures, the *Lactobacillus*-dominant microbiota found in the sourdough subculture owing to *F. sanfranciscensis* expansion was not validated (Oshiro et al. 2020). So, in final, the microbial succession stages leading up to the mature sourdough stage are spontaneous selection processes in which *Lactobacillus spp.* strains that can resist and adjust to severe acidity (final dough pH 3.49–3.57), temperature (50–60 °C), and additional specialized circumstances better control the ecosystem (Montemurro et al. 2020; Zhang et al. 2019a, b).

During the secondary fermentation stage, the lactic acid bacteria count gradually decreased, but the yeast count expanded (Zhang et al. 2019a, b). Since yeasts prefer maltose, low rates of maltose consumption by homofermentative LAB are likely to preserve the cohabiting *S. cerevisiae* population. The relationship between the LAB community's maltose degradation capability and yeast development shows the increased accessibility of maltose via maltose phosphorylase in the LAB community (*F. sanfranciscensis*, *L. brevis*, *L. paralimentarius*) triggers yeast growth in sourdough. Subsequently, glucose is transported to the surface and made available to grow yeast (Carbonetto et al. 2018; De Vuyst et al. 2002; Landis et al. 2021). During the fermentation phase, PCR-DGGE (polymerase chain reaction denaturing gradient gel electrophoresis) method revealed that the most common yeast species were *Saccharomyces spp.* and *Candida spp.* (Baek et al. 2021; Zhang et al. 2019a, b). On the other hand, it has been identified in certain types of sourdough, the dominant yeast is *W. anomalus*, but it began to decline after 21 backslopping phases. Thus, it results that *W. anomalus* could have a reduced chance of surviving in the sourdough community (Oshiro et al. 2019). Regarding yeast cell densities, a gradual increase in viable counts to log CFU/g was found in the artisan bakery, and laboratory sourdoughs studied, oscillating from mean values of 7.47 to 7.97 (Taccari et al. 2016). In sourdough, LAB is the most common bacteria, with a LAB to yeast proportion of 100:1. In mature sourdough, yeast counting is commonly found to be between 6 and 7 log CFU/g (Landis et al. 2021; Ozgolet et al. 2022).

The optimal fermentation period is commonly sought to achieve the most significant effects. Nevertheless, many bacteria from the environment contribute to the sourdough fermentation progression. Some of the issues were summarized by Oshiro et al. (2020), including the unexpected succession of LAB owing to spontaneous fermentation and the technological limits in measuring (Oshiro et al. 2020). The bacterial succession and metabolic characteristics and bacterial reproduction during sourdough fermentation are highly responsive to environmental biota, which can drive the stability of the sourdough fermentation ecosystem (Oshiro et al. 2020; Zhang et al. 2019a, b).

1.4 Benefits Associated with Sourdough Microbiota and Fermentation

1.4.1 *Bread Quality: Flavor, Aroma, Texture*

The metabolic activities of LAB, such as lactic acid fermentation, acidification, proteolysis, and flavor component formation, are primarily responsible for sourdough's distinctive benefits, which favorably affect nutritional, functional, and sensory qualities (Zhang et al. 2019a, b). The inclusion of sourdough in the bread fermentation improves finish product quality, and many of its structural characteristics are dependent on the sourdough resident bacteria and its biochemical processes (Fujimoto et al. 2019; Ripari et al. 2016).

Since the **flavor** is one of the most decisive factors influencing the customer acceptance of bread, considerable studies have been conducted on the volatile profiles of sourdough-fermented bread (Demirkesen-Bicak et al. 2021; Irakli et al. 2019; Liu et al. 2020; Olojede et al. 2022). In traditional sourdough, *F. sanfranciscensis*, a heterofermentative bacteria, has been regarded as an essential autochthonous microorganism (Raimondi et al. 2017; Zhang et al. 2019a, b). Ethanol was provisionally identified as the most abundant volatile component, followed by alcohols (1-propanol, 1-pentanol, 2-methylpentanol), aldehydes (hexanal, 2-methylbutanal, benzaldehyde, nonanal), esters (methyl acetate, 2-phenylethyl acetate, octyl acetate), and ketones (2-pentanone, 2-octanone) (Pétel et al. 2017; Van Kerrebroeck et al. 2018). Acetoin and ethyl acetate were also detected, although these compounds couldn't be separated (Van Kerrebroeck et al. 2018). Another study result suggests that adding sugars (pentose and hexose) to rye flour at 1% during sourdough fermentation forms a high level of propionic acid, which produces a strong odor like acetic acid in rye bread (Buksa 2020). Moreover, fermentation duration and temperature impact the volatile content (Xu et al. 2019). As an example, after 12 hours of fermentation, aldehydes and ketones were produced, including 2-nonenal, octanal, 2-octenal, 2-heptenal, 3-octen-2-one, 2,3-butanedione, and 1-octen-3-ol, 2-butoxyethoxy. The results suggest that most of these substances were formed from the oxidation of flour lipids (Diana et al. 2014; Liu et al. 2020). In another study, it was found that at 28 °C temperature for 24 h fermentation, carboxylic acids and alkanes predominated among the sourdough volatile components. At 35 °C, a similar production occurred, but the pH changed from 4.29 to 3.69, and the total titratable acidity increased from 11.78 to 17.42 mL 10 g^{-1} NaOH. Furthermore, peptides and free amino acids are involved in Maillard reactions and produce chemicals that affect bread's flavor. High temperatures in baking cause the bread to retain certain components from the Maillard reaction-based process, including acrylamide, glyoxal, methylglyoxal, heterocyclic aromatic hydrocarbons, and furfural (Ozgolet et al. 2022). The optimal retention of Maillard reaction products is 3%, resulting in bread with more baking flavor and the most significant assessment without changing the bread's characteristic volume, texture, or color (Wu et al. 2018). According to Pacyński et al. (2015), the absence of the principal bread odorants, such as pyrazines and 2-acetyl-1-pyrroline related to the Maillard reaction, is the primary factor that makes the taste of gluten-free bread more acceptable for a customer (Pacyński et al. 2015). The use of sourdough made with 8 log cfu/g *P. pentosaceus*, or *W. confusa*, added at a 20% for gluten-free sourdough fabrication, was tested to improve celiac disease products (Olojede et al. 2022; Rinaldi et al. 2017). The control sample was bread without sourdough. The *P. pentosaceus* sourdough bread had higher amounts of crude protein (5.31%), ash (2.44%), and dietary fiber (14.43%) than the control bread, crude protein (4.55%), ash (2.12%), and dietary fiber (13.25%). But, the sample created with sourdough with *W. confusa* had the most significant specific volume at 2.50 cm^3/g, followed by control bread without sourdough at 2.46 cm^3/g and *P. pentosaceus* at 2.16 cm^3/g. Taste, texture, scent, and overall acceptability were more remarkable in the sourdough samples than in control, with *P. pentosaceus* having the most significant general acceptance (6.56) (Olojede et al. 2022).

As previously mentioned, the main metabolic activities of sourdough include acidification (LAB), leavening (yeasts and heterofermentative LAB), and **aroma** production (yeasts and LAB). *Meyerozyma guilliermondii* and *Pichia kudriavzevii*, two yeast isolates, were mixed with *F. sanfranciscensis* for sourdough fermentation. The bread aroma increased when sourdough fermentation with a combination of starters occurred at 10 °C (Xu et al. 2020a). On the other hand, bread prepared using *Saccharomyces cerevisiae* and *Kazachstania exigua* yeasts in sourdough fermentation exhibited a more complex chemical aroma profile, a more varied crumb texture, and a darker color than bread made with industrial baker's yeast only. But by combining these yeasts with LAB, such as *L. plantarum*, *L. brevis*, *L. rossiae*, and *L. casei*, the bread aroma characteristics and crumb structure were more distinctive, with substances associated with sour aromas and chosen by consumers (Winters et al. 2019). Other bacteria, like *L. paracasei*, *L. lactis*, and *L. crustortun*, have shown a beneficial impact on bread aroma, according to consumer preference surveys and gas chromatography/mass spectrometry (Winters et al. 2019; Wu et al. 2022). This promoted the synthesis of aroma-active compounds like ethyl acetate (fruity), ethyl lactate (caramel-like), hexyl acetate (fruity), (E)-2-nonenal (fatty) and 2-pentylfuran (fruity) (Xu et al. 2020a, b; Yan et al. 2019). *P. pentosaceus* was shown to have a strong beneficial connection with the synthesis of aromatic compounds such ethyl caprylate, phenylethyl acetate, 2-methyl-1-propanol, isoamyl alcohol, and phenethyl alcohol in sourdough fermentation (Montemurro et al. 2020).

F. sanfranciscensis can also affect bread **texture** by affecting the ability of starch to gel and swell and the ability of flour to paste. The pasting characteristics of wheat flour slurry were primarily attributable to varied starch granule behaviors in the composite system (Zhang et al. 2018). The exopolysaccharides (EPS) independent impact of *F. sanfranciscensis* may be attributed to gluten modification mediated by glutathione reductase, whereas the interaction between high molecular weight EPS and starch-gluten or starch-starch connections may be responsible for delayed staling (Xu et al. 2020a, b). The generation of EPS by LAB variety in sourdough is another of their vital metabolic activities, which might impact the technical qualities of sourdough and sourdough bread (Chen et al. 2016; İspirli et al. 2020). The effect of α-glucan with (α1–3) and (α1–6) glycosidic linkages on sourdough rheology and textural characteristics of bread was investigated in the α-glucan producer *L. reuteri* E81 and the glucan deficient *L. reuteri* gtfA mutant strain. *In vivo*, glucan production during fermentation increased the rheological properties of sourdough and resulted in a more elastic sourdough, which was not observed in non-glucan sourdough (İspirli et al. 2020). Moreover, adding 30% sourdough from LAB EPS+ *Leuconostoc lactis* and *L. curvatus* extended bread shelf life, keeping a higher moisture content and increasing mechanical characteristics during storage (Torrieri et al. 2014). Also, increased proportions of organic acids, mainly lactic and acetic acids, which have antibacterial action and are efficient against fungal spoilage, may be responsible for raising the shelf life of sourdough bread (Axel et al. 2015; Quattrini et al. 2019). By incorporating 4% sucrose into sourdough fermentation with *L. brevis*, the mold-free shelf life of *A. niger*-challenged bread was extended by six days (Quattrini et al. 2019; Taglieri et al. 2021). The incorporation of *L. hammesii* in sourdough and ricinoleic acid (0.15%) extended the mold-free shelf-life of wheat sourdough bread by 7–8 days (Quattrini et al. 2019). Quinoa sourdough fermented with the antifungal *L. amylovorus* was compared to a non-antifungal strain. The 4-hydroxyphenyl acetic acid, phloretic acid, 3-phenyl lactic acid, and hydroferulic acid were significantly higher than in the quinoa sourdough. The inclusion of *L. amylovorus* fermented sourdough increased bread texture in terms of specific volume and crumb hardness, the main challenges in the quality of agglutinative bread (Axel et al. 2015). Another change between bread fermented using only yeasts and sourdough bread is the decreased pH, resulting in acidification or souring. Eleven hours at 25 °C and 9 h at 35 °C were necessary to produce sourdough with the optimal pH (4.2–4.3). An increased elastic modulus of dough at 25 °C was seen in sourdough (Alba et al. 2017). Chestnut flour prevented excessive acidification in dough and bread, resulting in decreased water holding capacity and increased crumb stiffness. Compared to the control, the volume of all loaves made with chestnut flour and/or sourdough was reduced. Still, the combination of chestnut flour and sourdough helped decrease crumb grain heterogeneity. Because of the reduced volume, adding sourdough and/or chestnut flour increased crumb hardness significantly. Even though the amylopectin fusion enthalpy was reduced, a significant decrease in the staling phenomena indicated by crumb hardness increase was detected only after five days of storage (Rinaldi et al. 2017).

The nutritional value and sensory qualities (aroma, taste, and texture) of bread may be significantly enhanced with the addition of sourdough.

1.4.2 Health Benefits Associated to Sourdough

Sourdough fermentation, in which bacteria and yeast work together, is thought to improve bread in several ways (including texture, taste, nutritional content, and shelf life) and reduce or eliminate the need for preservatives (Fig. 1.2) (Calasso

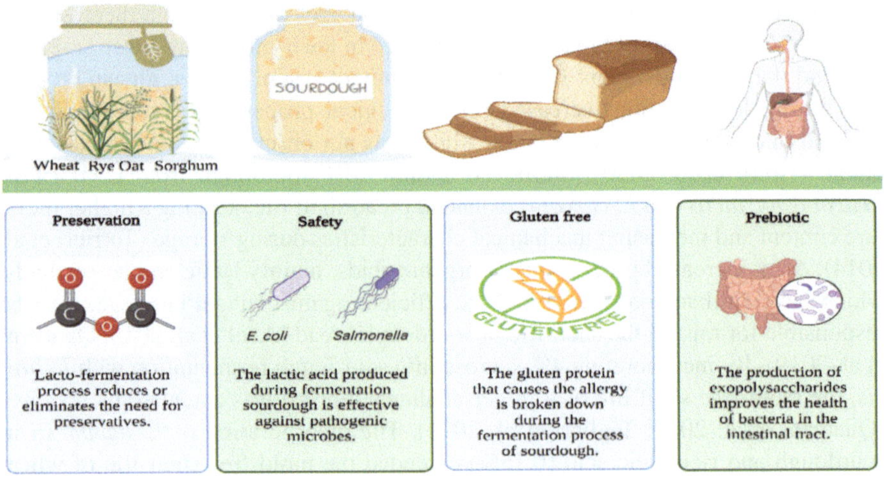

Fig. 1.2 Principle healthy benefits of sourdough

et al. 2018; De Vero et al. 2021; Diowksz et al. 2020; Modi et al. 2021). Sourdough bread's potential dietary value may be enhanced by the sourdough process's influence on the bread's beneficial components (Debonne et al. 2019; Nionelli et al. 2018). It has been shown that the sporangial development of *A. parasiticus*, *P. carneum*, *P. polonicum*, *P. paneum*, and *P. chermesinum* is significantly inhibited by the addition of 25% hop extract (*Humulus lupulus*) as a natural preservative with 7.0 log cfu/g LAB (*Enterococcus feacium*, *L. plantarum*, *L. brevis*, *L. helveticus*, *L. curvatus*, *P. pentosaceus*, and *P. acidilactici*). By use of hop-sourdough slowed the development of mold by 14 days, increased the bread's specific volume, and gave it a more attractive sensory profile, including the characteristic sourdough properties and a mild bitter/herbaceous sensation (Nionelli et al. 2018).

Research in this area has recently focused on how various preparation methods affect the bioactive qualities (total phenolic contents, total flavonoid contents, and antioxidant properties) of cereal products and their traditional role in improving gut flora (Lluansi et al. 2021; Smith et al. 2020). Examples include a study that looked at the impact of sourdough spelt fermentation on gut microbiota composition after exposing the grain to LAB for 0, 24, 48, 72, and 96 hours. The fermented spelt extracts added to the dough did not affect the development of *S. aureus*, *E. coli,* or probiotic bacteria *in vitro*. However, the spelt fermented for 96 hours was crucial in reducing the number of *S. typhimurium* and *L. innocua* (Pozzo et al. 2022). Antibacterial activity against *E. coli*, *Shigella*, and *S. typhimurium* was also discovered in LAB isolated from the fermented grain products *ogi* and *kunun-zaki* (Onipede and Fawole 2017). Bread's bioactive properties (total phenolic contents and antioxidant activities) were enhanced when hop (57.5%) and rice bran (13.5%) were added to the sourdough. Fortified sourdough bread with rice bran and hop had significantly higher antioxidant activity than sourdough bread (>1.2 times) and (>1.7 times) compared to the control (Irakli et al. 2019). Another two bread was compared, one made from 100% wheat flour, and another obtained from 70% wheat flour, 15% acha sourdough flour, and 15% bambara nut sourdough flour. The bread with acha and bambara nut sourdough flours showed higher DPPH: 143.65 mmol TEAC/100 g and ferric reducing antioxidant power FRAP: 512.91–565.19 mmol TE/ 100 g values than the wheat flour (DPPH: 102.47 mmol TEAC / 100 g and FRAP: 498.50 mmol TE/100 g) (Chinma et al. 2016). On the other hand, it has been suggested that a class of poorly digested carbohydrates known as fermentable oligo-, di-, mono-, and polyols (FODMAPs) may contribute to Irritable Bowel Syndrome (IBS) treatment (Laatikainen et al. 2017; Menezes et al. 2019). Menezes et al. (2019) results suggest that sourdough bread had more organic acids and polyols and fewer fructans, sucrose, fructose, and glucose than *S. cerevisiae*-fermented bread. The fructan reduction was between 69%–75%, showing that sourdough fermentation may be used to make low-FODMAP wheat baking goods. A pilot study was used to see how a lengthy dough fermentation (>12 hours) can impact IBS. However, neither sourdough or yeast-fermented bread significantly improves gastrointestinal symptoms or low-grade inflammatory indicators (Laatikainen et al. 2017). Their findings coincide with the results of other researchers (Calasso et al. 2018; Muir et al. 2019).

Following consumption of the bread, the starch is frequently swiftly digested and absorbed, resulting in hyperglycemia in those with insulin resistance syndrome (Amoah et al. 2021; Demirkesen-Bicak et al. 2021). The way glucans from oats lower cholesterol is by increasing the viscosity of digestive juices in the intestines. Oat bran sourdough bread might be a supply of β-glucan (0.5–1.3 g/100 g bread), but the baking process can degrade the β-glucanases. Cholesterol-lowering effects may be anticipated from the addition of 30% finely milled oat bran, which increases viscosity of digestive juices *in vivo* (Johansson et al. 2018). In addition, in a randomized crossover design, 12 people with diet-controlled type II diabetes ingested 50 g of available carbohydrates from four types of bread: 3-grain sprouted sourdough (SPR), whole-grain sourdough (WG), white sourdough (SD), and control white (WB). The glycemic response to SPR resulted in a significantly reduced iAUC (300.9 mmol/L 3 h) compared to all bread (WG-369.7 mmol/L 3 h; SD-378 mmol/L 3 h; WB-362.1 mmol/L 3 h). After the second meal, the iAUC of plasma insulin was significantly lower after SPR (9377 pmol/L 2 h) than SD (14,377 pmol/L 2 h) and WB (13,688 pmol/L 2 h) (Tucker et al. 2014). The treatment mechanism for diabetes using bread sourdough consists of inhibiting amylase activity (between 10% and 90%) and reducing starch. As a result, sourdough bread tends to be more suitable for diabetes (Diowksz et al. 2020).

As reported above, sourdough and bread's qualities, notably their nutritional properties, are significantly impacted by the temperature and time of fermentation. Sourdough, with a pH of 4.0 to 4.8, is more acidic than yeast bread, supporting the hydrolysis activity of the enzyme phytase (Mohammadi-Kouchesfahani et al. 2019; Nuobariene et al. 2015). The extracellular phytase activity of *W. confusa* and *P. pentosaceus* isolates in liquid medium ranged from 16.3 to 53.2 (U/ml) (Mohammadi-Kouchesfahani et al. 2019). Higher phytase activity (0.42 U/mg prot.) and cell growth (1.72 Log CFU/ml) were reported for *L. plantarum* CRL1964 isolated for sourdough. The enzyme activity was enhanced by pH 2.0–6.0, Co^{2+} (10 mmol/L), and temperature range (45–70 °C) (Sandez Penidez et al. 2020). Another healthful characteristic of sourdough was associated with prebiotic propriety (Baruah et al. 2017; Bockwoldt et al. 2021; Kim et al. 2022). The maximum dextran production (which indicated prebiotic properties) by *W. cibaria* 9 Log cfu/ml was found at 3.26% d.w. in sourdough made with 10% rye bran (Baruah et al. 2017; Kim et al. 2022). Consequently, the probiotic bacteria *L. plantarum*, *L. acidophilus*, *B. animalis*, *B. bifidum*, and *B. infantis*, all increased from 6.9, 6.8, 6.2, 5.4, and 6.6 to 9.7, 9.6, 9.8, 7.6, and 9.7 Log cfu/ml, respectively, after 24 hours in the presence of inulin (Baruah et al. 2017).

Sourdough bread has been discovered to provide various health benefits, including increased nutrient availability and protection against illness and cancer-causing free radicals, thanks to antioxidant and antihypertensive actions (Calasso et al. 2018; Demirkesen-Bicak et al. 2021; Galli et al. 2018).

References

Alba C, Alberto Á, Leticia G, Domingo F, José LM, María ET (2017) Effect of fermentation on microbiological, physicochemical and physical characteristics of sourdough and impact of its use on bread quality. Czech J Food Sci 35(6):496–506. https://doi.org/10.17221/68/2017-cjfs

Altilia S, Foschino R, Grassi S, Antoniani D, Dal Bello F, Vigentini I (2021) Investigating the growth kinetics in sourdough microbial associations. Food Microbiol 99:103837. https://doi.org/10.1016/j.fm.2021.103837

Amoah I, Cairncross C, Merien F, Rush E (2021) Glycaemic and appetite suppression effect of a vegetable-enriched bread. Nutrients 13(12). https://doi.org/10.3390/nu13124277

Arora K, Ameur H, Polo A, Di Cagno R, Rizzello CG, Gobbetti M (2021) Thirty years of knowledge on sourdough fermentation: a systematic review. Trends Food Sci Technol 108:71–83. https://doi.org/10.1016/j.tifs.2020.12.008

Axel C, Rocker B, Brosnan B, Zannini E, Furey A, Coffey A, Arendt EK (2015) Application of Lactobacillus amylovorus DSM19280 in gluten-free sourdough bread to improve the microbial shelf life. Food Microbiol 47:36–44. https://doi.org/10.1016/j.fm.2014.10.005

Baek HW, Bae JH, Lee YG, Kim SA, Min W, Shim S, Han NS, Seo JH (2021) Dynamic interactions of lactic acid bacteria in Korean sourdough during back-slopping process. J Appl Microbiol 131(5):2325–2335. https://doi.org/10.1111/jam.15097

Banik A, Mondal J, Rakshit S, Ghosh K, Sha SP, Halder SK, Ghosh C, Mondal KC (2019) Amelioration of cold-induced gastric injury by a yeast probiotic isolated from traditional fermented foods. J Funct Foods 59:164–173. https://doi.org/10.1016/j.jff.2019.05.039

Baruah R, Maina NH, Katina K, Juvonen R, Goyal A (2017) Functional food applications of dextran from Weissella cibaria RBA12 from pummelo (Citrus maxima). Int J Food Microbiol 242:124–131. https://doi.org/10.1016/j.ijfoodmicro.2016.11.012

Bockwoldt JA, Fellermeier J, Steffens E, Vogel RF, Ehrmann MA (2021) beta-Glucan production by levilactobacillus brevis and pediococcus claussenii for in situ enriched rye and wheat sourdough breads. Foods 10(3). https://doi.org/10.3390/foods10030547

Boudaoud S, Aouf C, Devillers H, Sicard D, Segond D (2021) Sourdough yeast-bacteria interactions can change ferulic acid metabolism during fermentation. Food Microbiol 98:103790. https://doi.org/10.1016/j.fm.2021.103790

Brandt MJ (2007) Sourdough products for convenient use in baking. Food Microbiol 24(2):161–164. https://doi.org/10.1016/j.fm.2006.07.010

Buksa K (2020) Effect of pentoses, hexoses, and hydrolyzed arabinoxylan on the most abundant sugar, organic acid, and alcohol contents during rye sourdough bread production. Cereal Chem 97(3):642–652. https://doi.org/10.1002/cche.10280

Calabrese FM, Ameur H, Nikoloudaki O, Celano G, Vacca M, Junior WJ, Manzari C, Vertè F, Di Cagno R, Pesole G, De Angelis M, Gobbetti M (2022) Metabolic framework of spontaneous and synthetic sourdough metacommunities to reveal microbial players responsible for resilience and performance. Microbiome 10(1):148. https://doi.org/10.1186/s40168-022-01301-3

Calasso M, Francavilla R, Cristofori F, De Angelis M, Gobbetti M (2018) New protocol for production of reduced-gluten wheat bread and pasta and clinical effect in patients with irritable bowel syndrome: a randomised, double-blind, cross-over study. Nutrients 10(12). https://doi.org/10.3390/nu10121873

Calvert MD, Madden AA, Nichols LM, Haddad NM, Lahne J, Dunn RR, McKenney EA (2021) A review of sourdough starters: ecology, practices, and sensory quality with applications for baking and recommendations for future research. PeerJ 9:e11389. https://doi.org/10.7717/peerj.11389

Canella-Rawls S (2012) Pão. Arte e ciência, 5th edn. SENAC São Paulo

Carbonetto B, Ramsayer J, Nidelet T, Legrand J, Sicard D (2018) Bakery yeasts, a new model for studies in ecology and evolution. Yeast 35(11):591–603. https://doi.org/10.1002/yea.3350

Chen XY, Levy C, Ganzle MG (2016) Structure-function relationships of bacterial and enzymatically produced reuterans and dextran in sourdough bread baking application. Int J Food Microbiol 239:95–102. https://doi.org/10.1016/j.ijfoodmicro.2016.06.010

Chinma CE, Anuonye JC, Ocheme OB, Abdullahi S, Oni S, Yakubu CM, Azeez SO (2016) Effect of acha and bambara nut sourdough flour addition on the quality of bread. LWT 70:223–228. https://doi.org/10.1016/j.lwt.2016.02.050

De Vero L, Iosca G, Gullo M, Pulvirenti A (2021) Functional and healthy features of conventional and non-conventional sourdoughs. Appl Sci 11(8). https://doi.org/10.3390/app11083694

De Vuyst L, Neysens P (2005) The sourdough microflora: biodiversity and metabolic interactions. Trends Food Sci Technol 16(1):43–56. https://doi.org/10.1016/j.tifs.2004.02.012

De Vuyst L, Schrijvers V, Paramithiotis S, Hoste B, Vancanneyt M, Swings J, Kalantzopoulos G, Tsakalidou E, Messens W (2002) The biodiversity of lactic acid bacteria in Greek traditional wheat sourdoughs is reflected in both composition and metabolite formation. Appl Environ Microbiol 68(12):6059–6069. https://doi.org/10.1128/AEM.68.12.6059-6069.2002

De Vuyst L, Van Kerrebroeck S, Leroy F (2017) Microbial ecology and process technology of sourdough fermentation. In: Sariaslani S, Gadd GM (eds) Advances in applied microbiology, vol 100. Academic, pp 49–160. https://doi.org/10.1016/bs.aambs.2017.02.003

De Vuyst L, Comasio A, Kerrebroeck SV (2021) Sourdough production: fermentation strategies, microbial ecology, and use of non-flour ingredients. Crit Rev Food Sci Nutr 1–33:2447–2479. https://doi.org/10.1080/10408398.2021.1976100

Debonne E, Baert H, Eeckhout M, Devlieghere F, Van Bockstaele F (2019) Optimization of composite dough for the enrichment of bread crust with antifungal active compounds. LWT 99:417–422. https://doi.org/10.1016/j.lwt.2018.10.020

Demirkesen-Bicak H, Arici M, Yaman M, Karasu S, Sagdic O (2021) Effect of different fermentation condition on estimated glycemic index, in vitro starch digestibility, and textural and sensory properties of sourdough bread. Foods 10(3). https://doi.org/10.3390/foods10030514

Diana M, Rafecas M, Quílez J (2014) Free amino acids, acrylamide and biogenic amines in gamma-aminobutyric acid enriched sourdough and commercial breads. J Cereal Sci 60(3):639–644. https://doi.org/10.1016/j.jcs.2014.06.009

Diowksz A, Malik A, Jaśniewska A, Leszczyńska J (2020) The inhibition of amylase and ACE enzyme and the reduction of immunoreactivity of sourdough bread. Foods 9(5). https://doi.org/10.3390/foods9050656

Fraberger V, Ammer C, Domig KJ (2020) Functional properties and sustainability improvement of sourdough bread by lactic acid bacteria. Microorganisms 8(12). https://doi.org/10.3390/microorganisms8121895

Fujimoto A, Ito K, Narushima N, Miyamoto T (2019) Identification of lactic acid bacteria and yeasts, and characterization of food components of sourdoughs used in Japanese bakeries. J Biosci Bioeng 127(5):575–581. https://doi.org/10.1016/j.jbiosc.2018.10.014

Galli V, Mazzoli L, Luti S, Venturi M, Guerrini S, Paoli P, Vincenzini M, Granchi L, Pazzagli L (2018) Effect of selected strains of lactobacilli on the antioxidant and anti-inflammatory properties of sourdough. Int J Food Microbiol 286:55–65. https://doi.org/10.1016/j.ijfoodmicro.2018.07.018

Garbacz K (2022) Anticancer activity of lactic acid bacteria. Semin Cancer Biol 86:356–366. https://doi.org/10.1016/j.semcancer.2021.12.013

Garnweidner-Holme L, Hallquist M, Watters SI, Gjøvik M, Pihl Fredriksen M, Jonassen S, Ravnanger I, Henriksen C, Myhrstad MCW, Telle-Hansen VH (2022) Attitudes towards and experiences with sourdough and baker's yeast bread amongst participants in a randomised controlled trial: a qualitative study. Food Nutr Res 66. https://doi.org/10.29219/fnr.v66.8839

Gobbetti M, Minervini F, Pontonio E, Di Cagno R, De Angelis M (2016) Drivers for the establishment and composition of the sourdough lactic acid bacteria biota. Int J Food Microbiol 239:3–18. https://doi.org/10.1016/j.ijfoodmicro.2016.05.022

Hansen A, Schieberle P (2005) Generation of aroma compounds during sourdough fermentation: applied and fundamental aspects. Trends Food Sci Technol 16(1):85–94. https://doi.org/10.1016/j.tifs.2004.03.007

Irakli M, Mygdalia A, Chatzopoulou P, Katsantonis D (2019) Impact of the combination of sourdough fermentation and hop extract addition on baking properties, antioxidant capacity and phenolics bioaccessibility of rice bran-enhanced bread. Food Chem 285:231–239. https://doi.org/10.1016/j.foodchem.2019.01.145

İspirli H, Özmen D, Yılmaz MT, Sağdıç O, Dertli E (2020) Impact of glucan type exopolysaccharide (EPS) production on technological characteristics of sourdough bread. Food Control 107:106812. https://doi.org/10.1016/j.foodcont.2019.106812

Jakobsson K (2021) Surdeg: lär dig hantverk. Polaris Fakta

Johansson DP, Andersson R, Alminger M, Landberg R, Langton M (2018) Larger particle size of oat bran inhibits degradation and lowers extractability of β-glucan in sourdough bread – potential implications for cholesterol-lowering properties in vivo. Food Hydrocoll 77:49–56. https://doi.org/10.1016/j.foodhyd.2017.09.016

Josephs-Spaulding J, Beeler E, Singh OV (2016) Human microbiome versus food-borne pathogens: friend or foe. Appl Microbiol Biotechnol 100(11):4845–4863. https://doi.org/10.1007/s00253-016-7523-7

Khlestkin VK, Lockachuk MN, Savkina OA, Kuznetsova LI, Pavlovskaya EN, Parakhina OI (2022) Taxonomic structure of bacterial communities in sourdoughs of spontaneous fermentation. Vavilovskii Zhurnal Genet Selektsii 26(4):385–393. https://doi.org/10.18699/vjgb-22-47

Kim G, Bae JH, Cheon S, Lee DH, Kim DH, Lee D, Park SH, Shim S, Seo JH, Han NS (2022) Prebiotic activities of dextran from Leuconostoc mesenteroides SPCL742 analyzed in the aspect of the human gut microbial ecosystem. Food Funct 13(3):1256–1267. https://doi.org/10.1039/d1fo03287a

Laatikainen R, Koskenpato J, Hongisto SM, Loponen J, Poussa T, Huang X, Sontag-Strohm T, Salmenkari H, Korpela R (2017) Pilot Study: comparison of sourdough wheat bread and yeast-fermented wheat bread in individuals with wheat sensitivity and irritable bowel syndrome. Nutrients 9(11). https://doi.org/10.3390/nu9111215

Lahue C, Madden A, Dunn R, Smukowski Heil C (2020) History and domestication of *Saccharomyces cerevisiae* in bread baking [Review]. Front Genet 11. https://doi.org/10.3389/fgene.2020.584718

Landis EA, Oliverio AM, McKenney EA, Nichols LM, Kfoury N, Biango-Daniels M, Shell LK, Madden AA, Shapiro L, Sakunala S, Drake K, Robbat A, Booker M, Dunn RR, Fierer N, Wolfe BE (2021) The diversity and function of sourdough starter microbiomes. eLife 10:e61644. https://doi.org/10.7554/eLife.61644

Liu T, Li Y, Yang Y, Yi H, Zhang L, He G (2020) The influence of different lactic acid bacteria on sourdough flavor and a deep insight into sourdough fermentation through RNA sequencing. Food Chem 307:125529. https://doi.org/10.1016/j.foodchem.2019.125529

Lluansi A, Lliros M, Oliver L, Bahi A, Elias-Masiques N, Gonzalez M, Benejam P, Cueva E, Termes M, Ramio-Pujol S, Malagon M, Amoedo J, Serrano M, Busquets D, Torreabla L, Sabat M, Buxo M, Cambra M, Serra-Pages M, Aldeguer X (2021) In vitro prebiotic effect of bread-making process in inflammatory bowel disease microbiome. Front Microbiol 12:716307. https://doi.org/10.3389/fmicb.2021.716307

Marklinder I (2015) Surdeg: brödbakning med rågad passion över gränserna. Reportage, Fakta & Recept

Martín-Garcia A, Riu-Aumatell M, López-Tamames E (2023) Influence of process parameters on sourdough microbiota, physical properties and sensory profile. Food Rev Intl 39(1):334–348. https://doi.org/10.1080/87559129.2021.1906698

Menezes LAA, Molognoni L, de Sá Ploêncio LA, Costa FBM, Daguer H, Dea Lindner JD (2019) Use of sourdough fermentation to reducing FODMAPs in breads. Eur Food Res Technol 245(6):1183–1195. https://doi.org/10.1007/s00217-019-03239-7

Meroth CB, Hammes WP, Hertel C (2003) Identification and population dynamics of yeasts in sourdough fermentation processes by PCR-denaturing gradient gel electrophoresis. Appl Environ Microbiol 69(12):7453–7461. https://doi.org/10.1128/aem.69.12.7453-7461.2003

Minervini F, De Angelis M, Di Cagno R, Gobbetti M (2014) Ecological parameters influencing microbial diversity and stability of traditional sourdough. Int J Food Microbiol 171:136–146. https://doi.org/10.1016/j.ijfoodmicro.2013.11.021

Modi B, Timilsina H, Bhandari S, Achhami A, Pakka S, Shrestha P, Kandel D, Gc DB, Khatri S, Chhetri PM, Parajuli N (2021) Current trends of food analysis, safety, and packaging. Int J Food Sci 2021:9924667–9924620. https://doi.org/10.1155/2021/9924667

Mohammadi-Kouchesfahani M, Hamidi-Esfahani Z, Azizi MH (2019) Isolation and identification of lactic acid bacteria with phytase activity from sourdough. Food Sci Nutr 7(11):3700–3708. https://doi.org/10.1002/fsn3.1229

Montemurro M, Celano G, De Angelis M, Gobbetti M, Rizzello CG, Pontonio E (2020) Selection of non-lactobacillus strains to be used as starters for sourdough fermentation. Food Microbiol 90:103491. https://doi.org/10.1016/j.fm.2020.103491

Muir JG, Varney JE, Ajamian M, Gibson PR (2019) Gluten-free and low-FODMAP sourdoughs for patients with coeliac disease and irritable bowel syndrome: a clinical perspective. Int J Food Microbiol 290:237–246. https://doi.org/10.1016/j.ijfoodmicro.2018.10.016

Nionelli L, Pontonio E, Gobbetti M, Rizzello CG (2018) Use of hop extract as antifungal ingredient for bread making and selection of autochthonous resistant starters for sourdough fermentation. Int J Food Microbiol 266:173–182. https://doi.org/10.1016/j.ijfoodmicro.2017.12.002

Novotni D, Gänzle M, Rocha JM (2021) Composition and activity of microbiota in sourdough and their effect on bread quality and safety. In: Galanakis CM (ed) Trends in wheat and bread making. Academic, pp 129–172. https://doi.org/10.1016/B978-0-12-821048-2.00005-2

Nuobariene L, Cizeikiene D, Gradzeviciute E, Hansen ÅS, Rasmussen SK, Juodeikiene G, Vogensen FK (2015) Phytase-active lactic acid bacteria from sourdoughs: isolation and identification. LWT Food Sci Technol 63(1):766–772. https://doi.org/10.1016/j.lwt.2015.03.018

Olojede AO, Sanni AI, Banwo K, Michael T (2022) Improvement of texture, nutritional qualities, and consumers' perceptions of sorghum-based sourdough bread made with pediococcus pentosaceus and weissella confusa strains. Fermentation 8(1). https://doi.org/10.3390/fermentation8010032

Onipede GO, Fawole AO (2017) Antibacterial activity of lactic acid bacteria isolated from fermented cereal products against organisms implicated in gastrointestinal tract infections. Niger J Microbiol 31(2):3979–3985

Oshiro M, Momoda R, Tanaka M, Zendo T, Nakayama J (2019) Dense tracking of the dynamics of the microbial community and chemicals constituents in spontaneous wheat sourdough during

two months of backslopping. J Biosci Bioeng 128(2):170–176. https://doi.org/10.1016/j.jbiosc.2019.02.006
Oshiro M, Tanaka M, Zendo T, Nakayama J (2020) Impact of pH on succession of sourdough lactic acid bacteria communities and their fermentation properties. Biosci Microbiota Food Health 39(3):152–159. https://doi.org/10.12938/bmfh.2019-038
Ozgolet M, Yaman M, Zeki Durak M, Karasu S (2022) The effect of five different sourdough on the formation of glyoxal and methylglyoxal in bread and influence of in vitro digestion. Food Chem 371:131141. https://doi.org/10.1016/j.foodchem.2021.131141
Pacyński M, Wojtasiak RZ, Mildner-Szkudlarz S (2015) Improving the aroma of gluten-free bread. LWT Food Sci Technol 63(1):706–713. https://doi.org/10.1016/j.lwt.2015.03.032
Pétel C, Onno B, Prost C (2017) Sourdough volatile compounds and their contribution to bread: a review. Trends Food Sci Technol 59:105–123. https://doi.org/10.1016/j.tifs.2016.10.015
Pozzo L, Alcántara C, Selma-Royo M, Garcia-Mantrana I, Bramanti E, Longo V, Collado MC, Pucci L (2022) The impact of sourdough fermentation of spelt (Triticum dicoccum) from Garfagnana on gut microbiota composition and in vitro activity. J Funct Foods 91:105007. https://doi.org/10.1016/j.jff.2022.105007
Qiao N, Wittouck S, Mattarelli P, Zheng J, Lebeer S, Felis GE, Gänzle MG (2022) After the storm—perspectives on the taxonomy of *Lactobacillaceae*. JDS Communications 3(3):222–227. https://doi.org/10.3168/jdsc.2021-0183
Quattrini M, Liang N, Fortina MG, Xiang S, Curtis JM, Ganzle M (2019) Exploiting synergies of sourdough and antifungal organic acids to delay fungal spoilage of bread. Int J Food Microbiol 302:8–14. https://doi.org/10.1016/j.ijfoodmicro.2018.09.007
Raimondi S, Amaretti A, Rossi M, Fall PA, Tabanelli G, Gardini F, Montanari C (2017) Evolution of microbial community and chemical properties of a sourdough during the production of Colomba, an Italian sweet leavened baked product. LWT 86:31–39. https://doi.org/10.1016/j.lwt.2017.07.042
Reese AT, Madden AA, Joossens M, Lacaze G, Dunn RR (2020) Influences of ingredients and bakers on the bacteria and fungi in sourdough starters and bread. mSphere 5(1). https://doi.org/10.1128/mSphere.00950-19
Rinaldi M, Paciulli M, Caligiani A, Scazzina F, Chiavaro E (2017) Sourdough fermentation and chestnut flour in gluten-free bread: a shelf-life evaluation. Food Chem 224:144–152. https://doi.org/10.1016/j.foodchem.2016.12.055
Ripari V, Cecchi T, Berardi E (2016) Microbiological characterisation and volatiles profile of model, *ex-novo*, and traditional Italian white wheat sourdoughs. Food Chem 205:297–307. https://doi.org/10.1016/j.foodchem.2016.02.150
Röcken W, Voysey PA (1995) Sour-dough fermentation in bread making. J Appl Bacteriol 79:38S–48S
Samuel D (1996) Investigation of ancient egyptian baking and brewing methods by correlative microscopy. Science 273(5274):488–490. https://doi.org/10.1126/science.273.5274.488
Sandez Penidez SH, Velasco Manini MA, Gerez CL, Rollan GC (2020) Partial characterization and purification of phytase from Lactobacillus plantarum CRL1964 isolated from pseudocereals. J Basic Microbiol 60(9):787–798. https://doi.org/10.1002/jobm.202000236
Seyed AM, Alireza S (2011) Investigating the sourdough potential for enhance microbiological shelf life and roasty aroma of traditional Lavash bread. Afr J Biotechnol 10(47):9668–9672. https://doi.org/10.5897/ajb10.2428
Smith C, Van Haute MJ, Rose DJ (2020) Processing has differential effects on microbiota-accessible carbohydrates in whole grains during in vitro fermentation. Appl Environ Microbiol 86(21). https://doi.org/10.1128/AEM.01705-20
Sugihara TF, Klina L, McCready LB (1970) Nature of San Francisco sour dough in French bread process. II Microbiological aspects. Baker's Digest 44:51–57
Sugihara TF, Kline L, Miller MW (1971) Microorganisms of the San Francisco sour dough bread process. I. Yeasts responsible for the leavening action. Appl Microbiol Biotechnol 21(3):456–458. https://doi.org/10.1128/am.21.3.456-458.1971

Suo B, Chen X, Wang Y (2021) Recent research advances of lactic acid bacteria in sourdough: origin, diversity, and function. Curr Opin Food Sci 37:66–75. https://doi.org/10.1016/j.cofs.2020.09.007

Taccari M, Aquilanti L, Polverigiani S, Osimani A, Garofalo C, Milanovic V, Clementi F (2016) Microbial diversity of Type I Sourdoughs prepared and back-slopped with wholemeal and refined soft (Triticum aestivum) wheat flours. J Food Sci 81(8):M1996–M2005. https://doi.org/10.1111/1750-3841.13372

Taglieri I, Macaluso M, Bianchi A, Sanmartin C, Quartacci MF, Zinnai A, Venturi F (2021) Overcoming bread quality decay concerns: main issues for bread shelf life as a function of biological leavening agents and different extra ingredients used in formulation. A review. J Sci Food Agric 101(5):1732–1743. https://doi.org/10.1002/jsfa.10816

Torrieri E, Pepe O, Ventorino V, Masi P, Cavella S (2014) Effect of sourdough at different concentrations on quality and shelf life of bread. LWT Food Sci Technol 56(2):508–516. https://doi.org/10.1016/j.lwt.2013.12.005

Tucker AJ, Vandermey JS, Robinson LE, Graham TE, Bakovic M, Duncan AM (2014) Effects of breads of varying carbohydrate quality on postprandial glycaemic, incretin and lipidaemic response after first and second meals in adults with diet-controlled type 2 diabetes. J Funct Foods 6:116–125. https://doi.org/10.1016/j.jff.2013.09.025

Van Kerrebroeck S, Comasio A, Harth H, De Vuyst L (2018) Impact of starter culture, ingredients, and flour type on sourdough bread volatiles as monitored by selected ion flow tube-mass spectrometry. Food Res Int 106:254–262. https://doi.org/10.1016/j.foodres.2017.12.068

Wang Y, Jian C, Salonen A, Dong M, Yang Z (2023) Designing healthier bread through the lens of the gut microbiota. Trends Food Sci Technol 134:13–28. https://doi.org/10.1016/j.tifs.2023.02.007

Winters M, Panayotides D, Bayrak M, Remont G, Viejo CG, Liu D, Le B, Liu Y, Luo J, Zhang P, Howell K (2019) Defined co-cultures of yeast and bacteria modify the aroma, crumb and sensory properties of bread. J Appl Microbiol 127(3):778–793. https://doi.org/10.1111/jam.14349

Wu F, Lv P, Yang N, Jin Y, Jin Z, Xu X (2018) Preparation of Maillard reaction flavor additive from germinated wheat and its effect on bread quality. Cereal Chem 95(1):98–108. https://doi.org/10.1002/cche.10019

Wu S, Peng Y, Xi J, Zhao Q, Xu D, Jin Z, Xu X (2022) Effect of sourdough fermented with corn oil and lactic acid bacteria on bread flavor. LWT 155:112935. https://doi.org/10.1016/j.lwt.2021.112935

Xu D, Zhang Y, Tang K, Hu Y, Xu X, Ganzle MG (2019) Effect of mixed cultures of yeast and lactobacilli on the quality of wheat sourdough bread. Front Microbiol 10:2113. https://doi.org/10.3389/fmicb.2019.02113

Xu D, Hu Y, Wu F, Jin Y, Xu X, Ganzle MG (2020a) Comparison of the functionality of exopolysaccharides produced by sourdough lactic acid bacteria in bread and steamed bread. J Agric Food Chem 68(33):8907–8914. https://doi.org/10.1021/acs.jafc.0c02703

Xu D, Zhang H, Xi J, Jin Y, Chen Y, Guo L, Jin Z, Xu X (2020b) Improving bread aroma using low-temperature sourdough fermentation. Food Biosci 37:100704. https://doi.org/10.1016/j.fbio.2020.100704

Yan B, Sadiq F, Cai Y, Fan D, Zhang H, Zhao J, Chen W (2019) Identification of key aroma compounds in Type I Sourdough-based chinese steamed bread: application of untargeted metabolomics analysisp. Int J Mol Sci 20(4). https://doi.org/10.3390/ijms20040818

Zhang Y, Guo L, Xu D, Li D, Yang N, Chen F, Jin Z, Xu X (2018) Effects of dextran with different molecular weights on the quality of wheat sourdough breads. Food Chem 256:373–379. https://doi.org/10.1016/j.foodchem.2018.02.146

Zhang G, Tu J, Sadiq FA, Zhang W, Wang W (2019a) Prevalence, genetic diversity, and technological functions of the lactobacillus sanfranciscensis in sourdough: a review. Compr Rev Food Sci Food Saf 18(4):1209–1226. https://doi.org/10.1111/1541-4337.12459

Zhang G, Zhang W, Sadiq FA, Arbab SH, He G (2019b) Microbiota succession and metabolite changes during the traditional sourdough fermentation of Chinese steamed bread. CyTA J Food 17(1):172–179. https://doi.org/10.1080/19476337.2019.1569166

Zhang Z, Wang K, Oh J-H, Zhang S, van Pijkeren J-P, Cheng CC, Ren D, Wei H, Gänzle MG, Walter J (2020) A Phylogenetic View on the Role of Glycerol for Growth Enhancement and Reuterin Formation in Limosilactobacillus reuteri. Front Microbiol 11

Zheng J, Wittouck S, Salvetti E, Franz CMAP, Harris HMB, Mattarelli P, O'Toole PW, Pot B, Vandamme P, Walter J, Watanabe K, Wuyts S, Felis GE, Gänzle MG, Lebeer S (2020) A taxonomic note on the genus Lactobacillus: description of 23 novel genera, emended description of the genus *Lactobacillus* Beijerinck 1901, and union of *Lactobacillaceae* and *Leuconostocaceae*. Int J Syst Evol Microbiol 70(4):2782–2858. https://doi.org/10.1099/ijsem.0.004107

Chapter 2
Sourdough Microbiota Diversity in Southern Europe

João Miguel Ferreira da Rocha, Pasquale Russo, Vittorio Capozzi, Mariagiovanna Fragasso, and Francisco Xavier Malcata

2.1 Introduction

Sourdough is a mixture of flour (or flours) and water fermented by endogenous complex microbial communities, mainly including yeasts and lactic acid bacteria (LAB), with a prevalence of heterofermentative lactobacilli (De Vuyst et al. 2016). These microorganisms are pivotal in shaping sourdough's unique sensory attributes, texture, nutritional features and derivative products (De Vuyst et al. 2023; Gänzle 2014; Ma et al. 2021). Noteworthy, it is also the microbial contribution that enhances the safety and shelf-life of sourdough-based foods by synthesizing several antimicrobial compounds useful to contrast the development of spoilage filamentous fungi and/or pathogens (Axel et al. 2016; Quattrini et al. 2019; Russo et al. 2017). More recently, the scientific interest has also moved towards the improvement of the

J. M. Ferreira da Rocha (✉)
Universidade Católica Portuguesa, CBQF – Centro de Biotecnologia e Química Fina – Laboratório Associado, Escola Superior de Biotecnologia, Porto, Portugal

LEPABE—Laboratory for Process Engineering, Environment, Biotechnology and Energy, Faculty of Engineering, University of Porto, Porto, Portugal

ALiCE—Associate Laboratory in Chemical Engineering, Faculty of Engineering, University of Porto, Porto, Portugal
e-mail: jmfrocha@fc.up.pt

P. Russo
Department of Food, Environmental and Nutritional Sciences, University of Milan, Milan, Italy
e-mail: pasquale.russo1@unimi.it

V. Capozzi
Institute of Sciences of Food Production, National Research Council (CNR), Foggia, Italy
e-mail: vittorio.capozzi@ispa.cnr.it

E. B. Ceresino et al. (eds.), *Sourdough Microbiota and Starter Cultures for Industry*, https://doi.org/10.1007/978-3-031-48604-3_2

functional properties suggesting the design of selected LAB and yeasts targeting the optimization of sourdough bioprocesses (Gobbetti et al. 2019; Graça et al. 2021; Ribet et al. 2023). Accordingly, the occurrence of postbiotic-like components synthesized through sourdough fermentation and/or released during the baking process is opening new perspectives in the effort of linking microbial resources to specific functions, with a particular interest in metabolites providing human health benefits (Akamine et al. 2023; Lau et al. 2021; Pérez-Alvarado et al. 2022).

Traditional sourdoughs are obtained by spontaneous fermentation further subjected to backslopping (type-I sourdough), thus offering a rich and dynamic ecosystem in terms of microbial diversity that is related, at different extent, to the raw matrices, the working environment, as well the conditions faced during food-processing (Calabrese et al. 2022; Novotni et al. 2021). In general, the most prevalent LAB species reported in sourdough are *Fructilactobacillus sanfranciscensis*, *Lactiplantibacillus plantarum*, *Levilactobacillus brevis*, *Companilactobacillus paralimentarius*, *Limosilactobacillus fermentum*, *Pediococcus pentosaceus*, *Leuconostoc* spp. and *Weissella* spp., while *Saccharomyces cerevisiae*, *Candida humilis*, *Kluyveromyces marxianus*, *Pichia* spp., *Wickerhamomyces anomalus* and *Torulaspora delbrueckii* are the most commonly found yeasts (recently reviewed by De Vuyst et al. 2023). Microbial region specificity is claimed to improve the unique qualities of traditional products (Capozzi et al. 2012). However, a clear relationship between a typical sourdough and its associated microbiota remains still an open question (De Vuyst et al. 2014). Elucidation can be deduced from a comprehensive systematic review analyzing 1230 studies based on 30 years of knowledge on sourdough fermentation (Arora et al. 2021). Moreover, a meta-analysis of 583 sourdough-related literature articles in the period 1999–2017 based on the microbial diversity and process condition revealed that sourdoughs were mainly characterized by the presence of *S. cerevisiae* or *C. humilis* and that the simultaneous presence of both yeasts seemed to be associated with the occurrence of the LAB *Fr. sanfranciscensis* (Van Kerrebroeck et al. 2017). Accordingly, sourdough often harbors only one or two yeast species. For instance, Italian bakery sourdoughs often displayed *C. humilis* and/or *S. cerevisiae* as dominating yeasts (De Vuyst et al. 2016). Recently, a comprehensive study has been performed to investigate how the microbial diversity of sourdough varies across and between continents. Interestingly, this study suggested that geographical location minimally influences microbial communities,

M. Fragasso
Department of Agriculture Food Natural Science Engineering (DAFNE), University of Foggia, Foggia, Italy
e-mail: mariagiovanna.fragasso@unifg.it

F. X. Malcata
LEPABE—Laboratory for Process Engineering, Environment, Biotechnology and Energy, Faculty of Engineering, University of Porto, Porto, Portugal

ALiCE—Associate Laboratory in Chemical Engineering, Faculty of Engineering, University of Porto, Porto, Portugal
e-mail: fmalcata@fe.up.pt

while the diversity likely depends on how the sourdough was obtained and maintained over time (Landis et al. 2021). In agreement, a recent study showed that the bacterial and fungal composition of wheat grain and flour was poorly represented in mature sourdough, indicating that selection by this ecological niche appears as the main driver of the mature sourdough microbial composition (von Gastrow et al. 2023).

Sourdough fermentation has played a pivotal role in the history and nutrition of humans. With its rich culinary history and diverse cultural heritage, Southern Europe offers a captivating landscape to explore the microbiota diversity of sourdoughs in this geographical region (Bonaccio et al. 2022).

This chapter aims to delve into the intricacies of sourdough microbiota in Southern Europe, unravelling the microbial players responsible for the distinct features of sourdough and bakery products by providing an overview of the state-of-art in Italy and Portugal. By examining the traditional features of sourdoughs from these countries, we can gain valuable insights into the microbial ecology and evolution of sourdough fermentation, as well as uncover potential connections between microbiota composition and sourdough quality.

2.2 Traditional Italian and Portuguese Sourdough Breads and Other Sourdough-Based Baking Products

The scientific, social and economic relevance of Southern Europe sourdough fermentation is widely testified by the importance of these products among the Protected Geographical Indications (PGI), a sign that allows associating unique quality features to a specific geographical origin in the framework of the intellectual property law (Capozzi et al. 2012). As listed in Table 2.1, three Italian breads have received the Protected Designation of Origin (PDO), namely *Pane di Altamura*, *Pagnotta del Dittaino*, and *Pane Toscano*, while a total of 18 Italian and Portuguese bakery products are recognized as Protected Geographical Indications (Table 2.1). Outside the European Community (EC) Regulation, Traditional Agri-Food Products (TAP) is an official approval for traditional Italian regional food products similar to the Protected Geographical Indication (PGI) status of the European Union (EU). TAPs encompass hundreds of bread, pastry, cakes, biscuits and other baker's wares diversified among all Italian Regions "*whose method of processing, preservation and maturation have been consolidated over time*" (https://www.politicheagricole.it/flex/cm/pages/ServeBLOB.php/L/IT/IDPagina/17979). The high number of sourdough-fermented products recorded in Italy is probably due to the existence of strong differences within the Italian territory, including geographical, historical and cultural concerns, which are reflected in a wide range of traditional products. Moreover, the artisanal features of this kind of food, the employment of traditional recipes, as well as the small-scale production are consistent with their local more than global availability.

Table 2.1 Protected Designation of Origin (PDO) and Protected Geographical Indication (PGI) in the field of bakery products in Italy and Portugal

Name	Product category	File Number	Type	Country	Status	Date of Registration
Coppia Ferrarese	Class 2.3. Bread, pastry, cakes, confectionery, biscuits and other baker's wares	PGI-IT-0120	PGI	Italy	Registered	18/10/2001
Pane di Altamura	Class 2.3. Bread, pastry, cakes, confectionery, biscuits and other baker's wares	PDO-IT-0136	PDO	Italy	Registered	19/07/2003
Ricciarelli di Siena	Class 2.3. Bread, pastry, cakes, confectionery, biscuits and other baker's wares	PGI-IT-0666	PGI	Italy	Registered	19/03/2010
Pastel de Chaves	Class 2.3. Bread, pastry, cakes, confectionery, biscuits and other baker's wares	PGI-PT-1126	PGI	Portugal	Registered	27/05/2015
Amêndoa Coberta de Moncorvo	Class 2.3. Bread, pastry, cakes, confectionery, biscuits and other baker's wares	PGI-PT-02235	PGI	Portugal	Registered	16/03/2018
Focaccia di Recco col formaggio	Class 2.3. Bread, pastry, cakes, confectionery, biscuits and other baker's wares	PGI-IT-0944	PGI	Italy	Registered	14/01/2015
Torrone di Bagnara	Class 2.3. Bread, pastry, cakes, confectionery, biscuits and other baker's wares	PGI-IT-1101	PGI	Italy	Registered	14/08/2014
Piadina Romagnola / Piada Romagnola	Class 2.3. Bread, pastry, cakes, confectionery, biscuits and other baker's wares	PGI-IT-1067	PGI	Italy	Registered	04/11/2014
Folar de Valpaços	Class 2.3. Bread, pastry, cakes, confectionery, biscuits and other baker's wares	PGI-PT-01392	PGI	Portugal	Registered	14/02/2017

(continued)

Table 2.1 (continued)

Name	Product category	File Number	Type	Country	Status	Date of Registration
Pane di Matera	Class 2.3. Bread, pastry, cakes, confectionery, biscuits and other baker's wares	PGI-IT-0372	PGI	Italy	Registered	22/02/2008
Ovos moles de Aveiro	Class 2.3. Bread, pastry, cakes, confectionery, biscuits and other baker's wares	PGI-PT-0518	PGI	Portugal	Registered	08/04/2009
Pagnotta del Dittaino	Class 2.3. Bread, pastry, cakes, confectionery, biscuits and other baker's wares	PDO-IT-0577	PDO	Italy	Registered	18/06/2009
Liquirizia di Calabria	Class 1.8. Other products of Annex I of the treaty (spices etc.), Class 2.3. Bread, pastry, cakes, confectionery, biscuits and other baker's wares	PDO-IT-0644	PDO	Italy	Registered	25/10/2011
Panforte di Siena	Class 2.3. Bread, pastry, cakes, confectionery, biscuits and other baker's wares	PGI-IT-0795	PGI	Italy	Registered	22/05/2013
Pastel de Tentúgal	Class 2.3. Bread, pastry, cakes, confectionery, biscuits and other baker's wares	PGI-PT-0938	PGI	Portugal	Registered	04/09/2013
Pastel de Tentúgal	Class 2.3. Bread, pastry, cakes, confectionery, biscuits and other baker's wares	PGI-PT-0938-AM01	PGI	Portugal	Applied	
Pane casareccio di Genzano	Class 2.3. Bread, pastry, cakes, confectionery, biscuits and other baker's wares	PGI-IT-1553	PGI	Italy	Registered	25/11/1997

(continued)

Table 2.1 (continued)

Name	Product category	File Number	Type	Country	Status	Date of Registration
Cantuccini Toscani / Cantucci Toscani	Class 2.3. Bread, pastry, cakes, confectionery, biscuits and other baker's wares	PGI-IT-01290	PGI	Italy	Registered	26/01/2016
Fogaça da Feira	Class 2.3. Bread, pastry, cakes, confectionery, biscuits and other baker's wares	PGI-PT-01342	PGI	Portugal	Registered	14/06/2016
Pane Toscano	Class 2.3. Bread, pastry, cakes, confectionery, biscuits and other baker's wares	PDO-IT-01016	PDO	Italy	Registered	04/03/2016
Pão de Ló de Ovar	Class 2.3. Bread, pastry, cakes, confectionery, biscuits and other baker's wares	PGI-PT-01341	PGI	Portugal	Registered	24/08/2016
Pampapato di Ferrara / Pampepato di Ferrara	Class 2.3. Bread, pastry, cakes, confectionery, biscuits and other baker's wares	PGI-IT-01323	PGI	Italy	Registered	08/12/2015
Südtiroler Schüttelbrot / Schüttelbrot Alto Adige	Class 2.3. Bread, pastry, cakes, confectionery, biscuits and other baker's wares	PGI-IT-02392	PGI	Italy	Registered	24/07/2020
Sebadas / Seadas / Sabadas / Seattas / Savadas / Sevadas di Sardegna	Class 2.3. Bread, pastry, cakes, confectionery, biscuits and other baker's wares, Class 2.5. Pasta	PGI-IT-02834	PGI	Italy	Published	---
Pampepato di Terni / Panpepato di Terni	Class 2.3. Bread, pastry, cakes, confectionery, biscuits and other baker's wares	PGI-IT-02467	PGI	Italy	Registered	23/10/2020
Caralhotas de Almeirim	Class 2.3. Bread, pastry, cakes, confectionery, biscuits and other baker's wares	PGI-PT-02622	PGI	Portugal	Applied	---

Source: eAmbrosia

For example, flat breads include a multitude of ancient and worldwide breads ranging from a few millimeters to a few centimeters in thickness. In a recent survey, a total of 143 different flat bread types were found to be produced in 9 countries of the Mediterranean area, of which 75 in Italy (Pasqualone et al. 2022). Apart from what above-reported, other reasons could explain the high occurrence of these products. Indeed, in Italy, they are considered a delicacy, thus admitting many variations on a regional basis, including pain, garnished and fried products, while, as a function of the segments/areas, they are also staple foods consumed on a daily basis (Pasqualone et al. 2022). Among these, *Focaccia* is a common Italian flat bread, typically consumed as street food. Even if some differences among different traditional recipes can be reported, the production process generally includes kneading a blend of soft and durum wheat flour with water, fatty substances, yeast and salt, as well as the addition of other ingredients, such as fresh cherry tomatoes, olives, dry oregano and some additional oil (Bavaro et al. 2021).

Italian traditional bakery goods also include some traditional sweet-leavened products mainly consumed for particular religious feasts (*i.e. Panettone* and *Pandoro* or *Colomba* for Christmas and Easter, respectively) and some snacks for breakfast like *Brioches* and *Cornetti* (Palomba et al. 2011). *Pandoro* originates from the city of Verona and is typically frustum shaped with a star section, whose main ingredients are flour, sugar, eggs and butter (Lattanzi et al. 2013). *Colomba,* which takes its name from the typical shape of a dove, is decorated with glaze, almonds and granulated sugar (Raimondi et al. 2017). The employment of sourdough characterizes these products, the so-called mother-sponge or mother-dough, kept active by continuous refreshments (backsloppings). Fermentation times are usually longer and performed at lower temperatures than what is used for other bakery products (Montanari et al. 2014). However, even if considered traditional Italian goods, they are produced at both industrial and artisanal levels and have a national and international diffusion (Vernocchi et al. 2004).

Although wheat is the most commonly used cereal in breadmaking in Europe, other cereals and pseudocereals (such as amaranth, quinoa and buckwheat) are used, and new formulations and baking products are continuously being developed worldwide to meet the needs of contemporary consumers (Novotni et al. 2023). Portugal has a wide multiplicity of traditional breads and *broa* or *broa de milho* – a traditional Portuguese bread made of regional maize and rye flours – is widely consumed in many regions of Portugal, especially in the Northern.

An inventory (DGADR 2001) of the types of traditional *broas* and other Portuguese breads and baking products (in addition to other traditional Portuguese agricultural products) was created by DGADR (2023) in partnership with FPCG (2023) and MINHA TERRA (2023). These traditional Portuguese specialties are summarized in Table 2.2, even though many other traditional bakery goods may be found around the country.

Broa de milho is widely produced at the farmer level in Northern Portugal and is the most scientifically studied. Other types of *broas* made of maize and rye flours (sometimes with some wheat flour) exist mainly in North and Center Portugal, for instance, *Broa à Lavrador*, *Broa de Avanca*, *Broa de Milho da Beira Alta*, *Broa de*

Table 2.2 Traditional breads and other baking goods from different geographical regions in Portugal (DGADR 2001)

Name (and other designations)	Flour or flours	General characteristics	Main geographical region or municipality
Bola com Torresmos (*Bolo de Torresmos*; *Torta de Torresmos*)	Wheat	More or less spherical-shaped balls, about the size of a clenched fist. Wheat, salt, mother-dough (*Massa-mãe*) and water, sometimes made with the addition of lard to the dough and stuffed with greaves. Variants: with dried fermented sausages (*linguiça*) instead of greaves.	Alentejo region
Bola de Carne à Lavrador and *Bola de Bacalhau*	Wheat	Rectangle shape, brownish color. Wheat, salt, mother-dough (*Massa velha*) and water stuffed with smoked meat or salted codfish (*bacalhau*) and soft texture.	Municipality of Marco de Canaveses. North region
Bola de Centeio de Barroso (*Bola Centeia*)	Rye	Flat base and round or oval shape. Rye, salt, mother-dough (*Massa-mãe*) and water stuffed with streaky pork (*entremeada*), dried meat sausage (*chouriça*) and onion.	North region
Bola de Lamego (*Bola de Carne de Lamego*; *Bola de Bacalhau de Lamego*; *Bola de Sardinha de Lamego*)	Wheat	Rectangle shape, with a weight of about 850 g. Wheat, salt, baker's yeast (or mother-dough), water and butter, lard or olive oil stuffed with meats, codfish or sardines.	North region
Bolo de Farinha de Milho com Carne	Maize, rye, wheat	Flat base and round shape. Maize, rye and wheat (3/1/1), salt, mother-dough (*Massa-mãe*) and water, stuffed with various pork meats, for instance, dried smoked sausages (*salpicão*), chorizo (*chouriça*), smoked bacon (*toucinho defumado*), smoked ham (*presunto*).	North region
Bolo do Caco	Wheat	Round and flat bread shape, 15 and 20 cm in diameter and 2 to 3 cm high, soft texture, baked in a tile fragment (*caco*) or clay skillet. Wheat, salt, starter (or mother-dough) and water. Long kneading, short fermentation and short baking.	Madeira archipelago
Bolo Folar da Guarda	Wheat	Folar cake. Long shape, with 0.5 to 2 kg weight. Wheat, salt, baker's yeast (or mother-dough), water, eggs, brandy and olive oil.	District of Guarda. Central region

(continued)

Table 2.2 (continued)

Name (and other designations)	Flour or flours	General characteristics	Main geographical region or municipality
Broa (*Broa de milho*; *Pão de milho*)	Maize, rye	Circular to ellipsoidal shape, round top and a flat basis, weight between 1 and 3.5 kg but can have *ca.* 5 kg. Bread made of regional maize and rye flours with varying proportions of maize and rye flours (from 50 to 85% of regional maize flour), salt, mother-dough (*Massa-mãe*), water.	North region
Broa à Lavrador	Maize, rye, wheat	Similar characteristics as *broa.*	Municipality of São João da Madeira, Santa Maria da Feira, Oliveira de Azeméis, *etc.*. District of Aveiro. North region
Broa de Avanca	Maize, rye	Similar characteristics as *broa,* weight between 1 and 5 kg, yellow corn flour. Yellow maize.	Municipality of Avanca. District of Aveiro
Broa de Avintes	Maize, rye, malt	Two typical sizes and shapes: ellipsoidal, a round top and a flat basis, 8–9 kg weight, about 30 cm diameter and 15 cm height; and cylindrical, a round top and a flat basis, 1 kg weight, about 10 cm diameter and 15 cm height. White maize, rye, malt, salt, mother-dough (*Massa-mãe*), water.	Municipality of Avintes. District of Porto. North region
Broa de Milho da Beira Alta	Maize, rye	Similar characteristics as *broa.* Circular shape, a round top and a flat basis. Two typical sizes: 12–15 cm diameter and 800 g weight; 10–12 cm diameter and 400 g weight.	Beira Alta region. Central region
Broa de Milho do Vale do Sousa	Maize, rye	Similar characteristics as *broa.* Circular shape, a round top and a flat basis.	Vale do Sousa sub-region. North region
Broa de Milho e Centeio	Maize, rye, wheat	Similar characteristics as *broa.*	Between Douro and Tâmega sub-region. North region
Broa de Vil Moinhos	Maize, whole rye	Similar characteristics as *broa.* Maize of high granularity.	District of Viseu. Central region
Broa Mimosa do Boco	Maize, rye	Similar characteristics as *broa.*	Municipality of Vagos. District of Aveiro. North region

(continued)

Table 2.2 (continued)

Name (and other designations)	Flour or flours	General characteristics	Main geographical region or municipality
Caralhotas de Almeirim	Wheat	Round shape, about 15 cm in diameter. Brittle and crunchy crust, brownish colour and rustic crumb. Wheat, salt, mother-dough (*Massa-mãe*) and water.	Municipality of Almeirim. District of Santarém. Central region
Fogaça da Feira	Wheat	Sweet bread with a light flavor and aroma of lemon and cinnamon, brownish hue and a conical shape with four nozzles at the top (representing the towers of Feira Castle). Loose and light dough, slightly yellowish in colour, with small holes and crispy outside.	Municipality of Santa Maria da Feira. District of Aveiro. North region
Folar de Trás-os-Montes	Wheat	Folar cake. Rectangular or round shape, 0.5–3 kg weight. It is only made at Easter. Wheat, salt, mother-dough (*Massa-mãe*), water, eggs, oil and butter stuffed with smoked pork meats of ham (*presunto*), dried smoked sausage (*salpicão*) and bacon (*toucinho*).	Trás-os Montes region. North region
Folar de Valpaços	Wheat	Folar cake. Rectangular shape. Wheat, salt, mother-dough (*Massa-mãe*), water, with eggs, olive oil and vegetable margarine and/or lard, stuffed with fat meat pork and/or salted and dried streaky (*entremeada*), salted and dried pork, dried smoked pork sausages (*salpicão* and *linguíça*), smoke-cured or naturally cured pork ham (*presunto*) and/or smoked pork shoulder (*pá de porco*).	Municipality of Valpaços. Trás-os Montes region. North region
Padas do Vale de Ílhavo	Wheat	Small wheat breads made up of two small units overlapping at one end. Golden and soft crust. Wheat, salt, mother-dough (*Massa-mãe*) and water.	Ílhavo valley. District of Aveiro. North region
Pão com Chouriço (*Pão Caseiro do Marco*)	Wheat	Bread in the shape of a small yellowish bagel, with an aroma of smoked meat and a fluffy texture. Wheat, salt, mother-dough (*Massa-mãe*) and water, stuffed with chorizo (*chouriço*)	Municipality of Marco de Canaveses. North region
Pão de Alfarroba	Wheat, carob	Circular, round top and a flat basis shape, sweet in flavour, with reminiscent chocolate aroma, and very dark in colour. Wheat (85%), carob (15%), salt, baker's yeast (or mother-dough) and water.	Algarve region. South region
Pão de Casa (*Pão regional*; *Pão de Trigo da Terra*; *Pão de Família*)	Wheat, sweet potato	Circular to ellipsoidal format, a round top and a flat basis, around 1 kg weight. Wheat (75%), sweet potato (17%), salt, mother-dough (*Massa-mãe*) and water.	Madeira archipelago

(continued)

Table 2.2 (continued)

Name (and other designations)	Flour or flours	General characteristics	Main geographical region or municipality
Pão de Centeio da Guarda	Rye	Ellipsoidal shape, a round top and a flat basis, with a relatively crispy and cracked crust. Three typical sizes: small, 10–12 cm diameter and 400 g weight; medium, 13–15 cm diameter and 800 g weight; and large, 16–18 cm diameter and 1200 g weight. Rye, salt, baker's yeast (or mother-dough), water.	District of Guarda. Central region
Pão de Centeio de Barroso	Rye	Circular to ellipsoidal shape, round top and a flat basis, 3 kg weight, dark colour. Rye, salt, mother-dough (*Massa-mãe*) and water.	Barroso sub-region. North region
Pão de Centeio de Castro Laboreiro	Rye, wheat or maize	Circular shape, a round top and a flat basis, 3–4 kg weight. Dark in colour, acidic, not very moist, with a thin crust. Rye, wheat or maize, mother-dough (*Massa-mãe*), salt and water.	Castro Laboreiro sub-region. Central region
Pão de Centeio do Sabugueiro	Rye, wheat	Circular shape, a round top and a flat basis, 40–50 cm diameter 1–1.5 kg weight. The interior appearance is porous with a relatively crispy crust. Dark pearl-white colour. Rye (> 60%) and wheat, mother-dough (*Massa-mãe*), salt and water.	Sabugueiro sub-region. Central region
Pão de Cornos (*Pão de azeite*)	Wheat	Shape of a half-moon. Rough texture, light brown colour, olive oil flavour and odor. Wheat, olive oil, mother-dough (*Massa-mãe*), salt and water.	Municipality of Vagos. District of Aveiro. North region
Pão de Escalhão	Rye, wheat	Circular shape, a round top and a flat basis, 40–50 cm diameter 1 kg weight. The interior appearance is dry and porous, with a relatively crispy and cracked crust. Pearl-white colour on the inside and light brown on the outside. Rye, wheat, baker's yeast (or mother-dough), salt and water.	Central region
Pão de Mafra (*Pão saloio*)	Wheat	Round, long or ball shape. Small size. Soft texture to the touch, both on the crust and inside. Normally, crumb with little density, large holes or alveoli. High humidity (32–40%) and low acidity. Low amounts of starter. Wheat, baker's yeast (or mother-dough), salt and water.	Municipality of Mafra. Central region
Pão de Padronelo (*Pão de Ovelhinha*)	Rye, wheat	Family bread, as its shape makes it easy to divide into four parts, rectangular shape, 10–15 cm diagonally, 5 cm height, 150 g weight. Yellowish-brown colour and firm and smooth consistency. Rye, wheat, mother-dough (*Massa-mãe*), salt and water.	Municipality of Amarante. North region

(continued)

Table 2.2 (continued)

Name (and other designations)	Flour or flours	General characteristics	Main geographical region or municipality
Pão de Trigo do Alentejo (*Pão Alentejano*)	Wheat	Forehead bread shape (higher at one end, resembling a fold), 1–1.5 kg weight. Light brown crust and a whitish crumb. Wheat, mother-dough (*Massa-mãe*), salt and water.	Alentejo. South region
Pão de Trigo em "Padas"	Wheat	Rustic-shaped bread, made up of two small units (about 50 g) overlapping at one end. Three variants: "white" (type 55 flour); "clean" (type 65 flour); and "dark" (type 80 flour). Floured crust, typically from a very soft dough, and irregular crumb. Sharp flavour and slightly acidic. Long baking, giving a characteristic crispy crust. Wheat, mother-dough (*Massa-mãe*), salt and water.	District of Aveiro. North region
Pão Doce	Wheat	Circular shape, a round top and a flat basis. Sweet bread with a soft texture, and lemon and cinnamon flavour. Characteristic crust from baking in a wood oven. Wheat, olive oil or butter, mother-dough (*Massa-mãe*), salt and water, eggs, sugar, lemon zest and cinnamon. Egg whites should only be half the amount of eggs.	Municipality of Vagos. District of Aveiro. North region
Pão Podre	Wheat	Ring shape, a round top and a flat basis, 1 kg weight. Sweet flavor and brownish colour. Wheat, baker's yeast (or mother-dough), salt and water, sugar, eggs and cinnamon.	Municipality of Marco de Canaveses. North region
Pão Santoro de Pêga	Wheat	Flat and elongated shape, with rounded ends, slightly concave in the longitudinal direction, approximately 40–50 cm long and 2 cm high, which can reach 4 cm high at the ends, 250–350 g weight. It is slightly concave in the longitudinal direction and has a relatively crispy crust with a floury and fatty appearance. It has a dark pearl-white colour. Relatively crispy crust with a floury and fatty appearance. Dark pearl-white color. Wheat, baker's yeast (or mother-dough), salt and water, sugar, brandy (*aguardente*) and local olive oil.	Municipality of Sabugal. District of Guarda. Central region

(continued)

Table 2.2 (continued)

Name (and other designations)	Flour or flours	General characteristics	Main geographical region or municipality
Pão Sovado (*Pão Arregueifado*)	Wheat	Different shapes and weights depending on the regions: B*icas* (spouts) – small units finished in a spout with a longitudinal cut; *Regueifinhas* – units with about 0.5 kg of dough in the shape of a snail; *Regueifas* – two elongated pieces with two palms long, braided, with the final shape of a braided ring and the surface decorated with shaped small rings and leaves; *Calo* – elongated mass whose ends are joined in the middle and creased with the hand in a cleaver; and *Redondo* (round) – flattened dough on the surface of which a square is made. Smooth crust and a very closed and regular crumb. Hard dough that needs to be bound by compressing with kneading cylinders. Variants: *Pão redondo* (round bread) (in the border areas of Beira Alta region) and *Pão de Calo* (in the Baixo Alentejo and Algarve regions). Wheat, mother-dough (*Massa-mãe*), salt and water (only 50% of the typical amount of water used).	North region. Other regions
Pãozinho (*Carcaça*; *Bijou*; *Molete*; *Papo-seco*; *Rosca*; *Viana*)	Wheat	Small bread with different shapes depending on the regions: *Carcaça* (carcass) (Lisbon and Tejo valley) – slightly elongated, creased in the middle, with a very fluffy crumb and crispy crust; *Bijou* or *Molete* (between Douro and Minho regions) – round bread, opened where the cut was made before baking; *Papo-seco* (Ribatejo, Alentejo and Southwest regions) – slightly elongated, creased in the middle and ending in small heads; *Rosca* (Lisbon region) – elongated, made from two twisted strips of dough, with a rough and crispy crust; Viana (Lisbon region) – very fluffy round bread. Wheat, baker's yeast, salt and water.	Lisboa and Tejo Valley. Central region. Other regions

Milho e Centeio, *Broa de Vil Moinhos*. *Broa de Avintes*, in Porto (Portugal) region, is one of the most known *broa* made of maize, rye and malt flours, and produced with traditional processes but at semi-industrial scale as a result of its commercial success (Novotni et al. 2021; Rocha 2011; Rocha et al. 2023).

The general breadmaking process of Portuguese sourdough *broa* (Fig. 2.1) is described in several publications (Kalo et al. 2006; Rocha 2011; Rocha and Malcata 1999, 2012, 2016a, b; Rocha et al. 2003, 2010a, b, 2011, 2012a, b, 2023) and varies according to the farmers and geographical locations. Briefly, traditional *broa*

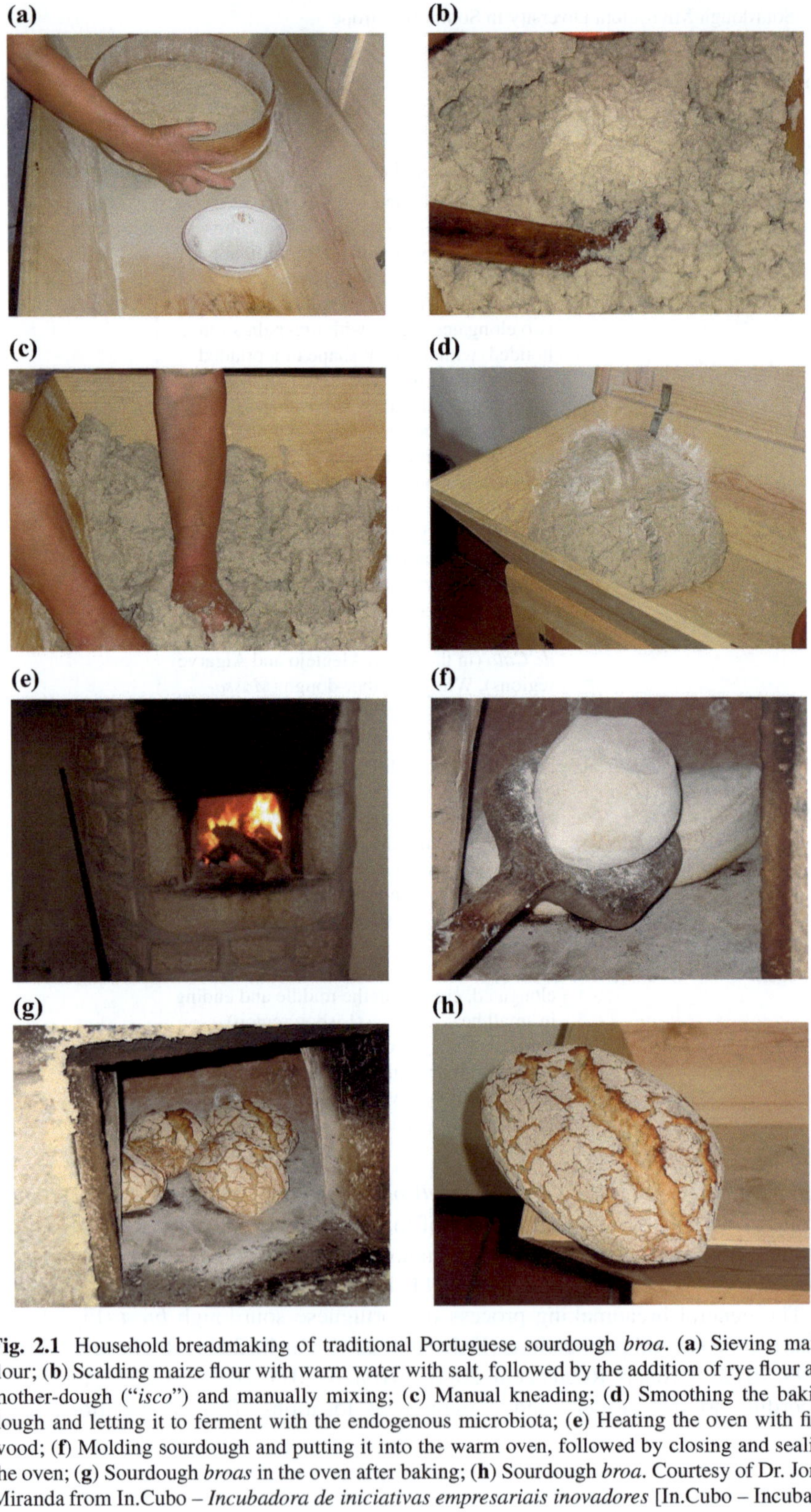

Fig. 2.1 Household breadmaking of traditional Portuguese sourdough *broa*. (**a**) Sieving maize flour; (**b**) Scalding maize flour with warm water with salt, followed by the addition of rye flour and mother-dough ("*isco*") and manually mixing; (**c**) Manual kneading; (**d**) Smoothing the baking dough and letting it to ferment with the endogenous microbiota; (**e**) Heating the oven with firewood; (**f**) Molding sourdough and putting it into the warm oven, followed by closing and sealing the oven; (**g**) Sourdough *broas* in the oven after baking; (**h**) Sourdough *broa*. Courtesy of Dr. Jorge Miranda from In.Cubo – *Incubadora de iniciativas empresariais inovadores* [In.Cubo – Incubator of innovative business initiatives] (https://www.incubo.eu/, accessed in 2023-08-30)

(Fig. 2.1g, h) is a type-I sourdough bread made of maize and rye flours by households and small farmers and following ancient procedures empirically transmitted between generations. Regional maize is still being milled in watermills by some farmers in the rural villages, but it is getting rarer. The percentage of maize and rye flours for sourdough *broa* also varies according to the producer, and may ranges between 50% and 85% of maize flour. *Broa* may have different formats, from circular to ellipsoidal, with a flat basis and a round top, and with an average weight of approximately 1.5 kg, but also varying between *ca.* 1 and 5 kg. Sourdough *broa* crust may have a thickness of approximately 1–2 cm. *Broa* is usually produced every 15 days in winter and weekly during summer. The sourdough *broa* fermentation is usually undertaken in two steps, the so-called first and second fermentation. On the day before baking it is prepared the sourdough, sometimes known as *crescente* (crescent): maize flour is sieved into a wooden kneader (to separate the bran from flour) and manually kneaded for 10–20 minutes, with salted warm water and rye flour. Rye flour is usually purchased in the local market but can also be produced by some farmers of certain Northern regions. Afterwards, mother-dough is added and mixed – *i.e.* a small portion of mature sourdough from the previous baking batch and acting as the microbial starter culture, which is a reservoir of adventitious microorganisms. This mixture is let to ferment usually overnight, at room temperature, but may vary as much as from 4.5 to 28 h. Early in the next morning, the largest amounts of sieved maize and rye flours are used for the breadmaking. Firstly, regional maize flour is scalded and kneaded with salted warm water, usually with a wooden shovel due to the high temperatures of the water, then rye flour and the sourdough from the previous day (first fermentation) is added and manually kneaded. This laborious and physically demanding kneading process usually takes 30–40 min, after which the baking dough is let to ferment. This second fermentation, undertaken at room temperature, also varies between householders from 1.5 to 3 h. When baking dough starts opening, the bakers know that the fermentation is successful and ready. The clay wall oven is heated with firewood during the second fermentation, and temperatures may reach approximately 250 °C. After the second fermentation, sourdough is molded with a wooden bowl (called "*gamelo*" or "*tigelão*") and, then, is taken into the oven, usually with a round or rectangular wooden shovel. The clay oven is closed and sealed, and *broas* are let to bake for approximately 2 hours.

One of the new trends in bakery production is the rediscovery of wheat landraces with particular nutritional and/or technological properties. Sicilian ancient durum wheats have been reported to have a different impact on the overall quality of bakery goods. In particular, Majorca flour showed the highest amount of glutenin, compared to other Sicilian durum wheat landraces, such as *Margherito*, *Perciasacchi*, *Russello* and *Timilia*, that improved the gluten strength and the technological properties of the dough (Visioli et al. 2021). Among the ingredients, water has also been reported to play a pivotal role in the traditional features of bakery goods since several correlations were found between sourdough microbiota and water features (Minervini et al. 2019). However, in spite of some differences in the technological aptitude of the flours and the production process, almost all traditional typical Italian breads are obtained by using sourdough as the natural starter, thus underlying the contribution of the microorganisms to the typicity of the final products (Minervini et al. 2012a).

In the last years, some biotechnological innovations have been reported in sourdough fermentation, mainly including the employment of innovative strains in order to improve the overall quality of these foodstuffs (*i.e.* technological, nutritional, functional sensorial and safety-related features), thus reducing the risk associated to spontaneous fermentation (Capozzi et al. 2017; Gobbetti et al. 2019).

For example, the incidence of intolerance to baker's yeast is increasing in the Western population, mainly due to some cell wall components of *Saccharomyces cerevisiae* that have been recognized as antigens in individuals with chronic inflammatory bowel disease (IBD). Therefore, a combination of unconventional yeast *Zymomonas mobilis*, isolated from fermented agave sap and *Fr. sanfranciscensis* has been proposed as a driver of innovation in the area of yeast-free leavened products (Musatti et al. 2016; Picozzi et al. 2022). In a similar way, a strain of *Leuconostoc citreum* inoculated in a liquid sourdough (type-II sourdough) has been employed for the production of a yeast-free and salt-free typical Apulian bread, known as *puccia* (De Bellis et al. 2019). In particular, salt was eliminated because it negatively impacted on the viability of *L. citreum*. However, the bread's sensory quality was acceptable, and this strain's suitability as a microbial starter was confirmed at the pilot scale in an industrial bakery (De Bellis et al. 2019). A similar protocol was applied to produce a yeast-free *focaccia* with reduced salt content. In particular, salt was replaced with seawater, thus introducing a technological innovation for producing a traditional baked good with added beneficial health effects due to the lowest predicted glycemic index (GI), salt reduction and magnesium cation (Mg^{2+}) fortification (De Bellis et al. 2020). According to the market's current trends, consumers are increasingly accepting bakery products with lower fat content. In this framework, the employment of exopolysaccharide (EPS)-producing strains of *Weissella cibaria* could be an attractive biotechnological solution to improve the rheological properties and substitute fat replacers (Bavaro et al. 2021). It is well-known that microbial dextrans increase the final volume, the softness of the crumb and the product's shelf-life, thus obtaining a special interest for application to the production of gluten-free baked goods (Ramos et al. 2021). On the other hand, dextrans synthesized by *W. cibaria* have been reported to be bioactive compounds with immuno-stimulants and anti-inflammatory properties, supporting their ability to improve the functionality of bakery products (Zarour et al. 2017). Thus, overproducing riboflavin strains of *W. cibaria* have been proposed to obtain multifunctional breads biofortified with both vitamin B_2 and dextrans (Hernández-Alcántara et al. 2022). Recently, it has been reported that sourdough fermentation encouraged the conversion of inorganic selenium into organic selenium compounds, resulting in a high content of bioaccessible selenium, suggesting that selenium-rich flours could be used for the production of a functional *piadina*, a typical flatbread of Emilia Romagna (Italy) (Di Nunzio et al. 2018).

Interestingly, a functional traditional sourdough, the so-called "*ciabatta*" bread, obtained by fermentation with selected strains of sourdough supplemented with insect powders, opens new perspectives in the field of novel foods applications and in the production of biofortification in the protein content (Gaglio et al. 2021).

2.3 Microbial Diversity of Traditional Italian and Portuguese Sourdoughs

Southern Europe boasts a multiplicity of typical breads and sourdough-based traditional products whose microbial communities and dynamics have been extensively characterized (Table 2.3). Recently, the microbiota of Tuscan sourdough bread has been investigated by both culture-independent and dependent methods. Though a total of 68 yeasts and 96 LAB isolates were molecularly identified, bacteria belonged only to the species *Fr. sanfranciscensis,* while yeasts were identified as *Candida milleri* and only three isolates as *S. cerevisiae* were found. The relative composition and specific physiological characteristics of such microbiota can potentially affect the nutritional features of PDO Tuscan bread, as suggested by the functional characterization of the strains (Palla et al. 2017). A low microbial diversity was also found in sourdoughs from the Apulian region, which revealed the occurrence of *Lp. plantarum* and *S. cerevisiae* as the dominant LAB and yeasts, ranging their concentration between 1.7×10^5 and 6.5×10^8 colony-forming units (CFU)/g, or 7.7×10^5 and 2.5×10^7 CFU/g, respectively. In particular, the isolated LAB strains showed interesting antimicrobial activity, suggesting their potential exploitation to extend the shelf-life of sourdough bread (Arena et al. 2019). In agreement, 28 typical sourdoughs of Irpinia, a large area of the Campania region (Southern Italy), were submitted to molecular approaches, revealing a high biodiversity in LAB community among the samples. The most abundant lactobacilli species belonged to *Lp. plantarum* (*ca.* 22% of total LAB isolates), *Fr. sanfranciscensis* (11%), *Companilactobacillus paralimentarius* (8%), and *Furfurilactobacillus rossiae* (6.5%), whereas LAB cocci included *Pediococcus pentosaceus* (9.5%), *Leuconostoc* spp. (7.8%), and *W. cibaria* (7.7%). However, most of the samples were characterized by the dominance of only one or two species, thus confirming the impact of the environment and sourdough fermentation on the development and persistence of few LAB better adapted to specific geographical, sourdough fermentation conditions and/or ecological niches (Reale et al. 2019). In another study, the microbiota of 19 Italian sourdoughs used for the manufacture of traditional breads allowed the identification, through a culture-dependent approach, of 20 LAB species, with a dominance of *Fr. sanfranciscensis*, *Lp. plantarum* and *Co. paralimentarius*, and four yeast species. Interestingly, *S. cerevisiae* was identified in 16 sourdoughs, while the yeasts *C. humilis*, *Kazachstania barnettii* and *Kazachstania exigua* were found as less represented species, thus indicating that differences in the microbial composition could contribute to enhancing the typicality of traditional breads (Minervini et al. 2012a). In consideration of the general occurrence of this species in traditional Italian sourdoughs, typing by combining conventional and multiplex RAPD–PCR profiles has been reported as an interesting tool to investigate the intra-specific variability among *Fr. sanfranciscensis* isolates (Venturi et al. 2012a). However, some sourdoughs, such as Maiorca, typical from Sicilia region, showed a higher microbial variability. Indeed, the LAB *Lp. plantarum* and *Lev. brevis* dominated this ecosystem, while *S. cerevisiae* and *W. anomalus* were the

Table 2.3 Microbial diversity in traditional sourdoughs and sourdoughs-based products in Italy and Portugal

Product	Geographical origin/Typical features of the product	Lactic Acid Bacteria (LAB)	Yeasts	Method of detection	Reference
Cornetto di Matera	Bread – Matera (Basilicata, Southern Italy)	*Lp. plantarum* (49%) *L. mesenteroides* (17%) *Latilactobacillus curvatus* (15%) *Lactiplantibacillus paraplantarum* (12%) *W. cibaria* (5%) *Lactiplantibacillus pentosus* (2%).	–	SDS-PAGE	Zotta et al. (2008)
Tuscan bread	Bread – Tuscan	*Fr. sanfranciscensis* (100%)	*C. milleri* (95%) *S. cerevisiae* (5%)	ARDRA (LAB) RFLP (yeasts)	Palla et al. (2017)
Traditional sourdoughs	Nineteen typical sourdoughs for bread production – Different regions of Italy	*Fr. sanfranciscensis* (28%) *Lp. plantarum* (16%) *Lp. paralimentarius* (14%)	*S. cerevisiae*, *C. humilis*, *K. barnettii*, *K. exigua*	RAPD-PCR	Minervini et al. (2012a)
Maiorca sourdough	Sourdough from ancient durum wheat landraces -Sicily, Southern Italy	*Lp. plantarum* *Lev. brevis*	*S. cerevisiae* *W. anomalus* *T. delbrueckii*, *Pichia kluyveri*, *Candida boidinii*, *Candida diddensiae*	rep-PCR (bacteria) RFLP (yeasts)	Pino et al. (2022)
Traditional sourdoughs	Six typical sourdoughs for bread production – Apulia (Southern Italy)	*Lp. plantarum*	*S. cerevisiae*	ARDRA (LAB) RFLP (yeasts)	Arena et al. (2019)

(continued)

Table 2.3 (continued)

Product	Geographical origin/Typical features of the product	Lactic Acid Bacteria (LAB)	Yeasts	Method of detection	Reference
Traditional sourdoughs	Twenty-eight typical sourdoughs for bread production – Irpinia (area of the Campania region, Southern Italy)	*Lp. plantarum* (22%) *Fr. sanfranciscensis* (11%) *Co. paralimentarius* (8%) *Fu. rossiae* (6.5%) *P. pentosaceus* (9.5%) *Leuconostoc* spp. (7.8%) *W. cibaria* (7.7%)	–	PCR-DGGE	Reale et al. (2019)
Traditional sourdoughs	Twenty typical sourdoughs for bread production – Abruzzo (Central Italy)	–	*S. cerevisiae,* (85%) *C. milleri* (11%), *C. krusei* (2.5%), *T. delbrueckii* (1%)	PCR-RFLP, RAPD-PCR, PCR-DGGE	Valmorri et al. (2010)
Traditional sweet leavened baked goods	Eighteen sourdoughs for traditional Italian sweet leavened baked goods – Different regions of Italy	*Fr. sanfranciscensis* (>80%) *Lp. plantarum* *L. citreum*	*S. cerevisiae* *C. humilis*	Pyrosequencing	Lattanzi et al. (2013)
Brioches and *cornetto*	Sweet baked goods – (Campania, Southern of Italy)	*Lactobacillus* spp. *Leuconostoc* spp. *Lactococcus* spp. *Fr. sanfranciscensis* *Lat. sakei* *W. groceries,* *S. thermophilus,*	*S. cerevisiae* *M. pulcherrima*	PCR-DGGE	Palomba et al. (2011)
Colomba	Easter sweet baked good – Italy	*Fr. sanfranciscensis*	*T. delbrueckii* *C. humilis* *S. cerevisiae* *K. marxianus*	NGS (MiSeq)	Raimondi et al. (2017)
Colomba	Easter sweet baked good - Italy	–	*C. milleri* *S. cerevisiae*	RAPD-PCR	Vernocchi et al. (2004)

(continued)

Table 2.3 (continued)

Product	Geographical origin/Typical features of the product	Lactic Acid Bacteria (LAB)	Yeasts	Method of detection	Reference
Panettone	Christmas sweet cake – Northern of Italy	*Fr. sanfranciscensis* *Lev. brevis*	*C. humilis*	PCR-DGGE	Garofalo et al. (2008)
Panettone	Christmas sweet cake – Northern of Italy	*Fr. sanfranciscensis* *Lp. plantarum* *Fur. rossiae* *Len. parabuchneri* *L. mesenteroides* *L. citreum*	*Kazachstania humilis* *S. cerevisiae*	rep-PCR (bacteria) RFLP (yeasts)	De Vero et al. (2021)
Lagaccio *Panettone*	Dry biscuit – Genoa (Liguria region, Italy) Christmas sweet cake – Northern of Italy	*Fr. sanfranciscensis*	*C. milleri*, *S. cerevisiae*	ARDRA (LAB) RFLP (yeasts)	Venturi et al. (2012b)
Broa de milho	Northern (and center) Portugal	*Lev. brevis*, *Lactobacillus delbrueckii* ssp. *lactis*, *Lp. plantarum*, *Lat. curvatus*, *Lactobacillus delbrueckii* ssp. *delbrueckii*, *Leuconoctoc* spp., *Lactococcus lactis* ssp. *lactis*, *Lactococcus lactis* ssp. *cremoris*, *Pediococcus* spp., *Gardnerella vaginalis*, *Streptococcus agalactiae*, *Streptococcus constellatus*, *Streptococcus equinus*	*Saccharomyces cerevisiae*, *Hansenula anomala*, *Kluyveromyces marxianus* var. *lactis*, *Pichia guilliermondii*, *Torulaspora delbrueckii*, *Pichia ohmeri*, *Pichia membranaefaciens*, *Issatchenkia orientalis*, *Pichia anomala*, *Lachancea kluyveri*, *Issatchenkia occidentalis*	Biochemical tests	Novotni et al. (2021); Rocha (2011); Rocha and Malcata (1999)

most frequently isolated yeasts. In addition, *T. delbrueckii*, *Pichia kluyveri*, *Candida boidinii*, and *Candida diddensiae* were also detected (Pino et al. 2022). A polyphasic approach was used to determine the yeast microbiota associated with spontaneous sourdough fermentations in the production of 20 traditional wheat sourdough breads of the Abruzzo region (Italy). PCR-RFLP analysis identified 85% of the isolates as *S. cerevisiae*, being other dominant species *Candida milleri* (11%), *C. krusei* (2.5%) and *T. delbrueckii* (1%), thus providing an exhaustive overview of the yeast populations in sourdoughs from this region (Valmorri et al. 2010). A metagenetic analysis was performed on *Focaccia* obtained by yeast or LAB starter-mediated fermentation. Interestingly, samples made with baker's yeast showed a dominance of *Proteobacteria* (synonym *Pseudomonadota*, which is a major phylum of Gram-negative bacteria) (82% of the bacterial population), associated with negative biochemical features of the dough. *Proteobacteria* population reduced to 43% when *L. citreum* was used as inoculum, indicating the ability of this starter to drive the fermentation, improving the microbiological quality of yeast-free bakery products (Ferrara et al. 2021). Indeed, such results underline the interest on using sourdough fermentation towards food safety.

To retrieve their origin in sourdough, LAB population have been monitored from ear harvest until the first step of fermentation in order to clarify the relationship among wheat variety, geographical area of production and processing (Alfonzo et al. 2017). In agreement, differences in dominant LAB have also been reported during artisanal pasta-making manufactory located in the Apulian Region (South Italy) (Russo et al. 2010).

Similar studies were performed to characterize the microbial ecosystems in sourdough employed for the production of sweet-leavened goods. A comprehensive study investigated the occurrence of LAB and yeasts by culture-dependent method and pyrosequencing in 18 sweet sourdoughs. Viable cells were about 8.05 and 7.03 log CFU/g for LAB and yeasts, respectively. *Fr. sanfranciscensis* was identified in all the sourdoughs, except for *Panaredda* and *Torcolo di San Costanzo*, by using culture-dependent approaches. Intriguingly, *Fr. sanfranciscensis* was the only species detected in 11 sourdoughs and dominated, with at least 80% of the LAB population, traditional sourdoughs, such as *Buccellato di Lucca*, *Mbriagotto*, *Pandoro*, and *Nadalin*. Other LAB found with a relatively high frequency belonged to *Lp. plantarum* and *L. citreum*. Molecular approaches confirmed these results, and, in a few cases, pyrosequencing complemented the culture-dependent methods, detecting *Fr. sanfranciscensis* also in *Panaredda* and *Torcolo di San Costanzo* sourdoughs. Among yeasts, *S. cerevisiae* was identified in all the sourdoughs, except for *Mbriagotto*, *Ciambella di Mosto* and *Pandolce Genovese*. These latter sourdoughs harbored strains of *C. humilis*, whereas five sourdoughs combined the presence of both yeast species (Lattanzi et al. 2013). Similarly, polymerase chain reaction denaturing gradient gel electrophoresis (PCR-DGGE) analysis allowed researchers to map different levels of microbial biodiversity in nine Italian sweet sourdoughs. Indeed, additional species than what detected by conventional culture-dependent methods were found in association with *S. cerevisiae*, including *Streptococcus thermophilus*, *Latilactobacillus sakei*, *Weissella groceries*, *Fr. sanfranciscensis* and

Metschnikowia pulcherrima, suggesting that typical microbial consortia could be selected for naturally fermented *brioche* and *cornetto* preparation (Palomba et al. 2011).

It has been reported that different technological parameters can impact on the final quality of the products (Novotni et al. 2021; Rocha 2011). For example, a higher temperature employed to propagate sourdough for *Pandoro* production resulted in a lower pH that was associated with higher LAB counts (about 1 log CFU/g higher) but without affecting yeast concentrations (Montanari et al. 2014). A similar result has been found investigating two sweet bakery goods, namely *Lagaccio,* a typical Genoese dry biscuit, and *Panettone*, over a period of three years. In particular, stable microbial associations were characterized by *Fr. sanfranciscensis*, *C. milleri* and *S. cerevisiae* dominance. However, these two products exhibited differences in the ratio LAB/yeasts, probably ascribed to moderate refrigeration conditions in *Panettone* manufacturing (Venturi et al. 2012b). A sourdough for the industrial production of *Colomba* showed a small biodiversity in terms of both LAB and yeasts. Culture-dependent analyses identified *Fr. sanfranciscensis* and *T. delbrueckii*, as dominant bacterial and yeast species, respectively. These results were confirmed by metataxonomic analysis only for the bacterial communities, whereas *C. humilis* was the most abundant yeast. A low microbial biodiversity has been considered an advantage in terms of stable and easier propagation because it is responsible for more reproducible lots of *Colomba* (Raimondi et al. 2017). Eighty-six yeast isolates from the productive process of *Colomba* were characterized by phenotypic methods and identified by random amplified polymorphic DNA (RAPD)-PCR as *C. milleri* and *S. cerevisiae*, being *C. milleri* the dominant species in the sourdough. In particular, this maltose-negative yeast seems to lead strict cooperation with heterofermentative LAB that improve some technological and organoleptic features of the products (Vernocchi et al. 2004).

Similarly, LAB and yeast dynamics during the production of three varieties of *Panettone* were investigated by PCR-DGGE analysis. Although molecular fingerprinting provides evidence of the dominance of *Fr. sanfranciscensis*, *Lev. brevis* and *C. humilis*, a shift in the final stages of two of the production processes suggested an impact of some technological conditions on the dough microbiota (Garofalo et al. 2008). More recently, the molecular characterization of 77 isolates from two doughs for the production of *Panettone* revealed the presence of the dominant yeast *K. humilis*, while *S. cerevisiae* was found in only one sample. Among the LAB species, *Fr. sanfranciscensis* was the most abundant in both sourdoughs, with a lower occurrence of *Lp. plantarum*, *Fu. rossiae*, *Lentilactobacillus parabuchneri*, *Leuconostoc mesenteroides* and *L. citreum*. Moreover, microbial strains correlated with the volatile organic compounds synthesized during the processing suggesting the employment of the best candidates in pure and/or mixed fermentation to contribute to the aromatic profile (De Vero et al. 2021).

Rocha and Malcata (1999) undertook a culture-dependent taxonomic identification of several groups of bacteria and yeasts in maize and rye flours and respective sourdough used for the manufacture of traditional sourdough *broa*. The novelty of this work lay in the study of this Portuguese traditional specialty for the first time and the broad spectrum of classes of bacteria scrutinized. Overall, 375 microbial isolates were identified, but only the results concerning sourdough are described and analyzed in this paragraph. Lactic acid bacteria and yeasts were the dominant species in sourdough. Moreover, it was reported a decrease in terms of microbial species diversity from maize and rye flours to the sourdough, as a result of the fermentation that drives a spontaneous competitive selection of the microorganisms existing in the raw-materials.

Facultative anaerobic Gram-negative rod-shaped Enterobacteriaceae were identified in maize and rye flours but not in sourdough. Aerobic Gram-negative rod-shaped Pseudomonadaceae were detected in sourdough, with *Sphingobacterium paucimobilis*, *Pseudomonas cepacia* and *Pseudomonas* spp. being the most commonly identified, in addition to other identified species, chiefly *Pseudomonas stutzeri*, *Xanthomonas maltophilia*, *Chryseomonas luteola* and *Agrobacterium radiobacter*. The Gram-negative rod *Achromobacter* spp. was also detected. Regarding the endospore-forming Gram-positive rods, it was identified several species belonging to the Bacillaceae family: *Bacillus pumilus*, *Bacillus cereus*, *Bacillus circulans*, *Bacillus laterosporus*, *Bacillus licheniformis*, *Bacillus macerans*, *Bacillus mycoides*, *Bacillus polymyxa*, *Bacillus stearothermophilus*, *Bacillus badius* and *Bacillus brevis*. Furthermore, a large diversity of Gram-positive catalase-positive cocci was detected in maize and rye flours, but such a diversity decreased in sourdough. Lastly, it was detected *Micrococcus* spp. and *Micrococcus kristinae*, belonging to the Micrococcaceae family, and *Staphylococcus lentus*, *Staphylococcus xylosus*, *Staphylococcus chromogenes*, *Staphylococcus schleiferi*, *Staphylococcus sciuri* and *Staphylococcus aureus*, belonging to the Staphylococcaceae family (Rocha and Malcata 1999).

With respect to the regular non-sporing Gram-positive rods belonging to the family of Lactobacillaceae, the main homofermentative or facultative heterofermentative *Lactobacillus* species identified were *Lev. brevis*, *Lactobacillus delbrueckii* ssp. *lactis*, *Lp. plantarum*, *Lat. curvatus* and *Lactobacillus delbrueckii* ssp. *delbrueckii*. It is interesting to note that other lactobacilli species were found in maize and rye flours, *viz. Lactobacillus acidophilus*, *Lacticaseibacillus paracasei* ssp. *paracasei* and *Lev. brevis*. In addition, the non-sporing Gram-positive rod *Listeria* spp. was also detected. Moreover, Gram-positive catalase-negative cocci belonging to lactic acid bacteria were also identified in maize, rye flours, and sourdough. Regarding the sourdough, the identified LAB species encompassed *Leuconoctoc* spp., *Lactococcus lactis* ssp. *lactis*, *Lactococcus lactis* ssp. *cremoris*, *Pediococcus* spp., *Gardnerella vaginalis*, *Streptococcus agalactiae*, *Streptococcus constellatus* and *Streptococcus equinus* (Rocha and Malcata 1999).

Concerning the yeasts, Rocha and Malcata (1999) identified species belonging to the Ascosporogenous and imperfect yeast groups. In sourdough, the main identified species were *S. cerevisiae* and *Hansenula anomala*, both belonging to the family Saccharomycetaceae. Other yeasts were the Cryptococcaceae *Kluyveromyces marxianus* var. *lactis*, the Cryptococcaceae *Pichia guilliermondii* and *Torulaspora delbrueckii*, and the Saccharomycetaceae *Pichia ohmeri*.

Almeida and Pais (1996a, b) used culture-dependent techniques to characterize the yeast population in 33 sourdoughs for *broa* production. *S. cerevisiae* was the species mostly identified, followed by *T. delbrueckii*, *Pichia membranaefaciens* and *Issatchenkia orientalis* (also known as *Pichia kudriavzevii* or *Candida krusei*). Other identified species were *Pichia anomala*, *Lachancea kluyveri* (a budding yeast related to *Saccharomyces cerevisiae*), *Issatchenkia occidentalis* and *K. marxianus* (Almeida and Pais 1996a). *S. cerevisiae* and *T. delbrueckii* isolated from *broa* sourdoughs were further subjected to studies on the leavening ability and freeze tolerance (Almeida and Pais 1996b) and, more recently, *T. delbrueckii* has been intensely studied by Alves-Araújo et al. (2005a, b, 2007) and Pacheco et al. (2020).

Aiming at characterizing the microbiota of maize and rye flours and respective *broa* sourdough and the effect of spontaneous sourdough fermentation by comparing the microbiota of flours and respective sourdoughs, Rocha and Malcata (2012) made a comprehensive study on the cell viable counts with samples of maize and rye flours and sourdough provided by 14 farmers and households from different regions and sub-regions from Northern Portugal and in two different time periods of the year (Autumn-Winter and Spring-Summer). The vegetative and (mesophilic and thermophilic) spore viable counts were undertaken for a large spectrum of microorganisms, encompassing total, mesophilic and thermophilic microorganisms, yeasts and molds (filamentous fungi), Gram-negative rods, endospore-forming and non-sporing Gram-positive rods, and catalase-positive and catalase-negative Gram-positive cocci. A total of 20 different general, selective and differential culture media were used and involved 26 different incubation conditions. The results unfolded no discrimination of the microbiological viable counts among samples from different regions or seasons, but spontaneous sourdough fermentation represented a key-factor affecting the profile of microbiological traits of the samples; it played a major impact in the spontaneous selection of microorganisms initially found in maize and rye cereals, chiefly the viable counts of yeasts, lactobacilli, streptococci, lactococci, enterococci and leuconostocs increased and, simultaneously, the viable counts of molds, Enterobacteriaceae, Pseudomonadaceae, staphylococci and micrococci decreased. Such results highlighted the importance of cereals as the main contributor to the microbial diversity found in sourdough, as well as the role of spontaneous sourdough fermentation towards the prevalence of desired microorganisms in the sourdough matrix via complex interactions, such as competition and synergism, but other interactions may exist such as symbiosis, neutralism, commensalism, mutualism, protocooperation, parasitism, amensalism, antagonism, *etc*.

The results in this study from Rocha and Malcata (2012) outlined that general total viable counts grown on tryptone soy agar (TSA) increased with sourdough fermentation, *i.e.* from cereal flours to sourdough, whereas total thermophilic viable counts were found in much fewer concentrations and were similar in the three types of studied food matrixes (maize and rye flours, and sourdoughs).

Facultative anaerobic Gram-negative rods were incubated in violet red bile dextrose agar (VRBDA) and MacConkey agar, whereas *Pseudomonas* agar base (PAB) was used for the aerobic Gram-negative rods. VRBDA is indicated for the growth of

Enterobacteriaceae, and MacConkey for *Salmonella*, *Shigella*, *Yersinia* and others, as well as for coliform bacteria. The undesirable presumptive viable counts of facultative anaerobic and aerobic Gram-negative rods decreased from flours to sourdough. The average number of viable microorganisms in flours growing on VRBDA, PAB and MacConkey were, respectively, 5.7–6.0, 6.2–6.6 and 5.8–6.4 log (CFU/g) in flours, whereas in sourdoughs were only 1.4, 1.5 and 1.2 log (CFU/g), respectively (Rocha and Malcata 2012).

Moreover, the viable counts of presumptive endospore-forming Gram-positive rods grown on *Bacillus cereus* medium (BCM) showed that *Bacillus* species are likely to find adequate growth conditions in sourdoughs and, thus, they persist after sourdough fermentation. *Bacillus* are linked with food spoilage and harmful effects on human health. The average number of vegetative cell forms grown on BCM was 6.1–6.6 log (CFU/g), whereas spore forms were 1.9–2.7 log (CFU/g). Although spores from Gram-positive rods may withstand the high baking temperatures, the lower pH due to lactic acid bacteria fermentation may inhibit spore germination (Rocha and Malcata 2012). Moreover, Baird-Parker medium base (BPM) was used to enumerate the presumptive Gram-positive catalase-positive cocci *Staphylococcus* and *Micrococcus*. The results showed a significative reduction of these microorganisms from flours to sourdoughs, on average from 3.6–5.2 to 2.5 log (CFU/g) (Rocha and Malcata 2012).

The regular non-sporing Gram-positive rods were incubated on de Man, Rogosa and Sharp agar (MRS); Gram-positive catalase-negative cocci belonging to LAB, such as *Pediococcus* and *Leuconostoc*, also grow on this culture medium. As a result of the fermentation, sourdough samples had higher viable counts than in cereal flours. The average viable counts were 5.4–5.8 log (CFU/g) in flours and 8.5 log (CFU/g) in sourdough (Rocha and Malcata 2012).

LAB Gram-positive catalase-negative cocci were studied in several specific culture media, chiefly M17 agar, lactic streak agar (LSA), Kenner faecal streptococcal agar (KFS), kanamycin esculin azide agar (KEA) and Mayeux, Sandine and Elliker agar (MSE). The presumptive *Streptococcus*, belonging to lactic acid bacteria (particularly, the homofermentative *Lactococcus*) obtained on M17 and LSA revealed similar or higher viable counts on sourdoughs when compared to cereal flours, ranging from 6.5 and 7.8 log (CFU/g), respectively. The presumptive *Streptococcus* (particularly, the homofermentative *Enterocococcus*) viable counts were attained by using the KFS and KEA culture media, and the results unfolded their increase after fermentation. The average viable counts on these two culture media were 3.2–4.3 in flours and 6.3–6.5 in sourdoughs. The average presumptive *Leuconostoc* viable counts obtained on MSE were 4.5–5.5 in flours and 6.1 in sourdoughs. Therefore, the LAB Gram-positive catalase-negative cocci viable counts showed that these microorganisms are present in the sourdoughs, which may play important roles when it comes 'slow' sourdough fermentations (Rocha and Malcata 2012).

Unlike mold viable counts, yeast viable counts were generally higher in sourdough than in flours – thus indicating the importance of sourdough fermentation towards the reduction of spoilage microorganisms and subsequent extension of the shelf-life of sourdough breads. Yeasts and molds were inoculated in two different culture media and under different incubation conditions, *viz.* yeast extract dextrose chloramphenicol agar (YEDCA) and Rose-Bengal chloramphenicol agar base (RBCAB). The average number of yeasts was 3.6–4.1 log (CFU/g) in flours and 7.2–8.1 log (CFU/g) in sourdough, whereas the average of molds was 4.1–6.2 log (CFU/g) in flours and 0.2–2.1 log (CFU/g) in sourdough (Rocha and Malcata 2012). Therefore, presumptive acid-tolerant yeasts and *Lactobacillus* dominated in sourdoughs after the overnight first fermentation, reaching an average value of 8.1 log (CFU/g), followed by the LAB Gram-positive catalase-negative cocci.

Overall, the study conducted by Rocha and Malcata (2012) pointed out the importance of fermentation towards the predominant microbiota found in sourdoughs and their distinction from the microbiota found in the main raw-materials, the cereal flours. The discrimination of microbiological profiles of maize and rye flours and sourdoughs between regions and seasons was not observed, indeed. The sourdough fermentation process employed by different local farmers and householders plays a major impact on the microbiological profile of sourdoughs and does not allow the discrimination of the microbiology of sourdoughs between geographical regions (Rocha and Malcata 2012). The process of manufacturing traditional sourdough seems to play a major impact, *i.e.*, the exogenous factors like fermentation time and temperature, dough yield, number and time of backslopping steps, percentage and time of the mother-dough used as inoculum. Other important exogenous factors also encompass, for instance, milling, sifting and kneading steps, sodium chloride addition, baking time and temperature. The endogenous factors in breadmaking also affect the final sourdoughs and sourdough breads, which are associated with the type and quality of the flours (for example, enzyme activity, buffering capacity, and composition of carbohydrate, peptide, amino acid and phenolic compounds) (Novotni et al. 2021; Rocha 2011). However, the processing factors seem to have a major impact on the metabolic activity of sourdough microorganisms and to constitute the major reason for the variability of the microbiota found in different sourdoughs, thus making unfeasible their microbial discrimination between geographical regions or seasons. As previously described, among different producers, the first fermentation of *broa* sourdough can be undertaken from 4.5 to 28 h, whereas the second fermentation can take 1.5–3 h, the mother-dough can be kept at room temperature or under refrigeration in a time variable, the ratio maize/rye flour varies, the frequency of *broa* breadmaking also varies significantly and holds a seasonality. These processing factors make of each *broa* a unique baking product.

Having in mind the importance of fermentation for the final microbiological profile of *broa* sourdough, a new study was conducted by Rocha and Malcata (2016a) to ascertain the microbial dynamics of type-I sourdough throughout a long-term spontaneous fermentation of 39 days (sampling in days 0, 1, 2, 3, 7, 9, 14, 29 and 39). The same large spectra of culture media and incubation conditions used by Rocha and Malcata (2012) were employed for the enumeration of cell viable counts. The pH substantially dropped in the first 24 h of fermentation, playing a key-role towards the growth inhibition of undesired microorganisms – in particular, the facultative anaerobic and aerobic Gram-negative rods (Enterobacteriaceae,

Pseudomonadaceae and coliforms) since the endospore-forming Gram-positive rods (*Bacillus*) were ubiquitous throughout the spontaneous fermentation of *broa* sourdough. The pH found in wheat and rye sourdoughs is typically lower (*ca.* 3.6–3.9) than in *broa* sourdough, mainly due to the relatively short-time farmers use in the second fermentation. Indeed, the average pH values in samples of *broa* sourdough and final *broa* were 4.2 and 5.2, respectively (Rocha and Malcata 2016a).

General total viable counts on TSA increased in the first 48 h of fermentation, and after a small decline in number in day 3 the values stabilized until the end of 39 days of fermentation. Facultative anaerobic Gram-positive rods grown on VRBDA decreased significantly after 24 h of fermentation and disappeared by 72 h, whereas those grown on MacConkey agar increased in the first 24 h of fermentation and were also vanished after 72 h of fermentation. Presumptive *Pseudomonas* counts on PAB increased substantially in the first 24 h and further decreased until they vanished after 7 days of fermentation (Rocha and Malcata 2016a). These microbial dynamics show that Gram-negative rods grow in the first 24 h of fermentation but are vanished by the pH drop and competition with the predominant LAB and yeasts. It also shows that increasing the time of the first fermentation of *broa* sourdough could be positive to decrease the number of Gram-negative bacteria. Yet, the time of the second fermentation might be prolonged.

Presumptive *Bacillus* grown on BCM increased rapidly in the first 24 h of fermentation and kept high count values during the entire fermentation period of 39 d, showing that better attention for these microorganisms should be assumed. These endospore-forming Gram-positive rods grow in pH values ranging from 2 to 11; their spores survive to baking temperatures (although they may be inhibited with low pH values), they cause ropiness in baking products and may be harmful to the consumers. Examples of species with this potential are *Bacillus cereus*, *Bacillus subtilis*, *Bacillus coagulans*, *Bacillus licheniformis* and *Bacillus megaterium*. Moreover, the presumptive Gram-positive catalase-positive cocci *Staphylococcus* and *Micrococcus* grown on BPM appeared only in the first 24 h of fermentation, being vanished thereafter (Rocha and Malcata 2016a).

The regular non-sporing Gram-positive rods (lactobacilli) grown on MRS increased significantly in the first 24 h of fermentation and kept in high values until day 9, after which decreased to values similar to the beginning of fermentation. It is important to mention that such a decrease was accompanied by the increase seen in yeasts. Presumptive *Lactobacillus* and LAB cocci were 10- to 100-fold higher than yeasts throughout a large time-period of 9 days of fermentation. Regarding the Gram-positive catalase-negative cocci grown on different culture media (M17, KFS, KEA and MSE), it was disclosed a significant increase in the first 24 h of fermentation and the high values were somehow kept until day 9, after which they vanished, maybe due to the drying and acid conditions. Such a long time of viability indicates that these microorganisms also play an important role in sourdough fermentations (Rocha and Malcata 2016a). Finally, yeast viable counts on YEDCA substantially increased in the first 24 h and, from day 9, on showing their synergistic interaction with lactic acid bacteria and adaptation to relatively low pH values. Regarding molds grown on RBCAB, high viable counts were detected between days 7 and 14 of fermentation, and remained almost in the same value as at the beginning (Rocha and Malcata 2016a).

From above, it was unveiled that presumptive *Lactobacillus*, yeasts and *Bacillus* dominated at the end of the long-term fermentation, although present in small counts after kneading. *Staphylococcus* and Gram-negative rods, present in the dough after kneading, were inhibited during fermentation. After 24 h of fermentation, LAB Gram-positive catalase-negative cocci were present at significant levels. This long-term fermentation reached a quasi-stationary state. The results may suggest the use of 2 to 3 days of the first fermentation of *broa* sourdough as a compromise between the development of endogenous desired microorganisms and the disappearance of most of the microorganisms associated with food spoilage (Rocha and Malcata 2016a).

Mother-dough – also known as sponge-dough, sour-ferment, mother-sponge or seed dough and known as "*isco*" (bait), "*massa-mãe*" (mother-dough), "*massa-azeda*" (sour dough) or "*crescente*" (crescent) in Northern Portugal (Rocha et al. 2023) – is the peace of leavened dough kept aside from baking batch to batch that constitutes a natural and spontaneous microbial starter culture for the traditional manufacture of sourdough *broa*. The addition of mother-dough accelerates spontaneous sourdough fermentations, and it is desirable to avoid deviations of quality between baking batches, which can be achieved by improving processing conditions, such as the storage time and amounts of mother-dough used. Mother-dough is usually stored at room temperature in the wooden kneader but nowadays is more often stored at refrigeration conditions (*ca.* 4 °C). Based on this, Rocha and Malcata (2016b) led a study to ascertain the microbial dynamics of mother-dough during a storage time of 6 days (sampling in days 0, 1, 2 and 6) in a plastic bag under refrigeration conditions using the same set of culture media and incubation conditions of Rocha and Malcata (2012, 2016a) and above described.

Total viable counts on TSA had a slight increase on days 2 and 6, indicating some slow activity of the adventitious microorganisms present in the mother-dough. The average viable counts on TSA were 8.6 log (CFU/g), and small changes in these value (6% maximum variation) indicates the presence of a steady-state in the microbiota of mother-dough. Moreover, a steady average of pH value of 4.1 was measured in the mother-dough, predicting the low microbial activity present in mother-doughs stored at refrigeration conditions.

Viable counts of presumptive facultative anaerobic Gram-negative rods grown on VRBDA (Enterobacteriaceae) and MacConkey (*Salmonella*, *Shigella*, *Yersinia* and other coliforms) disclosed the absence of Enterobacteriaceae and an average value of 4.6 log (CFU/g) in MacConkey, demonstrating the effect of fermentation in reducing these undesired microorganisms. Unexpectedly, *Pseudomonas* viable counts on PAB (presumptive aerobic Gram-negative rods) were found in relatively high concentrations, in average 7.0 log (CFU/g) (Rocha and Malcata 2016a). As a matter of fact, it is expected the disappearing of Gram-negative rods, as stated in the study of long-term sourdough fermentation by Rocha and Malcata (2016b).

Spore-forming Gram-positive rods (*Bacillus*) grown on BCM were maintained at high concentrations in mother-dough over storage time, with a steady average value of 8.6 log (CFU/g) (Rocha and Malcata 2016b). Such results are consistent with other works from the same authors (Rocha and Malcata 2012, 2016a). Furthermore, a steady and relatively low average value of Gram-positive catalase-positive cocci (*Staphylococcus* and *Micrococcus*), grown on BPM, of 4.3 log (CFU/g) was accounted in mother-dough (Rocha and Malcata 2016b). In sourdough, the disappearance of these microorganisms is expected at *ca.* 48 h (Rocha and Malcata 2016a), thus their presence in mother-dough may indicate that longer fermentation of sourdough could be used.

Presumptive lactobacilli in mother-dough (grown on MRS) were found in a steady average value of 8.8 log (CFU/g), which is ten-fold higher than yeast viable counts and shows their prevalence in sourdough fermentations. Gram-positive catalase-negative cocci LAB viable counts obtained on M17, KFS, KEA and MSE culture media reached a steady grand-average of 8.4 log (CFU/g), making apparent their importance in sourdough fermentations in addition to *Lactobacillus* and indicating the need to have deeper studies for this group of microorganisms in sourdough (Rocha and Malcata 2016b).

Sourdough LAB are usually sensitive to drying conditions. Therefore, keeping the mother-dough in plastic bags under refrigeration conditions seems to be a good strategy. LAB grow well at pH values ranging 4.0–4.5 but may still be active at pH values between 3.2 and 9.6. They are mesophilic microorganisms, but some strains may grow between 5 and 45 °C (Rocha and Malcata 2016b). Acetic and lactic acids are the main contributors to sourdough total titratable acidity (TTA) (Rocha 2011). The existence of both homofermentative and heterofermentative strains on sourdoughs are important because the first plays important roles in obtaining a final bread with a good grain and elastic crumb, whereas the last is responsible for shorter and harder gluten and is important to improve bread taste and aroma, as well as to indorse leavening (Rocha 2011; Rocha and Malcata 2016b). Organic acids produced by LAB have important anti-microbial properties, thus increasing the shelf-life of sourdough bread). Fungicidal traits are firstly attributed to acetic acid (Novotni et al. 2021; Rocha and Malcata 2016b). Conversely, lactic acid inhibits the growth of bacilli, which may survive to high baking temperatures and are related to ropy spoilage of bread (Novotni et al. 2021; Rocha and Malcata 2016b). Lactic acid is stronger than acetic acid in decreasing the pH but softer in flavor (Rocha and Malcata 2016b). Unlike lactic acid, acetic acid produced by heterofermentative LAB is a volatile with a sharp vinegar odor, thus producing different flavors and aromas in sourdough bread (Novotni et al. 2021; Rocha and Malcata 2016b). Heterofermentative LAB releases acetic acid which may suppress yeast growth. Such an effect may be avoided when slow acidification at low temperatures is employed (Novotni et al. 2021; Rocha and Malcata 2016b).

In the same study of the microbial dynamics of mother-dough (Rocha and Malcata 2016b), yeast (on YEDCA) and mold (on RBCAB) viable counts reached average values of 7.6 and 5.8 log (CFU/g), respectively. Unlike mounds, yeasts tended to increase slightly (13% maximum variation in both cases). The presence of yeasts and molds in mother-dough predicts their capacity to remain active at low temperatures, acidic conditions and high humidity. Additionally, the acidification of sourdough by LAB is slower at low temperatures, consequently favoring yeast activity – which optimal growth temperature is lower than that for lactobacilli. Conversely, low temperatures may reduce yeast activity because these conditions are favorable for the production of acetic acid by heterformentative LAB and yeasts are more sensitive to this type of organic acid. Furthermore, in the synergistic interactions between yeasts and LAB, yeasts produce growth factors, such as amino acids, peptides, and vitamins, that favors LAB growth and, on the other side, LAB produce anti-microbial compounds, such as organic acids, hydrogen peroxide, carbon dioxide, ethanol, diacetyl, bacteriocins and bacteriocin-like inhibitory substances, *etc*., that inhibit the growth of competitive microorganisms present in the mother-doughs and sourdoughs, including pathogenic and spoilage microorganisms present in cereal flours and baking doughs (Rocha and Malcata 2016b). In general, all the studied microbial groups by Rocha and Malcata (2016b) displayed stability throughout storage under refrigeration conditions, suggesting this process is suitable for the breadmaking of the Portuguese type-I sourdough *broa*.

Increasing the knowledge of the microbial communities during different sourdough fermentations paves the way for the selection of autochthonous beneficial strains further characterized by their technological properties. Accordingly, a large cohort of lactobacilli isolated from Italian sourdoughs have been screened for their proteolytic and peptidase activity in order to improve some functional properties of sourdough, such as antioxidant and anti-inflammatory activities (Galli et al. 2018). In another study, 41 LAB strains isolated from durum wheat sourdoughs for the production of *Cornetto di Matera* bread have been screened for their technological features (*i.e.* organic acids and exopolysaccharide production), as well as their antimicrobial potential (Zotta et al. 2008). Similarly, 100 LAB from Italian sourdoughs (Campania) were characterized for their tolerance to stress conditions commonly encountered within the production of leavened fermented bakery products, including the capability to face typical stress encountered during fermentation, such as acidic conditions, the occurrence of sodium chloride (NaCl) and ethanol, as well for some enzymatic activities (*i.e.* urease, amylase and proteolytic) or for the production of EPS. In particular, *Fr. sanfranciscensis*, *Fu. rossiae*, *Lev. brevis* and two strains of *Leuconostoc pseudomesentereoides* showed the highest survival to stress treatments and interesting technological properties (*i.e.* amino acid and exopolysaccharide production). Several strains exhibited high tolerance to the hardest stress conditions, suggesting their potential use for applications in the bakery industry (Reale et al. 2020).

In order to determine if the bakery environment and equipment are effective sources of LAB and yeasts during the propagation of sourdough, their origin has been investigated in sourdough submitted to back-slopping steps by both culture-dependent and molecular approaches. Interestingly, *Fr. sanfranciscensis* was found to be the most abundant in equipment and sourdoughs. In contrast, *W. cibaria* showed higher adaptability in sourdough than in bakery equipment, suggesting that flours are the main origin of this species. *Lp. plantarum* persisted only in the storage box, dough mixer and sourdough of two bakeries. *S. cerevisiae* was the dominant yeast in the house and sourdough microbiotas, except in one bakery dominated by *K. exigua* (Minervini et al. 2015).

In order to investigate the contribution of the manufacturing process on the microbial composition of sourdough, seven mature type-I sourdoughs were back-slopped for 80 days at artisan bakery and laboratory levels under constant technology parameters. Interestingly, *S. cerevisiae* was no longer detectable in several sourdoughs during late laboratory propagation. Among LAB, *L. plantarum*, *L. sakei* and *W. cibaria* dominated in only some sourdoughs back-slopped at artisan bakeries, while *Lat. citreum* seemed to be more persistent under laboratory conditions. In contrast, *Fr. sanfranciscensis* was found independently of the process. Similar studies could provide elucidation about the technological parameters leading to variations of the bacterial sourdough microbiota during the back-slopping, allowing better control of industrial processes and standardization of high-quality baked goods (Minervini et al. 2012b). A comparison between sourdoughs type-I and type-II obtained from stone-ground soft wheat revealed a similar stable microbiota characterized by a LAB concentration of about 9 log CFU/g, and the dominance of *Lev. brevis* and *Co. paralimentarius*. However, different volatile profiles were found among the experimental breads, suggesting that the sourdough type significantly impacted the bread aroma (Cardinali et al. 2022).

2.4 Conclusion

Traditional sourdough breads using spontaneous slow fermentation by the complex microbiota naturally present in cereal flours and environment, and which resorts to the know-how empirically transferred from generation to generation, play important socio-economic, environmental and cultural roles. From an economic point of view, traditional sourdough breads and other baking goods (*e.g.* biscuits, crackers, pastry, pizza and pasta) represent an extra income for many local small farmers and householders. In the social standpoint, the presence of population in the rural areas contributes to combating desertification. Furthermore, sourdough breads are enjoying a growing popularity among consumers since they are seen as convenient, nutritious, stable, natural and healthy food. In addition, sourdough-based products are highly appreciated for their distinct and unique flavors, aromas and texture, as well as for its nutritional attributes and extended shelf-life.

In terms of environment, the small local farmers constitute important and active environmental protective elements, contributing to sustainable agriculture and environmentally friendly agronomical practices, to the culture of autochthonous and ancestral cereal breeds, to the diversification of agriculture, and to the preservation of natural resources (soil, water, biodiversity) and plant genetic heritage. In fact, small farmers are often recognized for promoting best agronomical practices to improve productivity, quality, safety and sustainability of food, and to mitigate and adapt to climate change. They are also of utmost importance concerning the territorial cohesion and are vital to create agricultural stains of discontinuity to prevent rural fires – which are increasing in number and dimension over the years due to the climatic change and desertification of rural areas, among other factors. Finally, in the cultural and historical standpoint, traditional bakery products and their ancient manufacturing protocols represents a heritage worth preserving with immeasurable value.

Therefore, saving this immaterial cultural heritage and increasing hygiene and safety standards of the final sourdough-based products is important. Sourdough biotechnology is an opportunity to promote the public health and the sustainable manufacture of local, healthy, natural, low-processed baking goods. These high-added value baking goods may improve population's diet and bring opportunities and competitiveness to the industry and local economy. To that purpose, it is fundamental to get a thorough knowledge of the adventitious microbiota present in the flours and sourdough, as well as the dynamics of sourdough fermentations. Indeed, the knowledge of microbial dynamics of sourdough starter cultures in such natural and complex matrixes as sourdoughs is still embryonic. Therefore, dynamics of single and co-culture fermentations with microorganisms isolated from sourdoughs need to be better evaluated and modelled, and the environmental and growth conditions optimized to improve sourdoughs' technological, nutritional and health attributes. Moreover, the specific metabolic pathways and interactions between microorganisms and driving forces and mechanisms that keep sourdough-based microbial consortia stable and productive need to be better elucidated by interactomics. Increase the number of starter cultures (single or co-cultured) in the market is also a need so that greater diversity and quality of products in the global market may be achieved, and the consumption of sourdough baking products may be expanded and generalized.

In the last 45 years (since late 1970's), sourdough biotechnology has been intensively studied in different countries, but deepest scientific knowledge on the technological and functional impact of sourdough fermentation is still open. Research on sourdough technology is definitively a hotspot in biotechnology in Europe and worldwide, which is perceptible by the scientific activity and production and the number of research groups focused on this field. Such a valuable scientific knowledge is dispersed and the recent COST Action 18101 (2018–2023) (COST Action 18101 2018) entitled SOURDOMICS – *Sourdough biotechnology network towards novel, healthier and sustainable food and bioprocesses*, with hundreds of researchers and several companies from all the continents, is a good example of the importance of combining scientific and technological efforts between research groups and, also, highlights the importance given to the field of sourdough biotechnology. These kinds of networks are a driving force to aggregate and valorize the dispersed scientific and technological knowledge on sourdough biotechnology, to share knowledge and laboratory facilities, to promote multidisciplinary within research groups, to avoid duplication of research efforts, and to accelerate innovation and technological transfer.

Particularly, the studies on sourdough biotechnology have been mainly focused on the predominant yeasts and lactic acid bacteria, but other microorganisms exist in spontaneous sourdough fermentation, and their role and impact on breadmaking is still not well-known. Examples are the ubiquitous *Bacillus* and the Gram-positive catalase-negative cocci lactic acid bacteria present in sourdoughs. There is still room to continue the research work on sourdough biotechnology involving a comprehensive genotype characterization by metagenomics, and phenotype characterization of the whole microbiota present in cereals and sourdoughs, as well as involving the knowledge of the encrypted genetic potential and environmental expression of genetic traits and subsequent selection of microorganisms according to their health, nutritional and technological attributes. As a matter of fact, these microorganisms may be seen as cell factories to produce microbial functional (biological activity or technological functionality) metabolites and the metabolic pathways involved in converting of different substrates into high-added value functional metabolites need to be elucidated via metabolomics.

A deep scientifically sound characterization of sourdoughs and sourdough breads is essential to improve their quality, to better understand, control and optimize the sourdough fermentation and baking processes, and to support the health claims related to its consumption. Such knowledge will, consequently, help to expand market niches, contribute to the economic added-value, and trigger the integration of small farmers and households into the agri-food chain, as well as to enable agricultural activity throughout the whole territories. Such scientific knowledge will contribute to the expansion of this market niche and, therefore, will contribute to the increase of its economic value, to the baking companies better compete in a globalizing market, to influence consumers' preferences and market orientations, to the consumer's protection and public health, and to the preservation of the authenticity of traditional food products.

Acknowledgments The authors would like to acknowledge the COST Action 18101 SOURDOMICS – *Sourdough biotechnology network towards novel, healthier and sustainable food and bioprocesses* (https://sourdomics.com/; https://www.cost.eu/actions/CA18101/, accessed on August 30th 2023), where the author Pasquale Russo is member of the working groups 2, 3, 6, 7 and 8, and the author João Miguel Rocha is the Chair and Grant Holder Scientific Representative and is supported by COST (European Co-operation in Science and Technology) (https://www.cost.eu/, accessed on August 30th 2023). COST is a funding agency for research and innovation networks. Author João Miguel Rocha also acknowledges the Universidade Católica Portuguesa, CBQF - Centro de Biotecnologia e Química Fina – Laboratório Associado, Escola Superior de Biotecnologia, Porto, Portugal, as well as the support made by LA/P/0045/2020 (ALiCE) and UIDB/00511/2020-UIDP/00511/2020 (LEPABE) funded by national funds through FCT/MCTES (PIDDAC). Mariagiovanna Fragasso was supported by the European Union Next-Generation EU [Piano Nazionale di Ripresa e Resilienza (PNRR) – Missione 4 Componente 2, Investimento 1.4 – D.D. 1032 17/06/2022, CN00000022] within the Agritech National Research Centre for Agricultural Technologies.

References

Akamine IT, Mansoldo FRP, Vermelho AB (2023) Probiotics in the sourdough bread fermentation: current status. Fermentation 9(2):Article 2. https://doi.org/10.3390/fermentation9020090

Alfonzo A, Miceli C, Nasca A, Franciosi E, Ventimiglia G, Di Gerlando R, Tuohy K, Francesca N, Moschetti G, Settanni L (2017) Monitoring of wheat lactic acid bacteria from the field until the first step of dough fermentation. Food Microbiol 62:256–269. https://doi.org/10.1016/j.fm.2016.10.014

Almeida MJ, Pais CS (1996a) Characterization of the yeast population from traditional corn and rye bread doughs. Lett Appl Microbiol 23:154–158

Almeida MJ, Pais CS (1996b) Leavening ability and freeze tolerance of yeasts isolated from traditional corn and rye bread doughs. Appl Environ Microbiol 62:4401–4404

Alves-Araújo C, Almeida MJ, Sousa MJ, Leão C (2005a) Freeze tolerance of the yeast Torulaspora delbrueckii: celular and biochemical basis. FEMS Microbiol Lett 240:7–14. https://doi.org/10.1016/j.femsle.2004.09.008

Alves-Araújo C, Hernandez-Lopez MJ, Prieto JA, Randez-Gil F, Sousa MJ (2005b) Isolation and characterization of the LGT1 gene encoding a low-affinity glucose transporter from Torulaspora delbrueckii. Yeast 22:165–175. https://doi.org/10.1002/yea.1192

Alves-Araújo C, Pacheco A, Almeida MJ, Spencer-Martins I, Leão C, Sousa MJ (2007) Sugar utilization patterns and respiro-fermentative metabolism in the baker's yeast *Torulaspora delbrueckii*. Microbiology 153:898–904. https://doi.org/10.1099/mic.0.2006/003475-0

Arena MP, Russo P, Spano G, Capozzi V (2019) Exploration of the microbial biodiversity associated with North Apulian sourdoughs and the effect of the increasing number of inoculated lactic acid bacteria strains on the biocontrol against fungal spoilage. Fermentation 5(4):Article 4. https://doi.org/10.3390/fermentation5040097

Arora K, Ameur H, Polo A, Di Cagno R, Rizzello CG, Gobbetti M (2021) Thirty years of knowledge on sourdough fermentation: a systematic review. Trends Food Sci Technol 108:71–83. https://doi.org/10.1016/j.tifs.2020.12.008

Axel C, Brosnan B, Zannini E, Furey A, Coffey A, Arendt EK (2016) Antifungal sourdough lactic acid bacteria as biopreservation tool in quinoa and rice bread. Int J Food Microbiol 239:86–94. https://doi.org/10.1016/j.ijfoodmicro.2016.05.006

Bavaro AR, Di Biase M, Conte A, Lonigro SL, Caputo L, Cedola A, Del Nobile MA, Logrieco AF, Lavermicocca P, Valerio F (2021) Weissella cibaria short-fermented liquid sourdoughs based on quinoa or amaranth flours as fat replacer in focaccia bread formulation. Int J Food Sci Technol 56(7):3197–3208. https://doi.org/10.1111/ijfs.14874

Bonaccio M, Iacoviello L, Donati MB, de Gaetano G (2022) The tenth anniversary as a UNESCO world cultural heritage: an unmissable opportunity to get back to the cultural roots of the Mediterranean diet. Eur J Clin Nutr 76(2):Article 2–Articl183. https://doi.org/10.1038/s41430-021-00924-3

Calabrese FM, Ameur H, Nikoloudaki O, Celano G, Vacca M, Junior WJ, Manzari C, Vertè F, Di Cagno R, Pesole G, De Angelis M, Gobbetti M (2022) Metabolic framework of spontaneous and synthetic sourdough metacommunities to reveal microbial players responsible for resilience and performance. Microbiome 10(1):148. https://doi.org/10.1186/s40168-022-01301-3

Capozzi V, Russo P, Spano G (2012) Microbial information regimen in EU geographical indications. World Patent Inf 34(3):229–231. https://doi.org/10.1016/j.wpi.2012.04.001

Capozzi V, Fragasso M, Romaniello R, Berbegal C, Russo P, Spano G (2017) Spontaneous food fermentations and potential risks for human health. Fermentation 3(4) Scopus. https://doi.org/10.3390/fermentation3040049

Cardinali F, Garofalo C, Reale A, Boscaino F, Osimani A, Milanović V, Taccari M, Aquilanti L (2022) Liquid sourdough from stone-ground soft wheat (Triticum aestivum) flour: development and exploitation in the breadmaking process. Food Res Int 161:111796. https://doi.org/10.1016/j.foodres.2022.111796

COST Action 18101 (2018) SOURDOMICS – *Sourdough biotechnology network towards novel, healthier and sustainable food and bioprocesses*. Supported by the funding agency for research and innovation networks European Cooperation in Science and Technology (COST). COST is a funding agency for research and innovation networks (https://www.cost.eu/). SOURDOMICS: https://sourdomics.com/; https://www.cost.eu/actions/CA18101/. Accessed in 4th August 2022. Time period of the COST Action: 2018-2023

De Bellis P, Rizzello CG, Sisto A, Valerio F, Lonigro SL, Conte A, Lorusso V, Lavermicocca P (2019) Use of a selected leuconostoc citreum strain as a starter for making a "yeast-free" bread. Foods 8(2):70. https://doi.org/10.3390/foods8020070

De Bellis P, Montemurro M, D'Imperio M, Rizzello CG, Sisto A, Lavermicocca P (2020) Production of a yeast-free focaccia with reduced salt content using a selected Leuconostoc citreum strain and seawater. LWT 134:109918. https://doi.org/10.1016/j.lwt.2020.109918

De Vero L, Iosca G, La China S, Licciardello F, Gullo M, Pulvirenti A (2021) Yeasts and lactic acid bacteria for panettone production: an assessment of candidate strains. Microorganisms 9(5):Article 5. https://doi.org/10.3390/microorganisms9051093

De Vuyst L, Van Kerrebroeck S, Harth H, Huys G, Daniel H-M, Weckx S (2014) Microbial ecology of sourdough fermentations: diverse or uniform? Food Microbiol 37:11–29. https://doi.org/10.1016/j.fm.2013.06.002

De Vuyst L, Harth H, Van Kerrebroeck S, Leroy F (2016) Yeast diversity of sourdoughs and associated metabolic properties and functionalities. Int J Food Microbiol 239:26–34. https://doi.org/10.1016/j.ijfoodmicro.2016.07.018

De Vuyst L, Comasio A, Kerrebroeck SV (2023) Sourdough production: fermentation strategies, microbial ecology, and use of non-flour ingredients. Crit Rev Food Sci Nutr 63(15):2447–2479. https://doi.org/10.1080/10408398.2021.1976100

DGADR – Direcção-Geral de Agricultura e Desenvolvimento Rural [Directorate-General for Agriculture and Rural Development] (2023) Lisbon, Portugal https://www.dgadr.gov.pt/. Accessed in 2023-08-30

DGADR – Direcção-Geral de Agricultura e Desenvolvimento Rural [Directorate-General for Agriculture and Rural Development] (2001) Portuguese traditional Products – Agricultural Products, foodstuffs and prepared dishes. https://tradicional.dgadr.gov.pt/en/categories/bread-and-bakery-products. Accessed in 2023-08-30

Di Nunzio M, Bordoni A, Aureli F, Cubadda F, Gianotti A (2018) Sourdough fermentation favorably influences selenium biotransformation and the biological effects of flatbread. Nutrients 10(12):Article 12. https://doi.org/10.3390/nu10121898

Ferrara M, Sisto A, Mulè G, Lavermicocca P, De Bellis P (2021) Metagenetic analysis for microbial characterization of focaccia doughs obtained by using two different starters: traditional Baker's yeast and a selected Leuconostoc citreum strain. Foods 10(6):Article 6. https://doi.org/10.3390/foods10061189

FPCG – Federação Portuguesa das Confrarias Gastronómicas [Portuguese Federation of Gastronomic Confraternities] (2023) Santarém, Portugal. https://fpcggeral.wixsite.com/fpcg. Accessed in 2023-08-30

Gaglio R, Barbera M, Tesoriere L, Osimani A, Busetta G, Matraxia M, Attanzio A, Restivo I, Aquilanti L, Settanni L (2021) Sourdough "ciabatta" bread enriched with powdered insects: physicochemical, microbiological, and simulated intestinal digesta functional properties. Innovative Food Sci Emerg Technol 72:102755. https://doi.org/10.1016/j.ifset.2021.102755

Galli V, Mazzoli L, Luti S, Venturi M, Guerrini S, Paoli P, Vincenzini M, Granchi L, Pazzagli L (2018) Effect of selected strains of lactobacilli on the antioxidant and anti-inflammatory properties of sourdough. Int J Food Microbiol 286:55–65. https://doi.org/10.1016/j.ijfoodmicro.2018.07.018

Gänzle MG (2014) Enzymatic and bacterial conversions during sourdough fermentation. Food Microbiol 37:2–10. https://doi.org/10.1016/j.fm.2013.04.007

Garofalo C, Silvestri G, Aquilanti L, Clementi F (2008) PCR-DGGE analysis of lactic acid bacteria and yeast dynamics during the production processes of three varieties of Panettone. J Appl Microbiol 105(1):243–254. https://doi.org/10.1111/j.1365-2672.2008.03768.x

Gobbetti M, De Angelis M, Di Cagno R, Calasso M, Archetti G, Rizzello CG (2019) Novel insights on the functional/nutritional features of the sourdough fermentation. Int J Food Microbiol 302:103–113. https://doi.org/10.1016/j.ijfoodmicro.2018.05.018

Graça C, Lima A, Raymundo A, Sousa I (2021) Sourdough fermentation as a tool to improve the nutritional and health-promoting properties of its derived-products. Fermentation 7(4):Article 4. https://doi.org/10.3390/fermentation7040246

Hernández-Alcántara AM, Chiva R, Mohedano ML, Russo P, Ruiz-Masó JÁ, del Solar G, Spano G, Tamame M, López P (2022) Weissella cibaria riboflavin-overproducing and dextran-producing strains useful for the development of functional bread. Front Nutrit 9. https://www.frontiersin.org/articles/10.3389/fnut.2022.978831

Kalo PJ, Ollilainen V, Rocha JM, Malcata FX (2006) Identification of molecular species of simple lipids by normal phase liquid chromatography-positive electrospray tandem mass spectrometry, and application of developed methods in comprehensive analysis of low erucic acid rapeseed oil lipids. Int J Mass Spectrom 254:106–121. https://doi.org/10.1016/j.ijms.2006.05.022

Landis EA, Oliverio AM, McKenney EA, Nichols LM, Kfoury N, Biango-Daniels M, Shell LK, Madden AA, Shapiro L, Sakunala S, Drake K, Robbat A, Booker M, Dunn RR, Fierer N, Wolfe BE (2021) The diversity and function of sourdough starter microbiomes. elife 10:e61644. https://doi.org/10.7554/eLife.61644

Lattanzi A, Minervini F, Di Cagno R, Diviccaro A, Antonielli L, Cardinali G, Cappelle S, De Angelis M, Gobbetti M (2013) The lactic acid bacteria and yeast microbiota of eighteen sourdoughs used for the manufacture of traditional Italian sweet leavened baked goods. Int J Food Microbiol 163(2):71–79. https://doi.org/10.1016/j.ijfoodmicro.2013.02.010

Lau SW, Chong AQ, Chin NL, Talib RA, Basha RK (2021) Sourdough microbiome comparison and benefits. Microorganisms 9(7):Article 7. https://doi.org/10.3390/microorganisms9071355

Ma S, Wang Z, Guo X, Wang F, Huang J, Sun B, Wang X (2021) Sourdough improves the quality of whole-wheat flour products: mechanisms and challenges—a review. Food Chem 360:130038. https://doi.org/10.1016/j.foodchem.2021.130038

Minervini F, Di Cagno R, Lattanzi A, De Angelis M, Antonielli L, Cardinali G, Cappelle S, Gobbetti M (2012a) Lactic acid bacterium and yeast microbiotas of 19 sourdoughs used for traditional/typical Italian breads: interactions between ingredients and microbial species diversity. Appl Environ Microbiol 78(4):1251–1264. https://doi.org/10.1128/AEM.07721-11

Minervini F, Lattanzi A, De Angelis M, Di Cagno R, Gobbetti M (2012b) Influence of artisan bakery- or laboratory-propagated sourdoughs on the diversity of lactic acid bacterium and yeast microbiotas. Appl Environ Microbiol 78(15):5328–5340. https://doi.org/10.1128/AEM.00572-12

Minervini F, Lattanzi A, De Angelis M, Celano G, Gobbetti M (2015) House microbiotas as sources of lactic acid bacteria and yeasts in traditional Italian sourdoughs. Food Microbiol 52:66–76. https://doi.org/10.1016/j.fm.2015.06.009

Minervini F, Dinardo FR, De Angelis M, Gobbetti M (2019) Tap water is one of the drivers that establish and assembly the lactic acid bacterium biota during sourdough preparation. Sci Rep 9(1):Article 1. https://doi.org/10.1038/s41598-018-36786-2

MINHA TERRA – Federação Portuguesa de Associações de Desenvolvimento Local [Portuguese Federation of Local Development Associations] (2023) Lisbon, Portugal https://www.minhaterra.pt/. Accessed in 2023-08-30

Montanari C, Bargossi E, Lanciotti R, Chinnici F, Gardini F, Tabanelli G (2014) Effects of two different sourdoughs on the characteristics of Pandoro, a typical Italian sweet leavened baked good. LWT Food Sci Technol 59(1):289–299. https://doi.org/10.1016/j.lwt.2014.04.045

Musatti A, Mapelli C, Foschino R, Picozzi C, Rollini M (2016) Unconventional bacterial association for dough leavening. Int J Food Microbiol 237:28–34. https://doi.org/10.1016/j.ijfoodmicro.2016.08.011

Novotni D, Gänzle M, Rocha JM (2021) Chapter 5. Composition and activity of microbiota in sourdough and their effect on bread quality and safety. In: Galanakis CM (ed) Trends in wheat and bread making. Elsevier-Academic Press, Cambridge, MA, pp 129–172. https://doi.org/10.1016/B978-0-12-821048-2.00005-2, 469 pages. Galanaksis-TWBM-1632435, ISBN 978-0-12-821048-2

Novotni D, Gamel TH, Helou C, Rocha JM (2023) Chapter 17. Transferring theoretical principles into practical applications: Cereals, pseudocereals, and their applications in breadmaking and other agri-food. In: Rakszegi M, Papageorgiou M, Rocha JM (eds) Developing sustainable and health promoting cereals and pseudocereals. Subtitle: conventional and molecular breeding. Elsevier, Academic Press, London, pp 400–432. eBook ISBN: 9780323906890, Paperback ISBN: 9780323905664. Published: 27 March 2023. No. of pages: 506

Pacheco A, Donzella L, Hernandez-Lopez MJ, Almeida MJ, Prieto JA, Randez-Gil F, Morrissey JP, Sousa MJ (2020) Hexose transport in Torulaspora delbrueckii: identification of Igt1, a new dual-affinity transporter. FEMS Yeast Res 20:1–10. https://doi.org/10.1093/femsyr/foaa004

Palla M, Cristani C, Giovannetti M, Agnolucci M (2017) Identification and characterization of lactic acid bacteria and yeasts of PDO Tuscan bread sourdough by culture dependent and independent methods. Int J Food Microbiol 250:19–26. https://doi.org/10.1016/j.ijfoodmicro.2017.03.015

Palomba S, Blaiotta G, Ventorino V, Saccone A, Pepe O (2011) Microbial characterization of sourdough for sweet baked products in the Campania region (southern Italy) by a polyphasic approach. Ann Microbiol 61(2):307–314. https://doi.org/10.1007/s13213-010-0140-2

Pasqualone A, Vurro F, Summo C, Abd-El-Khalek MH, Al-Dmoor HH, Grgic T, Ruiz M, Magro C, Deligeorgakis C, Helou C, Le-Bail P (2022) The large and diverse family of Mediterranean flat breads: a database. Foods 11(15):Article 15. https://doi.org/10.3390/foods11152326

Pérez-Alvarado O, Zepeda-Hernández A, Garcia-Amezquita LE, Requena T, Vinderola G, García-Cayuela T (2022) Role of lactic acid bacteria and yeasts in sourdough fermentation during breadmaking: evaluation of postbiotic-like components and health benefits. Front Microbiol 13:969460. https://doi.org/10.3389/fmicb.2022.969460

Picozzi C, Clagnan E, Musatti A, Rollini M, Brusetti L (2022) Characterization of two Zymomonas mobilis wild strains and analysis of populations dynamics during their leavening of bread-like doughs. Foods 11(18):Article 18. https://doi.org/10.3390/foods11182768

Pino A, Russo N, Solieri L, Sola L, Caggia C, Randazzo CL (2022) Microbial consortia involved in traditional Sicilian sourdough: characterization of lactic acid bacteria and yeast populations. Microorganisms 10(2):Article 2. https://doi.org/10.3390/microorganisms10020283

Quattrini M, Liang N, Fortina MG, Xiang S, Curtis JM, Gänzle M (2019) Exploiting synergies of sourdough and antifungal organic acids to delay fungal spoilage of bread. Int J Food Microbiol 302:8–14. https://doi.org/10.1016/j.ijfoodmicro.2018.09.007

Raimondi S, Amaretti A, Rossi M, Fall PA, Tabanelli G, Gardini F, Montanari C (2017) Evolution of microbial community and chemical properties of a sourdough during the production of Colomba, an Italian sweet leavened baked product. LWT 86:31–39. https://doi.org/10.1016/j.lwt.2017.07.042

Ramos L, Alonso-Hernando A, Martínez-Castro M, Morán-Pérez JA, Cabrero-Lobato P, Pascual-Maté A, Téllez-Jiménez E, Mujico JR (2021) Sourdough biotechnology applied to gluten-free baked goods: rescuing the tradition. Foods 10(7):Article 7. https://doi.org/10.3390/foods10071498

Reale A, Di Renzo T, Boscaino F, Nazzaro F, Fratianni F, Aponte M (2019) Lactic acid bacteria biota and aroma profile of Italian traditional sourdoughs from the Irpinian area in Italy. Front Microbiol 10. https://www.frontiersin.org/articles/10.3389/fmicb.2019.01621

Reale A, Zotta T, Ianniello RG, Mamone G, Di Renzo T (2020) Selection criteria of lactic acid bacteria to be used as starter for sweet and salty leavened baked products. LWT 133:110092. https://doi.org/10.1016/j.lwt.2020.110092

Ribet L, Dessalles R, Lesens C, Brusselaers N, Durand-Dubief M (2023) Nutritional benefits of sourdoughs: a systematic review. Adv Nutr 14(1):22–29. https://doi.org/10.1016/j.advnut.2022.10.003

Rocha JM (2011) Microbiological and lipid profiles of broa: contributions for the characterization of a traditional portuguese bread. Ph.D thesis dissertation. Instituto Superior de Agronomia [Higher Institute of Agriculture], Universidade de Lisboa [University of Lisbon] (ISA-UL), Lisbon, Portugal, 705 pages. Thesis available at http://hdl.handle.net/10400.5/3876

Rocha JM, Malcata FX (1999) On the microbiological profile of traditional Portuguese sourdough. J Food Prot 62(12):1416–1429. ISSN: 0362-028X. url: http://www.scopus.com/inward/record.url?eid=2-s2.0-0345201643&partnerID=MN8TOARS

Rocha JM, Malcata FX (2012) Microbiological profile of maize and rye flours, and sourdough used for the manufacture of traditional Portuguese bread. Food Microbiol 31:72–88. https://doi.org/10.1016/j.fm.2012.01.008

Rocha JM, Malcata FX (2016a) Microbial ecology dynamics in Portuguese broa sourdough. J Food Qual 39(6):634–648. https://doi.org/10.1111/jfq.12244

Rocha JM, Malcata FX (2016b) Behavior of the complex micro-ecology in maize and rye flour and mother-dough for broa throughout storage. J Food Qual 39:218–233. https://doi.org/10.1111/jfq.12183

Rocha JM, Brás A, Malcata FX (2003) Pão de milho: Caracterização do produto tradicional de produção e melhoramento tecnológico [Bread of maize: Characterization of the traditional product and technological improvement], ESB-UCP (Editor), ISBN: 972-98476-9-X, 76 pp.

Rocha JM, Kalo PJ, Malcata FX (2010a) Neutral lipids in non-starch lipid and starch lipid extracts from portuguese sourdough bread. Eur J Lipid Sci Technol 112:1138–1149. https://doi.org/10.1002/ejlt.201000101

Rocha JM, Kalo PJ, Ollilainen V, Malcata FX (2010b) Separation and identification of neutral cereal lipids by normal phase high-performance liquid chromatography, using evaporative light-scattering and electrospray mass spectrometry for detection. J Chromatogr A 1217:3013–3025. https://doi.org/10.1016/j.chroma.2010.02.034

Rocha JM, Kalo PJ, Malcata FX (2011) Neutral lipids in free, bound and starch lipid extracts of flours, sourdough and portuguese sourdough bread, determined by NP-HPLC-ELSD. Cereal Chem 88(4):400–408. https://doi.org/10.1094/CCHEM-11-10-0157

Rocha JM, Kalo PJ, Malcata FX (2012a) Fatty acid composition of non-starch and starch neutral lipid extracts of portuguese sourdough bread. J Am Oil Chem Soc 89(11):2025–2045. https://doi.org/10.1007/s11746-012-2110-2

Rocha JM, Kalo PJ, Malcata FX (2012b) Composition of neutral lipid classes and content of fatty acids throughout sourdough breadmaking. Eur J Lipid Sci Technol 114(3):294–305. https://doi.org/10.1002/ejlt.201100208

Rocha JM, Brás A, Miranda J, Malcata FX (2023) Chapter 13. Broa – a Portuguese traditional sourdough bread, made of maize and rye flours. In: Garcia-Vaquero M, Pastor K, Orhun GE, McElhatton A, Rocha JMF (eds) Traditional european breads – an illustrative compendium of ancestral knowledge and cultural heritage. Springer Nature, Cham, Switzerland, pp 251–293. https://doi.org/10.1007/978-3-031-23352-4. Hardcover ISBN 978-3-031-23351-7 (Published: 17 May 2023), Softcover ISBN 978-3-031-23354-8 (Due: 31 May 2024), eBook ISBN 978-3-031-23352-4 (Published: 16 May 2023). Number of Pages VIII, 422

Russo P, Beleggia R, Ferrer S, Pardo I, Spano G (2010) A polyphasic approach in order to identify dominant lactic acid bacteria during pasta manufacturing. LWT Food Sci Technol 43(6):982–986. https://doi.org/10.1016/j.lwt.2010.01.013

Russo P, Fares C, Longo A, Spano G, Capozzi V (2017) Lactobacillus plantarum with broad antifungal activity as a protective starter culture for bread production. Foods 6(12):Article 12. https://doi.org/10.3390/foods6120110

Valmorri S, Tofalo R, Settanni L, Corsetti A, Suzzi G (2010) Yeast microbiota associated with spontaneous sourdough fermentations in the production of traditional wheat sourdough breads of the Abruzzo region (Italy). Antonie Van Leeuwenhoek 97(2):119–129. https://doi.org/10.1007/s10482-009-9392-x

Van Kerrebroeck S, Maes D, De Vuyst L (2017) Sourdoughs as a function of their species diversity and process conditions, a meta-analysis. Trends Food Sci Technol 68:152–159. https://doi.org/10.1016/j.tifs.2017.08.016

Venturi M, Guerrini S, Granchi L, Vincenzini M (2012a) Typing of Lactobacillus sanfranciscensis isolates from traditional sourdoughs by combining conventional and multiplex RAPD–PCR profiles. Int J Food Microbiol 156(2):122–126. https://doi.org/10.1016/j.ijfoodmicro.2012.03.011

Venturi M, Guerrini S, Vincenzini M (2012b) Stable and non-competitive association of Saccharomyces cerevisiae, Candida milleri and Lactobacillus sanfranciscensis during manufacture of two traditional sourdough baked goods. Food Microbiol 31(1):107–115. https://doi.org/10.1016/j.fm.2012.02.011

Vernocchi P, Valmorri S, Gatto V, Torriani S, Gianotti A, Suzzi G, Guerzoni ME, Gardini F (2004) A survey on yeast microbiota associated with an Italian traditional sweet-leavened baked good fermentation. Food Res Int 37(5):469–476. https://doi.org/10.1016/j.foodres.2004.01.004

Visioli G, Giannelli G, Agrimonti C, Spina A, Pasini G (2021) Traceability of Sicilian durum wheat landraces and historical varieties by high molecular weight Glutenins footprint. Agronomy 11(1):Article 1. https://doi.org/10.3390/agronomy11010143

von Gastrow L, Michel E, Legrand J, Amelot R, Segond D, Guezenec S, Rué O, Chable V, Goldringer I, Dousset X, Serpolay-Bessoni E, Taupier-Letage B, Vindras-Fouillet C, Onno B, Valence F, Sicard D (2023) Microbial community dispersal from wheat grains to sourdoughs: a contribution of participatory research. Mol Ecol 32(10):2413–2427. https://doi.org/10.1111/mec.16630

Zarour K, Llamas MG, Prieto A, Rúas-Madiedo P, Dueñas MT, de Palencia PF, Aznar R, Kihal M, López P (2017) Rheology and bioactivity of high molecular weight dextrans synthesised by lactic acid bacteria. Carbohydr Polym 174:646–657. https://doi.org/10.1016/j.carbpol.2017.06.113

Zotta T, Piraino P, Parente E, Salzano G, Ricciardi A (2008) Characterization of lactic acid bacteria isolated from sourdoughs for Cornetto, a traditional bread produced in Basilicata (Southern Italy). World J Microbiol Biotechnol 24(9):1785–1795. https://doi.org/10.1007/s11274-008-9671-0

Chapter 3
Sourdough Microbiota Diversity in Central Europe

Katarzyna Dybka-Stępień, Anna Otlewska, and Anna Diowksz

3.1 Introduction

In the face of extremely strong market competition, the success of a product is determined by its high quality, maintained at a constant level. From the consumer's perspective the most important factor in the evaluation of a food product is its sensory quality, with palatability being the main component of it. Very broad diversity of regional breads, mostly based on rye sourdough, famous for their outstanding taste and aroma is behind not fading popularity of sourdough bread in Central Europe.

In Central Europe the dominant type of bread is a mixed wheat-rye and rye bread produced with the traditional technology of sourdough fermentation. Craft bakers continue the centuries-old art of baking, once performed at home. The main secret for sourdough performance lies in microbial diversity and the existence of complex microbial consortia colonizing this ecosystem. A wide range of metabolites produced by lactic acid bacteria and yeast selected in a multistage fermentation process makes taste and smell much richer compared to yeast bread (Gänzle et al. 2007; Diowksz 2011; Arora et al. 2021).

The definition widely used in the literature describes sourdough as mixture of flour and water, fermented by lactic acid bacteria (LAB) and yeasts, which produce the metabolites and activate a cascade of enzymes, as a result of which flour ingredients are transformed (Poutanen and Katina 2014; Arora et al. 2021). Today, there is an increasing use of sourdough starters in the production of cereal-based foods (bread, crackers, and cake) both in Central Europe and beyond, due to the technological and nutritional quality, as well as the safety of the products obtained. In addition, the balanced and natural nature of sourdough allows to obtain healthy

K. Dybka-Stępień (✉) · A. Otlewska (✉) · A. Diowksz
Institute of Fermentation Technology and Microbiology, Faculty of Biotechnology and Food Sciences, Lodz University of Technology, Lodz, Poland
e-mail: katarzyna.dybka@p.lodz.pl; anna.otlewska@p.lodz.pl

E. B. Ceresino et al. (eds.), *Sourdough Microbiota and Starter Cultures for Industry*, https://doi.org/10.1007/978-3-031-48604-3_3

products with a clean label (Poutanen and Katina 2014). Sourdough-based baking products are considered to be healthier alternatives to those made using regular dough. The presence of *Lactobacillus* and wild yeast helps neutralize the acid in the leaven (phytic acid), thus making these products easily digestible unlike those made from regular dough. Furthermore, the presence of other vital nutrients, such as folic acid, magnesium, iron, B vitamins, and zinc, make these products a preferred choice among consumers owing to health benefits.

Research conducted in 2019 by Federation of European Manufacturers and Suppliers of Ingredients to the Bakery, Confectionery and Patisserie Industries (FEDIMA) among 5111 respondents indicates that 42% of people knew what sourdough is, 53% knew the name, and only 5% had not heard of sourdough. What's more, two of the three respondents consciously choose sourdough bread (17% always, 52% time to time, 21% never, 10% do not know). It should be emphasized that the knowledge of sourdough bread and the rate of its consumption differ in individual European countries. Among the Central European countries, the highest purchase rate is in Poland and Germany, 90% and 76% respectively. Here, too, 61% and 40% of respondents are familiar with sourdough (FEDIMA 2019).

In 2021, the size of the sourdough market was estimated at USD 2.9 billion. The sourdough industry is expected to grow from $1.98 billion in 2022 to $5.37 billion by 2030. The industry will grow exponentially over the forecast period due to increasing consumption of sourdough bread worldwide, with Europe dominating the global sourdough market by 2027, while Germany will be the largest market in the region (Albagli et al. 2023; Singh 2023; Report GVR-3-68038-510-6 2023).

3.2 Factors Determining the Diversification of Sourdough Microbiota

Sourdough bread making is an ancient biotechnological process and regardless of whether the type of product and implemented protocol, the biodiversity of microbiota creating its unique properties depends on many factors. Quantity and biodiversity of sourdough microbiota along with their metabolic abilities, strongly impact on proper dough and bread quality. The microbial ecosystem of sourdough is peculiar for the presence of yeasts and lactic acid bacteria (LAB), as well as other microorganisms originated (e.g., acetic acid bacteria AAB, spoilage causing bacteria and moulds) from raw materials used for sourdough preparation (before stabilization of microbiota). Microorganisms identified in sourdough are responsible for its acidification (LAB), flavour formation (LAB, AAB, and yeasts), and leavening (yeasts and heterofermentative LAB species). The interactions between microorganisms tend to be complex in nature but are essential for sourdough properties.

In context of four sourdough's types, their ecology is well documented, especially the initial development of the sourdough seems to be well recognized. In-depth sourdoughs' characteristic was made by Pérez-Alvarado et al. (2022) in the relationship to closer look into re-feeding practices. Type I of the sourdough is a result of spontaneous fermentation, but nowadays it is prepared mostly as artisanal baking for local, small-scale production of bread (Calvert et al. 2021). Type II sourdough starters are made of LAB and/or yeasts in a flour-water mixture. What is characteristic for Type II starters they are richer in organic acids and ensure adequate flavour and technological properties for industrial scale production. Young sourdough starters are rich in carbohydrates, whereas, the mature one, are characterized by limitation of nutrients and high content of CO_2 (De Vuyst and Neysens 2005). There have been many scientific articles describing the effects of numerous factors on sourdough ecosystems (Calvert et al. 2021; Gänzle and Ripari 2016; Minervini et al. 2014). Simultaneously, there are different ways to classify factors affecting the biodiversity of sourdough microbiota. In general, they can be categorized as of chemical, physical, and biological origin or summarized as being external and internal. Endogenous factors impacting sourdough cover quality of the raw materials. It refers to the type of the flour related with the ash content as well as the plant origin and enzymatic activity. Microbial consortium (being a consequence of microorganisms present in the ingredients and the in-house microbiota) and its composition change during sourdough maturation. Of course, such traditional classification of listed parameters does not fully reflect the situation of spontaneous fermentations, especially when sourdough is prepared manually (Minervini et al. 2014; De Vuyst et al. 2009).

3.2.1 Technological Parameters

The ingredients and the recipe used for sourdough preparation drive the ecology of the microbiota. Extrinsic factors impacting biodiversity of microbes refer mostly to the practices used for sourdough propagation. Amongst such factors the well-recognized are the temperature of sourdough propagation and storage, the number and duration of propagation steps (both for fermentation and/or refreshment), water activity (a_w), the presence of oxygen, the value of initial pH, as well as the dough yield (reflecting the flour to water ratio). Not only the ratio between lactic acid bacteria and yeasts impacts the dough yield. Both leavening capacities and flavour may be improved by introduction of proper amount of heterofermentative LAB. The ratio between lactic acid bacteria and yeasts can be influenced by the fermentation temperature, pH, redox potential, NaCl concentration, dough yield and the dose of sourdough (% content) used as inoculum. Depending on pH not only yeasts or LAB will grow preferably but also different LAB genera were found to dominate the sourdough in such situation as well. Simultaneously, the temperature around 25–27 °C will favourable the development of yeasts in comparison to higher, being in the range of 30 and 40 °C, which is more suitable for LAB growth. The different fermentation temperatures can influence both: growth of the microorganisms and formation of secondary metabolites, organic acids, and aromatic compounds. Yeast grows better at lower temperatures, whereas at higher temperatures releases more ethanol and CO_2. Similarly, homofermentative lactic acid bacteria proliferate better at relatively low temperatures, while lactic acid is produced most intensively in the temperature range of 30–35 °C. They become dominant at higher temperatures but have little effect on the characteristic bread flavour. In turn, the acidic environment promotes the development of yeast, which enhances alcoholic fermentation and gas release. On the other hand, colder dough favours heterofermentative lactic acid bacteria and the formation of acetic acid. Since the pH value strongly influences lactic acid bacteria and yeast performance, the formation of the sourdough microbiota is also influenced by refreshments in the subsequent fermentation stages. With a high percentage of the previous stage, the pH remains low, and yeast develops better, whereas with high multiplication the dough acidity gets lower and lactic bacteria find better conditions for development. The fermentation time also affects the condition of the microorganisms. The longer the dough is fermented the more lactic acid bacteria grow, which is reflected by a significant contribution of their metabolites in bread flavour. To shorten the process time, the temperature should be increased, the dough should have a stiff consistency and low multiplication is necessary (Decock and Capelle 2005; De Vuyst et al. 2009; Martín-Garcia et al. 2023). More details are discussed by Calvert et al. (2021), Minervini et al. (2014), De Vuyst et al. (2009) or De Vuyst and Neysens (2005).

3.2.2 Flour and Other Ingredients

Numerous reports show differences in the taxonomic structure of sourdough communities as a result of the different flours usage. Flours may serve as a habitat for many microorganisms which are essential for proper quality of the sourdough or are considered as unwanted contaminants. The flour, which is the basic ingredient for sourdough, can come from milling whole or refined grains. Different types of flour have different value in the preparation of sourdough. Flour types differ in their chemical compositions, nutrient value and microbiological quality depending on the crop, its variety, geographical site, and the bath. Thus, flour determines the consortium of microorganisms by the varying content of characteristic carbohydrates, nitrogen sources, minerals, or enzymatic activity (e.g., proteases, amylases) as well as the presence of microorganisms (Gobbetti et al. 2016). Some sourdoughs are enriched with additional ingredients such as sugar, salt, fruit must, resulting with changes of the stability of sourdough together with biodiversity of the microbiota (Calvert et al. 2021). The amount of volatile compounds produced during sourdough fermentation depends not only on the type of flour (wheat, rye) and its ash content (type of flour), but also on the fermentation temperature and dough yield. These factors determine the quantity and quality of the microorganisms active in the sourdough. The type of flour and the consistency of the dough have a particular effect on the development of lactic bacteria, and thus on the acidification of the dough. The higher type of flour promotes lactic acid bacteria over yeast development, thus in consequence the production of lactic and acetic acid. Similarly, dough yield influences the ratio between lactic acid bacteria and yeast, as well as between homo- and heterofermentative lactic acid bacteria. The thick consistency increases the acidification of the dough, because in the stiff dough lactic acid bacteria develop better than yeast. Additionally, at denser consistencies a better development of heterofermentative bacteria then homofermentative ones is observed, which promotes acetic acid production. The taste of bread is perceived pleasant when lactic acid prevails over acetic acid. Therefore, for a milder-tasting bread a yield of 170–180 is recommended, while a more pronounced sharp taste of bread results from fermentation of the dough prepared with the yield below 160. On the other hand, at looser consistencies yeast propagation is enhanced and yeast-produced aromatic compounds are accentuated along with increase in isoalcohols content (Decock and Capelle 2005; Hansen 2005; Minervini et al. 2014; Van Kerrebroeck et al. 2017; Martín-Garcia et al. 2023).

3.2.3 Production Environment

The presence of in-house specific microorganisms also drives the succession and biodiversity of sourdough microorganisms as well as the stability of sourdough over time. Household microorganisms of the bakery are distributed by the air as well as by the employees and by the usage of vessels and other equipment required in production process. The modern consumer is more and more demanding and expects a product of the highest, constant quality. However, traditional spontaneous sourdough is difficult to manage as the microbiota consists of numerous strains and is highly diverse, thus the control over its composition depends entirely on the skill of the baker (Decock and Capelle 2005; Van Kerrebroeck et al. 2017). Several authors highlighted the distinguished microbiota depending on the manual vs. machine preparation of the sourdough and the changes resulting from the sex of the employees feeding the sourdough. Women are considered as the reservoir of LAB that colonize their genitourinary tract. On the other hand, some unexpected issues (causing so called drift) may also drastically impact on the sourdough biodiversity (Calvert et al. 2021 and references therein).

3.2.4 Interactions Between Microorganisms

Depending on the type of the sourdough different interactions between microorganisms may or may not be "as expected". Traditionally LAB and yeasts interact each other through metabolisms of the carbohydrates and the nitrogen sources connected with the biosynthesis of some inhibitory or stimulatory chemical compounds (De Vuyst et al. 2009). The nature of microbial interactions in the sourdough may vary from mutualistic, through commensal to fully antagonistic. The succession of LAB and yeasts relates to the metabolisms of specific carbohydrates (like maltose used by *Fructilactobacillus sanfranciscensis*), was very well described in the review by Minervini et al. (2014). Metabolites produced by different LAB (including inter alia lactic acid, acetic acid, phenyllactic acid, short-chain fatty acids, bacteriocins) may inhibit the growth of fungi and ropiness spore-forming bacteria (Mohsen et al. 2016).

It is important to be aware that biodiversity of the sourdough is a result of a combination of multiple parameters at the same time. In case of endogenous factors, the chemical and microbiological composition of the sourdough is the most important, whereas exogenous factors are mainly affected by temperature and redox potential (oxygen concentration).

3.3 Microbial Diversity of Sourdough

Hundreds of sourdoughs have been investigated in the last decades. Up to now, more than 90 different LAB species and more than 40 different yeast species have been isolated from sourdoughs worldwide (De Vuyst et al. 2017).

Lactic acid bacteria active in sourdoughs are characterized by a great diversity. Production of lactic acid is their common feature, however, depending on the species, some strains also produce significant amounts of other acids and volatile compounds. The most important by-products of this fermentation are acetic acid, ethyl alcohol, diacetyl, and acetaldehyde. Other volatile acids (propionic acid, formic acid) and alcohols (butanol, propanol) are also released in smaller amounts. Yeast present in sourdoughs, apart from ethanol and carbon dioxide, also produces a number of fermentation by-products, such as aldehydes, aliphatic and isoaliphatic alcohols, acids, keto acids and esters. The final effect is therefore determined by the combination of the activities of these two groups of microorganisms. This wide pool of metabolites creates a specific, unique taste and smell of sourdough bread incomparably more interesting than those of yeast bread (Hansen and Schieberle 2005).

Sourdoughs fermented by heretofermentative lactic acid bacteria are not only higher in acetic acid and ethanol, but also in ethyl acetate and ethyl hexanoate compared to sourdoughs fermented by homofermentative bacteria, which contain more diacetyl and other carbonyl compounds. In the presence of yeast significantly more ethanol, isoalcohols, esters and diacetyl are found in the dough (Corsetti and Settani 2007).

The best way to overcome the disadvantages of a potentially elusive spontaneous process and to ensure that fermentation proceeds properly is to use a bakery starter culture. Standardised bacteria/yeast population guarantees consistent performance throughout a time. The bakery industry is interested in defined starter cultures that have strong stabilizing effect and facilitate the production procedure, while maintaining the quality typical of artisan bread. In order to achieve such benefits from the use of fermentation starters, appropriate strains selection is essential. This effect is strictly dependent on the restoration of a natural microbiota, which is a combination of autochthonous strains previously isolated from high-quality spontaneous sourdough. Therefore, to obtain an aroma and pleasant taste of sourdough bread, mixed cultures of lactic acid bacteria and yeast are recommended. A bakery starter culture must at the same time meet certain technological requirements to be suitable for industrial applications. The strains forming the starter consortium should demonstrate the ability to coexist in synergy. This refers in particular to the end products of the metabolism of the sourdough-active microorganisms that shape the organoleptic properties of bread of traditional artisan style (De Vuyst and Neysens 2005; Minervini et al. 2014; Arora et al. 2021; Petkova et al. 2021).

The final flavour of bread depends on many factors, including recipe and processing parameters, but there is no doubt that fermentation is the crucial stage for the creation of the unique flavour in sourdough bread. The biodiversity of microorganisms found in sourdoughs is reflected by the wide variety of baked goods obtained with sourdough fermentation throughout the world. The composition of sourdough microbiota developed during spontaneous fermentation, very typical for a given region, determines the unique, exceptionally attractive flavour of sourdough bread (Gänzle and De Vuyst 2004; Diowksz and Ambroziak 2006; Hansen 2005; De Vuyst et al. 2014; Fraberger et al. 2020c; Petkova et al. 2021; Sluková et al. 2021).

In order to achieve the desired effect, bakers have learned to manage the fermentation process by conducting it in multiple stages, adding a fresh batch of flour and water in successive stages. The goal is achieved by applying the appropriate consistency of the dough and maintaining optimal temperatures and fermentation time at each stage of the process. The combination of these parameters is responsible for the selection of the sourdough microbiota, and thus affects the volatile composition of the sourdough and the resulting bread. However, this requires not only great skills, but also a lot of effort, as the process is high labour-intensive and time-consuming (Minervini et al. 2014; Martín-Garcia et al. 2023).

The aroma of unprocessed flour is very weak. Some of the flour compounds, such as amino acids, fatty acids, phenolic acids, are aroma precursors, being transformed into flavour compounds during the baking step. However, the key stage in shaping the taste and smell of bread is sourdough fermentation. Although a significant amount of volatiles formed during the fermentation stage is lost during baking, enough volatile substances remain to contribute directly to the bread flavour or are thermally transformed into a unique flavour of fresh bread (Hansen and Schieberle 2005; Arora et al. 2021).

3.3.1 Microbial Diversity of Sourdough Origin from Central Europe

Until recently, it was believed that there is a close relationship between the diversity of sourdough microbiota (both bacteria and yeast) and the geographical origin of sourdough starters. However, based on taxonomic analysis of sourdoughs, it was found that geographical origin does not directly affect the species structure of sourdoughs (De Vuyst et al. 2023). The factors described above in Sect. 3.2 have a more significant impact. Therefore, the sourdough contains its own unique cocktail of microbes, which varies in composition and number depending on the artisanal tradition and region-dependent processes (De Vuyst and Neysens 2005).

3.3.1.1 Sourdough Microbiome Origin from Germany

Germany is a bread country, in the German Bread Register more than 3000 types of bread are listed (Göttlich et al. 2023). According to TasteAtlas (www.tasteatlas.com), among the four most popular sourdough breads in Central Europe is pumpernickel - a typical Westphalian specialty bread listed as food with protected geographical indication. Traditionally, production of all bread in Germany is based only on sourdough (Göttlich et al. 2023).

Meroth et al. (2003a, b) in sourdough obtained with rye bran flour in traditional process (temperature 25 °C) revealed presence of bacteria *Fructilactobacillus sanfranciscensis, Lactobacillus mindensis* and yeast *Candida humilis* (Table 3.1). In turn, Vogelmann and Hertel (2011) determined the effect of technological parameters (refreshment time, temperature, backslopping number) on two microbial communities *Fructilactobacillus sanfranciscensis, Candida humilis* and *Limosilactobacillus reuteri, Lactobacillus johnsonii* and *Issatchenkia orientalis* in the rye and wheat sourdoughs from Germany. The results demonstrated that both microbial communities were stable under different combination of the process conditions.

During the fermentation of type II sourdough different parameters are used, which results in a different microbiological composition of the sourdough. Müller et al. (2001) employed randomly amplified polymorphic DNA PCR (RAPD-PCR) to analyse microbiota of two rye sourdough of type II. The lactic acid bacteria present in these sourdoughs belonged to four different species. The predominant species under the three fermentations were *Lactobacillus amylovorus* and *Limosilactobacillus pontis/panis* (Table 3.1). *Lim. reuteri* were detected sporadically, only in some sourdough. *Lim. pontis, Lim. panis,* and *Lim. reuteri* were also detected during the continuous propagation of type II sourdoughs (at 30 °C and 40 °C), in studies conducted by Meroth et al. (2003a). In addition, *Lb. crispatus, Lb. frumenti, Lb. johnsonii* and *Limosilactobacillus fermentum* were detected in the same sourdough. Among yeast species *Debaryomyces hansenii, Saccharomyces cerevisiae,* and *S. uvarum* were found. The microbial composition of sourdoughs depended on the temperature and time of fermentation (Meroth et al. 2003b).

Bockwoldt et al. (2021) used two lactic acid bacteria strains *Levilactobacillus brevis* (heterofermentative) and *Pediococcus claussseni* (homofermentative) for fermentation rye and wheat sourdoughs. Interestingly, both strains were isolated from the brewery and produced O_2-substituted (1,3)-β-D-glucan. It was shown that the fermentation process can significantly affect the concentration of β-D-glucan in sourdough starters.

Table 3.1 Microorganisms identified in sourdoughs origin from Central Europe

Flour type	Microorganism	Origin	References
Bacteria			
Rye	*Companilactobacillus farciminis* *Fructilactobacillus sanfranciscensis* *Lacticaseibacillus casei* *Lactobacillus amylovorus* *Lb. crispatus* *Lb. frumenti* *Lb. johnsonii* *Lb. mindensis* *Lb. acidophilus* *Lb. alimentarius* *Lb. fructivorans* *Levilactobacillus brevis* *Lentilactobacillus buchneri* *Limosilactobacillus pontis* *Lim. panis* *Lim. reuteri* *Lim. fermentum* *Pediococcus claussseni*	Germany	Brandt (2014), Bockwoldt et al. (2021), Huys et al. (2013) and references therein Meroth et al. (2003a, b), Müller et al. (2001), and Vogelmann and Hertel (2011)
	Companilactobacillus kimchii *Co. paralimentarius* *Furfurilactobacillus rossiae* *Fructilactobacillus sanfranciscensis* *Lactiplantibacillus plantarum* *Lactobacillus parabrevis* *Lb. gallinarum* *Lb. harbinensis* *Lentilactobacillus diolivorans* *Len. otakiensis* *Len. parabuchneri* *Len. kisonensis* *Levilactobacillus brevis* *Lev. hammesii* *Lev. senmaizukei* *Lev. spicheri* *Limosilactobacillus pontis* *Pediococcus pentosaceus* *Weissella cibaria* *Weissella viridescens*	Austria	Fraberger et al. (2020c)
	Companilactobacillus kimchii *Co. paralimentarius* *Fructilactobacillus sanfranciscensis* *Lactiplantibacillus plantarum* *Levilactobacillus brevis* *Lactobacillus delbruecki* *Pediococcus pentosaceus* *Weissella confuse* *W. cibaria*	Poland	Boreczek et al. (2020), Diowksz and Włodarczyk (2003), Zaręba and Ostasiewicz (2009), and Ostasiewicz and Taglińska (2011)

(continued)

Table 3.1 (continued)

Flour type	Microorganism	Origin	References
	Fructilactobacillus sanfranciscensis *Lactiplantibacillus plantarum* *Lactobacillus delbrueckii* *Lb. leichmannii* *Lentilactobacillus buchneri* *Levilactobacillus brevis* *Limosilactobacillus fermentum* *Lim. panis*	Czech Republic	Sluková et al. (2021)
Wheat	*Companilactobacillus farciminis* *Fructilactobacillus sanfranciscensis* *Lactobacillus hoilgardii* *Lb. homohiochii* *Lb. johnsonii* *Lb. debrueckii* *Lactiplantibacillus plantarum* *Lacticaseibacillus casei* *Lentilactobacillus buchneri* *Levilactobacillus brevis* *Limosilactobacillus reuteri* *Pediococcus claussseni*	Germany	Bockwoldt et al. (2021), Huys et al. (2013) and references therein Vogelmann and Hertel (2011)
	Fructilactobacillus sanfranciscensis *Furfurilactobacillus rossiae* *Lactobacillus perolens* *Lactobacillus sakei* *Lactobacillus senmaizukei* *Lactobacillus spicheri* *Lacticaseibacillus paracasei* *Lactiplantibacillus plantarum* *Latilactobacillus curvatus* *Lentilactobacillus parabuchneri* *Limosilactobacillus pontis* *Loigolactobacillus coryniformis* *Paucilactobacillus xiangfangensis* *Leuconostoc citreum* *Lc. mesenteroides* *Lc. pseudomesenteroides* *Pediococcus parvulus* *P. pentosaceus* *Streptococcus dentisani* *S. salivarius* *Weissella cibaria* *W. viridescens*	Austria	Fraberger et al. (2020a, b)
	Companilactobacillus kimchi *Co. paralimentarius* *Lactiplantibacillus plantarum* *Levilactobacillus brevis* *Pediococcus pentosaceus* *Weissella confuse* *W. cibaria*	Poland	Boreczek et al. (2020), Kawka and Górecka (2010), Kawka et al. (2010), Litwinek et al. (2022) and Piasecka-Jóźwiak et al. (2011)

(continued)

Table 3.1 (continued)

Flour type	Microorganism	Origin	References
	Lactiplantibacillus plantarum *Lacticaseibacillus casei* *Latilactobacillus curvatus* *Lactobacillus delbrueckii* *Lb. sake* *Lb. rhamnosus* *Limosilactobacillus fermentum* *Lactibacillus acidophilus* *Lactococcus lactis subsp. lactis*	Hungary	Zalán et al. (2015)
Spelt	*Lactiplantibacillus plantarum* *Levilactobacillus brevis*	Poland	Boreczek et al. (2020)
Oat	*Lactiplantibacillus plantarum* *Lacticaseibacillus casei* *Latilactobacillus curvatus* *Lactobacillus delbrueckii* *Lb. sake* *Lb. rhamnosus* *Limosilactobacillus fermentum* *Lactibacillus acidophilus* *Lactococcus lactis subsp. lactis*	Hungary	Zalán et al. (2015)
Yeast			
Rye	*Candida humilis* *Debaryomyces hansenii, Saccharomyces cerevisiae* *S. uvarum* *S. pastorianus* *Issatchenkia orientalis* *Pichia kudriavzevii* *Kazachstania exigua* *Torulaspora delbrueckii*	Germany	Brandt (2014), Meroth et al. (2003a, b), and Vogelmann and Hertel (2011)
	Kazachstania humilis *Pichia fermentans* *Saccharomyces cerevisiae*	Austria	Fraberger et al. (2020a, b)
	Saccharomyces cerevisiae	Poland	Diowksz and Włodarczyk (2003) and Zaręba and Ostasiewicz (2009)
	Saccharomyces cerevisiae *Saccharomyces minor* *Lachancea fermentati* *Candida miller* *Candida humilis*	Czech Republic	Sluková et al. (2021)
Wheat	*Saccharomyces cerevisiae* *Saccharomyces uvarum* *Issatchenkia orientalis* *Candida glabrata*	Germany	Vogelmann and Hertel (2011)
	Kazachstania humilis *Kazachstania unispora* *Saccharomyces cerevisiae* *Saccharomyces uvarum* *Torulaspora delbrueckii*	Austria	Fraberger et al. (2020c)
	Saccharomyces chevalieri *Saccharomyces cerevisiae*		Kawka and Górecka (2010)

3.3.1.2 Austrian Sourdough Microbial Diversity

Fraberger et al. (2020c) evaluated twenty traditional sourdoughs originated from small-scale bakeries and households from different location in Austria (mainly Vienna, Salzburg, and Lower Austria). In 12 samples of wheat sourdoughs, 26 species of lactic acid bacteria and 5 yeasts were detected. In more than half of the tested samples, *Latilactobacillus curvatus* (8 samples), *Lactiplantibacillus plantarum* (7), and *Pediococcus pentosaceus* (6) were identified. *Loigolactobacillus coryniformis, Leuconostoc mesenteroides, Levilactobacillus brevis* were found in 3–5 samples, while the remaining species (see Table 3.1) were rather specific for individual sourdoughs. Among the yeasts *Saccharomyces cerevisiae, S. uvarum*, and *Kazachstania unispora* were the predominant species. *Kazachstania unispora* and *Torulaspora delbrueckii* were present in only four starters from Vienna. In leavens with rye flour revealed presence of 21 LAB species and only 3 yeasts. The species composition of the dominant microorganisms differed from wheat sourdoughs. *Lactiplantibacillus plantarum* (4 samples), *Fructilactobacillus sanfrancisiciencis,* and *Lacticaseibacillus paracasei* were detected in the highest number of rye sourdoughs. Furthermore, *Lentilactobacillus diolivorans, Len. otakiensis, Lactobacillus gallinarum, Lb. parabrevis, Companilactobacillus kimchii, Co. paralimentarius*, and *Paucilactobacillus xiangfangensis* were exclusively detected in rye sourdoughs. However, the strains belonging to the *Streptococcus* spp. and *Leuconostoc* spp., as well as some *Lactobacillus* spp. were not found (Table 3.1). In rye sourdough samples revealed coexistence *S. cerevisiae* with *K. humilis*, species frequently dominant in both rye and wheat sourdoughs. In further studies, Fraberger et al. (2020b) showed that sourdough-related LAB exhibited the potential to degrade immunogenic proteins like gluten and alpha-amylase-trypsin inhibitors (ATIs). Moreover, strains isolated from Austrian rye and wheat starters, namely *Lp. planatrum Fr. sanfrancisiciencis, L. paracasei, Len. diolivorans, Len. otakiensis, Len. kisonensis, Len. parabuchneri, Lb. gallinarum, Lb. parabrevis, Lev. brevis, Lo. coryniformis,* and *Paucilactobacillus xiangfangensis* showed antimicrobial activity against moulds *Aspergillus brasiliensis, A. fumigatus, A. flavus, Fusarium graminearum, Penicillium roqueforti* and bacteria *Bacillus subtilis, B. licheniformis, and B. cereus* (Fraberger et al. 2020c).

3.3.1.3 Microorganisms in the Production of Traditional Polish Bread

Sourdough bread is deeply rooted in Polish culture. In Poland, there are many types of regional bread, but 70% is produced from rye sourdough (Diowksz and Ambroziak 2006). Nowadays wheat-rye bread baked by traditional method is most popular type of bread. This is due to climate change in the Central European region. Rye is a plant that is very resistant to climate change, has low habitat requirements and can be grown on infertile and dry soils (Diowksz 2023). Obtaining good quality sourdough bread takes time. Traditionally, a mixture of rye flour and water is left for a suitable time to initiate the development of microbiota. As more flour and water are added, the microbiota in the flour is influenced by environmental factors and mutual interactions that induce natural selection (Diowksz and Ambroziak 2006). In addition, the artisanal, local nature of production means that the microbiota of the sourdoughs is characterized by great biodiversity, which directly affects the variety of the bread (Diowksz 2011).

Typical sourdough starters contain in 1 g from 10^7 to 10^{11} lactic acid bacteria and 10^5 to 10^6 yeasts. Single-species or multi-species systems of LAB (*Lactiplantibacillus plantarum, Lev. brevis, Fructilactobacillus sanfranciscensis*) and/or yeast (*S. cerevisiae, Candida milleri*) are commonly used, very often enriched with probiotic strains of lactic acid bacteria (e.g., *Lacticaseibacillus casei, L. paracasei, Lactibacillus acidophilus, Lb. rhamnosus, Pediococcus acidilactici, Propionibacterium freudenreichii*) (Zaręba and Ostasiewicz 2009).

Diowksz and Włodarczyk (2003) determined the suitability of homo- and heterofermentative strains of LAB for the construction bacterial consortium and evaluated their physiological activity in two-population systems in coexistence with *Saccharomyces cerevisiae* yeast. Five different combinations of bacteria containing *Fr. sanfranciscensis*, *Levilactobacillus brevis*, *Lp. plantarum*, and *Lactobacillus delbrueckii* were tested (Table 3.1). All consortia were characterized by high stability, good growth dynamics and production of appropriate metabolites. However, the starter containing *Fr. sanfranciscensis, Lev. brevis, Lp. plantarum* and *S. cerevisiae* was found to be optimal. Moreover, the combination of homofermentative and heterofermentative LAB reduced the content of volatile acids in the final metabolites. Also, Ostasiewicz andEglińska (2011) studied the effect of sourdough containing *Fr. sanfranciscensis, Lev. brevis,* and *Lp. plantarum* (without yeast) on the rye bread parameters and compared them with those for bread obtained as a result of spontaneous fermentation. Kawka and Górecka (2010) proved the positive effect of oat and barley sourdoughs containing *Lev. brevis* and *Saccharomyces chevalieri* on the chemical composition of wheat-oat and wheat-barley breads. In turn, Piasecka-Jóźwiak et al. (2011) used the single-species *Lp. plantarum* amylolytic system for baking wheat bread. The bread obtained in this way was characterized by a higher score in sensory tests, but worse extensographic parameters.

Boreczek et al. (2020) analysed, using culture-depended and independent techniques, spontaneous sourdoughs prepared from wholemeal flours: rye, wheat and spelt. Among the bacteria present at each stage of wheat sourdough fermentation, *Lp. plantarum, Lev. brevis, Companilactobacillus paralimentarius, Co. kimchii, Pediococcus pentosaceus, Weissella confusa,* and *W. cibaria* were identified. The taxonomic composition of rye sourdough was similar to that of wheat, the same species of lactic fermentation bacteria were detected at each stage of fermentation. Interestingly, in sourdough with spelt flour, only two species were predominant, namely *Lp. plantarum* and *Lev. brevis.*

An innovative approach to the preparation of starters was proposed by Koj and Pejcz (2023). In the study of the effect of rye dietary fiber components on the fermentation process, the sourdough was inoculated with potential probiotic microorganisms *Saccharomyces boulardii, Lactiplantibacillus plantarum, Lacticaseibacillus rhamnosus*, and *Bacillus coagulans*. The use of this microbial cocktail, and above all of *S. boulardii* may seem promising, as it allowed for the largest reduction of fructans in rye bread.

3.3.1.4 Traditional Production of Sourdough Bread in Czech Republic

In the traditional production of bread in Czech Republic, the rye flour is used to prepare sourdough. Significantly, the rye sourdough is a key player in this technology as the flour is not subjected to any processes beforehand (Sluková et al. 2021). Active sourdough is added in the process and is refreshed so that sourdough bread is made without yeast addition (Šedivý and Albrecht 2014). In typical Czech bread, sourdough makes up 40–50%, and the whole amount of rye flour undergoes a fermentation process. For such addition of sourdough to be possible in the process, it must have a low titratable acidity in the range of 100–130 mmol/kg. This causes that sourdough is unstable and requires the continuity of the process. Otherwise, the leaven needs to be refreshed under the optimal conditions before each use to ensure multiplication of all microorganisms, namely homofermentative and heterofermentative LAB: *Lactiplantibacillus plantarum, Lactobacillus delbrueckii, Lb. leichmannii, Fructilactobacillus sanfranciscensis, Levilactobacillus brevis, Limosilactobacillus fermentum, Lim. panis, Lentilactobacillus buchneri* and yeasts *Saccharomyces cerevisiae, Saccharomyces minor, Lachancea fermentati, Candida milleri, C. humilis* (Table 3.1). In Czech terminology, the sourdough obtained in this way is referred to as active sourdough refreshed to produce acids and fermentation gas (Sluková et al., 2021).

Another method in baking technology is the use of a starter culture with a known microbiological composition containing selected strains of LAB and sometimes yeast. In Czech terminology is defined as active refreshed sourdough for the acids production (Šedivý and Albrecht 2014). Sourdough preparation or its refreshing it is easier in this case and focused on production of enough amount of acids (mainly lactic and acetic acid). The titratable acidity of the prepared sourdough is above 200 mmol/kg, so it is more acidic, and its addition cannot be as high as in the previously described case. In addition, sourdough does not contain enough yeast to evolve fermentation gas (alcoholic fermentation is secondary process) (Chavan and Chavan 2011).

3.3.1.5 Bacterial Communities Used for Production of Hungarian Sourdough Bread

As in most Central European countries, the consumption of sourdough bread in Hungary has a long tradition. According to Tömösközi (2023), in 2019 over 63% of all grain products consumed by Hungarians were sourdough bakery products.

Zalán et al. (2015) conducted selection of twelve lactic acid bacteria strains belonging to *Lactiplantibacillus plantarum, Lacticaseibacillus casei, Latilactobacillus curvatus, Lactobacillus delbrueckii* subsp. *bulgaricus, Lb. sakei,*

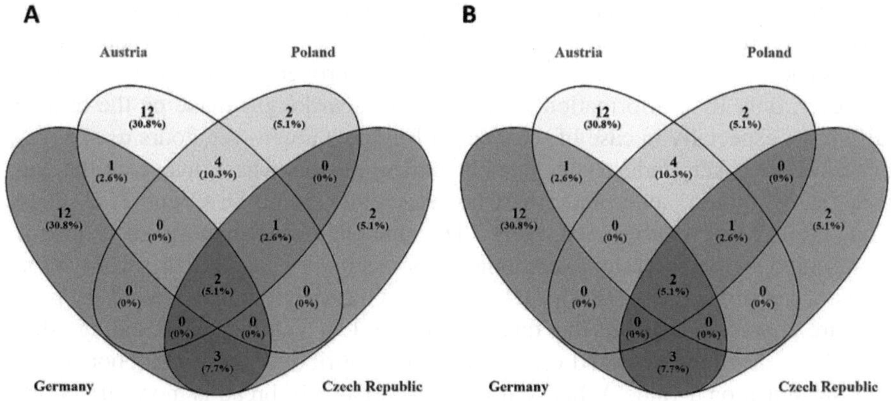

Fig. 3.1 The comparison of microbial diversity of rye (**a**) and wheat (**b**) sourdoughs in Central Europe countries

Lb. rhamnosus, Limosilactobacillus fermentum, *Lactibacillus acidophilus* and *Lactococcus lactis* subsp. *lactis* for antifungal properties and their suitability for fermentation of sourdough containing wheat and oat brans. The authors proved that the selected strains could be used for what and oat bran fermentation. Moreover, nine tested strains inhibited the growth of *Aspergillus* spp. and *Penicillium* spp., with the strongest mould-inhibiting properties of strains *Latilactobacillus curvatus* and *Limosilactobacillus fermentum.*

Venn's diagram (Fig. 3.1, Venny 1.2) illustrates the similarities and differences in the microbial biodiversity of rye (Fig. 3.1a) and wheat (Fig. 3.1b) sourdoughs from the abovementioned countries of Central Europe**.** Take together all data, two species of lactic acid bacteria *Fructilactobacillus sanfranciscensis* and *Levilactobacillus brevis* were typical for rye sourdough, regardless of the origin country. In wheat sourdoughs, only one species - *Lactiplantibacillus plantarum* was detected in all the examined countries. The most unique bacteria (12 species) are found in rye sourdoughs from Germany and Austria. Similar dependencies were observed for wheat sourdoughs (Fig. 3.1b), namely 18 specific species revealed in Austrian sourdoughs, 12 in German and 6 in Hungarian.

3.4 The Role of Sourdough Microbiota in Production of Gluten-Free Bread and Bakery Products in Central Europe

Nowadays, preparation of bakery products is not limited only to the usage of wheat or rye flours, but many other types of flours might be used for their production. A trend that is increasingly attracting bakery industry as well as consumers in the countries of Central Europe, is the use of gluten-free (GF) non-conventional flours.

It is very important especially in case of the people suffering from gluten intolerance. Currently, the production process to obtain gluten-free bakery products is not well standardized with variable effects on the properties of bakery products. However, only little information is available for sourdoughs made on the basis of GF flours, especially in case of the Central Europe. Gluten-free flours used as raw materials for bakery industry include i.e., amaranth, buckwheat, maize, millet, quinoa, rice, sorghum, spelt, and teff (Chochkov et al. 2022; Di Cagno et al. 2014; Coda et al. 2014; Moroni et al. 2009). The challenge in fermenting such flours is represented by the necessity to combine proper technology as well as sensory properties of the final product (Ramos et al. 2021). Gluten-free bread and bakery products are characterized with low quality and low loaf volume, rapid staling (short shelf-life), dry mouthfeel and crumbling texture attributes as well as poor flavour and low nutritional value. What is also characteristic GF bread is more susceptible to the microbiological spoilage (Šmídová and Rysová 2022). Gluten possesses unique viscoelastic properties which are essential for the water holding capacity of the dough and the gas retention during fermentation. To avoid quality defects of products resulting from undesired technological changes of gluten-free products related with gluten absence, many different strategies are implemented. Addition of ingredients such as hydrocolloids proteins (e.g. from soybean, peas, eff), or the usage of enzymes (e.g. cyclodextrin glycosyl transferase or glucose oxidase) is beneficial to overcome of some GF bread defects but at the same time generate high-cost of the final product. The alternative to solve many problems related to the production of GF bread is the usage of the sourdough. The role of the sourdough microbiota for the production of such products seems to be crucial. Both, lactic acid bacteria and yeast, positively impact many properties of the bakery products, and are present in the mature sourdough in the number higher than 10^8 cfu/g and around 10^6 cfu/g or 10^7 cfu/g, respectively. The usage of sourdough to prepare GF bakery goods results with improved technological and nutritional properties. The biodiversity of microbiota of the sourdough is a consequence of the quantity and the variety of microorganisms present in raw materials as well as the bakery house microbiota impacted by environmental factors described previously in this chapter. Microbiota of many different gluten-free cereals and pseudo-cereals has been investigated. The biodiversity of the species characteristic for GF sourdough depends on the type of the flour and other ingredients, number of the refreshment steps, time and temperature of propagation, and dough yield (Zannini et al. 2012; Ramos et al. 2021; Nionelli and Rizzello 2016). Table 3.2 summarizes microorganisms isolated from different GF sourdoughs.

Gluten-free bread has been found "problematic" in the relationship to its structure parameters, gas holding abilities, poor aroma quality and susceptibility to staling and microbial spoilage. What is more, the nutritional value of GF bread depends on the chemical composition of raw materials (flours and functional ingredients such as hydrocolloids, fiber etc.) used for its preparation (Šmídová and Rysová 2022; Wronkowska et al. 2013). Axel et al. (2015) have proven that presence and activity of antifungal compounds of microbial origin in sourdough is not only substrate-specific but also strain-specific. It means that the final quality of GF products is strongly impacted by the properties of GF flours and abilities of microorganisms used for their fermentation. Mentioned in the Table 3.2 species of microorganisms were also abundant as components of the starter cultures for GF sourdough development. The cultures used for fermentation of GF sourdough differ from the classical starter cultures used typical for rye or wheat due to a specific composition of the flour and the huge number of used ingredients. The choice of suitable strains is extremely important when different types of flour are mixed for dough preparation (Šmídová and Rysová 2022). Thus, the high quality of gluten-free sourdough -bread remains a challenge.

Table 3.2 Microorganisms identified in GF sourdoughs

Flour type	Bacteria	Yeast	Reference
Amaranth	*Companilactobacillus paralimentarius* *Lactiplantibacillus plantarum* *Lb. helveticus* *Lb. sakei* *Pediococcus pentosaceus*	*Saccharomyces cerevisiae*	Gobbetti et al. (2016) and references therein Vogelmann et al. (2009), Ruiz Rodríguez et al. (2016), Różyło et al. (2014), and Sterr et al. (2009)
Buckwheat	*Companilactobacillus paralimentarius* *Lc. argentinum* *Lb. helveticus* *Lactiplantibacillus plantarum* *Limosilactobacillus pontis Lim. fermentum*	*Kazachstania barnetti*	Arendt et al. (2010), Moroni et al. (2011a, b), and Różyło et al. (2015)
Maize	*Companilactobacillus paralimentarius* *Lb. alimentarius* *Lb. helveticus* *Limosilactobacillus. fermentum* *Lim. pontis* *Levilactobacillus brevis* *Lacticaseibacillus casei* *Lactiplantibacillus plantarum* *Lc. mesenteroides* *Lc. dextranicum* *P. acidilactici* *P. pentosaceus* *Enterococcus casseliflavus* *E. durans* *E. faecium* *Streptococcus constellatus* *S. equinus*	*S. cerevisiae* *Schizosaccharomyces pombe* *Issatchenkia orientalis* *C. krusei* *Torulaspora delbrueckii,* *Wickerhamoyces anomalus*	Annan et al. (2003), Gobbetti et al. (2016) and references therein Huys et al. (2013), Vogelmann et al. (2009)

(continued)

Table 3.2 (continued)

Flour type	Bacteria	Yeast	Reference
Millet	*Lb. helveticus* *Limosilactobacillus fermentum* *Lim. pontis*	*S. cerevisiae*	Vogelmann et al. (2009)
Quinoa	*Companilactobacillus paralimentarius* *Levilactobacillus brevis* *Limosilactobacillus fermentum* *Lactiplantibacillus plantarum*		Chavan and Chavan (2011), Huys et al. (2013), Rizzello et al. (2016), and Vogelmann et al. (2009)
Rice	*Companilactobacillus kimchii* *Co. paralimentarius* *E. faecium* *E. durans* *Lb. helveticus* *Lb. gallinarium* *Lb. perolens* *Lacticaseibacillus paracasei* *Limosilactobacillus plantarum* *Lim. fermentum* *Pediococcus pentosaceus*	*S. cerevisiae* *S. fibuligera* *T. delbrueckii* *W. anomalus*	Gobbetti et al. (2016), Huys et al. (2013), Meroth et al. (2004), Moroni et al. (2011a), and Vogelmann et al. (2009)
Sorghum	*E. faecalis* *Levilactobacillus brevis* *Lentilactobacillus parabuchneri* *Lb. amylovorus* *Lb. cellobiosus* *Lb. confuses* *Lb. helveticus* *Lb. herbinensis* *Lb. vaginali,* *Lb. reuteri* *Lactiplantibacillus plantarum* *Pediococcus pentosaceus* *Lc. lactis* *Limosilactobacillus fermentum*	*C. intermedia* *C. norvegensis* *Debaromyces hansenni* *P. kudriavzevii* *Rhodotorula glutinis*	Aguilar et al. (2016), Gobbetti et al. (2016), Huys et al. (2013), and Różyło et al. (2016)

(continued)

Table 3.2 (continued)

Flour type	Bacteria	Yeast	Reference
Teff	*Lentilactobacillus parabuchneri* *Lb. amylovorus* *Lb. crispatus* *Lb. helveticus* *Lb. gallinarum* *Lb. graminis* *Lb. pentosaceus* *Lb. sakei* *Lb. sanfrancisiciensis* *Lb. vaginalis* *Levilactobacillus brevis* *Limosilactobacillus pontis* *Lim. fermentum* *Lactiplantibacillus plantarum* *Lc. holzapfelii* *Pediococcus pentosaceus* *P. acidilactici* *Weissella cibaria* *E. faecalis* *E. casseliflavus* *Lc. mesenteroides* *Lc. Raffinolactis*	*S. cerevisiae* *Kazachstania exigua* *K. barnetti* *C. humilis* *C. tropicalis* *C. glabrata* *Pichia norvegensis*	Arendt et al. (2010), Gobbetti et al. (2016), Huys et al. (2013), Meroth et al. (2004), and Sterr et al. (2009)

During studies performed in Austria, the quality of buckwheat bread was improved by sourdough fermentation combined with the addition of arabinoxylan (AX) and pyranose 2-oxidase (POx) treatment (Zand et al. 2018). Researchers found that wholemeal buckwheat flour with sourdough (added concentration set at 82%) and a moderate amount of AX (up to 3%) along with POx (in dose up to 1 nkat/g flour) resulted with highly acceptable: specific volume, crumb firmness, crumb colour and a good pore structure.

Adamek et al. (2023), in Czech Republic, have developed the monitoring system (thermodynamic sensor combined with an electronic nose) for gluten-free dough leavening fortified with insect meal (*Tenebrio molitor*). Additionally, it was demonstrated that the fermentation activity was increased in corn dough by the addition of insects. In case of rice dough supplemented with insects, the fermentation activity was similar to that of the flour without the addition (no positive or negative result was noticed).

In Germany, which is considered as a country of sourdough-bread, the intensive work on GF bread is also conducted. For example, the impact of the sourdough introduction on the technological quality of bread based on buckwheat, millet and amaranth flours was investigated (Bender et al. 2018; Houben et al. 2010). A very interesting approach was made by Bender et al. (2018), who determined the influence of seven different *Lactobacillus* spp. strains in comparison to commercial starter culture on the functional properties of gluten-free sourdough breads. It was seen that the fermentation significantly improved the crumb firmness and porosity of the prepared GF breads. It should be highlighted that the sourdough technology alone is able to improve the quality of GF breads only when using the appropriate lactobacilli strains. The authors suggested that *Fructilactobacillus sanfranciscensis* was the most versatile species for sourdough production from tested flours in conducted studies.

The idea of using microorganisms to improve sourdough' parameters is in agreement with the results of research conducted by Houben et al. (2010), who confirmed a positive effect of fermentation on amaranth sourdough and dough properties. Amaranth flour and its potential for fermentation was also assessed in the research made by Sterr et al. (2009). Autochthonous microbiota of five amaranth flours was investigated for the selection of microorganisms used as a starter culture. Amongst detected lactic acid bacteria, strains belonging to *P. pentosaceus, Latilactobacillus sakei,* and *Lactiplantibacillus plantarum,* were identified. As a starter cultures, strains *Lp. plantarum* RTa12, *Lb. sakei* RTa14 and *P. pentosaceus* RTa11, were chosen to ferment six different amaranth flours. The doughs prepared on the laboratory scale were inoculated with 1×10^7 cfu/g of the respective strains and fermented with daily back-slopping at 25 °C, 30 °C and 35 °C for 10 days. From the technological point of view, strain: *Lp. plantarum* RTa12 (high volatile fatty acid concentration and total titratable activity) and *P. pentosaceus* RTa11 (fast growth and high number of cells over autochthonous microbiota in dough) occurred as the most suitable for amaranth flour fermentation. The fermentation properties and sensory attributes of the amaranth sourdoughs were also explored by Jekle et al. (2010). In the study, fermentation properties of *Lb. plantarum, Lb. helveticus* and *Companilactobacillus paralimentarius* were examined in reference to production of lactic and acetic acids, pH and total titratable activity. Starter cultures used for preparation of amaranth-based sourdough showed different abilities impacting on sensory and technological quality of the amaranth sourdough.

Breads from buckwheat flour were under the investigation of the researchers from the Technical University of Munich (Föste et al. 2013). They characterized the dough rheological properties as well as baking technological potential of the buckwheat milling fractions (flour, bran) against the background of substrate availability, varying endogenous enzyme activity and the growth of lactobacilli. The use of buckwheat flour revealed with increased of the specific bread volume by 7% compared to the control flour mixture. The replacement of the part of the control flour mixture by buckwheat bran, reduced the bread volume by 3.6%, and by 18.1% after fermentation, respectively. Darker crust colour and a more compact crumb structure were noticed. What is more, the application of fermented buckwheat bran attributed the reduction of baking loss compared to the control. Employment of buckwheat for sourdoughs preparation was also evaluated by Capuani et al. (2013). The authors highlighted the significance of pH, incubation time and LAB (*E. faecalis*, *P. pentosaceus, W. cibaria*) metabolism for the enhancement of proteolysis in buckwheat sourdoughs. Proteolysis is extremely important in context of GF bread structure.

The use of levan-producing acetic acid bacteria (AAB) was examined in buckwheat sourdoughs by Ua-Arak et al. (2016). The results showed dominant growth of all three AAB strains (*Gluconobacter albidus* TMW 2.1191, *Kozakia baliensis* NBRC 16680 and *Neoasaia chiangmaiensis* NBRC 101099) over pathogenic contaminants, the pH reduction through organic acids production and the high amounts of levan produced *in situ* in the buckwheat doughs. In another study, breads based on gluten-free buckwheat, quinoa, sorghum and teff flours, were prepared with addition of 20% sourdough fermented with exopolysaccharide producing *Weissella cibaria* MG1 (Wolter et al. 2014). Although the aroma of most baked GF breads was not improved by sourdough addition, the staling rate was significantly reduced in buckwheat and teff sourdough breads. In addition, crumb hardness was significantly reduced in buckwheat (−122%), teff (−29%) and quinoa (−21%) sourdough breads in comparison to "sourdough non-containing" control breads. Optimization of impact of exopolysaccharides to improve properties of gluten-free sourdough by *in situ* application of EPS-forming strains was investigated (Rühmkorf et al. 2012). The highest yields were obtained with *Lb. animalis* TMW 1.971 (17.93 g EPS /flour kg) and *Latilactobacillus curvatus* TMW 1.624 (16.28 g EPS /flour kg), respectively.

In Poland the theme of gluten-free bread is quite well recognized. The studies conducted by Diowksz and Kordialik-Bogacka (2014) demonstrated the possibility to shorten the process of gluten-free sourdough fermentation by simple modification of bread composition. Addition of glucose (0.5%), casein hydrolysate and soy sprouts as fermentation enhancers to mixtures of maize and soy flours as well as maize and potato starches, shortens the fermentation time.

Piasecka-Jóźwiak et al. (2016) demonstrated the utility of two *Pediococcus pentosaceus* strains (labelled with symbols ZFG2, ZFG4) for buckwheat flour fermentation. Both strains were previously isolated from buckwheat sourdough. Diowksz and Sadowska (2021) prepared GF bread using *Lactiplantibacillus plantarum* ŁOCK 860 and *Levilactobacillus brevis* ŁOCK 1267 and buckwheat flour applicated along with transglutaminase (TG). Mechanical properties of the crumb were influenced by the sourdough. The use of LAB resulted with the improvement of overall sensory quality of experimental bread. The characteristic bitter aftertaste of buckwheat was less noticeable. The total acceptability of the buckwheat bread increased. The addition of buckwheat flour can also enrich the protein and microelements content in gluten-free bread (Krupa-Kozak et al. 2011).

Korus et al. (2022) reported the possibility of mesquite (*Prosopis* L.) flour usage for the replacement of starch in gluten-free formulation. Sensory evaluation exposed an increase in the acceptability of bread's crumb colour, smell, and taste. In addition, the crumb with 7.5% and 10% mesquite flour had considerably lower hardness during storage compared to the control sample, indicating the reduction of the staling process. Summarizing, mesquite flour not only strengthened the structure of the gluten-free dough but also it was found that mesquite flour improve shelf life and sensory acceptability of GF bread. Różyło et al. (2017) demonstrated beneficial effects of the addition of carob fibre to the GF bread, whereas the addition of chestnut flour to the recipe was investigated by Marciniak-Lukasiak et al. (2022). The usage of chestnut flour in an amount of up to 10% increased significantly loaves volume and positively impacted the texture of the final product by reducing the hardness and increasing the elasticity and cohesiveness of the bread crumb. In the same studies, the vacuum packaging was pointed out as a better way to protect gluten-free bread from spoilage and to prolong its shelf life.

Literature search on the topic of inulin-type fructans (ITFs) introduction into GF products was conducted by Drabińska et al. (2016). In general, ITFs improve technological and sensory value of gluten-free products and affect rheological and physicochemical properties of bread.

According to Wieczorek et al. (2022), cricket powder (CP) can be valuable additive to GF products. The results of presented studies showed not only significant improvement in nutritional value but also in flavour (enriched with caramel, roasty, and cooked potatoes notes) when compared to the traditional GF bread without CP. The trend of utilization of waste materials in food processing is widely observed. Waste products from chia seed oil pressing can be considered as valuable functional additive in the production of both wheat and gluten-free breads (Zdybel et al. 2019). It was found that the usage of chia seed waste and chia seeds (5% of the total flour) increased the bread volume with no deterioration in its sensory qualities. The microbial stability and shelf life of gluten-free bread can be prolonged by the addition of flaxseed oil cake extract (Łopusiewicz et al. 2023). Pecyna et al. (2023) revealed that the addition of hemp inflorescence significantly increased the total amount of polyphenols, flavonoids, and antioxidant activity in the bread. Researchers found that up to 2% of this byproduct did not affect tested sensory properties of bread.

The comparison of the fresh and freeze-dried amaranth sourdough was performed by Różyło et al. (2014) in model gluten-free bread. GF bread was prepared from corn, rice, and amaranth flour in two variants (45:45:10% and 40:40:20%). Addition of the amaranth sourdough resulted in significant changes in the characteristics of bread texture. It was noticed that up to 10% replacement of the amaranth flour with sourdough was acceptable for the consumers. Simultaneously it was demonstrated that the freeze-dried amaranth sourdough may be a good substitute for freshly prepared sourdough.

Slovenian researchers analysed the impact of chickpea fermentation on the production of nutritionally undesirable biogenic amines as well as beneficial gamma-aminobutyric acid (GABA) during spontaneous and inoculated sourdoughs (Polak et al. 2021). Spontaneous fermentation was proven to be related with high accumulation of cadaverine, putrescine, and tyramine for certain examined flours. What is more, commercial starter culture was not effective in the reduction of biogenic amines content in tested sourdoughs. Whereas, the addition of fenugreek sprouts to the sourdough along with pH raised to 6.5 resulted in reduction of biogenic amines content.

As demonstrated, production of gluten-free bread, including the usage of sourdough, is a matter of investigation in Central Europe, especially in Germany and Poland. On the other hand, it should be highlighted that the study of the sourdough microbiota affecting the properties of gluten-free baked goods, requires an increase in intensity in many countries included to the Central Europe to fill the gaps in this area of knowledge.

3.5 A Deep Insight into the Microflora of Sourdough – A Review of Methods

For over 30 years, the sourdough microbiota has been the subject of research by many scientists around the world. At that time, according to data collected by Arora et al. (2021) the taxonomic structure of the unique microbial cocktail was investigated both by classical culture-dependent methods, culture-independent methods, and a combination of the previously mentioned.

For a long time, a culture-based approach has been used to isolate and identify both desirable and contaminating microorganisms in sourdough. Most often, the microbiological analysis of sourdoughs includes determination of the total number of bacteria and fungi, the presence of spore-forming bacteria, and above all lactic acid bacteria and yeast. In general, for the enumeration and isolation of lactic acid bacteria, MRS medium is commonly used, and incubation is carried out at 30–37 °C for 2–3 days (Meroth et al. 2003; Vancanneyt et al. 2005; Ehrmann et al. 2007). Yeasts from sourdough are cultivated on Sabouraud dextrose agar (SDA), yeast extract–peptone–glucose (YPG) medium, malt extract agar (MEA) with chloramphenicol used as antibacterial agent at 25–30 °C for 2–3 days (Huys et al. 2013; De Vuyst et al. 2016). The culture-dependent methods allow for obtaining pure cultures of microorganisms, their identification, and also understanding the physiological and biochemical potential of isolates. Literature data indicate that biochemical tests, i.e., API 50 CH system (bioMérieux) or PCR-based methods are commonly used to identify bacteria or yeasts originating from sourdoughs (Wiese et al. 1996; Vancanneyt et al. 2005; Ehrmann et al. 2007). The most common molecular markers for bacteria are genes encoding small subunit of ribosomal RNA (rRNA) – 16S rRNA (Vogel et al. 1994; Wiese et al. 1996; Meroth et al. 2003; Vancanneyt et al. 2005; Ehrmann et al. 2007), while for yeast nucleotide sequence of region ITS and 28S rRNA (Gatto and Torriani 2004; Ripari et al. 2016). PCR-based methods are employed for microorganisms' identification, among others, in randomly amplified polymorphic DNA (RAPD), amplified/restriction fragment length polymorphism (AFLP/RFLP) (Corsetti et al. 2003; Ehrmann and Vogel 2005; De Vuyst and Vancanneyt 2007; Scheirlinck et al. 2009; Ripari et al. 2016). Without a doubt, culture-based approach is powerful tool for the isolation and characterization of microorganisms, however it provides limited information on the diversity of complex microbial communities in the sourdough. This is due to the fact that significantly over 90% of microorganisms in the natural environment belong to viable, non-culturable on synthetic microbiological media under laboratory conditions. This phenomenon is called the VBNC stage of microorganisms, and it occurs in all environments, especially those with stressogenic conditions.

For the first time, the culture-independent approach in the research on the microbiota of sourdoughs was applied by Ehrmann et al. (1994). They identified lactic acid bacteria direct in fermented foods, including sourdoughs, without cultivation step, based on the reverse dot blot hybridization technique. According to the authors, in addition to the LAB detected in the sourdough (in this case it was *Fructilactobacillus sanfranciscensis*, syn. *Lb. sanfranciscensis* and *Lb. sanfrancisco*; *Limosilactobacillus pontis,* syn. *Lb. pontis* and *Levilactobacillus brevis*, known as *Lb. brevis*), using the above-mentioned method, it is also possible to detect the contaminating microflora. The following approaches used denaturing gradient gel electrophoresis (DGGE) and temporal temperature gel electrophoresis (TTGE) of PCR-amplified fragments of 16S rRNA gene to detect the microbiota of sourdoughs type I and II (Meroth et al. 2003a, b; Randazzo et al. 2005; Ferchichi et al. 2007; Vrancken et al. 2011). DGGE and TTGE are based on electrophoretic separation under denaturation conditions of DNA molecules having the same length but different nucleotide sequences. Importantly, the migration of bands in the gel depends on the degree of DNA denaturation, and the separation products can be excised and sequenced (Ferchichi et al. 2007). Meroth and co-authors (2003a) used DGGE not only to identify LAB, but also to monitor their changes during sourdough propagation. Research revealed different species of the *Lactobacillus* genus (*Fructilactobacillus sanfranciscensis, Limosilactobacillus pontis, Lb. mindensis, Lb. crispatus, Lb. panis, Lb. johnsonii* and *Lb. reuteri*) in sourdoughs. Moreover, the detection results obtained by the PCR-DGGE method were consistent with the results of the microbiological analysis, apart from *Lim. fermentum* which was detected only based on the culture methods. In turn, Gatto and Torriani (2004) investigated changes in bacterial and yeast population dynamics during traditional sourdough fermentation. A similar methodological approach (combination of DGGE and culture-based methods) was used by Van Kerrebroeck et al. (2016) in monitoring the relationship between pH changes of sourdough and dynamics of LAB (mainly *Lim. fermentum, Lp. paraplantarum, Lb. pentosus, Lp. plantarum*) and yeast (*Kazachstania exigua, K. bulderi, K. barnettii*) community. As with DGGE, the TTGE technique was mainly employed for monitoring the dynamics of microbial population. First time was applied by Ferchichi et al. (2007) to analyse microbiota of French sourdoughs. Overall, all authors agree that the above-mentioned methods are an excellent tool for tracking changes in microbiota dynamics under sourdough propagation and detecting undesirable microorganisms. The methods described above are labour intensive and time consuming, and more importantly, some of them allow the detection and identification of only cultivable microorganisms.

Lin and Gänzle (2014) and Ripari et al. (2016) developed methods based on high-resolution melting curves (HRM-qPCR) for monitoring of sourdough microbiota during fermentation. Additionally, Lin and Gänzle (2014) applied this technique to determine the effect of cereal substrate on sourdough microbiota. The essence of the method is the possibility to detect differences in single nucleotide in the target gene. DNA is extracted directly from the sample without the need for microorganisms' cultivation. This method is a suitable tool in the taxonomic analysis of dominant microorganisms in sourdoughs, however, the authors point out that there are some limitations for very complex microbiota containing many genotypes. Analysis of sourdough microbiota requires a combination of culture dependent and culture-independent methods.

For over 10 years, high throughput sequencing (HTS) approaches have been widely used to the study of sourdough microbial consortia and monitor changes in their dynamics under fermentation process. Phylobiomics also allows for deep insights into undesirable microbiota in sourdoughs and tracking their entry routes. HTS is a more sensitive approach compared to traditional culture-independent techniques, aside from quantification (De Filippis et al. 2017; Lhomme et al. 2015a, b). Lhomme et al. (2016) determined microbial diversity (bacterial and yeast) over 68 French sourdoughs during bread-making process and their stability. Also, Michel et al. (2016) used HTS technology (MiSeq Illumina) to study the phylogenetic structure of 16 French sourdoughs from different bakeries (farmer-bakers, artisan-bakers, and industrial-bakers).

In many cases, the research on sourdough microbiota is focused on relationships between microbial and technological parameters (time, pH, temperature, water activity, raw materials, and additional compounds). Mendez et al. (2020) investigated whether the temperature determines the structure sourdough bacterial community. These approaches in the analysis of the sourdough consortia revealed not only the lactobacilli dynamics but also undesirable microbiota. The authors observed, after one step propagation, that *Clostridium* and *Bacillus* genera were dominated, while *Lactobacillus* spp. was in the minority. After two days, *Clostridium* spp. was displaced by *Lactobacillus* spp., and after 6 days no contaminating microflora was found. The temperature of 30 °C was conducive to bacterial growth of *Pseudomonas* and *Enterobacteriaceae,* while in 21 °C their growth was inhibited. Further studies also indicate the usefulness of HTS technology in dynamics analysing of the entire microbiota and displacement of atypical bacteria by *Lactobacillus* spp. In studies by De Angelis et al. (2019) after 8 h of sourdough propagation bacteria belonging to *Enterobacter, Serratia, Erwinia* genera (*Enterobacteriaceae*) as well as *Clostridium* and *Enterococcus* were detected. In contrary to research by Ercolini et al. (2013) which revealed that after 8 h of propagation at 25 °C the undesired bacteria were completely eliminated.

Currently, sourdoughs, unique in their properties, with an unimaginable variety of lactic acid bacteria and yeast, are used both in small, artisan bakeries and in a large industrial plants. The interactions between microorganisms, as well as between the microbiota and the raw material, and the technological parameters used significantly affect the properties of sourdough. To understand the complexity of these processes, highly specialized fields of knowledge such as genomics, metabolomics and proteomics have developed (De Vuyst et al. 2009; Gänzle and Ripari 2016; Gänzle and Zheng 2019).

References

Adamek M, Zvonkova M, Buresova I, Buran M, Sevcikova V, Sebestikova R, Adamkova A, Skowronkova N, Mlcek J (2023) Use of a thermodynamic sensor in monitoring fermentation processes in gluten-free dough proofing. Sensors 23, 534. https://doi.org/10.3390/s23010534

Aguilar N, Albanell E, Miñarro B, Capellas M (2016) Chestnut flour sourdough for gluten-free bread making. Eur Food Res Technol 242:1795–1802

Albagli G, Finotelli PV, Ferreira TF, Amaral PFF (2023) Toward Sourdough Microbiome Data: A Review of Science and Patents. Foods 12:420. https://doi.org/10.3390/foods12020420
Annan NT, Poll L, Plahar WA, Jakobsen M (2003) Aroma characteristics of spontaneously fermented Ghanaian maize dough for kenkey. Eur Food Res Tech 217(1):53–60
Arendt EK, Moroni AV, Morrissey JP, Dal Bello F (2010) Development of buckwheat and teff sourdoughs with the use of commercial starters. Int J Food Microbiol 142(1–2):142–148
Arora K, Ameur H, Polo A, Di Cagno R, Rizzello CG, Gobbetti M (2021) Thirty years of knowledge on sourdough fermentation: a systematic review. Trends Food Sci Technol 108:71–83
Axel C, Röcker B, Brosnan B, Zannini E, Furey A, Coffey A, Arendt EK (2015) Application of *Lactobacillus amylovorus* DSM19280 in gluten-free sourdough bread to improve the microbial shelf life. Food Microbiol 47:36–44
Bender D, Fraberger V, Szepasvári P, D'Amico S, Tömösközi S, Cavazzi G, Jäger H, Domig KJ, Schoenlechner R (2018) Effects of selected lactobacilli on the functional properties and stability of gluten-free sourdough bread. Eur Food Res Technol 244(6):037–1046
Bockwoldt JA, Fellermeier J, Steffens E, Vogel RF, Ehrmann MA (2021) β-Glucan production by *Levilactobacillus brevis* and *Pediococcus claussenii* for in situ enriched Rye and Wheat Sourdough Breads. Foods 10(3):547
Boreczek J, Litwinek D, Żylińska-Urban J, Izak D, Buksa K, Gawor J, Gromadka R, Bardowski JK, Kowalczyk M (2020) Bacterial community dynamics in spontaneous sourdoughs made from wheat, spelt, and rye wholemeal flour. MicrobiologyOpen 9:e1009
Brandt JM (2014) Starter cultures for cereal based foods. Food Microbiol 37:41–43
Calvert MD, Madden AA, Nichols LM, Haddad NM, Lahne J, Dunn RR, McKenney EA (2021) A review of sourdough starters: ecology, practices, and sensory quality with applications for baking and recommendations for future research. PeerJ 9:e11389
Capuani A, Behr J, Vogel R (2013) Influence of lactic acid bacteria on redox status and on proteolytic activity of buckwheat (*Fagopyrum esculentum* Moench) sourdoughs. Int J Food Microbiol 165(2):148–155
Chavan R, Chavan R (2011) Sourdough technology – a traditional way for wholesome foods: a review. Compr Rev Food Sci Food 10:169–182
Chochkov R, Savova-Stoyanova D, Papageorgiou M, Rocha JM, Gotcheva V, Angelov A (2022) Effects of teff-based sourdoughs on dough rheology and gluten-free bread. Quality Foods 11:1012
Coda R, Di Cagno R, Gobbetti M, Rizzello CG (2014) Sourdough lactic acid bacteria: exploration of non-wheat cereal-based fermentation. Food Microbiol 37:51–58
Corsetti A, Settanni L (2007) Lactobacilli in sourdough fermentation. Food Res Int 40:539–558
Corsetti A, De Angelis M, Dellagio F, Paparella A, Fox PF, Settanni L, Gobbetti M (2003) Characterization of sourdough lactic acid bacteria based on genotypic and cell-wall protein analyses. J Appl Microbiol 94:641–651
De Angelis M, Minervini F, Siragusa S, Rizzello CG (2019) Wholemeal wheat flours drive the microbiome and functional features of wheat sourdoughs. Int J Food Microbiol 302:35–46
De Filippis F, Parente E, Ercolini D (2017) Metagenomics insights into food fermentations. Microb Biotechnol 10(1):91–102
De Vuyst L, Neysens P (2005) The sourdough microflora: biodiversity and metabolic interactions. Trends Food Sci Technol 16:43–56
De Vuyst L, Vancanneyt M (2007) Biodiversity and identification of sourdough lactic acid bacteria. Food Microbiol 24:120–127
De Vuyst L, Vrancken G, Ravyts F, Rimaux T, Weckx S (2009) Biodiversity, ecological determinants, and metabolic exploitation of sourdough microbiota. Food Microbiol 26:666–675
De Vuyst L, Van Kerrebroeck S, Harth H, Huys G, Daniel HM, Weckx S (2014) Microbial ecology of sourdough fermentations: diverse or uniform? Food Microbiol 37:11–29
De Vuyst L, Harth H, Van Kerrebroeck S, Leroy F (2016) Yeast diversity of sourdoughs and associated metabolic properties and functionalities. Int J Food Microbiol 239:26–34

De Vuyst L, Van Kerrebroeck S, Leroy F (2017) Microbial ecology and process technology of sourdough fermentation. Adv Appl Microbiol 100:49–160

De Vuyst L, Comasio A, Kerrebroeck SV (2023) Sourdough production: fermentation strategies, microbial ecology, and use of non-flour ingredients. Crit Rev Food Sci Nutr 63(15):2447–2479

Decock P, Capelle S (2005) Bread technology and sourdough technology. Trends Food Sci Technol 16:113–120

Di Cagno R, Pontonio E, Buchin S, De Angelis M, Lattanzi A, Valerio F, Gobbetti M, Calasso M (2014) Diversity of the lactic acid bacterium and yeast microbiota in the switch from firm- to liquid-sourdough fermentation. Appl Environ Microbiol 80(10):3161–3172

Diowksz A (2011) Quality and specificity of polish bread. Polish Food 4:29–32

Diowksz A (2023) Traditional European breads. In: Garcia-Vaquero M, Pastor K, Orhun GE, McElhatton A, Rocha JMF (eds) Traditional European breads. Springer, Cham, pp 233–249

Diowksz A, Ambroziak W (2006) Sourdough. In: Hui YH (ed) Bakery products. Science and Technology. Blackwell Publishing, Ames, pp 365–380

Diowksz A, Kordialik-Bogacka E (2014) Stimulation of gluten-free sourdough fermentation. Acta Aliment 43(2):225–231

Diowksz A, Sadowska A (2021) Impact of sourdough and transglutaminase on gluten-free buckwheat bread quality. Food Biosci 43:101309

Diowksz A, Włodarczyk M (2003) Bakery starter cultures: I. Selection of lactic acid bacteria for sourdough fermentation. Zeszyty Naukowe. Chemia Spożywcza i Biotechnologia, Politechnika Łódzka 67:17–24 (in Polish)

Drabińska N, Zieliński H, Krupa-Kozak U (2016) Technological benefits of inulin-type fructans application in gluten-free products – a review. Trends Food Sci Technol 56:149–157

Ehrmann MA, Vogel RF (2005) Molecular taxonomy and genetics of sourdough lactic acid bacteria. Trends Food Sci Technol 16:31–42

Ehrmann M, Ludwig W, Schleifer KH (1994) Reverse dot blot hybridization: a useful method for the direct identification of lactic acid bacteria in fermented food. FEMS Microbiol Lett 117:143–149

Ehrmann MA, Brandt M, Stolz P, Vogel RF, Korakli M (2007) *Lactobacillus secaliphilus* sp. nov., isolated from type II sourdough fermentation. Int J Syst Evol Microbiol 57:745–750

Ercolini D, Pontonio E, De Filippis F, Minervini F, La Storia A, Gobetti M, Di Cagno R (2013) Microbial ecology dynamics during rye and wheat sourdough preparation. Appl Environ Microbiol 79(24):7827–7836

FEDIMA (2019) How bread consumers perceive sourdough in Europe. FED/19/082 https://www.fedima.org/resources. Accessed 14 Jan 2023

Ferchichi M, Valcheva R, Prévost H, Onno B, Dousset X (2007) Molecular identification of the microbiota of French sourdough using temporal temperature gradient gel electrophoresis. Food Microbiol 24:678–686

Föste M, Elgeti D, Jekle M, Becker T, Nordlohne S, Linden M, Heinz V (2013) Manufacture of gluten-free breads – a question of the substrate? Baking Biscuit 6:46–49

Fraberger V, Unger C, Kummer C, Domig KJ (2020a) Insights into microbial diversity of traditional Austrian sourdough. LWT Food Sci Technol 127:109358

Fraberger V, Ladurner M, Nemec A, Grunwald-Gruber C, Call LM, Hochegger R, Domig KJ, D'Amico S (2020b) Insights into the potential of sourdough-related lactic acid bacteria to degrade proteins in wheat. Microorganisms 8:1689

Fraberger V, Ammer C, Domig KJ (2020c) Functional properties and sustainability improvement of sourdough bread by lactic acid bacteria. Microorganisms 8(12):1895

Gänzle M, De Vuyst L (2004) Sourdough fermentation: from fundamentals to applications. Cereal Foods Rep 49:42–43

Gänzle M, Ripari V (2016) Composition and function of sourdough microbiota: from ecological theory to bread quality. Int J Food Microbiol 239:19–25

Gänzle MG, Zheng J (2019) Lifestyles of sourdough lactobacilli – do they matter for microbial ecology and bread quality? Int J Food Microbiol 302:15–23

Gänzle MG, Vermuelen N, Vogel RF (2007) Carbohydrate, peptide and lipid metabolism of lactic acid bacteria in sourdough. Food Microbiol 24:128–138

Gatto V, Torriani S (2004) Microbial population changes during sourdough fermentation monitored by DGGE analysis of 16S and 26S rRNA gene fragments. Ann Microbiol 54(1):31–42

Gobbetti M, Minervini F, Pontonio E, Di Cagno R, De Angelis M (2016) Drivers for the establishment and composition of the sourdough lactic acid bacteria biota. Int J Food Microbiol 239:3–18

Göttlich M, Brown JA, Brandt MJ (2023) Traditional breads from Germany. In: Garcia-Vaquero M et al (eds) Traditional European breads. Springer, Cham, pp 115–133

Hansen A (2005) Sourdough bread. In: Hui YH (ed) Handbook of food science, technology, and engineering, 1st edn. Taylor & Francis

Hansen A, Schieberle P (2005) Generation of aroma compounds during sourdough fermentation: applied and fundamental aspects. Trends Food Sci Technol 16:85–94

Houben A, Götz H, Mitzscherling M, Becker T (2010) Modification of the rheological behavior of amaranth (*Amaranthus hypochondriacus*) dough. J Cereal Sci 51:350–356

Huys G, Daniel HM, De Vuyst L (2013) Taxonomy and biodiversity of sourdough yeasts and lactic acid bacteria. In: Gobbetti M, Gänzle M (eds) Handbook on sourdough biotechnology. Springer, New York, pp 105–154

Jekle M, Houben A, Mitzscherling M, Becker T (2010) Effects of selected lactic acid bacteria on the characteristics of amaranth sourdough. J Sci Food Agric 90:2326–2332

Kawka A, Górecka D (2010) Comparison of chemical composition of wheat-oat and wheat-barley bread with sourdoughs fermented by 'LV2' starter. Food Sci Technol Quality 3(70):44–55 (in Polish)

Kawka A, Rausch P, Budna A (2010) Starters in production of wheat-oat bread. Sci Nat Technol 4(2):1–14 (in Polish)

Koj K, Pejcz E (2023, 1910) Rye dietary Fiber components upon the influence of fermentation inoculated with probiotic microorganisms. Molecules 28(4)

Korus J, Witczak M, Korus A, Juszczak L (2022) Mesquite (*Prosopis* L.) as a functional ingredient in gluten-free dough and bread. LWT Food Sci Technol 168:113957

Krupa-Kozak U, Wronkowska M, Soral-Śmietana M (2011) Effect of buckwheat flour on microelements and proteins contents in gluten-free bread. Czech J Food Sci 29:103–108

Lhomme E, Lattanzi A, Dousset X, Minervini F, De Angelis M, Lacaze G, Onno B, Gobbetti M (2015a) Lactic acid bacterium and yeast microbiotas of sixteen French traditional sourdoughs. Int J Food Microbiol 215:161–170

Lhomme E, Orain S, Courcoux P, Onno B, Dousset X (2015b) The predominance of *lactobacillus sanfranciscensis* in French organic sourdoughs and its impact on related bread characteristics. Int J Food Microbiol 213:40–48

Lhomme E, Urien C, Legrand J, Dousset X, Onno B, Sicard D (2016) Sourdough microbial community dynamics: an analysis during French organic bread-making processes. Food Microbiol 53:41–50

Lin XB, Gänzle MG (2014) Quantitative high-resolution melting PCR analysis for monitoring of fermentation microbiota in sourdough. Int J Food Microbiol 186:42–48

Litwinek D, Boreczek J, Gambuś H, Buksa K, Berski W, Kowalczyk M (2022) Developing lactic acid bacteria starter cultures for wholemeal rye flour bread with improved functionality, nutritional value, taste, appearance and safety. PLoS One 17(1):e0261677

Łopusiewicz Ł, Kowalczewski PŁ, Baranowska HM, Masewicz Ł, Amarowicz R, Krupa-Kozak U (2023) Effect of flaxseed oil cake extract on the microbial quality, texture and shelf life of gluten-free bread. Foods 12:595

Marciniak-Lukasiak K, Lesniewska P, Zielinska D, Sowinski M, Zbikowska K, Lukasiak P, Zbikowska A (2022) The influence of chestnut flour on the quality of gluten-free bread. Appl Sci 12:8340

Martín-Garcia A, Riu-Aumatell M, López-Tamames E (2023) Influence of process parameters on sourdough microbiota, physical properties and sensory profile. Food Rev Int 39(1):334–348

Mendez LAA, Savo Sardaro ML, Duarte RTD, Mazzon RR, Neviani E, Gatti M, De Dea LJ (2020) Sourdough bacterial dynamics revealed by metagenomic analysis in Brazil. Food Microbiol 85:103302

Meroth CB, Hammes WP, Hertel C (2004) Characterisation of the microbiota of rice sourdoughs and description of *lactobacillus spicheri* sp. nov. Syst Appl Microbiol 27(2):151–159

Meroth CB, Walter J, Hertel C, Brandt MJ, Hammes WP (2003a) Monitoring the bacterial population dynamics in sourdough fermentation processes by using PCR-denaturing gradient gel electrophoresis. Appl Environ Microbiol 69(1):475–482

Meroth CB, Hammes WP, Hertel C (2003b) Identification and population dynamics of yeasts in sourdough fermentation processes by PCR-denaturing gradient gel electrophoresis. Appl Environ Microbiol 69(12):7453–7461

Michel E, Monfort C, Deffrasnes M, Guezenec S, Lhomme E, Barret M, Sicard D, Dousett X, Onno B (2016) Characterization of relative abundance of lactic acid bacteria species in French organic sourdough by cultural, qPCR and MiSeq high-throughput sequencing methods. Int J Food Microbiol 239:35–43

Minervini F, De Angelis M, Di Cagno R, Gobbetti M (2014) Ecological parameters influencing microbial diversity and stability of traditional sourdough. Int J Food Microbiol 171:136–146

Mohsen SM, Aly MH, Attia AA, Osman DB (2016) Effect of sourdough on shelf life, freshness and sensory characteristics of Egyptian Balady bread. J Appl Environ Microbiol 4:39–45

Moroni AV, Dal Bello F, Arendt EK (2009) Sourdough in gluten-free bread-making: an ancient technology to solve a novel issue? Food Microbiol 26:676–684

Moroni AV, Arendt EK, Bello FD (2011a) Biodiversity of lactic acid bacteria and yeasts in spontaneously-fermented buckwheat and teff sourdoughs. Food Microbiol 28:497–502

Moroni AV, Bello FD, Zannini E, Arendt EK (2011b) Impact of sourdough on buckwheat flour, batter and bread: biochemical, rheological and textural insights. J Cereal Sci 54:195–202

Müller MRA, Wolfrum G, Stolz P, Ehrmann MA, Vogel RF (2001) Monitoring the growth of *lactobacillus* species during a rye four fermentation. Food Microbiol 18:217–227

Nionelli L, Rizzello CG (2016) Sourdough-based biotechnologies for the production of gluten-free foods. Foods 5(3):65

Oliveros JC (2007–2015) Venny. An interactive tool for comparing lists with Venn's diagrams. https://bioinfogp.cnb.csic.es/tools/venny/index.html. Accessed 20 June 2023

Ostasiewicz A,Ieglińska A (2011) The comparison of quality of rye bread prepared on spontaneous fermenting leaven and with starter cultures application. Adv Food Process Technol 1:18–23 (in Polish)

Pecyna A, Buczaj A, Rózyło R, Kobus Z (2023) Physical and antioxidant properties of innovative gluten-free bread with the addition of hemp inflorescence. Appl Sci 13:4889

Pérez-Alvarado O, Zepeda-Hernández A, Garcia-Amezquita LE, Requena T, Vinderola G, García-Cayuela T (2022) Role of lactic acid bacteria and yeasts in sourdough fermentation during breadmaking: evaluation of postbiotic-like components and health benefits. Front Microbiol 13:969460

Petkova M, Stefanova P, Gotcheva V, Angelov A (2021) Isolation and characterization of lactic acid bacteria and yeasts from typical Bulgarian sourdoughs. Microorganisms 9:1346

Piasecka-Jóźwiak K, Rozmierska J, Słowik E, Chabłowska B, Stecka KM, Kotryba D (2011) Evaluation of usefulness of amylolytic strain *lactobacillus plantarum* for obtaining wheat bread. Adv Sci Technol Agri-Food Ind 66(2):5–20 (in Polish)

Piasecka-Jóźwiak K, Słowik E, Rozmierska J, Chabłowska B (2016) Development of organic buckwheat gluten-free bread, characterized by a high level of bioactive compounds. J Res Appl Agricult Eng 61(4):110–116

Polak T, Mejaš R, Jamnik P, Kralj Cigić I, Poklar Ulrih N, Cigić B (2021) Accumulation and transformation of biogenic amines and gamma-aminobutyric acid (GABA) in chickpea sourdough. Foods 10:2840

Poutanen K, Katina K (2014) Preface: sourdough in multifunctional process technology for future food challenges. J Microbiol 37:1

Ramos L, Alonso-Hernando A, Martínez-Castro M, Morán-Pérez JA, Cabrero-Lobato P, Pascual-Maté A, Téllez-Jiménez E, Mujico JR (2021) Sourdough Biotechnology Applied to Gluten-Free Baked Goods: Rescuing the Tradition. Foods 10:1498. https://doi.org/10.3390/foods10071498
Randazzo CL, Heilig H, Restuccia C, Giudici P, Caggia C (2005) Bacterial population in traditional sourdough evaluated by molecular methods. J Appl Microbiol 99(2):251–258
Report Sourdough Market Size, Share & Trends Analysis Report By Type, By Application (Breads, Cookies, Cakes, Waffles, Pizza), By Region, Competitive Landscape, And Segment Forecasts, 2019–2025 Report ID: GVR-3-68038-510-6. https://www.grandviewresearch.com/industry-analysis/sourdough-market. Accessed 14 Jan 2023
Ripari V, Gänzle MG, Berardi E (2016) Evolution of sourdough microbiota in spontaneous sourdoughs started with different plant materials. Int J Food Microbial 232:35–42
Rizzello CG, Lorusso A, Montemurro M, Gobbetti M (2016) Use of sourdough made with quinoa (Chenopodium quinoa) flour and autochthonous selected lactic acid bacteria for enhancing the nutritional, textural and sensory features of white bread. Food Microbiol 56:1–13
Różyło R, Rudy S, Krzykowski A, Dziki D (2014) Novel application of freeze-dried Amaranth sourdough in gluten-free bread production. J Food Process Eng 38:135–143
Różyło R, Rudy S, Krzykowski A, Dziki D, Gawlik-Dziki U, Różyło K, Skonecki S (2015) Effect of adding fresh and freeze-dried buckwheat sourdough on gluten-free bread quality. Int J Food Sci Technol 50:313–322
Różyło R, Rudy S, Krzykowski A, Dziki D, Siastała M, Polak R (2016) Gluten-free bread prepared with fresh and freeze-dried Rice sourdough-texture and sensory evaluation. J Texture Stud 47:443–453
Różyło R, Dziki D, Gawlik-Dziki U, Biernacka B, Wójcik M, Ziemichód A (2017) Physical and antioxidant properties of gluten-free bread enriched with carob fibre. Int Agrophys 31:411–418
Rühmkorf C, Jungkunz S, Wagner M, Vogel RF (2012) Optimization of homoexopolysaccharide formation by lactobacilli in gluten-free sourdoughs. Food Microbiol 32(2):286–294
Ruiz Rodríguez L, Vera Pingitore E, Rollan G, Martos G, Saavedra L, Fontana C, Hebert EM, Vignolo G (2016) Biodiversity and technological potential of lactic acid bacteria isolated from spontaneously fermented amaranth sourdough. Lett Appl Microbiol 63:147–154
Scheirlinck I, Van der Meulen R, De Vuyst L, Vandamme P, Huys G (2009) Molecular source tracking of predominant lactic acid bacteria in traditional Belgian sourdoughs and their production environments. J Appl Microbiol 106:1081–1092
Šedivý P, Albrecht J (2014) Bakery technology II.: Bread production (Pekařská technologie II.: Výroba chleba). Praha, Pekař a cukrář s.r.o.: 27–51 (in Czech)
Singh S (2023) Sourdough Market Research Report Information By Type (Type I, Type II, and Type III), By Ingredients (Wheat, Rye, Oats, Barley, and Others), By Application (Bread & Buns, Pizza, Cakes, and Others), And By Region (North America, Europe, Asia-Pacific, And Rest Of The World) – Market Forecast Till 2030. Report ID: MRFR/F-B & N/1291-HCR. https://www.marketresearchfuture.com. Accessed 20 Jan 2023
Sluková M, Jurkaninová L, Gillarová S, Horáčková Š, Skřivan P (2021) Rye – nutritional and technological evaluation in Czech cereal technology – a review: sourdoughs and bread. Czech J Food Sci 39:65–70
Šmídová Z, Rysová J (2022) Gluten-Free Bread and Bakery Products Technology. Foods 11(3):480. https://doi.org/10.3390/foods11030480
Sterr Y, Weiss A, Schmidt H (2009) Evaluation of lactic acid bacteria for sourdough fermentation of amaranth. Int J Food Microbiol 136:75–82
Tömösközi S, Németh R, Farkas A, Rakszegi M (2023) Hungarian history of cereal cultivation, processing and sourdough making. In: Garcia-Vaquero M, Pastor K, Orhun GE, McElhatton A, Rocha JMF (eds) Traditional European breads. Springer, Cham, pp 161–180
Ua-Arak T, Jakob F, Vogel R (2016) Characterization of growth and exopolysaccharide production of selected acetic acid bacteria in buckwheat sourdoughs. Int J Food Microbiol 239:103–112
Van Kerrebroeck S, Bastos FCC, Harth H, De Vuyst L (2016) A low pH does not determine the community dynamics of spontaneously developed backslopped liquid wheat sourdoughs but does influence their metabolite kinetics. Int J Food Microbiol 239:54–64

Van Kerrebroeck S, Maes D, De Vuyst L (2017) Sourdoughs as a function of their species diversity and process conditions, a meta-analysis. Trends Food Sci Technol 68:152–159

Vancanneyt M, Neysens P, De Wachter M, Engelbeen K, Snauwaert C, Cleenwerck I, Van der Meulen R, Hoste B, Tsakalidou E, De Vuyst L, Swings J (2005) *Lactobacillus acidifarinae* sp. nov. and *lactobacillus zymae* sp. nov., from wheat sourdoughs. Int J Syst Evol Microbiol 66:615–620

Vogel RF, Böcker G, Stolz P, Ehrmann M, Fanta D, Ludwig W, Pot B, Kersters K, Schleifer KH, Hammes WP (1994) Identification of *lactobacilli* from sourdough and description of *lactobacillus pontis* sp. nov. Int J Syst Bacteriol 44(2):223–229

Vogelmann SA, Hertel C (2011) Impact of ecological factors on the stability of microbial associations in sourdough fermentation. Food Microbiol 28:583–589

Vogelmann SA, Seitter M, Singer U, Brandt MJ, Hertel C (2009) Adaptability of lactic acid bacteria and yeasts to sourdoughs prepared from cereals, pseudocereals and cassava and use of competitive strains as starters. Int J Food Microbiol 130(3):205–212

Vrancken G, Rimaux T, Weckx S, Leroy F, De Vuyst L (2011) Influence of temperaturę and backslopping time on the microbiota of a type I propagated laboratory wheat sourdough fermentation. Appl Environ Microbiol 77:2716–2726

Wieczorek MN, Kowalczewski PŁ, Drabińska N, Różańska MB, Jeleń HH (2022) Effect of cricket powder incorporation on the profile of volatile organic compounds, free amino acids and sensory properties of gluten-free bread. Pol J Food Nutr Sci 72:431–442

Wiese BG, Strohmar W, Rainey FA, Diekmann H (1996) *Lactobacillus panis* sp. nov., from sourdough with a long fermentation period. Int J Syst Bacteriol 46(2):449–453

Wolter A, Hager AS, Zannini E, Czerny M, Arendt E (2014) Influence of dextran-producing *Weissella cibaria* on baking properties and sensory profile of gluten-free and wheat breads. Int J Food Microbiol 172:83–91

Wronkowska M, Haros M, Soral-Śmietana M (2013) Effect of starch substitution by buckwheat flour on gluten-free bread quality. Food Bioprocess Technol 6:1820–1827

Zalán Z, Hegyi F, Szabó EE, Maczó A, Baka E, Du M, Liao Y, Jianquan K (2015) Bran fermentation with *lactobacillus* strains to develop a functional ingredient for sourdough production. Int J Nutr Food Sci 4(4):409–419

Zand E, Bender D, D'Amico S, Tömösközi S, Jaeger H, Schoenlechner R (2018) Improving gluten-free buckwheat bread by sourdough fermentation and addition of arabinoxylan and pyranose 2-oxidase. Die Bodenkultur: J Land Manage Food Environ 69(4):227–237

Zannini E, Pontonio E, Waters DM (2012) Applications of microbial fermentations for production of gluten-free products and perspectives. Appl Microbiol Biotechnol 93:473–485

Zaręba D, Ostasiewicz A (2009) The technological and health-promoting functionality of dough fermentation proces – tradition and scientific progres. Adv Food Process Technol 1:70–74 (in Polish)

Zdybel B, Różyło R, Sagan A (2019) Use of a waste product from the pressing of chia seed oil in wheat and gluten-free bread processing. J Food Process Pres 43(8):e14002

Chapter 4
Traditional and Novel Applications of Sourdough Microbes in Northern Europe and the Baltic Countries

Clarissa Schwab and Maria Florencia Bambace

4.1 A Short History on the Arrival of Cereals in Northern Europe Including Iceland and the Faroese Island

Agriculture first appeared in Europe in around 7000 BC and reached Scandinavia around 4000–3500 BC (Bjørnstad 2012). Grains were grown around 3200 BC in Denmark, Southern Sweden and Norway. The oldest cereal crop used in Northern Europe is likely barley (*Hordeum vulgare* L., Table 4.1) with first observations dating back to the Bronze Age. Barley has the advantage to mature grains in a colder climate with short growing season and to tolerate frost and cold (Reykdal et al. 2014). Oat (*Avena sativa* L.) first appeared as weed around the Bronze Age contaminating the fields cultivated with barley, and began to be itself cultivated on larger scale in the eighteenth century. Similar to oat, rye (*Secale cereal* L.) arrived as a weed contaminant of other grains (Behre 1992). Being adapted to the prevailing climate, rye had become a major cultivated crop by 400–600 AD in Northern Europe (Bjørnstad 2012; Behre 1992). Other grains that were cultivated early were the wheat varieties emmer (*Triticum turgidum* subsp. *Dicoccum* L.), einkorn (*Triticum monococcum* L.) and spelt (*Triticum aestivum* subsp. *spelta* L.) (Bjørnstad 2012).

On the Farø islands, barley arrived possibly as early as 300–500 CE (Curtin et al. 2021). On Iceland, a warm period lasting from the ninth to sixteenth century allowed to cultivate and harvest rye, barley, and wheat, which was abolished by a following cold spell, also reducing grain consumption, while on the Farø islands barley cultivation continued (Bjørnstad 2012). Greenland traditionally relied on a protein-based diet consisting of marine mammals, game, fish and birds. Similarly, cereals

C. Schwab (✉) · M. F. Bambace
Functional Microbe Technology Group, Department of Biological and Chemical Engineering, Aarhus University, Aarhus, Denmark
e-mail: schwab@bce.au.dk

E. B. Ceresino et al. (eds.), *Sourdough Microbiota and Starter Cultures for Industry*, https://doi.org/10.1007/978-3-031-48604-3_4

Table 4.1 Composition of major grains of flours used in Northern Europe and the Baltic countries

Composition	Water (%)	Carbohydrates (%)	Major carbohydrates (proportion in %)	Proteins (%)	Major gluten storage proteins (proportion in %)	Lipids (%)
Rye	13–14	70–73	Starch (52–60)	9–11	Secalins as prolamins, secalinin as glutelins	2
			Non-starch (13–17) (arabinoxylan 1–8, β-glucan 2)			
Barley	12	67–72	Starch (52–63)	11	Hordeins as prolamins, hordenin as glutelins	2
			Non-starch (10–20) (arabinoxylan 0–4, β-glucan 3–7)			
Wheat	13–14	70–75	Starch (60)	11–12	70–80, gliadins as prolamin, glutenins as glutelins	2
			Non-starch (10–13) (arabinoxylan 0–3, β-glucan 1)			
Oat	13	58–66	Starch (40–56)	10–11	Avenins as prolamins, avenalins as glutelins	5–7
			Non-starch (10–13) (arabinoxylan 0–2, β-glucan 3.5–6)			

Data collected from De Vuyst et al. (2017), Koehler and Wieser (2013), and Sullivan et al. (2013)

were not part of the traditional diet in the Northern Sami regions leading to the statement of the 'land of no bread' (Kylli 2014).

4.2 The Beginning of Bread Production in Northern Europe

With the arriving of milling techniques after 1000 AD, flour could be prepared allowing to bake throughout the year. Barley was mostly used in the north and west of Northern Europe to produce unleavened thin, flatbread with low water content that could be stored for up to two years. Further south, hard flatbread was popular, which was often produced using rye fermented with sourdough (see also Sect. 4.3).

In the southern part of Northern Europe and the Baltics, soft loaves prepared from sourdough fermented rye were popular similar as in Germany. The merchants of the Hanseatic League brought along their own brewers and possibly bakers, which might have supported the development of local Scandinavian baking traditions (Bjørnstad 2012; Unger 1988). From around 1500 AD, rye bread was consumed daily in Denmark, Sweden, Finland and the Baltics. Wheat was considered a luxury grain for special festive occasions. On Iceland, grain was often imported and frequently used to prepare cereal porridges instead of bread. Leavened sourdough bread appeared first in the eighteenth century (Steingrímsdóttir et al. 2018).

Sourdough was used to increase shelf life, to create a specific flavour profile, and to nutritionally improve the grains (see also Sect. 4.9), and was likely the main leavening agent until the use of brewer's became common. Already in the eighth and ninth century, monasteries in Scandinavia built brewhouses next to their bakeries (Unger 1988). After the Middle age, barm, which is *Saccharomyces cerevisiae* containing foam from local breweries, started to be used as alternative or additional leavening agent (Catzeddu 2011) similarly as in Germany, where brewer's yeast was introduced from the fifteenth and sixteenth century (Cappelle et al. 2013). Since the nineteenth century and the beginning of the industrial production of yeast, the major contribution of sourdough to rye bread production became acidification.

Currently, the most commonly harvested temperate cereals in Northern Europe are wheat and barley, and to a lesser extent rye and oat (Bjørnstad 2012), which are all used for production of bread or other cereal-based products including beer and animal feed. On Iceland, Farø Islands and Greenland, most food grains are imported (Reykdal et al. 2014). Cultural differences prevail with rye contributing most to the whole-grain intake in Denmark (around 70%) and less in Sweden (about 50%) and Norway (about 20%) (Kyrø et al. 2012). In a Swedish survey from 2012, about 1/3 of Swedes were classified as white or whole grain bread consumers. The consumption of dry crisp bread remains to be higher in the north (Kyrø et al. 2012). There was a difference between age groups, with white bread being preferred by the younger population, while older age groups more often ate dry crisp bread, whole-grain bread, and whole-grain rye sourdough bread (Sandvik et al. 2014).

4.3 Sourdough Fermentation Processes in Northern European Countries and the Baltics

Bread production in Northern Europe has been characterized by the widespread use of flours other than wheat, i.e. rye and barley. A distinctive feature of bread of Northern European bread is the use of sourdough for acidification and leavening. Sourdough is a mixture of flour and water fermented with a microbial community mainly consisting of few different genera or species of *Lactobacillaceae* (or lactic acid bacteria, LAB) (Fig. 4.1). Homofermentative LAB theoretically use 1 M glucose to produce 2 M lactate gaining 2 ATP, while heterofermentative LAB ferment 1 M glucose to 1 M lactate, 1 M ethanol and 1 M CO_2. Heterofermentative LAB can gain additional energy when fermenting maltose, which is abundant in cereal fermentations, or when reducing fructose to mannitol (see also Sect. 4.9.4) (Gänzle 2015).

Scalding, i.e. the pre-soaking of milled grains in boiling/hot water with consequent cooling down is used in some regions (Table 4.2). Scalding solubilizes and gelatinizes the starches rendering it more accessible for degradation (Li et al. 2022).

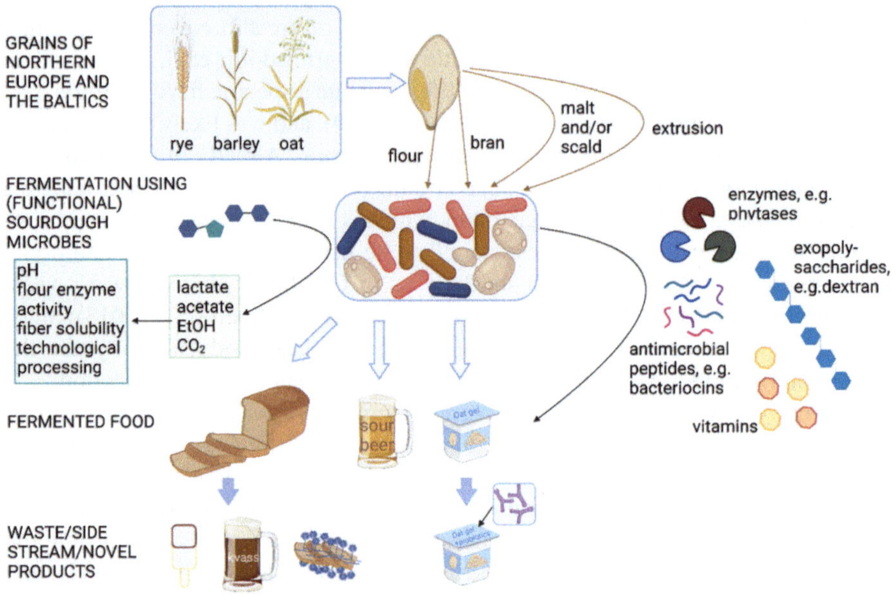

Fig. 4.1 Examples of products of sourdough fermentation of grains grown in Northern Europe, and the activities and compounds that are transferred by the souring microbes. (Figure was prepared using Biorender)

Table 4.2 Examples of Northern European bread varieties

Country	location	bread	Grain	type	Leavening style	Additional information
Northern Europe	Sweden	Tunnbröd	Barley, rye wheat	Flour	Unleavend, yeast or chemical	Hard and dry, long storage
	Norway	Flatbrød				Hard and dry
	Finland	Rieska				Soft
	Iceland	Flatkaka				Soft
Northern Europe	Finland	Näkkileipä				
	Estonia	Näkileib				
	Sweden	Knäckebröd	Rye	Flour	Sourdough, yeast	Low water content, long storage
	Denmark	Knækbrød				
	Norway	Knekkebrød				
	Iceland	Hrökkbrauð				
	The Farø islands	Knekkbreyð				

(continued)

Table 4.2 (continued)

Country	location	bread	Grain	type	Leavening style	Additional information
Finland	Easter Finland	Ruislimppu (Finnish Rye)	Rye (100%)	Flour	Sourdough	
	Western Finland	Ruisreikäleipä (Halme Bakery's Ring Rye)	Whole grain rye (90%)	Flour	Sourdough and yeast	
		Mallas Leipää (Finnish Malt Bread)	Rye malt and wheat	Flour	Yeast	
	Central Finland	Tavastian Rye Hämäläiset Ruiskorput (Rusks)	Whole grain rye (90%)	Flour	Yeast	Sweetened with molasse, spiced
Estonia		Leib (Estonian Black Bread)	Whole rye	Flour	Sourdough	e.g. Kihnu Rukkileib, Rukkipüülileib prepared with sugar and cumin or caraway seeds; possible addition of potatoes, fatty pork or Baltic herrings
		Karask	Barley	Flour	Baking powder	Addition of sour milk
Lithuania		Ruginė Duona iš Palangos (Palanga Rye)	Whole rye and malt	Flour	Sourdough and yeast	Series and combination of sourdough sponge, scald, scald-sponge, yeast sponge; addition of caraway seed
		Juoda Ruginė Duona (Black Rye Bread)	Whole rye and malt	Flour	Sourdough	Combinations of sourdough sponge, scald-sponge; addition of honey
Latvia		Latgaliešu Maize (Latgalian Rye)	Whole rye and malt	Flour		Combination of sponge, scald, 2-stage sponge-scald; addition of caraway seed
		Rudzu Rupjā Maize (Latvian Coarse Rye)	Whole rye	Flour and coarse meal		Sponge, final dough

(continued)

Table 4.2 (continued)

Country	location	bread	Grain	type	Leavening style	Additional information
Sweden	Southern and Western Sweden	Kavring	Rye	Flour	Sourdough	Sponge and scald; spices can be added
	Western Sweden	Rågkakor (Anise-Fennel Loaves)	Rye (100%)	Flour	Yeast	Addition of spices, sweet cake
Denmark		Rugbrød	Whole rye	Flour and kernels	Sourdough	Series of stage 1 sponge, stage 2 sponge, final dough; addition of other seeds and grains
Norway	North West	Sunnmørsbrød	Rye, whole wheat	Flour (+oats)	Yeast	
Iceland		Rugbrauð baked as Pottbrauð (pot bread) or Hverabrauð (hot-spring bread)	Rye	Flour	Sourdough	Traditionally baked in pots or in geothermal hot springs

References: tasteatlas.com, theryebaker.com, visitestonia.com/en/bread, swedishfood.com

Additionally, the temperature during scalding might direct influence amylase activity (Hopek et al. 2006), and might reduce cell counts of flour indigenous bacteria, yeast and fungi (Li et al. 2022). In some countries such as Denmark, whole grains are added to the sourdough bread.

Sourdoughs are distinguished based on inoculation style and process parameters (De Vuyst et al. 2017). Type I and II sourdoughs are commonly used in domestic and industrial settings in Northern Europe and the Baltic countries (Table 4.3). Type 1 sourdoughs initiate spontaneously; consequently backslopping, i.e. the inoculation of new batches with fermented dough material from previous fermentation in multiple consequent cycles, is employed (De Vuyst et al. 2017). For type 2 sourdoughs, starter cultures are added to initiate the fermentation. In addition, there are different alternatives in running the fermentation process (De Vuyst et al. 2017). A type 0 process generates a pre-dough/pre-ferment/sponge that form when flour and water are incubated, yeast might be added. In a type I process, doughs with a dough yield of ≤200 are fermented at 20–30 °C for 6–24 h with sourdough as main leaving agent; sourdoughs are firm while the sourdoughs of type 2 processes are more liquid (dough yield >200). Type 2 processes are conducted at temperatures >30 °C (De Vuyst et al. 2017).

As an example, traditional Estonian rye sourdoughs can be classified as semi-solid type 1 processed sourdoughs, because the consistency of doughs is relatively dense. Estonian sourdoughs are fermented at ambient temperature, which may vary between 18 and 27 °C depending on the season (Bessmeltseva et al. 2014).

Table 4.3 Examples of microbiota composition of Northern European sourdoughs

Country	Cereal	Sourdough process	Type	Temperature (°C)/Time	Final pH	Type of analysis	Bacterial species	Cell counts (log g^{-1})	Yeast species	Cell counts	Reference
Finland	Rye	Home made	NA	NA	NA	16S rRNA gene/ ITS sequencing	*F. sanfranciscensis*^{he} *C. paralimentarius*^{ho} *F. rossiae*^{he}	NA	*S. cerevisiae* *T. delbrueckii*	NA	Landis et al. (2021)
		Industrial, daily backslopping	NA	27–30 °C, 12 h	3.8	PCR-RFLP	NA	NA	*C. humilis* *K. exigua* *S. cerevisiae*	NA	Mäntynen et al. (1999)
Lithuania	Rye	Industrial		NA	3.9–4.1		*L. curvatus*^{he} *C. farciminis*^{ho} *L. sakei*^{ho}				Juodeikiene (2016)
		Laboratory spontaneous	Type I & II, DY 200 or 300	25–35 °C, 2–3 d	3.7–4.1	Phenotypic, ITS sequencing	*P. acidilactici*^{ho} *P. pentosaceus*^{ho}	8.3–8.6			
		Laboratory spontaneous	Type II, DY 250	30 °C	NA	Cultivation, species-level PCR	*L. plantarum*^{ho} *L. casei*^{he} *L. paracsei*^{he} *L. curvatus*^{he} *L. farraginis* ^{he} *L. coryniformis* ^{he} *L. brevis* ^{he} *L. uvarum*^{ho} *P. pentosaceus*^{ho} *P. acidilactici*^{ho} *E. pseudoavium*^{ho}	NA	NA		Bartkiene et al. (2020)

(continued)

Table 4.3 (continued)

Country	Cereal	Sourdough process	Type	Temperature (°C)/Time	Final pH	Type of analysis	Bacterial species	Cell counts (log g^{-1})	Yeast species	Cell counts	Reference
Latvia	Rye	Laboratory spontaneous 3 cycles	Type I DY 200	28/34 °C	3.8	Phenotypic (API CH50, ID32C)	*L. brevis*^{he} *L. fermentum*^{he}	6.8	*S. cerevisae*	6.7	Kozlinskis et al. (2008, 2010)
	Barley	Laboratory spontaneous 3 cycles	Type I	28.5–31 °C	3.8	16S rRNA gene/ ITS sequencing	*P. pentosaceus*^{ho}	8.2	Divers	7.3	Reidzane et al. (2021)
Estonia	Rye	Laboratory spontaneous 56 cycles	Type I DY 200	20 °C	3.5–4.0	Cultivation, DGGE 16S rRNA gene sequencing	*L. brevis*^{he} *C. crustorum*^{ho} *L. plantarum*^{ho} *C. paralimentarius*^{ho}	8.7–9.6	NA		Bessmeltseva et al. (2014)
			Type I DY 200	30 °C	3.5–4.0		*C. paralimentarius*^{ho} *L. plantarum*^{ho}	8.8–9.4	NA		
	Rye	Industrial A backslopping	Type II, DW 250	32 °C	3.5–3.7	Cultivation, 16S rRNA gene sequencing	*L. helveticus*^{ho} *L. pontis*^{ho} *L. panis*^{he}	7.1–9.1	NA		Viiard et al. (2012, 2016)
		Industrial B backslopping	Type II, DW 400	42–44 °C	3.4–3.6	Cultivation, 16S rRNA gene sequencing	*L. amylovorus*^{ho} *L. frumenti*^{he}	8.1–8.9	NA		Viiard et al. (2016)
		Industrial C backslopping	Type I DY 200	<28 °C	4.0–4.2	Cultivation, 16S rRNA gene sequencing	*F. sanfranciscensis*^{he}	6.6–8.3	NA		Viiard et al. (2016)
		Industrial D backslopping	Type I DY 200	19–30 °C	3.8–4.3	Cultivation, 16S rRNA gene sequencing	*L. pontis*^{ho} *L. helveticus*^{ho}	8.0–9.1	NA		Viiard et al. (2016)

Sweden	Rye	Laboratory spontaneous 2 cycles		NA	4.1	Cultivation	*Lactobacillus* sp. *P. pentosaceus*[ho]	9.0	*S. delbrueckii*	6	Lönner et al. (1986)
	Rye / whole wheat	Industrial		NA	NA	16S rRNA gene / ITS sequencing	L. *xiangfangensis*[ho], *A. pasteurians*[AAB] *C. paralimentarius*[ho]		*P. membrani-faciens, Alternaria spp., K. bulderi*		Landis et al. (2021)
	Wheat	Industrial A Spontaneous	Type I DY 150	30 °C, 6 h	5.0	Phenotypic	*L. plantarum*[ho] *L. brevis*[he]	9.0	NA		Spicher and Lönner (1985)
	Rye / wheat	Industrial B, backslopping	DY 160	40 °C, 18 h	3.7		*F. lindneri*[he] *L. delbrueckii* [ho]	9.5	NA		
	Rye / wheat	Industrial C, backslopping	TypeI DY 170	28 °C -30, 10–12 h	3.9		*L. fermentum*[he] *F. lindneri*[he]	9.4	NA		
	Rye	Industrial D, backslopping	Type II DY 250	33 °C, 15 h	3.5		*L. fermentum*[he]	9.4	NA		
	Rye	Industrial E backslopping	Type II DY 250	30–32 °C, 18–20 h	3.4		*L. fermentum*[he] *W. viridescens*[he]	9.6	NA		
	Rye	Industrial F Commercial starter	Type III DY 250	53 °C, 16 h	3.2		*L. delbrueckii* [ho] *L. bulgaricus*[ho] *C. farciminis*[ho]	9.0	NA		
	Rye	Industrial G backslopping	Type I DY 150	28–30 °C, 12 h	3.7		*L. fermentum*[he]	9.5	NA		
	Rye	Industrial H Commercial starter	Type III DY 180	29–30 °C, 44 h	3.3		*L. casei/ rhamnosus*[he] *L. acidophilus*[ho] *C. farciminis*[ho] *W. viridescens*[he]	9.5	NA		
	Rye	Industrial I backslopping	DY 200	35–37 °C, 3 × 8 h	3.4			9.5	NA		
	Barley	Laboratory spontaneous 2 cycles		30 °C	3.5	Cultivation	NA	8.9	NA		Marklinder and Johansson (1995)

(continued)

Table 4.3 (continued)

Country	Cereal	Sourdough process	Type	Temperature (°C)/Time	Final pH	Type of analysis	Bacterial species	Cell counts (log g^{-1})	Yeast species	Cell counts	Reference
Denmark	Rye	Industrial	Type II DY 225	30–34 °C	3.4–3.7	Phenotypic (API) 50CH and ID32C and 16S rRNA gene sequencing	*L. amylovorus*^{ho} *L. panis*^{he} *L. reuteri* group^{he}	8.4–9.1	*S. cerevisiae*	6.5–8.0	Rosenquist and Hansen (2000)
	Organic rye		Type II DY 240	28–31 °C	3.3–3.5		*L. amylovorus*^{ho} *L. panis*^{he} *L. reuteri* group^{he}	8.4–9.1	*S. cerevisiae*	6.0–7.1	
Norway	Rye	Home made		NA	NA	16S rRNA gene/ ITS sequencing	*C. paralimentarius*^{ho}, *L. brevis*^{he}	NA	*S. cerevisiae*	NA	Landis et al. (2021)

NA not available, *he* heterofermentative, *ho* homofermentative, *AAB* acetic acid bacteria

4.4 Microbes Present in Sourdoughs from Northern Europe and the Baltic Countries

During the last four decades, a variety of studies investigated the microbiota of industrial, homemade and laboratory sourdoughs in Northern European and Baltic countries (Table 4.3). In the earlier days of microbiota analysis, identification of the sourdough microbes relied entirely on cultivation and phenotypic analysis based on for example enzymatic activity profiles and fermentation pattern. From the 1990s, the 16S rRNA gene started to be used as marker gene for strain identification often in combination with culture-based techniques. More recently, 16S rRNA gene sequencing was used to profile sourdough communities which reports the relative abundance of certain taxa (Landis et al. 2021). The taxonomic assignment presented in this chapter takes into consideration the recent reclassifications that occurred for example within the genus *Lactobacillus* (Zheng et al. 2020).

In general, LAB dominate the microbial community of sourdoughs and are responsible for acid production while yeasts work to leaven the dough. The species *Fructilactobacillus sanfranciscensis, Furfurilactobacillus rossiae, Companilactobacillus paralimentarius, Levilactobacillus brevis, Lactiplantibacillus plantarum, Limosilactobacillus pontis,* and *Limosilactobacillus panis* were frequently identified in homemade and industrial sourdoughs from Northern Europe and the Baltic Countries (Table 4.3). This is in accordance with *F. sanfranciscensis, F. rossiae, C. paralimentarius, L. brevis,* and *L. plantarum* being associated with type I sourdoughs while *L. pontis, L. panis,* the *Limosilactobacillus reuteri* group including *Limosilactobacillus reuteri* and *Limosilactobacillus fermentum* and the *Lactobacillus delbrueckii* group dominate type II sourdoughs that are fermented at higher temperatures (De Vuyst et al. 2017; Huys et al. 2013).

The *L. delbrueckii* group includes *Lactobacillus johnsonii* and *Lactobacillus amylovorus*, which were isolated from Danish and Estonian sourdoughs, and also the species *Lactobacillus helveticus* and *L. delbrueckii* subsp. *bulgaricus* with isolates that were recovered from sourdoughs from Estonia and Sweden (Viiard et al. 2016; Spicher and Lönner 1985; Rosenquist and Hansen 2000) (Table 4.3).

C. paralimentarius was isolated from several sourdoughs (Table 4.3) and belongs to the *Companilactobacillus alimentarius* group which often co-occur with other heterofermentative lactobacilli in cereal and vegetable fermentations (Zheng et al. 2020).

Latilactobacillus sakei and *Latilactobacillus curvatus* were isolated from an industrial dough in Lithuania (Juodeikiene 2016). These species frequently occur on meat products (Zagorec and Champomier-Vergès 2017; Zheng et al. 2020). Pediococci and enterococci, namely *Pediococcus pentosaceus* and *Pediococcus acidilactici,* and *Enterococcus pseudoavium,* were only recovered from spontaneous laboratory fermentations using rye and barley (Table 4.3). Acetic acid bacteria, which are rarely isolated from sourdough, were not recovered from Scandinavian doughs by cultivation, but were recently identified in a Swedish sourdough using 16S rRNA gene sequencing (Landis et al. 2021).

Yeast profiles were analysed in fewer studies; *S. cerevisiae* was frequently detected, with *Candida humilis* and *Kazaschtania exigua* also being observed among other yeasts (Landis et al. 2021; Mäntynen et al. 1999; Kozlinskis et al. 2008, 2010; Rosenquist and Hansen 2000) (Table 4.3).

When a variety of potential starter cultures was tested for the ability to grow and acidify rye sourdoughs under different conditions in regard to inoculum size, dough yield, fermentation temperature and time (Spicher and Lönner 1985), *Lactobacillus acidophilus, C. alimentarius, L. brevis* and *Fructilactobacillus lindneri* decreased the pH in the shortest time. Bartkiene et al. (2017b) showed that a strain of *L. plantarum* grew in rye, wheat, barley and oat flour to 7–9 log CFU g^{-1} while decreasing the pH to 3.4–3.9 in 48 h. Mixed rye-wheat breads produced with sourdough lost less mass during baking and had a higher specific volume compared to bread without sourdough. Sourdough mixed rye-wheat breads were overall acceptable (Bartkiene et al. 2017b).

4.5 Impact of Sourdough Fermentation on Texture and Flavour of Breads Produced from Grains Used in Northern Europe

In contrast to wheat flour, rye does not contain the gluten proteins gliadin and glutenin and instead mainly the structurally different secalins and glutelins (Table 4.1). High and low molecular glutentins are mostly responsible for the formation of a protein network structure during wheat dough preparation and baking, secalins show only limited network formation (Verwimp et al. 2007). Additionally, rye flour is rich in β-glucan and arabinoxylan (Table 4.1) with only a small proportion of water soluble arabinoxylan. Bread made with a major proportion of rye flour is usually produced with sourdough. The lowering of the pH during sourdough fermentation supports the solubilization of rye pentosans including arabinoxylan through activation of cereal xylanases (Boskov Hansen et al. 2002), which is beneficial for gluten network formation and gas cell destabilization (De Vuyst and Neysens 2005; Gänzle 2014; Galle 2013). Yet, increasing arabinoxylan content decreased network formation of wheat and rye doughs as observed by rheology and visual analysis (Döring et al. 2015).

Starch determines the quality of the crumb structure of bread while amylopectin retrogradation causes staling. Rye flours harbour α-amylases, β-amylases and glucoamylases that release maltodextrins, maltose, and glucose, respectively, from starch during sourdough fermentation (Gänzle 2014). A decrease to pH <4.5 during sourdough fermentation by *Lactobacillaceae* inactivates α-amylases but not glucoamylases present in rye flour reducing starch degradation. It should be taken notice that there were observations of an overall decrease in amylase content of rye flours in recent decades (Neumann et al. 2006), which might reduce the role of sourdough fermentation for enzyme inactivation.

Taken together, rye dough acidification through the fermentative activity of *Lactobacillaceae* enables pentosane solubilization, reduces starch degradation, which allows for the formation of a crumb structure and can increase bread volume through increased CO_2 retention. Nevertheless, rye doughs are less capable to retain the CO_2 produced during fermentation compared to wheat dough, which leads to denser doughs.

In mixed wheat–rye breads, yeast alone can be used for leavening to obtain an increase in loaf volume. Similarly to rye, barley and oat have a different protein content compared to wheat and lack gluten (Table 4.1). Bread prepared solemnly or with a substantial proportion of barley and oat flour are characterized by low loaf volume and dense structure. In addition, breads prepared from both flours have characteristic sensory properties. Marklinder et al. (1996) showed that barley bread (40% barley and 60% wheat) fermented with sourdough starter cultures had lower volume yield but a less distinctive barley flavour compared to bread prepared without sourdough.

Concurrently, sourdough fermentations contribute to flavour development (Pétel et al. 2017) as bread produced with fermented rye (or other flours) sourdough has a sour, intense flavour and an aftertaste. More than 70 volatile compounds were identified in rye sourdough including hexanal, nonanal, 1-pentanol, 1.4-butanediol, 6-methyl-5-hepten-2-one, octyl acetate, 2-methylbutanal, 3-methylbutanal, 1-hexanol, 2-pentanone and 2-acetylfuran. LAB and yeast fermentations contributed to volatile compound production (Pétel et al. 2017). Rye bread produced with sourdough fermented with a mix of starter cultures consisting of *Candida milleri, L. plantarum* and *L. brevis* contained 110 compounds identified with LC–QTOF–MS whose levels were significantly increased compared to bread produced only with yeast. The most prominent components were amino acids and derivatives (Koistinen et al. 2018).

4.6 Bread Variations in Northern Europe and the Baltics

Throughout the North of Europe, sourdough containing (rye) breads are produced and local bread varieties have developed (Table 4.2, Fig. 4.2). Hard crispbreads breads are consumed as staple food throughout Scandinavia and vary depending on the grains that are used, whether sourdough fermentation is employed, and whether spices have been added (Frølich et al. 2013; Jonsson et al. 2018; Steingrímsdóttir et al. 2018).

Especially in Finland, a variety of locally different rye bread is available (Table 4.2, sources: theryebaker.com, tasteatlas.com). Ruislimpuu (Finnish Rye) is a representative bread from Eastern Finland, while Ruisreikäleipä (Halme Bakery's Ring Rye) is more popular in Western Finland. The weekly baked Ruislimpuu (Mäkelä and Rautavirta 2018) is a thick loaf bread, while Ruisreikäleipä consists of a flat loaf with a hole in the middle and low moisture that can be kept for longer periods. Traditionally Ruisreikäleipä was only baked twice a year and was stored on

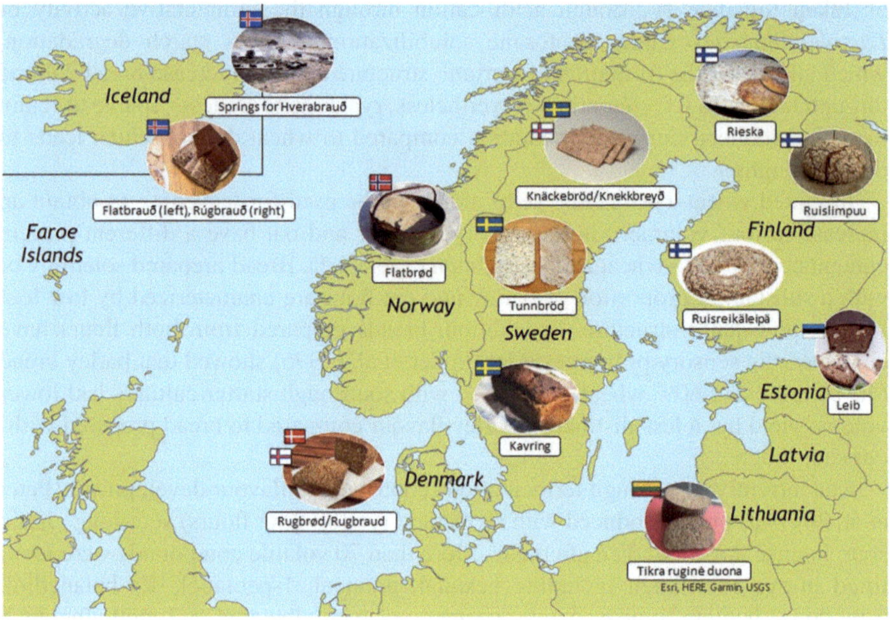

Fig. 4.2 Local varieties sourdough-containing breads in Northern Europe and the Baltics. (The map was generated with ArcGis Pro (courtesy of Dr. Jirí Hošek). Corresponding licenses disclaimer is found under "Image references" section)

horizontal wooden poles (The Rye Baker 2019; Mäkelä and Rautavirta 2018). Mallas Leipää (Finnish Malt Bread) combines rye malt and wheat, and is only leavened using yeast (The Rye Baker 2018). Similarly, Hämäläiset Ruiskorput (Tavastian Rye Rusks) is only leavened with yeast. Made with rye, crispbread is another type of bread that can be found in Finland in different formats: yeast fermented, sourdough fermented, and cold crispbread (no yeast added). The final product offers an extended shelf life due to its low water content (5–7%) (Poutanen et al. 2014).

The traditional Estonian rye bread, Leib, employs sourdough fermentation, while the barley based Karsak uses baking soda. Two basic kinds of rye Leib are prepared in Kihnu Island, a small Estonian community that keeps a strong bread heritage: Rukkileib for daily consumption and Rukkipüülileib for special occasions with better rye flour, sugar and cumin or caraway addition. Rukkipüülileib is usually left overnight in the oven to form a dark crust. Small pieces of potatoes, fatty pork or Baltic herrings could be added to the bread (Wu 2021). The Lithuanian and Latvian bread types Ruginė Duona iš Palangos (Palanga Rye), Juoda Ruginė Duona (Black Rye Bread) and Latgaliešu Maize (Latgalian Rye) are originally initiated through a scalding and saccharification process by mixing the flour with hot water or steam, and fermentation of gelatinized fluor. To reduce production time, many bakeries have replaced this steps by using technology-treated flours, like extruded flours (Juodeikiene et al. 2011). Optionally caraway seed, glucose syrup or other sweetening options can be added to generate a sweet-sour taste (Taste Atlas 2020).

Compared to Finland and the Baltic countries, rye bread is less popular in other Northern European countries. Nevertheless, the main whole grain employed in baking in Denmark and Sweden is rye, while it is wheat in Norway (Engeset et al. 2015). Kavring, a rye loaf bread from Southern and Western Sweden, combines a sourdough sponge and scalding. The Swedish rågkakor uses rye flour, is leavened by yeast and contains spices, i.e. anis and fennel. Danish rye bread (Rugbrød) uses sourdough fermented rye flour and is rich in unmilled whole grains. A similar type of rye bread is also available on the Farø islands (Rugbraud) and has been exported to Greenland as Rugbrød. In Iceland, sourdough fermented rye bread (Rugbrauð) was traditionally baked in iron pots on peat stoves while Hverabrauð breads were baked underground using the steam from natural geothermal springs (Steingrímsdóttir et al. 2018).

4.7 Novel Applications of Souring Microbes for Cereal-Based Products in Northern Europe and the Baltic Countries

4.7.1 Oat Based Fermented Beverages Using Sourdough Microbes as a Base to Produce Yoghurt-Type Gels

In addition to traditional bread, which is among the most consumed food categories in Northern Europe (Engeset et al. 2015), the use of souring LAB has extended to other cereal-based food products (Fig. 4.1). One approach is oat-based fermentations. Oat with its high content in β-glucans and a unique profile of proteins, lipids and carbohydrates (Table 4.2) is considered nutritionally relevant. However, oat is difficult to process technologically, challenging the development of fermented oat-based functional food. In a series of studies, Mårtensson et al. (2000, 2001, 2002b) compared and evaluated the fermentation of oat beverages by LAB. Three commercial oat beverages marketed by Ceba AB (Lund, Sweden) were studied on their suitability to support microbial growth for the development of fermented vegetarian beverage/yoghurt as alternative to products based on cow's milk: Mill Milk, which is a 100% oat beverage produced by wet or dry milling and enzymatic hydrolysis (does not contain glucose, higher maltose, higher fat), Adavena M40 (does not contain glucose, higher maltose, higher fat), and Adavena G40 (added glucose, lower maltose, lower fat).

Nine strains of LAB, which were species associated with dairy and cereal fermentations (Table 4.4) were able to reach cell counts of 7–8 log CFU mL^{-1} in Mill Milk, confirming the suitability of the matrix to support LAB growth. Among the strains tested only *Leuconostoc mesenteroides* developed a pleasant flavour and a good taste, while samples fermented with *L. bulgaricus* showed the highest viscosity (Mårtensson et al. 2000). Adavena M40, was fermented using both the common starters involved in yoghurt production (*L. bulgaricus* and *Streptococcus salivarius* subsp. *thermophilus*) and a combination of maltose-fermenting LAB and

Table 4.4 Novel products fermented with sourdough microbes

Type of product	Cereal	Microorganisms	Fermentation conditions	Aim of study	Reference
Oat-based fermented beverage	Oat	*L. bulgaricus* NCFB 2772[ho] *L. dextranicum* NCFB 2706[he] *L. brevis* DSMZ 1269[he] *L. kefiri* DSM 20485[he] *L. bulgaricus* DSM 20081[ho] *S. thermophilus* DSM 20259[ho] *P. damnosus* DSM 20331[ho] *A. acidipropionici* DSM 20272[he] *L. mesenteroides* DSM 20240[he]	Inoculum size: 5% 30 and 37 °C for 24 h	Evaluation of LAB fermentation of commercial plant-based beverage	Mårtensson et al. (2000)
	Oat	*L. bulgaricus* [ho] *S. thermophilus* [ho] *L. acidophilus* [ho] *Bifidobacterium* spp.[he]	Inoculum size: 0.02% 43 °C for 16 h	Ability of cultures to growth in oat-base drink and evaluation of the main physicochemical characteristics and sensory acceptance	Mårtensson et al. (2001)
	Oat	*L. bulgaricus* NCFB 2772[ho] *L. brevis* DSM 1269[he] *L. bulgaricus* DSM 20081[ho] *S. thermophilus* DSM 20259[ho]	Inoculum size: 5% 25, 30, 37 and 43 °C for 24 h	Exopolysaccharide production and texture enhancement	Mårtensson et al. (2002b)
	Oat	*L. reuteri* ATCC 55730[he] *L. acidophilus* DSM 20079[ho] *B. bifidum* DSM 20456[he]	Inoculum size: 5% 37 °C for 16 h	Ability of oat-based beverage to support bacterial growth and its viability	Mårtensson et al. (2002a)

(continued)

Table 4.4 (continued)

Type of product	Cereal	Microorganisms	Fermentation conditions	Aim of study	Reference
Oat-based yoghurt alternatives	Oat	*L. rhamnosus*[he]	37 °C for 24 h	Study of changes and factors implicated in oat proteins during oat bran fermentation	Loponen et al. (2007)
	Oat	*L. bulgaricus*[ho] *S. thermophilus*[ho]	42 °C for 24 h	Development, physicochemical characterization and consumer acceptance of fermented oat-based gels	Brückner-Gühmann et al. (2019)
Kvass	Rye	*L. sakei KTU05-6*[he] *P. pentosaceus KTU05-10*[ho] *S. cerevisiae*	Inoculum size: 10% 30 and 37 °C for 48–60 h	Development of functional properties of kvass made from extruded rye, enzymes and LAB	Basinskiene et al. (2016)
Sour beer	Malted barley, malted wheat, wheat flakes, oat flakes	*L. plantarum*[ho] *L. rhamnosus*[he] *L. brevis*[he] *L. buchneri*[he] *L. delbrueckii* [ho] *C. alimentarius* [ho] *S. cerevisiae* US-05	Yeast inoculum size: 0.7 g L^{-1} LAB: 33 °C for 48 h plus 21 °C for 2 weeks approx. plus 21 °C for 1 week	Effects of LAB species on sensory and chemical properties of sour beer	Tenhovirta (2019)
Cold smoked pork sausages	Rye	*P. acidilactici* KTU05-7[ho] *P. pentosaceus* KTU05-9[ho] *L. sakei* KTU05-6[he]	Inoculum: fermented potato juice with 9.6 log CFU g^{-1} 18 °C for 1 h	Polycyclic aromatic hydrocarbons and biogenic amines development	Bartkiene et al. (2017a)
Extrudes for breakfast cereals	Rye	Baker's yeast *L. brevis*[he] *L. plantarum*[ho]	Inoculum: 0.4% dough weight 30 °C for 22 h	Evaluation of sensory profile and flavour-active volatile compounds	Heiniö et al. (2003)
	Rye and barley	*W. confusa* A16[he] *L. pseudomesenteroides* 20193[he]	25 °C for 24 h	Evaluation of fermented BSG as a novel food ingredient to make extruded breakfast cereals	Thorvaldsson (2020)

(continued)

Table 4.4 (continued)

Type of product	Cereal	Microorganisms	Fermentation conditions	Aim of study	Reference
Fermented bran	Oat, rye	Commercial rye starter containing hetero- and homofermentative lactobacilli *C. milleri*	30 °C for 1, 2, 3, and 4 h	Effect of fermentation on the soluble β-glucan and protein content, and evaluation of β-glucan hydrolysis	Degutyte-Fomins et al. (2002)
	Rye	Baker's yeast Indigenous rye LAB community	Baker's yeast inoculum: 7 log CFU g^{-1} 20, 27.5 and 35 °C for 6, 13 20 h	Improvement of technological properties and bioactivity of rye bran	Katina et al. (2007)

he heterofermentative, *ho* homofermentative

bifidobacteria (Mårtensson et al. 2001). An acceptable yoghurt-like product was obtained by oat-based beverage fermentation with the help of stabilizer (xanthan gum), fat and flavour agents enhancing texture and organoleptic characteristics. When Mill Milk, Adavena M40 and Adavena G40 were fermented with four LAB strains including *L. brevis* (Mårtensson et al. 2002b), higher viscosity was obtained in samples fermented by co-cultures compared to individual strain emphasizing the functionality of cooperative microbial communities. In addition, exopolysaccharide-forming starter cultures contributed to product viscosity.

These studies indicated that microbial consortia must be carefully selected in order to achieve a product with desired organoleptic (taste, aroma, acidity), physicochemical (texture properties, viscosity) and nutritional (no β-glucan degradation) characteristics when using oat as a base matrix.

4.7.2 *Sourdough Fermentation to Produce Novel Breakfast Cereals?*

Snacking and eating on the go have been shown to be on the rise, opening new opportunities for traditional bakery and novel cereal products (Jonsson et al. 2018). Despite the nutritional benefits and acceptability of rye bread, the intense flavour and aroma of this cereal hinder its application for the development of new functional food products. Extrusion technology can be applied together with fermentation to produce nutritional enhanced snack options with altered texture, structure and sensory attributes.

Different volatile compounds were identified in extruded breakfast cereals produced from germinated, sourdough fermented, and native rye, while rye cultivars did not have influence on any of the attributes studied (Heiniö et al. 2003). The masking of the intense rye flavour was greater when non pretreated rye was extruded. Higher levels of furfural, ethyl acetate, 3-methylbutanol and 2-methylbutanol were recovered from fermented rye, which might relate to the activity of the sourdough microbes. Breakfast cereals produced with sourdough obtained the highest scores in most of the sensory attributes, but were characterized by a strong sour and intense flavour and aftertaste perception (Heiniö et al. 2003).

Thorvaldsson (2020) developed a cereal breakfast product by extruding a dough containing fermented brewer's spent grains collected from breweries in Sweden (barley), Finland (barley) and Italy (barley and corn), rye flour, and potato starch as major ingredients. The product fermented with exopolysaccharide-producing strains *Weissella confusa* and *Leuconostoc pseudomesenteroides* developed a pleasant taste and texture according to a costumer panel evaluation. Samples made using brewer's spent grains that were fermented after sucrose was added were the consumer's favorite in combination with milk, yoghurt or 'filmjölk', a traditional Scandinavian sourmilk.

4.7.3 The Use of Sourdough LAB in the Production of Sour Beer in Northern Europe

Different from conventional beer where sour taste is perceived as an off-flavour, the acidification by LAB in sour beer is crucial to confer it the characteristic flavour and aroma profile that is linked to lactate and acetate formation during fermentation. Traditional manufacture of sour beer involves a natural (spontaneous) fermentation process driven by LAB, acetic acid bacteria, and yeasts such as *Saccharomyces* and *Brettanomyces*. Sour brewing culture has a long tradition in Belgium (Dysvik et al. 2020a), an increasing interest in sour beers has emerged particularly in Finland, where sour ales have been gaining popularity (Sahlberg 2020). In Sweden and Denmark, sour beer is brewed in local breweries (www.uglyduckbrewing.dk/sours). It is even possible to find breweries that are exclusively dedicated to the production of sour beer (www.brekeriet.se). In addition, a number of recent studies from Norway optimized the sour beer production process (Dysvik et al. 2019, 2020a, b, c).

The raising interest in sour beers in recent decades initiated a search for fermentation techniques alternative to natural fermentation to allow shorter manufacturing times and higher process controllability to obtain product uniformity even in different production batches. One option is to conduct pre-fermentation with LAB prior to yeast inoculation (Dysvik et al. 2019). Candidate sour beer cultures have been *Lactobacillaceae* species such as *L. plantarum* and *L. brevis* that are commonly associated with cereal fermentations (also see Sect. 5) or the silage-associated *Lentilactobacillus buchneri*. LAB pre-fermentation can occur at different stages of the brewing process, i.e. in the mashing kettle (sour mashing), in the brewing kettle (kettle souring) and after the wort has been transferred from the brew kettle (sour wort) (Dysvik et al. 2019); pre-fermentation avoids exposing the LAB to ethanol-producing yeast. When wort was pre-fermented with *L. buchneri,* which increased 100-fold in cell counts, less 2-methyl-1-butanol but more short chain carboxylic acids (e.g. acetate) were detected in beer. Yet, few significant differences were found between LAB fermented and chemically acidified sour beers (Dysvik et al. 2019). Tenhovirta (2019) monitored the effect of six LAB on the sensory and chemical properties of sour beer produced in a six-week process. Through sensory evaluation tests, which included consumer acceptance and descriptive analysis by a trained panel, it was observed that higher lactate concentrations were related to increased overall preference with a positive correlation with raspberry aroma, apple flavour and vinous flavour. Beer fermented with heterofermentative *L. buchneri* and *L. brevis* contained acetate. *C. alimentarius* was evaluated for the first time for sour beer production and reached higher cell counts and acidified wort faster than the commonly used *L. plantarum* (Tenhovirta 2019).

As alternative to a two-step fermentation process, Dysvik et al. (2020b) tested the potential of co-fermentation of *S. cerevisiae* and strains of *L. brevis, L. buchneri* and *L. plantarum,* and of secondary LAB fermentation in the presence of xylooligosaccharides from birch wood (Dysvik et al. 2020c). Sour beers produced by different co-cultures in a short production time of only 21 days had different flavour profiles, and were perceived as comparable to commercial sour beer highlighting the possibility of process improvements with controlled fermentations using cereal-associated LAB. The process with secondary fermentation took only 2–4 weeks (Dysvik et al. 2020c). These studies together suggest the potential to produce a fast and controlled sour beer with defined starter cultures.

4.7.4 Nordic Grains and 'Translated Fermentation'

Beyond the products available at points of sale such as supermarkets or kiosks, it is possible to find innovative preparations from sourdough such as pasta, pizza, pastries in the menu of restaurants, coffee shops, and food trucks. These emerging opportunities are part of the development of the New Nordic Cuisine, a larger culinary movement founded in the local terroir that exploit processing methods such as pickling, fermenting, and smoking of local raw ingredients and generates several business opportunities (Bech-Larsen et al. 2018). An example is the Danish restaurant Noma, which was rated the best restaurant in the world for five times between 2010 and 2021 by Restaurant Magazine and obtained three stars by the Guide Michelin in 2021. Rene Redzepi, its founder, and its chef David Zilber, wrote 'The Noma Guide to Fermentation' (2018) that includes recipes for preparing fermented products from Scandinavian grains, including miso from rye bread (or as it is called in the book 'ryeso') and koji from rye ('breadso'). Miso and koji are traditionally used in the production of soy sauce. Miso is defined as a semisolid paste made from soybeans or rice, barley and from mixed ingredients, which are combined, fermented, and matured with soybeans and/or grains cultured with koji mold

(*Aspergillus oryza*) and salt (Hutkins 2006). The difference between these two types of adapted miso was that ryeso manufacturing involved a fermentation of pre-sliced Danish rye bread with koji and salt for 3–4 months, while for breadso, the koji was grown directly on the bread (Redzepi and Zilber 2018).

With such an approach, the Noma chefs fermented Nordic varieties of protein-rich, leguminous substrates and cereals using traditional Japanese recipes. From the authors' own words: 'Ryeso marries Nordic ingredients with Japanese techniques'. Evans and Lorimer (2021) called this combination of microorganisms and techniques from other parts of the world using Nordic raw materials 'translated fermentation'. Translated fermentation has emerged within the New Nordic Cuisine, which aims at exploring the biodiversity and cultural identity of the Nordic region through cooking (Evans and Lorimer 2021). With a similar local ingredient philosophy, the Domestic restaurant (https://restaurantdomestic.dk/) located in Aarhus (Jylland, Denmark) refers to a fermentation process as an important foundation of the restaurant, which led to the opening of their own 'domestic. Lab' in 2021. Beyond the application of the concept of translated fermentation, which is used to obtain miso from beans and barley, the chefs of Domestic aim to integrate sustainability practices and to reduce food waste, for example by using the surplus sourdough from bread production for kvass preparation (see also Sect. 4.9.3).

4.7.5 Rye Sourdough Microbes as Biopreservatives and to Reduce Undesired Compounds

In addition to being used to manufacture fermented products with enriched organoleptic properties and preservability, sourdough microbes can be applied as part of a natural food preservation strategy by the development of antimicrobial metabolites such as short chain carboxylic acids, ethanol, and bacteriocins; souring microbes can contribute to an increase bioavailability or remove undesirable compounds (Kårlund et al. 2020). For example, bacteriocin-like substances were produced by strains of *L. sakei*, *P. acidilactici* and *P. pentosaceus* isolated from rye sourdoughs, with additional antimicrobial activity against bread spoilage fungi (Digaitiene et al. 2012). Bartkiene et al. (2020) reported that strains of *L. plantarum*, *Lacticaseibacillus casei*, *L. curvatus*, *Lacticaseibacillus paracasei*, and *Loigolactobacillus coryniformis* isolated from rye sourdough showed *in vitro* growth inhibition properties when tested against fifteen pathogens. Moreover, the thirteen LAB isolates exerted antifungal activities against *Aspergillus nidulans*, *Penicillium funiculosum*, and *Fusarium poae*.

These antimicrobial properties an be harnessed to other food products in order to reduce biological or chemical contamination (see also Sect. 4.8.1). Particularly in Northern Europe, smoking of fermented sausage is important for its distinctive flavour and lactic acid is believed to be the main taste component (Leroy et al. 2006). However, smoke particles can transfer polycyclic aromatic hydrocarbons, which are considered carcinogenic. During fermentation, biogenic amines can be formed. In 2017, Bartkiene et al. treated Lithuanian cold smoked pork meat sausages before or after the smoking step with *L. sakei* and *P. acidilactici* isolated from spontaneous rye sourdough (Table 4.4). The addition of sourdough cultures before smoking reduced biogenic amine levels of cadaverine and spermidine, and lowered the content of putrescine by about 50% when *L. sakei* and *P. acidilactici* were applied. Levels were below detection limited when *P. pentosaceus* was added after smoking. Moreover, the content of benzo[α]pyrene and chrysene decreased compared to controls without added cultures. The application of LAB before smoking step resulted in significantly higher acceptability highlighting the suitability of such an approach.

4.8 Use of Northern Europe Cereal Side Streams and Bread Waste for Novel Fermented Products

4.8.1 Modifications of the Rye and Oat Bran Fraction Using Sourdough Fermentation

The rapidly growing market for health-promoting foods, as well as the search for alternatives usages for side stream products of grains from Northern Europe has initiated research on the possibilities to increase technological and nutritional properties. Cereal bran is a milling side product rich in phytochemicals including phenolic acids, lignans, benzoxazinoids, and alkylresorcinols and dietary fibre (Jonsson et al. 2018). The diverse composition of bran makes it an interesting ingredient to enhance nutritional content as well as functionality of conventional products. However, bran addition can negatively affect textural and organoleptic properties of baked goods, such as taste or mouthfeel. In wheat bread, the addition of bran may cause a disruption of the gluten network structure reducing textural properties and volume. Therefore, novel processing approaches are needed to produce bran containing foods that are attractive to consumers.

Degutyte-Fomins et al. (2002) investigated the impact of a rye sourdough starter and additional rye flour as a source of enzymes on the technological properties of rye and oat bran (Degutyte-Fomins et al. 2002). The rye starter reduced the pH of the oat bran; the addition of rye flour increased protein concentrations. Addition of rye as flour or sourdough did not modify overall β-glucan content but increased solubility of β-glucan and reduced overall viscosity of the rye-oat bran mixes. These changes in rheological properties might improve process integration of oat bran (Degutyte-Fomins et al. 2002).

In combination with sourdough fermentation, the milling process was demonstrated to have a significant impact on rye bran processability (Katina et al. 2007) increasing the growth of starter culture LAB. Growth, bacteriocin and short chain carboxylic acid production of the LAB community during fermentation reduced the count of aerobic bacteria conferring a biopreservative effect especially when whole grain rye was used. Lower LAB abundance was linked to reduced acidity levels of fermented peeled grain rye bran, which could be an advantage in terms of organoleptic characteristics of the product (Katina et al. 2007). Levels of bioactive compounds and complex carbohydrates, such as folates, phenolic acids and soluble pentosans, were higher in fermented whole grain rye bran compared to peeled grain rye, highlighting the contribution of the outer layer of the grain to nutritional content and bioactivity. However, the concentration of free ferulic acid was pH-dependent in connection with cinnamoyl esterase activity of wholegrain rye flour (Katina et al. 2007). Bacteriocin-like substance produced of sourdough microbes play a role in biopreservation, although their antimicrobial activity can be affected by environmental conditions. Narbutaite et al. (2008) demonstrated that the use of microbially-obtained baking enzymes did not affect the antimicrobial activity of bacteriocins produced by LAB isolated from Lithuanian sourdough against *Bacillus subtilis*, the main responsible for bread ropiness and unpleasant fruity odour.

4.8.2 Fermented Oat-Based Yoghurt Alternatives Produced from Oat Bran or Oat Protein

In addition to the studies carried out on oat beverages (Sect. 4.7.1), oat bran, which is a side product after grinding and separating of oat groats or rolled oats by sieving or bolting, has been used in the manufacture of plant-based yoghurt-type products. One example is Yosa, the first oat bran-based yoghurt product fermented with a combination of LAB that also contains bifidobacteria. Yosa was firstly placed on the Finnish market in October 1995 by Bioferme, a family-owned company located in Kaarina near Turku (Finland) six years after initial development. More than 20 years later -in 2017- the Fazer Group bought Bioferme, with the ambitious objective of strengthening the segment of oat-based products. The health beneficial effects associated with Yosa consumption are claimed based on the probiotic properties of the strains present in product as well on the presence of β-glucan from oat bran (Salovaara 1996). Currently, the Yosa line includes oat-based items for multiple uses such as Yosa Oat Drinks, Yosa Oat Snacks; Classics, and Muesli on top. These products are commercialized in several Scandinavian countries, such as Finland, Sweden, Denmark, and Norway.

The procedure to obtain Yosa involves the cooking of oat bran, the use of enzymes, and addition of microorganisms responsible of fermentation with final addition of flavouring agents (Salovaara and Kurka 1997). The microbes used in Yosa are reported to be *L. acidophilus* LA5 and *Bifidobacterium lactis* Bb12 (Rozada-Sánchez et al. 2008; Nyanzi and Jooste 2012). At pilot plant scale, Salovaara (1996), who led the product development, reported that the cultures were able to remain cultivable at more than 7 log CFU g^{-1} after weeks of refrigerated storage recommending a product shelf life of four weeks.

Yosa is a pudding-like gel with a consistency and texture similar to yoghurt. Loponen et al. (2007) investigated how fermentation with a freeze-dried starter culture of *Lacticaseibacillus rhamnosus* impacted the globulins, the major protein group of oats, and gel fermentation during oat bran fermentation. While protein hydrolysis did not occur during cooking or fermentation of oat bran, the acidic conditions shifted the globulin solubility from salt-soluble to residual protein fraction causing the unfolding of globulins and also apparently inducing protein aggregation. The gel structure of oat bran-based products depended on starch gelatinization during the pre-fermentation heating step (100 °C) and on the presence of other non-starch polysaccharides. In contrast to sourdough fermentation, proteolytic or other cereal-derived enzymes were not active in oat bran fermentation due the cooking step (Loponen et al. 2007).

In 2019, Brückner-Gühmann et al. (2019) determined the rheological properties and sensory response of an alternative oat-based gel with a fermentation matrix consisting of a mixture of oat extracted protein and starch added to water or commercial oat drink plus 1% sugar. *S. thermophilus* and *L. bulgaricus* were used as inoculum and a heat treatment was applied after fermentation in order to gelatinize the starch. The fermented oat drink had higher total solid content and higher elasticity compared to a water-based control sample, while the higher native starch content in the water-based fermented oat gel conferred a higher water holding capacity. Interestingly, these differences in rheological properties were recognized by consumers, which described the oat-based gel as creamy and sweet. When information on protein content was given prior to sensory testing, acceptability was significantly increased, highlighting how health claim information can affect purchase intention.

4.8.3 Rye Bran Fermentation with Exopolysaccharide Producing Starter Cultures to Improve Extrusion Behaviour

Extrusion of bran can improve the nutritional value through an increase in dietary fibre content. On the same time, bran addition negatively impacts extrudate expansion and texture due to a reduction of starch levels. Competition for water with the starch matrix reduces melt viscoelasticity and hinders bubble formation (Roye et al. 2021). As pre-fermentation might improve extrusion processing, an exopolysaccharide -producing *Weissella* strain was used together with *K. exigua* and *L. brevis* to ferment rye bran (Nikinmaa et al. 2017). Strains of *Weissella* produce technologically significant amounts of exopolysaccharides, particularly dextran, which is glucose polymer with mainly α-D-(1–6) and few α-D-(1–3) linkages through the activity of extracellular dextransucrases using sucrose as substrate. In cereal food systems, dextran acts as a hydrocolloid.

Improved expansion and texture were achieved when rye bran was fermented in the presence of dextran producing *W. confusa* suggesting exopolysaccharide producing LAB for the production of high-fibre extruded products. However, a follow-up study identified acidification as a major contributor to for the improved structure and texture of fermented rye bran used in extrusion (Nikinmaa et al. 2020).

In addition, these results indicated that rye bran might also act a suitable matrix to produce dextran, which is usually commercially produced by *Leuconostoc* strains in sucrose rich medium. Kajala et al. (2016) pre-fermented wheat and rye bran with dextran-producing *Weisella* strains. Both strains reached cell counts of 9.4–9.9 log CFU g^{-1} and produced more dextran when grown in rye (0.9% without added sucrose and 2.6% with sucrose supplementation) than in wheat bran (0.2% and 2.1%). The production of up 25 g kg^{-1} dextran suggests the use of a rye bran matrix with added sucrose for dextran production.

4.8.4 Reuse of Sourdough Bread Waste to Produce Ice Cream Bars

The search for alternatives for waste products within the new concept of circular (bio-)economy has led to the development of novel food products based on rye sourdough bread. Beside traditional products that reuse leftover sourdough bread are kvass (see Sect. 4.9.3) or the Danish øllebrød, a soup or porridge made of old rye sourdough bread and beer that is eaten for breakfast. The sugared and fried ymerdrys from Rugbrød bits was first produced by Struerbrød (http://www.struer-brod.dk/sortiment/ymerdrys.aspx) and is eaten with dairy products. Novel food creations have been brought to the market recently. An example is 'Salty Caramel Sourdough', the result of a successful transformation of old sourdough bread into ice cream bars (sources: jalmogb.dk/bread-ice-cream, foodpeople.dk/en/salty-caramel-sourdough-from-bread-to-ice-cream). The project started when the medium-sized Danish bakery Jalm&B collaborated with Amass restaurant, Copenhagen, Denmark, with the idea to upcycle leftover bread. Loaves of stale sourdough bread were soaked in warm water to initiate starch degradation producing a sweet syrup, which was integrated in the dairy ice cream matrix. The 'Bread Ice Cream' was first served at the Copenhagen Cooking Food Festival, where it obtained positive feedback and caught the attention of the food industry. The product was brought to market by Hansens, an ice cream manufacturer in Copenhagen. With this approach, Amass, Jalm&B and Hansens were able to transform food waste into marketable ice cream bars; in the first batch they used around 320 kg food waste and upcycled in 15.000 ice cream bars.

4.9 Sourdough Fermentation to Modify Biofunctionalities and Health Properties of Products Using Grains from Northern Europe

4.9.1 *Oat Beverages or Gels as a Matrix for the Cultivation of Probiotic Strains*

In addition to serving as a substrate for fermentation (Sect. 4.7), oat-based beverages or yoghurt-type gels can serve as a matrix for probiotic strains. For example, Mårtensson et al. (2002a) reported that cell counts of *L. reuteri*, *L. acidophilus*, and *Bifidobacterium bifidum* were lower when they were co-inoculated with yoghurt starters due to lower final pH and the more acidic conditions generated in oat-based beverages. In fact, *B. bifidum* completely lost viability in all the co-cultured oat-based gels after 10 days of storage. The samples fermented with *B. bifidum* had lower final β-glucan content, which was associated with β-glucanase activity of the *B. bifidum*.

In Sweden, Proviva was the first fermented oat-based functional beverage launched on the market. Proviva was developed by researchers at the Lund Institute of Technology in the 1980s and started being marketed in Sweden in 1994. Initially, Proviva drink was intended for hospitalized patients as a post-operative recovery drink containing oats and probiotics. Proviva preparation includes a mixture of water, oat flour, and malted barley flour, which contributes amylases, proteinases, and β-glucanases. Fruit juices (e.g. black currant, strawberry, blueberry, rose hip, or other tropical fruit) are added. Cultivable counts of the added probiotic *L. plantarum* 299v were maintained at around 10.6 log CFU L^{-1} for more than one month despite the acidic environment generated by the addition of fruit juice (pH about 2.8–3.4) (Basinskiene and Cizeikiene 2020; Nyanzi and Jooste 2012). In 2010, Danone acquired 51% of the shares of Proviva AB. The Proviva beverage is currently sold in Sweden and Finland with both classic and innovative offerings such as no added sugar and 'low carb'.

4.9.2 *Fermented Oat Fibre Concentrates or Bran as a Novel Prebiotic?*

Oat fibre concentrates are rich in dietary fibre (around 25%) and especially in β-glucans, which could act as prebiotic, i.e. 'a substrate that is selectively utilized by host microorganisms conferring a health benefit' (Gibson et al. 2017). *In vitro*, oat dietary fibre addition increased the formation of the short chain fatty acid (SCFA) butyrate by fecal microbiota compared to control fermentations supplied with resistant starch (Casterline et al. 1997). Oat bran, which is also rich in β-glucans, increased fecal levels of major SCFA acetate, propionate, butyrate, and of isobutyrate and isovalerate in a human cohort during a 12-week intervention (Nilsson et al.

2008). Lambo-Fodje et al. (2006) fermented oat fibre concentrates with a commercial yoghurt culture consisting of *L. bulgaricus* and *S. thermophilus*, *Pediococcus damnosus* and a strain of *L. reuteri* from Biogaia before feeding to rats. The impact on oat fibre consumption on SCFA formation depended on gut segment, with fermented oat fibres generally increasing the levels of acetate in the cecum and of propionate in the distal column compared to an unfermented control, while in the proximal colon, butyrate levels were lower when fermented oat fibres. There were also differences in the response to oat fibres fermented by different bacterial cultures (Lambo-Fodje et al. 2006) highlighting the potential to support intestinal SCFA formation with a choice of fermentation cultures and fermentation substrate.

4.9.3 New Approaches to Manufacture Kvass for Higher Content of Non-digestible Oligosaccharides

Grains used in Northern Europe and the Baltic countries are rich in biofunctional and nutritional compounds including starch and cell wall polysaccharides, and proteins that account for approx. 90% of the dry weight, and lipids, terpenoids, phenolics, minerals, and vitamins (Shewry et al. 2013). Functional activities of souring LAB or yeast can enrich the nutritional properties of fermented cereal products.

Kvass is a traditional non-alcoholic beverage that is popular in eastern European and Baltic countries, were it can be found as kali (Estonia), dzersis (Latvia), or gira (Lithuania) (Table 4.4). Kvass is usually made from rye and barley malt, rye flour, and stale rye bread that are fermented with *S. cerevisiae* and LAB. The two main kvass making processes use either stale sourdough bread or malt as main ingredients. Basinskiene et al. (2016) developed a new technology to manufacture kvass by using extruded rye wholemeal. During extrusion, overall dietary fibre content that included arabinoxylans, β-glucans and arabinogalactans, was preserved when performed under mild or moderate conditions, while the soluble fraction increased. Prior to fermentation, the extruded rye was treated with enzymes to obtain fermentable monosaccharides and prebiotic xylooligosaccharides and arabinoxylooligosaccharides through the degradation of arabinoxylan. Fermentation was carried out using *L. sakei* or *P. pentosaceus* isolated from spontaneous rye sourdough (Basinskiene et al. 2016). The addition of LAB increased the content of flavour compounds and lowered the pH value compared to kvass fermented only with yeast. Starter cultures used glucose and fructose, while prebiotic oligosaccharides were not degraded. Both strains maintained the cell counts above 6 log CFU mL^{-1} for at least 30 days of refrigerated storage. Moreover, fermentation with *L. sakei* reduced the cell numbers of the native background microbiota. In sensory trials, both kvass fermented by LAB were less accepted than commercial kvass, mainly due to differences in organoleptic attributes, i.e. increased odour and decreased brown colour intensity of the product (Basinskiene et al. 2016) requiring further process development.

4.9.4 Sourdough Fermentation to Change FODMAPS in Rye Bread

FODMAPs are **F**ermentable **O**ligosaccharides, **D**isaccharides, **M**onosaccharides, and **P**olyols (FODMAPs), which are rich in cereal flours, i.e. oligosaccharides of the raffinose family and fructans. Rye flour contains 4.3–5% fructans and 0.1–0.7% rye (Loponen and Gänzle 2018). FODMAPS are not easily digested carbohydrates and have been linked to the development of pathologies, such as irritable bowel syndrome (IBS) or non-celiac gluten sensitivity (NCGS). The relationship of NCGS and FODMAPs has not yet well been supported by clinical trials (Loponen and Gänzle 2018).

Conventional sourdough fermentations using LAB can decrease levels of FODMAPs (e.g. fermentable oligosaccharides) but may also change FODMAP content as heterofermentative LAB generate the FODMAP mannitol from sucrose and fructose (Sect. 4.3), Gänzle 2015). *S. cerevisiae* does not possess fructanase activity, the addition of fructanase-positive *Kluyveromyces marxianus* allows the production of low-FODMAP bread through the degradation of fructans (Struyf et al. 2017) suggesting that overall FODMAP reduction might be achieved through a combination of the choice of fermentation organisms, the fermentation process itself, and the cereal used.

Two clinical trials tested how low FODMAP rye bread influences the gastrointestinal symptoms including intestinal hydrogen formation. In a randomized double-blind controlled crossover study, low FODMAP rye bread caused less flatulence, less abdominal pain, fewer cramps, and less stomach rumbling compared to regular rye bread (Laatikainen et al. 2016). In a follow up analysis, Laatikainen et al. (2019) suggested that microbiota composition at baseline might be associated with the extent of symptom relief of the patients. In a randomized double-blind controlled crossover study, consumption of low FODMAP rye bread reduced levels of breath hydrogen (Pirkola et al. 2018) linking diet to gut microbial fermentative activity.

4.9.5 Formation of B-Vitamins during Sourdough Fermentation of Rye

Cereal flours are rich in B-vitamins, whose levels might be reduced during sourdough fermentation or baking (Mihhalevski et al. 2013). Folates are B-vitamins (vitamin B9), and derivatives of folic acid. Rye has high folate content (65–140 μg 100 g^{-1}). Kariluoto et al. (2004, 2006) showed that in fermentation medium, in sourdough model systems, and in rye sourdough fermentations, yeast and not LAB were the main responsible for folate formation. Rye sourdough fermentation in the presence of baker's yeast increased the content of 10-HCO-H2folate, 5-CH3-H4folate, and 5-HCO-H4folate. Folate synthesis by yeast compensated for folate loss that commonly occurs during baking (around 25%) (Kariluoto et al. 2004).

A second B-vitamin of interest in a bread context is vitamin B12. Vitamin B12 exists naturally in foods of animal origin and is synthesized by few bacteria, one example is the species *Propionibacterium freudenreichii* which is a traditional starter culture of Swiss Emmental cheese. The potential to produce vitamin B12 differs between isolates of *P. freudenreichii* and depends on the availability of the lower ligand 5,6-dimethylbenizimidazole (DMBI) and the precursors of DMBI riboflavin (RF) and nicotinamide (NAM) (Chamlagain et al. 2016). Edelmann et al. (2016) compared the stability of added chemical and natural vitamin B12 variants to vitamin B12 produced *in situ* in a barley malt medium by *P. freudenreichii*. During baking, up to 25% of vitamin B12 was lost in a wheat bread with added rye sourdough, nevertheless this study showed that vitamin B12 can be bacterially produced in a cereal matrix, which can be added during the bread making process (Edelmann et al. 2016). In follow up studies, Xie et al. (2018, 2021) incubated *P. freudenreichii* together with *L. brevis* combining a vitamin B12 producer with an acidifying microbe to control growth of background microbiota using different non-sterile flours. Production was higher in bran and whole wheat flour than in durum flour (Xie et al. 2018) possibly due available nutrients. Vitamin B12 formation was also observed in rye and oat bran (Xie et al. 2018) albeit at lower levels than for example when buckwheat was used. Taken together, these studies indicate that through a suitable choice of starter and adjunct culture, the functional properties related to B-vitamin content of cereal fermented food can be enhanced.

4.9.6 Reduction of Rye Phytate Content Through Fermentation

Whole grain rye is rich in minerals including P, Mg, Fe, Cu, and Zn, which mostly form salts with phytate (IP6, myo-inositol 1,2,3,4,5,6 hexaphosphate), a storage compound of phosphor. In rye, the phytate content is around 0.5–1.5%. Phytate bound minerals are not hydrolyzed at neutral pH, which possibly translates to low bioavailability upon human (or animal) consumption. Phytases (*myo*-inositol hexakis (dihydrogenphosphate) phosphohydrolase) catalyze the degradation of phytic acid to lower inositols increasing the availability of phosphor, minerals and myo-inositol.

In whole wheat flour, endogenous phytase activity is considered not sufficient to reduce the amount of phytate, while phytate was fully degraded in commercial rye bread made from milled rye with the addition of sourdough possibly due to longer fermentation times (Nielsen et al. 2007, 2008). The addition of whole grain increased levels of non-degraded IP6 phytate (Nielsen et al. 2008). A combination of adding scalded rye bran and 10% sourdough reduced phytate content of wheat bread by 96–97% (Larsson and Sandberg 1991). Sourdough microbes can directly contribute phytase activity. High extracellular phytase activity was identified for yeast strains of *S. cerevisiae*, *C. humilis*, and *Pichia kudriavzevii* isolated from Danish and

Lithuanian sourdough (Nuobariene et al. 2012). Strains of L. panis, L. reuteri, L. fermentum, and P. pentosaceus isolated from Lithuanian sourdough showed both extra- and intra-cellular phytase activity (Nuobariene et al. 2015).

Dietary phytate content has been previously hypothesized to be related to iron bioavailability. A study from Sweden (Hoppe et al. 2019) using a free-living cohort of young women that consumed either high or low-phytate wholegrain bread contradicted this assumption. Hoppe et al. revealed that the consumption of low phytate bread for 12 weeks reduced markers of iron status, suggesting that modifying iron bioavailability from dephytinization of cereal products might not work for adults under free-living conditions (Hoppe et al. 2019).

4.10 Summary

Early in the Bronze Age, eating possibilities in Northern Europe were dictated by climate conditions, which were followed by the domestication of contaminating weed for cultivation as new crops and the development of milling as processing techniques. Extended preservation and improved organoleptic properties were obtained by spontaneous sourdough fermentation. The consumption of sourdough fermented rye and to a lesser extent barley and oat containing bread has been and still is a distinguishable cultural characteristic and eating habit of societies of Northern Europe and the Baltic Countries.

The high nutritional content of the rye, oat, wheat, barley grains that have been traditionally cultivated in Northern Europe, and the health-promoting effects that have been related to cereal consumption as well as the trend of a vegan style renders the development of novel cereal products highly attractive. However, the processing of these grains remains technologically challenging and the sensory profiles are distinctive.

Exploiting cereal-associated microbes do not only ferment a raw product to a food, but can add value by enhancing the nutritional profile, re-using side and waste streams. Additionally cereal-based products can act as matrix for probiotics. During the last years, novel and innovative cereal-based (fermented) food has been successfully launched in Northern Europe and the Baltic Countries in response to the demand for non-dairy products, the need to add value to local crops and the profit associated with the popularity of health-promoting food. During the last decades, research on old Northern European grains has indicated new application possibilities especially in regard to current transitions from animal protein to plant diets, and the use of fermented products as functional food.

Acknowledgements CS acknowledges funding from the Aarhus Universitet Forsknings Fonden (AU FF-F-2020-7-2) and the Novo Nordisk Foundation (BIOFUNC, NNF21OC0066725). Dr. Jiři Hošek is acknowledged for providing the map which was created usins ArcGISPro (Fig. 4.2). Danielius Muchinas is acknowledged for the picture of "Tikra ruginė duona" bread (Fig. 4.2).

References

Bartkiene E, Bartkevics V, Mozuriene E, Krungleviciute V, Novoslavskij A, Santini A et al (2017a) The impact of lactic acid bacteria with antimicrobial properties on biodegradation of polycyclic aromatic hydrocarbons and biogenic amines in cold smoked pork sausages. Food Control 71:285–292

Bartkiene E, Bartkevics V, Pugajeva I, Krungleviciute V, Mayrhofer S, Domig K (2017b) Parameters of rye, wheat, barley, and oat sourdoughs fermented with *Lactobacillus plantarum* LUHS135 that influence the quality of mixed rye-wheat bread, including acrylamide formation. Int J Food Sci Technol 52:1473–1482

Bartkiene E, Lele V, Ruzauskas M, Domig KJ, Starkute V, Zavistanaviciute P et al (2020) Lactic acid bacteria isolation from spontaneous sourdough and their characterization including antimicrobial and antifungal properties evaluation. Microorganisms 8:64

Basinskiene L, Cizeikiene D (2020) Cereal-based nonalcoholic beverages. In: Galanakis CM (ed) Trends in non-alcoholic beverages. Academic, London, pp 63–99

Basinskiene L, Juodeikiene G, Vidmantiene D, Tenkanen M, Makaravicius T, Bartkiene E (2016) Non-alcoholic beverages from fermented cereals with increased oligosaccharide content. Food Technol Biotechnol 54:36–44

Bech-Larsen T, Mørk T, Bundgaard L (2018) Future outlooks: Nordic gastronomy in food marketing and catering. In: Nutritional and health aspects of food in Nordic countries. Academic, pp 245–255

Behre K-E (1992) The history of rye cultivation in Europe. Veget Hist Archaeobot 1:41–156

Bessmeltseva M, Viiard E, Simm J, Paalme T, Sarand I (2014) Evolution of bacterial consortia in spontaneously started rye sourdoughs during two months of daily propagation. PLoS One 9:e95449

Bjørnstad Å (2012) Our daily bread: a history of cereals. Vidarforlaget AS, Oslo

Boskov Hansen H, Andreasen MG, Nielsen MM, Larsen LM, Bach Knudsen KE, Meyer AS et al (2002) Changes in dietary fibre, phenolic acids and activity of endogenous enzymes during rye bread-making. Eur Food Res Technol 214:33–42

Brückner-Gühmann M, Banovic M, Drusch S (2019) Towards an increased plant protein intake: rheological properties, sensory perception and consumer acceptability of lactic acid fermented, oat-based gels. Food Hydrocoll 96:201–208

Cappelle S, Guylaine L, Gänzle M, Gobbetti M (2013) History and social aspects of sourdough. In: Gobbetti M, Gänzle M (eds) Handbook on sourdough biotechnology. Springer, New York, pp 1–10

Casterline JL Jr, Oles CJ, Ku Y (1997) *In vitro* fermentation of various foods fibre fractions. J Agric Food Chem 45:2463–2467

Catzeddu P (2011) Sourdough breads. In: Preedy V, Watson R, Patel V (eds) Flour and breads and their fortification in health and disease prevention. Academic, London, pp 177–187

Chamlagain B, Deptula P, Edelmann M, Kariluoto S, Grattepanche F, Lacroix C et al (2016) Effect of the lower ligand precursors on vitamin B12 production by food-grade Propionibacteria. LWT-Food Sci Technol 72:117–124

Curtin L, D'Andrea WJ, Balascio NL, Shirazi S, Shapiro B, de Wet GA et al (2021) Sedimentary DNA and molecular evidence for early human occupation of The Faroe Islands. Comm Eart Environ 2:253

De Vuyst L, Van Kerrebroeck S, Leroy F (2017) Microbial ecology and process technology of sourdough fermentation. Adv Appl Microbiol 100:49–160

De Vuyst L, Neysens P (2005) The sourdough microflora: biodiversity and metabolic interactions. Trends Food Sci Technol 16(1–3):43–56.

Degutyte-Fomins L, Sontag-Strohm T, Salovaara H (2002) Oat bran fermentation by rye sourdough. Cereal Chem 79:345–348

Digaitiene A, Hansen ÅS, Juodeikiene G, Eidukonyte D, Josephsen J (2012) Lactic acid bacteria isolated from rye sourdoughs produce bacteriocin-like inhibitory substances active against *Bacillus subtilis* and fungi. J Appl Microbiol 112:732–742

Döring C, Nuber C, Stukenborg F, Jekle M, Becker T (2015) Impact of arabinoxylan addition on protein microstructure formation in wheat and rye dough. Food Eng 154:10–16

Dysvik A, Liland KH, Myhrer KS, Westereng B, Rukke E-O, de Rouck et al (2019) Pre-fermentation with lactic acid bacteria in sour beer production. J Inst Brew 125:342–356

Dysvik A, La Rosa SL, De Rouck G, Rukke E-O, Westereng B, Wicklund T (2020a) Microbial dynamics in traditional and modern sour beer production. Appl Environ Microbiol 86:e00566–e00520

Dysvik A, La Rosa SL, Liland KH, Myhrer KS, Østlie HM, De Rouck G et al (2020b) Co-fermentation involving *Saccharomyces cerevisiae* and *lactobacillus* species tolerant to brewing-related stress factors for controlled and rapid production of sour beer. Front Microbial 11:279

Dysvik A, La Rosa SL, Buffetto F, Liland KH, Myhrer KS, Rukke EO et al (2020c) Secondary lactic acid bacteria fermentation with wood-derived xylooligosaccharides as a tool to expedite sour beer production. J Agric Food Chem 68:301–314

Edelmann M, Chamlagain B, Santin M, Kariluoto S, Piironen V (2016) Stability of added and in situ-produced vitamin B12 in breadmaking. Food Chem 204:21–28

Engeset D, Hofoss D, Nilsson L, Olsen A, Tjønneland A, Skeie G (2015) Dietary patterns and whole grain cereals in the Scandinavian countries – differences and similarities. The HELGA project. Public Health Nutr 18:905–915

Evans J, Lorimer J (2021) Taste-shaping-natures making novel miso with charismatic microbes and new Nordic fermenters in Copenhagen. Curr Anthropol 62:S24–S375

Frølich W, Åman P, Tetens I (2013) Whole grain foods and health–a Scandinavian perspective. Food Nutr Res 57(1):18503

Galle S (2013) Sourdough: a tool to improve bread structure. In: Gobbetti M, Gänzle M (eds) Handbook on sourdough biotechnology. Springer, New York, pp 217–228

Gänzle MG (2014) Enzymatic and bacterial conversions during sourdough fermentation. Food Microbiol 37:2–10

Gänzle MG (2015) Lactic metabolism revisited: metabolism of lactic acid bacteria in food fermentations and food spoilage. Curr Op Food Sci 2:106–117

Gibson GR, Hutkins R, Sanders ME, Prescott SL, Reimer RA, Salminen SJ et al (2017) Expert consensus document: the International Scientific Association for Probiotics and Prebiotics (ISAPP) consensus statement on the definition and scope of prebiotics. Nat Rev Gastroenterol Hepatol 14:491–502

Heiniö RL, Katina K, Wilhelmson A, Myllymäki O, Rajamäki T, Latva-Kala K et al (2003) Relationship between sensory perception and flavour-active volatile compounds of germinated, sourdough fermented and native rye following the extrusion process. LWT-Food Sci Technol 36:533–545

Hopek M, Ziobro R, Acremowicz B (2006) Comparison of the effects of microbial alpha-amylases and scalded flour on bread quality. Acta Sci Pol Technol Aliment 5:97–106

Hoppe M, Ross AB, Svelander C, Sandberg A-S, Hulthén L (2019) Low-phytate wholegrain bread instead of high-phytate wholegrain bread in a total diet context did not improve iron status of healthy Swedish females: a 12-week, randomized, parallel-design intervention study. Eu J Nutr 58:853–864

Hutkins RW (2006) Fermentation of foods in the orient. Microbiology and technology of fermented foods. Blackwell Publishing Ltd, Oxford, pp 419–455

Huys G, Daniel H-M, de Vuyst (2013) Taxonomy and biodiversity of sourdough yeasts and lactic acid bacteria. In: Gobbetti M, Gänzle M (eds) Handbook on sourdough biotechnology, 11. Springer, New York, pp 105–154

Jonsson K, Andersson R, Knudsen KEB, Hallmans G, Hanhineva K, Katina K et al (2018) Rye and health-where do we stand and where do we go? Trends Food Sci Technol 79:78–87

Juodeikiene G (2016) Traditional rye sourdough bread in the Baltic region. In: Kristbergsson K, Oliveira J (eds) Traditional foods: general and consumer aspects. Springer, Boston, pp 173–187

Juodeikiene G, Salomskiene J, Eidukonyte D, Vidmantiene D, Narbutaite V, Vaiciulyte-Funk L (2011) The impact of novel fermented products containing extruded wheat material on the quality of wheat bread. Food Technol Biotechnol 49(4):502–510

Kajala I, Mäkelä J, Coda R, Shukla S, Shi Q, Maina ND et al (2016) Rye bran as fermentation matrix boosts in situ dextran production by *Weissella confusa* compared to wheat bran. Appl Microbiol Biotechnol 100:3499–3510

Kariluoto S, Vahteristo L, Salovaara H, Katina K, Liukkonen K-H, Piironen V (2004) Effect of baking method and fermentation on folate content of rye and wheat breads. Cereal Chem 81:134–139

Kariluoto S, Aittamaa M, Korhola M, Salovaara H, Vahteristo L, Piironen. (2006) Effects of yeasts and bacteria on the levels of folates in rye sourdoughs. Int J Food Microbiol 106:137–143

Kårlund A, Gómez-Gallego C, Korhonen J, Palo-Oja OM, El-Nezami H, Kolehmainen M (2020) Harnessing microbes for sustainable development: food fermentation as a tool for improving the nutritional quality of alternative protein sources. Nutrients 12:1020

Katina K, Laitila A, Juvonen R, Liukkonen KH, Kariluoto S, Piironen V et al (2007) Bran fermentation as a means to enhance technological properties and bioactivity of rye. Food Microbiol 24:175–186

Koehler, Wieser (2013) Chemistry of cereal grains. In: Gobbetti M, Gänzle M (eds) Handbook on sourdough biotechnology. Springer, New York, pp 11–45

Koistinen VM, Mattila O, Katina K, Poutanen K, Aura A-M, Hanhineva K (2018) Metabolic profiling of sourdough fermented wheat and rye bread. Sci Rep 8:5684

Kozlinskis E, Skudra L, Klava D, Kunkulberga D (2008) Characterization of rye sourdough microflora. Latv Univ Agric Fac Food Technol:89–93

Kozlinskis E, Skudra L, Rakcejeva T, Kunkulberga D (2010) Changes in the chemical and microbiological properties of spontaneous rye sourdough during fermentation. Spontano Rudzu Maizes Ieraugu Kim Mikrobiol Rad Izmainas Ferment Laik 25:67–77

Kylli RM (2014) Bread and power in the "land of no bread"—low-carbohydrate Sami diet in transition. Acta Boreal 31:176–197

Kyrø C, Skeie G, Dragsted LO, Christensen J, Overvad K, Hallmans G et al (2012) Intake of whole grain in Scandinavia: intake, sources and compliance with new national recommendations. Scand J Public Health 40:76–84

Laatikainen R, Koskenpato J, Hongisto SM, Loponen J, Poussa T, Hillilä M et al (2016) Randomised clinical trial: Low-FODMAP rye bread vs. regular rye bread to relieve the symptoms of irritable bowel syndrome. Aliment Pharmacol Ther 44:460–470

Laatikainen R, Jalanka J, Loponen J, Hongisto S-M, Hilliä M, Koskenpato J et al (2019) Randomised clinical trial: effect of low-FODMAP rye bread versus regular rye bread on the intestinal microbiota of irritable bowel syndrome patients: association with individual symptom variation. BMC Nutr 5:12

Lambo-Fodje AM, Oste R, Nyman MEG-L (2006) Short-chain fatty acid formation in the hindgut of rats fed native and fermented oat fibre concentrates. Br J Nutr 96:47–55

Landis EA, Oliverio AM, McKenney EA, Nichols LM, Kfoury N, Biango-Daniels M et al (2021) The diversity and function of sourdough starter microbiomes. elife 10:e61644

Larsson M, Sandberg A-S (1991) Phytate reduction in bread containing oat flour, oat bran or rye bran. J Cereal Sci 14:141–149

Leroy F, Verluyten J, De Vuyst L (2006). Functional meat starter cultures for improved sausage fermentation. Int J Food Microbiol 106(3):270–285.

Li H, Dai F, Zhang L, Li Z (2022) Characterisation of scalded dough and its impact on the growth of mixed yeasts originating from Jiaozi. Food Biosci 49:101920

Lönner C, Welander T, Molin N, Dostalek M (1986) The microflora in a sour dough started spontaneously on typical Swedish rye meal. Food Microbiol 3:3–12

Loponen J, Gänzle MGG (2018) Use of sourdough in low FODMAP baking. Foods 7:96

Loponen J, Laine P, Sontag-Strohm T, Salovaara H (2007) Behaviour of oat globulins in lactic acid fermentation of oat bran. Eu Food Res Technol 225:105–110

Mäkelä J, Rautavirta K (2018) Food, nutrition, and health in Finland. In Nutritional and health aspects of food in Nordic countries. Academic, pp 127–143

Mäntynen VH, Korhola M, Gudmundsson H, Turakainen H, Alfredsson GA, Salovaara H et al (1999). A polyphasic study on the taxonomic position of industrial sour dough yeasts. Syst Appl Microbiol 22(1):87–96.

Marklinder I, Johansson L (1995) Sour dough fermentation of barley flours with varied content of mixed-linked (1→ 3),(1→ 4) β-d-glucans. Food Microbiol 12:363–371

Marklinder I, Haglund Å, Johansson L (1996) Influences of lactic acid bacteria on technological, nutritional, and sensory properties of barley sour dough bread. Food Qual Prefer 7:285–292

Mårtensson O, Öste R, Holst O (2000) Lactic acid bacteria in an oat-based non-dairy milk substitute: fermentation characteristics and exopolysaccharide formation. LWT-Food Sci Technol 33:525–530

Mårtensson O, Andersson C, Andersson K, Öste R, Holst O (2001) Formulation of an oat-based fermented product and its comparison with yoghurt. J Sci Food Agric 81:1314–1321

Mårtensson O, Öste R, Holst O (2002a) The effect of yoghurt culture on the survival of probiotic bacteria in oat-based, non-dairy products. Food Res Int 35:775–784

Mårtensson O, Öste R, Holst O (2002b) Texture promoting capacity and EPS formation by lactic acid bacteria in three different oat-based non-dairy media. Eu Food Res Technol 214:232–236

Mihhalevski A, Nisamedtinov I, Hälvin K, Ošeka A, Paalme T (2013) Stability of B-complex vitamins and dietary fiber during rye sourdough bread production. J Cereal Sci 57:30–38

Narbutaite V, Fernandez A, Horn N, Juodeikiene G, Narbad A (2008) Influence of baking enzymes on antimicrobial activity of five bacteriocin-like inhibitory substances produced by lactic acid bacteria isolated from Lithuanian sourdoughs. Lett Appl Microbiol 47:555–560

Neumann H, Stephan H, Brümmer J-M (2006) Roggen als Rohstoff und Technik der Roggensauerteigführung. In: Brandt MJ, Gänzle MG (eds) Handbuch Sauerteig. Behr's Verlag, Hamurg, pp 185–284

Nielsen MM, Damstrup ML, Dal Thomsen A, Rasmussen SK, Hansen Å (2007) Phytase activity and degradation of phytic acid during rye bread making. Eu Food Res Technol 225:173–181

Nielsen MM, Damstrup ML, Hansen Å (2008) An optimised micro-titer plate method for characterisation of endogenous rye phytase under industrial rye bread making conditions. Eu Food Res Technol 227:1009–1015

Nikinmaa M, Alam SA, Raulio M, Katina K, Kajala I, Nordlund E et al (2017) Bioprocessing of bran with exopolysaccharide producing microorganisms as a tool to improve expansion and textural properties of extruded cereal foams with high dietary fibre content. LWT– Food Sci Technol 77:170–177

Nikinmaa M, Kajala I, Liu X, Nordlund E, Sozer N (2020) The role of rye bran acidification and *in situ* dextran formation on structure and texture of high fibre extrudates. Food Res Int 137:109438

Nilsson U, Johansson M, Nilsson A, Björck I, Nyman M (2008) Dietary supplementation with beta-glucan enriched oat bran increases faecal concentration of carboxylic acids in healthy subjects. Eur J Clin Nutr 62:978–984

Nuobariene L, Hansen ÅS, Arneborg N (2012) Isolation and identification of phytase-active yeasts from sourdoughs. LWT – Food Sci Technol 48:190–196

Nuobariene L, Cizeikiene D, Gradzeviciute E, Hansen ÅS, Rasmussen SK, Juodeikiene G et al (2015) Phytase-active lactic acid bacteria from sourdoughs: isolation and identification. LWT – Food Sci Technol 63:766–772

Nyanzi R, Jooste PJ (2012) Cereal-based functional foods. In: Rigobelo E (ed) Probiotics. InTech, pp 161–197

Pétel C, Onno B, Prost C (2017) Sourdough volatile compounds and their contribution to bread: a review. Trends Food Sci Technol 59:105–123

Pirkola L, Laatikainen R, Loponen J, Hongisto SM, Hillilä M, Nuora A et al (2018) Low-FODMAP vs regular rye bread in irritable bowel syndrome: randomized SmartPill® study. World J Gastroenterol 24:1259–1126

Poutanen K, Katina K, Heiniö RL (2014) Rye. In: Zhou W, Hui YH, De Leyn I, Pagani MA, Rosell CM, Selman JD, Therdthai N (eds) Bakery products science and technology. Wiley, pp 75–87

Redzepi R, Zilber D (2018) The Noma guide to fermentation: including koji, kombuchas, shoyus, misos, vinegars, garums, lacto-ferments, and black fruits and vegetables. Artisan Books

Reidzane S, Kruma Z, Kazantseva J, Traksmaa A, Klava D (2021) Determination of technological parameters and characterization of microbiota of the spontaneous sourdough fermentation of hull-less barley. Foods 10:2253

Reykdal O, Kristjánsdóttir PA, Hermannsson J, Martin P, Dalmannsdóttir S, Djurhuus R et al (2014) Status of cereal cultivation in the North Atlantic region. Skýrsla Matis:23–14

Rosenquist H, Hansen Å (2000) The microbial stability of two bakery sourdoughs made from conventionally and organically grown rye. Food Microbiol 17:241–250

Roye C, Henrion M, Chanvrier H, Loret C, King R, Lamothe L et al (2021) Changing wheat bran structural properties by extrusion-cooking on a pilot and industrial scale: a comparative study. Foods 10:472

Rozada-Sánchez R, Sattur AP, Thomas K, Pandiella SS (2008) Evaluation of *Bifidobacterium* spp. for the production of a potentially probiotic malt-based beverage. Process Biochem 43:848–854

Sahlberg A (2020) Exporting Finnish craft beer to The Netherlands

Salovaara HO (1996) The time of cereal based functional foods is here: introducing Yosa®, a vellie., Haugesund, Norway. In Proceedings, G Skrede & EM Magnus, (eds.), Haugesund, pp 195–202

Salovaara H, Kurka AM (1997) Food product containing dietary fiber and method of making said product. European Patent Specication EP 0568530B1

Sandvik P, Kihlberg I, Lindroos AK, Marklinder I, Nydahl M (2014) Bread consumption patterns in a Swedish national dietary survey focusing particularly on whole-grain and rye bread. Food. Nutr Res 58:24024

Shewry PR, Hawkesford MJ, Piironen V, Lampi A-M, Gebruers K, Boros D et al (2013) Natural variation of grain composition of wheat and related cereals. J Agric Food Chem 61:8295–8303

Spicher G, Lönner C (1985) Die Mikroflora des Sauerteiges. XXI. Mitteilung: Die in Sauerteigen schwedischer Bäckereien vorkommenden Lactobacillen, vol 181. Z Lebensm Unters und Forsch, pp 9–13

Steingrímsdóttir L, Thorkelsson G, Eythórsdóttir E (2018) Food, nutrition, and health in Iceland. In: Nutritional and health aspects of food in Nordic countries. Academic, pp 145–177

Struyf N, Laurent J, Verspreet J, Verstrepen KJ, Courtin CM (2017) *Saccharomyces cerevisiae* and *Kluyveromyces marxianus* co-cultures allow reduction of fermentable oligo-, di-, and mono-saccharides and polyols levels in whole wheat bread. J Agric Food Chem 65:8704–8871

Sullivan P, Arendt E, Gallagher E (2013) The increasing use of barley and barley by-products in the production of healthier baked goods. Trends Food Sci Technol 29:124e134

TasteAtlas (2020) Top 50 Most Popular Baltic Food and Beverages. What to Eat and Drink in the Baltic? August 31, 2020. Available at: https://www.tasteatlas.com/50-most-popular-foods-in-baltic

Tenhovirta S (2019) The effects of lactic acid bacteria species on properties of Sour Beer. Master thesis at the University of Helsinki, Faculty of Agriculture and Forestry, Finland. https://helda.helsinki.fi/handle/10138/303018

The Rye Baker (2018) By Stanley Ginsberg. Finnish Malt Bread/Mallas Leipää November 01, 2018. https://theryebaker.com/mallas-leipaa/

The Rye Baker (2019) By Stanley Ginsberg. Halme Bakery's Ring Rye/Ruisreikäleipä (Finland). November 07, 2019. https://theryebaker.com/halme-bakerys-ring-rye-ruisreikaleipa-finland/

Thorvaldsson M (2020) Biotransformation of brewer's spent grain and application as a food ingredient in extruded breakfast cereals. Master thesis at Chalmers University, Department of Biology and Biotechnology, Sweden. https://odr.chalmers.se/handle/20.500.12380/300764

Unger RW (1988) The trade in beer to medieval Scandinavia. Deutsches Schiffahrtsarchiv 11:249–258

Verwimp T, Courtin CM, Delcour JA (2007) Rye constituents and their impact on rye processing. Food Biochem Food Process 654:567–592

Viiard E, Mihhalevski A, Rühka T, Paalme T, Sarand I (2012) Evaluation of the microbial community in industrial rye sourdough upon continuous back-slopping propagation revealed *Lactobacillus helveticus* as the dominant species. J Appl Microbiol 114:404–412

Viiard E, Bessmeltseva M, Simm J, Talve T, Aaspõllu A, Paalme T et al (2016) Diversity and stability of lactic acid bacteria in rye sourdoughs of four bakeries with different propagation parameters. PLoS One 11:e0148325

Wu MS (2021) The heritage bread: Rural food tradition, heritage-making and community-building on Kihnu Island, Estonia. HeritageForAll. Posted in: Articles, Heritage Preservation and Rural Communities. March 16, 2021

Xie C, Coda R, Chamlagain B, Edelmann M, Deptula P, Varmanen P et al (2018) In situ fortification of vitamin B12 in wheat flour and wheat bran by fermentation with *Propionibacterium freudenreichii*. J Cereal Sci 81:133–139

Xie C, Coda R, Chamlagain B, Edelmann M, Varmanen P, Piironen V et al (2021) Fermentation of cereal, pseudo-cereal and legume materials with *Propionibacterium freudenreichii* and *Levilactobacillus brevis* for vitamin B12 fortification. LWT – Food Sci Technol 137:110431

Zagorec M, Champomier-Vergès M-C (2017) *Lactobacillus sakei*: a starter for sausage fermentation, a protective culture for meat products. Microorganisms 5:56

Zheng J, Wittouck S, Salvetti E, Franz CM, Harris H, Mattarelli P et al (2020) A taxonomic note on the genus *Lactobacillus*: description of 23 novel genera, emended description of the genus Lactobacillus Beijerinck 1901, and union of *Lactobacillaceae* and *Leuconostocaceae*. Int J Syst Evol Microbiol 70:2782–2858

Images References

Flatbrød in Figure 2 is adapted from "Brødkorg.JPG" by Knut, March 20, 2008, available at https://commons.wikimedia.org/wiki/File:Br%C3%B8dkorg.JPG under an Attribution-Share Alike 3.0 Unported license. To view a copy of this license, visit https://creativecommons.org/licenses/by-sa/3.0/legalcode

Kavring in Figure 2 is adapted from "Kavring" by Erik Forsberg, July 25, 2015, available at https://www.flickr.com/photos/hepp/19385887814/in/photolist-aqcC2h-ASJsS6-AQrEHs-ARxsjh-ot1sER-osQERR-wrL5M3-vx4K45-8oQyAp-5KsY1t-5EgQna-zVR9H4-6eMgEq-zVH1Xb-ATHTyH-zVH67o-zVH3zQ-ARxF1j-AeLyWa-r7ToVR-ouSJvF-rpksHt-qtBT5V-r8PNTS-jEvip-7hi71A-2i3tMM1-oujjZV-xqNDiD under an Attribution-NonCommercial 2.0 Generic (CC BY-NC 2.0) license. To view a copy of this license, visit https://creativecommons.org/licenses/by-nc/2.0/legalcode

Leib in Figure 2 is adapted from "Today we had a first bite of this year's Christmas bread/Täna võtsime esimese ampsu selle aasta jõululeivast" by Marju Randmer, December 23, 2010, available at https://www.flickr.com/photos/isutekitaja/5285863875/in/photolist-bWH9si-ce5vpY-bWH9ga-bWH8Ji-ce5vM1-bWH8QB-FPEuxo-FRYxVg-FxHEXu-F3zx4B-FxHEMQ-FRYypH-FPEtR3-8ZnHh1-ctL3cL-949uXC-946yxi-946rb4-946r3n-ctL6VG-ctL4MN-ctL5w3-ctL4eN-ctL541-ctL6nC-ctL61s-ctL3zy-ctL3Ps under an Attribution-NonCommercial 2.0 Generic license. To view a copy of this license, visit https://creativecommons.org/licenses/by-nc/2.0/legalcode

Mallas Leipää in Figure 2 is adapted from "Creamy mushroom soup and malt bread at restaurant Frans & Amélie.jpg" by JIP, February 4, 2020, available at https://commons.wikimedia.org/wiki/File:Creamy_mushroom_soup_and_malt_bread_at_restaurant_Frans_%26_Am%C3%A9lie.

jpg under an Attribution-Share Alike 4.0 International license. To view a copy of this license, visit https://creativecommons.org/licenses/by-sa/4.0/legalcode

Rieska in Figure 2 is adapted from "Rye and Potato rieska – 1.JPG" by Hellahulla, November 23, 2007, available at https://commons.wikimedia.org/wiki/File:Rye_and_Potato_rieska_-_1.JPG under an Attribution-Share Alike 4.0 International, 3.0 Unported, 2.5 Generic, 2.0 Generic and 1.0 Generic license. To view a copy of these licenses, visit https://creativecommons.org/licenses/by-sa/4.0/legalcode, https://creativecommons.org/licenses/by-sa/3.0/legalcode, https://creativecommons.org/licenses/by-sa/2.5/legalcode, https://creativecommons.org/licenses/by-sa/2.0/legalcode, https://creativecommons.org/licenses/by-sa/1.0/legalcode, respectively

Rúgbrauð and Flatbrauð in Figure 2 is adapted from "Thorramatur.jpg" by The blanz, February 2, 2010 available at https://commons.wikimedia.org/wiki/File:Thorramatur.jpg under an Attribution-Share Alike 3.0 Unported license. To view a copy of this license, visit https://creativecommons.org/licenses/by-sa/3.0/legalcode

Ruislimppu in Figure 2 is adapted from "Fazer_Leipurit_972950_iso_ruislimppu_leivinhiivaton_vaaka.jpg" by Fazer Public Image Bank, September 25, 2015, available at https://fast.fazer.com/f/K6kw

Ruisreikäleipä in Figure 2 is adapted from "FB_Oululainen_RuisSata.jpg" by Fazer Public Image Bank, October 16, 2014, available at https://fast.fazer.com/f/G9KV

Spings for hverabrauð in Figure 2 is adapted from "Hverabrauð - Forni per Pane/Bread Ovens" by Francesco Giordano, July 18, 2008 available at https://www.flickr.com/photos/francescog/2803463147/in/photolist-5gJtVv under an Attribution-Share Alike 2.0 Generic (CC BY-SA 2.0) license. To view a copy of this license, visit https://creativecommons.org/licenses/by-sa/2.0/legalcode

Tunnbröd in Figure 2 is adapted from "Tunnbröd-2007-07-14.jpg" by WikiPedant, September 19, 2020, available at https://commons.wikimedia.org/wiki/File:Tunnbr%C3%B6d-2007-07-14.jpg under an Attribution-Share Alike 2.5 Generic license. To view a copy of this license, visit https://creativecommons.org/licenses/by-sa/2.5/legalcode

Chapter 5
Microbiological and Biomolecular Methods Applicable to the Isolation, Characterization and Identification of Microbial Sourdough Strains

Antonio A. Câmara Jr, Larissa P. Margalho, Wilson J. F. Lemos Jr, Vanessa S. de Oliveira, André F. Guerra, and Anderson S. Sant'Ana

5.1 Introduction

Sourdough fermentation is one of the oldest biotechnological processes extensively applied in food production and traditionally used in bread-making. Since approximately 1990, there has been a growing interest in discovering the potential of sourdough fermentation that started focusing on aspects such as flavor, rheology, and shelf life, and was extended to knowledge about microbial diversity and interactions (Arora et al. 2021).

Sourdough has been rediscovered as an alternative to consumers' demand for high-quality bread and bakery products (Montemurro et al. 2019; Arora et al. 2021). In recent years, sourdough containing new ingredients has been increasingly used worldwide, providing products with improved sensory attributes and prolonged shelf life (Rizzello et al. 2016; Melini and Melini 2018; Korcari et al. 2021; Olojede et al. 2022). Moreover, more natural, nutritious, and healthy products can be

A. A. Câmara Jr · L. P. Margalho · A. S. Sant'Ana (✉)
Department of Food Science and Nutrition, Faculty of Food Engineering, University of Campinas, Campinas, SP, Brazil
e-mail: and@unicamp.br

W. J. F. Lemos Jr
Bioresources Unit, Center for Health & Bioresources, AIT Austrian Institute of Technology, Tulln, Austria

V. S. de Oliveira
Department of Food Technology, Institute of Technology, University Federal Rural of Rio de Janeiro, Seropédica, RJ, Brazil

A. F. Guerra
Department of Food Engineering, Federal Center of Technological Education Celso Suckow da Fonseca, Valença, RJ, Brazil

E. B. Ceresino et al. (eds.), *Sourdough Microbiota and Starter Cultures for Industry*, https://doi.org/10.1007/978-3-031-48604-3_5

obtained, meeting the trend in the food industry represented by functional foods (Fernández-Peláez et al. 2020; Canesin and Cazarin 2021; Graça et al. 2021).

Therefore, the increasing popularity of sourdough highlights the importance of improving the control of fermentation mechanisms involved in sourdough products. As microbial sourdough strains play a role in the characteristics and quality of the final product, their proper isolation, identification, and characterization are pivotal.

The sourdough microbiota is mainly composed of lactic acid bacteria (LAB) and yeasts (Yu et al. 2019). It shows the predominance of bacteria, where the genera *Secundilactobacillus, Levilactobacillus, Lactiplantibacillus, Lactobacillus, Lacticaseibacillus,* and *Fructilactobacillus* are the most abundant (Martorana et al. 2018; Arena et al. 2019; Oshiro et al. 2021). However, other species, such as *Pediococcus*, *Enterococcus*, *Lactococcus*, and *Weissella*, can also be found in sourdough (Ferchichi et al. 2007; Oshiro et al. 2021). The dominant yeasts belong to *Saccharomyces* and *Candida* genus (Martorana et al. 2018; Arena et al. 2019). Nevertheless, the presence of *Kazachstania*, *Candida* (*Torulopsis*), *Yarrowi*, and *Pichia* genera has been also reported in sourdough (Xu et al. 2019; Carbonetto et al. 2020; Palla et al. 2020a).

Yeasts and bacteria present in the sourdough are part of a symbiotic microbiome with different species coexisting in a certain harmony level. Therefore, the success of fermented products like sourdough bread, cheeses, and wines lies in the microbial interaction and diversity of these complex ecosystems (Lau et al. 2021). This must be taken into account, especially if the intention is to use individual strains in processes involving sourdough. In other words, the original characteristics of the product should not be expected to be the same using only one purified culture compared to those from products obtained using mature sourdoughs, which are characterized by strains well acclimatized and adapted to the formulation conditions (De Vuyst et al. 2014). Indeed, a common practice of industrial processes consists in adding adjunct cultures to the sourdough, where strains capable of boosting one or more technological parameters are inoculated during the formulation, for example, to increase the acidification rate (Arora et al. 2021).

For this reason, the isolation and characterization of these species are essential to comprehend sourdough fermentation. Firstly, it must be considered that the key to select the best isolation and biomolecular approach may be affected by process parameters, such as fermentation time or temperature, cereal and flour composition, water microbiota, dough yield, pH, and temperature, as well as geographic location, microclimate, and region where sourdoughs are produced (Gänzle and Ripari 2016; De Vuyst et al. 2017; Landis et al. 2021). It is important to highlight that the type of sourdough, 0 (pre-dough), I (natural sourdough fermentation), II (industrially produced sourdough, generally liquid), or III (dried), contributes to the sourdough microbiome and determines the isolation or biomolecular method that will be used.

Traditional microbial methods (e.g. morphological, physiological, and biochemical) have been used to isolate and characterize fermenting microorganisms. Although these approaches are relatively fastidious, they are the cheapest and easiest to apply both in industrial and laboratory environments. Thus, molecular technologies have been highlighted, showing advantages such as accuracy, specificity, reproducibility, ease of use, sensitivity, and speed (Cocolin and Ercolini 2008; Sharma et al. 2020; El Sheikha and Hu 2020).

Microbial enumeration, identification, and characterization performed by traditional methods cannot monitor specific strains in complex ecosystems. While biochemical and physiological techniques present limitations in the differentiation of microbial isolates with comparable physiological characteristics, molecular methods allow the proper exploration of the microbial diversity and the knowledge of the dynamic of microbial communities (Cocolin and Ercolini 2008; El Sheikha et al. 2018; El Sheikha and Hu 2020).

Molecular methods include proteomics and metabolomics-based technologies. Studies concerning proteomics approaches reported the use of liquid chromatography tandem-mass spectrometry (LC–MS/MS) and matrix-assisted laser desorption ionization time-of-flight (MALDI-TOF-MS) (Shevchenko et al. 2014; Sevgili et al. 2021) in sourdough. Regarding metabolomics, Fourier transform infrared (FTIR) has been used (Arici et al. 2018; Arora et al. 2021; Cakir et al. 2021). In addition, biomolecular techniques have contributed to the selection of new starter cultures with desirable competitive abilities and metabolic properties (Mantzourani et al. 2019; Holt et al. 2022).

Most studies on sourdough describe both microbial and biomolecular methods, which can be divided into culture-dependent and culture-independent techniques. While the former may contribute increasing the knowledge about the taxonomic diversity of the food-related microflora and the selection of cultivable species, culture-independent techniques allow studying a microbial population without the need to isolate and culture its distinct components (Giraffa and Carminati 2008), which include next-generation sequencing (NGS) methods (not described in this chapter). The description and comparisons of NGS technologies, as well as the advantages and disadvantages, have been extensively reviewed (Shendure and Ji 2008; Glen 2011; Liu et al. 2012; Loman et al. 2012; Danilo 2013).

Therefore, this chapter aims to evidence some of the current microbiological and biomolecular methods applied for the isolation, characterization and identification of microorganisms from the sourdough microbiome.

5.2 Microbiological Classical Approaches for Isolation of Sourdough Strains

5.2.1 Methods for Yeast Isolation and Culture Media

The most common yeast genera found in sourdough belong to *Saccharomyces*, *Kazachstania*, *Candida* (*Torulopsis*), *Yarrowia*, and *Pichia* (De Vuyst et al. 2017). The presence of yeasts in sourdough is associated with several positive nutritional and technological effects such as increased bioavailability of nutrients (Fernández-Peláez et al. 2020), enhancement of antioxidant activity (Palla et al. 2020a) and hydrolysis of short-chain carbohydrates (Gélinas et al. 2016; Graça et al. 2021), biosynthesis of esters and aromatic compounds (Guerzoni et al. 2007), and antifungal activity (Coda et al. 2013; Arora et al. 2021). Therefore, it is crucial to know some classical methods frequently used to isolate yeasts.

The microbiological methods usually applied for the isolation of yeasts are the culture-dependent ones. Basically, these methods consist of dissolving a considerable amount of the sourdough sample, typically 1:9 proportion, on sterile physiological solution, like peptone water or 0.85% NaCl solution, followed by homogenization in a stomacher (or paddle blender) during 2–5 min.

After obtaining the microbial suspension, serial dilutions must be carried out at 1:10 proportion in order to isolate individual cells. Then, an aliquot of 100 μL is spread on the surface of the culture medium (Yarrow 1998; Kurtzman et al. 2011).

At this stage, the general idea is to select a growth medium that meets the basic nutritional requirements for most yeasts, i.e. generally containing yeast extract, glucose, and peptone; however, this composition can undergo some variations.

Several culture media have already been described in the literature and some of them were reported by Kurtzman et al. (2011) and Yarrow (1998). In sourdough microbiota that also contains bacteria, an antibiotic is normally added to the culture medium to limit the spread of "undesirable" microorganisms and increase the probability of isolating only yeasts. The most common antibiotic applied is chloramphenicol at a concentration range between 0.1 and 0.3 g/L. This antibiotic is thermostable and can be added to the culture medium before the sterilization cycle at 121 °C without affecting its antibacterial activity (Kurtzman et al. 2011). Regarding the media, YPD (yeast extract peptone dextrose agar) and MEA (Malt extract agar) are commonly used. However, other growth media can be considered depending on the purpose. For example, strategies such as varying the composition of sugars or adding sodium chloride may be employed to select osmophilic species. Table 5.1 presents a non-exhaustive list of culture-dependent methods recently employed to isolate sourdough yeasts, including information on the sourdough raw material, predominant species isolated, the culture medium, and a summary of the methods used to identify yeasts species.

5.2.2 *Methods for Bacteria Isolation and Culture Media*

LAB play a fundamental role in steering the spontaneous fermentation that occurs during the preparation of sourdough in the manufacture of bread, steamed bread, and other cereal-based products. Among the most recurrent isolated LAB species from sourdough in studies conducted over the last 70 years, the following stand out: *Fructilactobacillus sanfranciscensis, Lactiplantibacillus plantarum, Levilactobacillus brevis, Pediococcus pentosaceus, Limosilactobacillus fermentum,* and *Companilactobacillus paralimentarius* (De Vuyst et al. 2017), which are distributed along the entire production chain (threshing, milling, or as part of wheat, handler, and the environmental microbiota) (Suo et al. 2021). The ratio of LAB to

Table 5.1 Non-exhaustive list of microbiological and molecular methods used to isolate and identify yeasts in sourdoughs

Country	Sourdough composition	Isolated dominant species	Culture medium and growth conditions	Summary of biomolecular method used to identify at species level	Reference
Austria	Wheat, rye	*Saccharomyces cerevisiae, Kazachstania humilis* (*Candida humilis*), *S. uvarum, K. unispora*	Yeast glucose chloramphenicol (YGC), Dichloran-Bengalrot-chloramphenicol (DRBC) or malt extract agar (MEA) agar, all supplemented with chloramphenicol 0.1 g/L. 25 °C for 72 h.	26S rDNA gene amplification with primers NL1 and NL4.	Fraberger et al. (2020b)
Belgium	Wheat, spelt, rye	*Saccharomyces cerevisiae, Wickerhamomyces anomalus* (*Pichia anomala*), *Candida glabrata*	Yeast extract-glucose agar (YGA) supplemented with chloramphenicol 0.1 g/L. 30 °C for 48 h.	M13 for initial fingerprinting, followed by Ribosomal DNA regions, ITS, D1/D2 LSU, and *ACT1* gene fragments amplification.	Vrancken et al. (2010)
Brazil	Grape and wheat flour	*Saccharomyces cerevisiae*	Potato dextrose agar (PDA) supplemented with 10% tartaric acid. 25 °C for 72 h.	ITS regions using primers ITS4 and ITS5.	Aplevicz et al. (2014)
Bulgaria	White, wholegrain wheat	*Saccharomyces cerevisiae, Kazachstania barnettii, Kluyveromyces marxianus*	Malt extract agar (MEA) and Sabouraud dextrose agar (SDA) supplemented with chloramphenicol 0.1 g/L. 30 °C for 48 h.	ITS1-5.8S-ITS2 rRNA gene partial sequence amplification with primers ITS4 and ITS5.	Petkova et al. (2021)

(continued)

Table 5.1 (continued)

Country	Sourdough composition	Isolated dominant species	Culture medium and growth conditions	Summary of biomolecular method used to identify at species level	Reference
China	Wheat and wheat/corn/rice for Jiaozi	*Saccharomyces cerevisiae, Kazachstania humilis* (*Candida humilis*), *Wickerhamomyces anomalus*	YPD agar supplemented with chloramphenicol 0.1 g/L. 30 °C for 48 h.	28S rDNA gene amplification with primers P1 and P2.	Yan et al. (2019)
France	Organic wheat	*Saccharomyces cerevisiae Kazachstania bulderi, K. barnettii, K. saulgeensis, K. unispora, K. humilis*	Yeast extract (YE) agar. 48 h–72 h at room temperature.	Amplification of rDNA Non-Transcribed Spacer 2 (NTS2). Also confirmed by sequencing the D1/D2 region and the actin coding gene, *ACT1*.	Urien et al. (2019)
France	Organic wheat or organic wheat + organic rye	*Kazachstania bulderi, Kazachstania humilis* (*Candida humilis*), *Kazachstania unispora, Torulaspora delbrueckii, Kazachstania exigua, Rhodotorula mucilaginosa*	Yeast extract (YE) agar. Room temperature for 48 h–72 h.	3 molecular markers were used: the rDNA NTS2 using primers SR21 and LR13; sequencing the rDNA D1/D2 region with primers NL1 and NL4; the gene *ACT1* with primers CA5R and CA14.	Lhomme et al. (2016b)
Italy	Maiorca wheat	*Saccharomyces cerevisiae, Wickerhamomyces anomalus, Torulaspora delbrueckii, Pichia kluyveri, Candida boidinii, Candida diddensiae*	Sabouraud dextrose Agar (SDA) and Wallerstein laboratory nutrient agar (WLN), both supplemented with chloramphenicol 0.1 g/L. 30 °C for 48 h–72 h.	Amplification of 5.8S rRNA gene and the ITS1 and ITS2 regions using primers ITS1 and ITS4. One representative selected to amplification of D1/D2 region of the 26S rRNA gene using primers NL1 and NL4.	Pino et al. (2022)

(continued)

Table 5.1 (continued)

Country	Sourdough composition	Isolated dominant species	Culture medium and growth conditions	Summary of biomolecular method used to identify at species level	Reference
Italy	Wheat	*Saccharomyces cerevisiae*, *Kazachstania humilis* (*Candida milleri*), *Pichia kudriavzevii* (*Candida krusei*), *Torulaspora delbrueckii*	YPD supplemented with chloramphenicol 0.1 g/L. 28 °C for 48 h.	Amplification of 5.8S ITS rRNA region using primers ITS1 and ITS4. Products digested with the restriction endonucleases *Cfo*I, *Hae*III, and *Hinf*I. Senquencing of 26S rRNA D1/ D2 domain using primers NL1 and NL4.	Valmorri et al. (2010))
Italy	Spelt wheat flour	*Saccharomyces cerevisiae*, *Wickerhamomyces anomalus*, *Clavispora lusitaniae*, *Pichia fermentans*	YGC with chloramphenicol. 30 °C for 48 h.	Amplification of ITS1-5.8S-ITS2 using primers ITS1 and ITS4, and/or partial 26S rRNA gene using NL1 and NL4 primers.	Korcari et al. (2020)
Japan	Wheat and rye	*Saccharomyces cerevisiae*, *Kazachstania humilis* (*Candida humilis*)	YPD agar supplemented with chloramphenicol 0.1 g/L. 30 °C for 48 h.	Senquencing performed using FAST MicroSEQ D2 LSU rDNA fungal PCR kit and ABI PRISM device.	Fujimoto et al. (2019)
Nigeria	Sorghum flour	*Saccharomyces cerevisiae*	Glucose yeast peptone agar (GYPA) supplemented with chloramphenicol 0.1 g/L. 37 °C for 72 h.	18S rRNA for yeasts typing.	Ogunsakin et al. (2017)

(continued)

Table 5.1 (continued)

Country	Sourdough composition	Isolated dominant species	Culture medium and growth conditions	Summary of biomolecular method used to identify at species level	Reference
Thailand	Wheat	*Candida tropicalis, Pichia stipitis, Candida parapsilosis, Pichia kudriavzevii (Issatchenkia orientalis) and Saccharomyces cerevisiae*	Yeast malt extract (YM) agar. 25 °C for 72 h.	Senquencing of 26S rRNA D1/D2 variable domains using primers NL1 and NL4.	Luangsakul et al. (2009)
Turkey	Whole wheat, rye	*Saccharomyces cerevisiae, Kazachstania bulderi, Pichia fermentans, Pichia membranifaciens, K. servazzii, K. unispora, Hanseniaspora valbyensis*	YPD agar supplemented with oxytetracycline 0.1 g/L. 28 °C for 48 h to 72 h.	ITS region amplification of the 5.8S rRNA gene, RFLP analysis, and the D1/D2 domain of the 26S rDNA gene. Primers used: ITS1 and ITS4.	Boyaci-Gunduz and Erten (2020)
Turkey	Wheat flour	*Saccharomyces cerevisiae, Torulaspora delbrueckii* and *Kluyveromyces marxianus*	DRBC agar. 28 °C for 24 h to 48 h.	26S rDNA sequencing using primers NL1 and LS2.	Arici et al. (2018)

It describes some isolation and biomolecular methods used in the sourdough microbiomes context and includes the country, sourdough composition, and the isolated and identified dominant species

the yeast population is usually recorded as 100:1 and 10:1, but in some cases, it can reach up to 1:1000 values (Comasio et al. 2020; Petkova et al. 2021).

The literature has reported a great interest in isolating and characterizing LAB due to their antimicrobial, nutraceutical, functional, and sensorial properties, as well as their impact on the quality and shelf life extension of baked goods (Arora et al. 2021; Lau et al. 2021). In addition, there was an increase in people's interest in resuming the sourdough bread trend due to the covid-19 pandemic in 2020 and social confinement (Lau et al. 2021). Studies recently published have intended to apply new starter cultures present in the autochthonous microbiota of different sourdough matrices and geographic locations worldwide (Landis et al. 2021; Shen et al. 2022; Zhou et al. 2022; Pino et al. 2022).

Thus, to assess their technological and functional traits, it is necessary to isolate pure cultures of LAB by culture-based methods using a specific medium, as previously described for yeasts (Item 5.2.1). The preliminary steps of weighing, homogenization, and serial dilution of sourdoughs are generally the same for counting yeasts and lactic acid bacteria. The main difference is related to the "pour plate" technique applied for LAB instead of the "spread plate", which is currently applied for yeasts. A recent review showed that most studies involving sourdough combined microbiological and biochemical characterization, and 75% of research articles used culture-dependent methods for microbiological characterization, while 9% dealt with culture-independent techniques only (Arora et al. 2021).

The most commonly applied culture medium in the recovery and isolation of lactobacilli and other LAB from food matrices is MRS agar since it presents a complex nutritional composition (amino acids, peptides, fatty acids, nucleic acid derivatives, vitamins, salts, and fermentable carbohydrates) as required by this fastidious bacterial group, which is classified according to the carbon metabolism as homofermentative (lactic acid is the dominant product formed) and heterofermentative (acetate, carbon dioxide, and ethanol are formed as products) (Meruvu and Harsa 2022). However, several modifications in the MRS formula are applied. Therefore, variations of carbon sources present in foods and their assimilation by the sourdough microbiota must be taken into account (Comasio et al. 2020; Petkova et al. 2021; Baek et al. 2021). Furthermore, there is no universal medium that attends to the great diversity of LAB present in the sourdough microbiota, but the modified MRS (mMRS) medium may be rendered more selective by adding inhibitory compounds (mostly sodium acetate, sorbic acid, and antibiotics), increasing their concentration (different sugar contents), or by reducing the pH to values below 5.7 (Schillinger and Holzapfel 2003). To avoid yeast and mold growth (highly present in fermented food matrices), an antifungal is added to the final medium, usually cycloheximide, amphotericin, and nystatin at concentrations varying from 0.05 to 0.2 g/L in a pre-filtered solution.

The most commonly used media for LAB enumeration and isolation are: maltose (1–5%) and yeast extract (1–5%) added MRS agar (2–5%); Sabouraud dextrose broth (SDB) agar; MRS agar containing bromophenol blue (mMRS-BPB); M17 agar; M17-glucose (containing 0.5% w/v glucose instead of lactose); MRS5 agar (10 g of tryptone, 5 g of meat extract, 5 g of yeast extract, 10 g of maltose, 5 g of fructose, 5 g of glucose, 5 g of $C_2H_3NaO_2 \cdot 3H_2O$, 3 g of ammonium chloride, 2.6 g of $K_2HPO_4 \cdot 3H_2O$, 4 g of KH_2PO_4, 0.1 g of $MgSO_4 \cdot 7H_2O$, 0.05 g of $MnSO_4 \cdot 4H_2O$, 0.5 g of cysteine-HCl, 1 mL of Tween 80, and 1 mL of a vitamin mixture, pH 5.8); Deoxycholate-mannitol-morbitol (mDMS) agar; modified deoxycholate-mannitol-morbitol (mDMS) agar; MRS agar supplemented with 2% of sugars, sucrose (Su-MRS), glucose (Glu-MRS), or raffinose (Raf- MRS); mannitol-MRS agar; MRS-starch agar (in which glucose was replaced by starch at 0.5%) (Nuobariene et al. 2015; Ispirli et al. 2018; Çakır et al. 2020; Galli et al. 2020; Dentice Maidana et al. 2020; Fraberger et al. 2020b; Comasio et al. 2020; Petkova et al. 2021; Baek et al. 2021; Pino et al. 2022). Table 5.2 presents a non-exhaustive list of culture-dependent methods recently employed to isolate sourdough LAB, including information on the sourdough raw material, predominant species isolated, the culture medium, and a summary of the methods used to identify LAB species.

Table 5.2 Non-exhaustive list of microbiological methods used to isolate and identify bacteria in sourdoughs

Country	Sourdough composition	Isolated dominant species	Culture medium and growth conditions	Summary of biomolecular method used to identify at species level	Reference
Argentina	Chia	*Enterococcus casseliflavus, E. faecium, E. hirae, E. mundtii, E. devriesei, E. italicus, Lactococcus lactis, E. hermanniensis Lacticaseibacillus rhamnosus, and Weisella cibaria*	MRS-5 agar, MRS-starch (glucose was replaced by starch at 0.5%), and MRS-5.5 (pH adjusted to 5.5) + cycloheximide at 100 µg/mL. 30 °C for 48 h; anaerobiosis.	rep-PCR and taxonomic identification by sequence analysis of 16S ribosomal RNA gene.	Dentice Maidana et al. (2020)
Austria	Rye and wheat	*Levilactobacillus brevis, Latilactobacillus curvatus, Lacticaseibacillus paracasei, Lactiplantibacillus plantarum, Fructilactobacillus sanfranciscensis, Pediococcus pentosaceus*	SDM, MRS, manitol MRS, M17 and glucose M17. 30 °C for 72 h; anaerobiosis (10% CO2, 10% H_2, and 80% N_2).	MALDI-TOF MS	Fraberger et al. (2020b)
Bulgaria	Wheat, whole wheat, rye and chickpea	*Lactiplantibacillus plantarum, Pediococcus pentosaceus, Levilactobacillus brevis, Pediococcus acidilactici, Enterococcus faecium, Enterococcus durans and Pediococcus parvulus*	MRS, MRS-5, M17 and M17-glucose (containing 0.5% w/v glucose instead of lactose) + cycloheximide (0.1 g/L). 37 °C for 48 h; anerobiosis.	PCR amplification of the 16S rRNA gene. Primers: LacbF and LacbR.	Petkova et al. (2021)
China	Tartary buckwheat	*Weissella cibaria, Lactococcus taiwanensis; Pediococcus pentosaceus* and *Latilactobacillus curvatus*	MRS agar. 37 °C for 48 h; anaerobiosis.	Amplification of the V3-V4 region of bacterial 16S rRNA genes. Primers: 338-F and 806-R.	Zhou et al. (2022)

China	Whole wheat	*Lactobacillus johnsonii, Pediococcus pentosaceus, Levilactobacillus brevis, Lacticaseibacillus casei, Lacticaseibacillus paracasei, Pediococcus acidilactici, and Acetobacter pasteurianus*	MRS agar. 37 °C for 48–72 h.	Amplification of 16S rRNA gene. Primers: 1492-R and 27F.	Shen et al. (2022)
Iran	Oat Bran	*Levilactobacillus brevis, Latilactobacillus sakei, Enterococcus hirae, and Pediococcus pentosaceus*	MRS agar. 37 °C for 48 h; in the presence of 10% CO_2.	16S rRNA gene sequencing. Primers: F44 and R1543.	Masoomshahi et al. (2020)
Italy	Triticum vulgare Host. var. albidum Koern (Maiorca grain)	*Levilactobacillus brevis, Lactiplantibacillus plantarum, Lacticaseibacillus rhamnosus*	mMRS agar (maltose 1%, fresh yeast extract 5%). Fiinal pH = 5.6. Anaerobiosis; 30 ± 2 °C for 48 h.	PCR amplification of the 16S rRNA gene. Primers: 7-f and 1510-r.	Pino et al. (2022)
Italy	Chickpea flour	*Weissella confusa*	MRS agar and MRS agar +2% of sugars: sucrose (Su-MRS), glucose (Glu-MRS), and raffinose (Raf- MRS). 30 °C for 24–48 h, under micro-aerophilic conditions.	RAPD-PCR analysis using the following primers: P1 and P4; Amplification by Ribosomal DNA Restriction Analysis (ARDRA) and 16S rDNA gene. Primers: FD1 and RD1 .	Galli et al. (2020)
Japan	Rye and wheat	*Levilactobacillus brevis, Companilactobacillus alimentarius, Lactiplantibacillus pentosus, Paucilactobacillus vaccinostercus, Fructilactobacillus sanfranciscensis, and Latilactobacillus sakei*	mMRS agar + cycloheximide (10 mg/L). Final pH: 5.4. 30 °C for 48 h.	FAST MicroSEQ 500 16S rDNA bacterial polymerase chain reaction (PCR) kit.	Fujimoto et al. (2019)

(continued)

Table 5.2 (continued)

Country	Sourdough composition	Isolated dominant species	Culture medium and growth conditions	Summary of biomolecular method used to identify at species level	Reference
Korea	Whole wheat (with bran) sourdough, named *Nuruk* (traditional Korean fermentation starter)	*Le. mesenteroides, Levilactobacillus brevis, Latilactobacillus curvatus, Pediococcus pentosaceus and Fructilactobacillus sanfranciscensi*	mMRS (2% maltose instead of glucose), mMRS-BPB, and SDB agar + cycloheximide (0.2 g/L). 37 °C for 24 h; anaerobiosis.	Multiplex-RAPDPCR. Primers for OPL-05 + RD1 and P1 + RD1 combinations were used.	Baek et al. (2021)
Lithuania	Wheat rye	*Limosilactobacillus panis, Limosilactobacillus reuteri, Limosilactobacillus fermentum, and Pediococcus pentosaceus*	MRS + D-glucose (7 g/L), Fructose (7 g/L), Maltose (7 g/L). 30 °C for 72 h; anaerobic jar.	Amplification of the 16S rRNA gene. Primers: 16S-27F and 16S- 1492R.	Nuobariene et al. (2015)
Nigeria	Sorghum	*Lactiplantibacillus pentosus, Lactiplantibacillus plantarum, Companilactobacillus alimentarius,* and *Micrococcus varians*	MRS agar. 30 °C for 48 h; anaerobiosis.	–	Ewuoso et al. (2020)
Switzerland and Austria	Wheat/rye and brewer's spent grain.	*Lc. citreum, Lc. mesenteroides, Lc. pseudomesenteroides, Lc. palmae, Lac. paracasei subsp. paracasei, L. lactis, Llb. curvatus, F. fructivorans, Lo. coryniformis subsp. coryniformis, and W. confusa.*	MRS agar. 30 °C for 72 h.	MALDI-TOF MS	Müller et al. (2021)

Turkey	Rye and rye bran	*Leuconostoc citreum, Lactiplantibacillus plantarum, Latilactobacillus graminis, Lactococcus lactis subsp. cremoris, Weissella confusa, and Weissella cibaria*	MRS-5 agar and SDB agar + 0.1 g/L of cycloheximide. 30 °C for 48 h; anaerobiosis.	Amplification of the 16S rRNA gene. Primers: AMP_F and AMP_R	Ispirli et al. (2018)
Turkey	Hull-less barley	*Pediococcus, Latilactobacillus curvatus, Levilactobacillus brevis, Lactiplantibacillus plantarum, Limosilactobacillus fermentum, Companilactobacillus musae, Companilactobacillus paralimentarius, Leuconostoc mesenteroides and Limosilactobacillus equigenerosi*	MRS, mMRS (1% w/v maltose and 5% v/v fresh yeast extract), and RA agar + cycloheximide. Final pH = 5.6. 34 °C for 72 h; anaerobic jar.	FTIR and 16S rDNA gene sequencing	Çakır et al. (2020)

(continued)

Table 5.2 (continued)

Country	Sourdough composition	Isolated dominant species	Culture medium and growth conditions	Summary of biomolecular method used to identify at species level	Reference
United Kingdom, Belgium, France, USA	Rye, buckwheat, kamut, spelt, wheat and multigrain based	*Levilactobacillus brevis, Latilactobacillus curvatus, Levilactobacillus hammesii, Lacticaseibacillus paracasei subsp. paracasei, Lacticaseibacillus paracasei subsp. tolerans/paracasei, Companilactobacillus paralimentarius, Lactiplantibacillus plantarum, Fructilactobacillus sanfranciscensis, Lactiplantibacillus xiangfangensis, Leuconostoc citreum, Pediococcus parvulus, Weissella cibaria/confuse, Acetobacter fabarum, Acetobacter cerevisiae, Acetobacter pasteurianus subsp. pasteurianus, Acetobacter oryzifermentans, Acetobacter sicerae, Acetobacter senegalensis, Komagataeibacter xylinus, Komagataeibacter sucrofermentans*	mMRS-5 and mDMS + 0.1 g/L of cycloheximide and 0.005 g/L of amphotericin B. 30 °C for 72 h.	(GTG)5-PCRfingerprinting analysis of genomic DNA and 16S rRNA gene sequencing, using the primers pA and pH.	Comasio et al. (2020)

It includes the country of origin of the sourdough, as well as its composition and the dominant species isolated

[1]Some studies did not report the antifugal used. *mMRS* modified MRS, *MRS agar* Man, Rogosa and Sharp agar, *BPB* bromophenol blue, *SDB* Sabouraud dextrose broth, *mDMS* modified deoxycholate-mannitol-sorbitol

The mMRS-BPB selection medium is usually applied to isolate *Latilactobacillus curvatus* and *Levilactobacillus brevis* from sourdough. The detection of lactococci species is commonly carried out with M 17. SDB agar was set to isolate *F. sanfranciscensis*. Moreover, the replacement of glucose with starch in MRS effectively promoted the growth of Enterococcus and Lactobacillus strains from fermented food samples (Baek et al. 2021; Meruvu and Harsa 2022). When plated on Sucrose-MRS, the dominant microbial LAB population produced slimy colonies, suggesting the prospective production of dextran enabled by the sucrose addition, which indicates a potential exopolysaccharide producer strain, as reported by Galli et al. (2020). These authors have found a great EPS producer, identified as *Weissella confusa*.

Although most studies have focused on LAB-specific culture media, recent studies concerning molecular techniques have shown that acetic acid bacteria (AAB) also play an important role in the composition of sourdough microbiota (Li et al. 2021; Landis et al. 2021; Shen et al. 2022). AAB are gram-negative, strictly aerobic, and capable of producing acetic acid from sugars and alcohol compounds. In addition, they improve sensory and quality attributes of buckwheat bread mostly due to their high potential to form exopolysaccharides (Ua-arak et al. 2016; Ua-Arak et al. 2017; Anguluri et al. 2022). However few studies involving sourdoughs have used culture-based mediums targeting the isolation of AAB due to their fastidious nature in food matrices (Comasio et al. 2020). Moreover, the absence of an effective selective medium for the recovery and differentiation of these microorganisms may explain their misidentified and underrepresented status. All the restrictions above represent a barrier to investigating the dynamics and physiology of the AAB population, as well as the search for novel high-functional AAB starters.

The most common culture media used to study AAB in sourdough and bread matrices are: modified gluconate medium (20 g/L of sodium gluconate, 3 g/L of yeast extract, 2 g/L of peptone, 3 g/L of glycerol, 10 g/L of mannitol, 80 g/L of sucrose, and 20 g/L of agar, pH 6.0); glucose-yeast extract-calcium carbonate agar (10 g/L of glucose, 1 g/L of yeast extract, 2 g/L of $CaCO_3$, and 1.5 g/L of agar); Hestrin-Schramm (HS) (2 g/L of glucose, 0.5 g/L of yeast extract, 0.5 g/L of polypeptone, 0.73 g/L of Na_2HPO_4, 0.115 g/L of citric acid, and 0.05 g/L of $MgSO_4$) and modified HS agar containing sucrose (25 g/L); Sodium gluconate (NaG) medium (20 g/L of sodium gluconate, 3 g/L of yeast extract, 2 g/L of peptone from casein, 3 g/L of glycerol, 0.2 g/L of $MgSO_4.7H_2O$, and 10 g/L of mannitol, pH 6.0) (Jakob et al. 2012; Ua-Arak et al. 2017; Anguluri et al. 2022). None of these media are selective, being used mostly to maintain and preserve AAB.

To the best of our knowledge, only one selective culture method that was recently developed by Kim et al. (2019) is available for AAB isolation, detection, and enumeration from complex foods matrices. This medium was named acetic acid bacteria-selective agar (ABS) (D-(+) 50 g/L of glucose, 10 g/L of yeast extract, 20 mg/L of bromophenol blue, 20 g/L of bacteriological agar 20 g, 0.1% of glacial acetic acid, 5% of pure ethanol, and 5000 U penicillin, final pH = 4.8 ± 0.1). Culture incubation of AAB was determined at 30 °C for 72 h, under aerobiosis, using spread plate method. This media enabled the selection of AAB based on the following phenotype: green colonies with a yellow halo formed by typical AAB growth were regarded as positive. In addition to this formula, the selectivity of ABS could be further enhanced by the supplementation of 0.1–5 mg/L of cycloheximide to inhibit the growth of yeasts and molds, especially in samples with high levels of yeasts and molds (Liu et al. 2016; Kim et al. 2019; Comasio et al. 2020). This medium was proven to provide an accurate and reliable method for AAB detection and quantification in fermented food production. Therefore, it can facilitate the effective isolation of novel functional AAB strains from various sources in future studies involving sourdough.

In conclusion, although several molecular assays have been developed and opened new frontiers for the understanding of the microbial communities involved during sourdough production processes, the culture-based methods are still essential to identify and characterize the complex microbial populations in sourdough matrices for further industrial application. Thus, the best way to obtain a more accurate and reliable result on the composition of the complex microbiota present in sourdoughs is the simultaneous application of dependent and independent cultivation methods by using recent biomolecular techniques and multivariate statistical tools.

5.3 Characterization Methods Applied to Sourdough Strains

5.3.1 Characterization of Sourdough Yeasts

There are numerous methods to characterize the metabolic or physiological activity of yeasts (Yarrow 1998; Kurtzman et al. 2011). However, for practical application purposes in fermentation processes using sourdough, it is expected that the isolated yeast strains have at least a considerable fermentative activity and resistance to low pH to improve the leavening of baking goods and resist the acidification caused by lactic acid bacteria metabolism, respectively (Heitmann et al. 2018). Besides, other interesting studies have also characterized yeast antioxidant and antimicrobial activities (Palla et al. 2020a; Arora et al. 2021).

5.3.1.1 Leavening Ability

The leaving ability is perhaps the most frequently reported characteristic in studies concerning sourdough yeasts. In yeasts, it is mainly evaluated in kinetic terms, i.e., mass expansion/leavening speed, CO_2 production, and substrate consumption, among others. The dough expansion rate is a qualitative rheological parameter related to the metabolic activity of strains and microbial growth and can be easily measured in laboratory or industrial environments. The specific volume of the expansion is calculated by the ratio of the dough expansion to the apparent volume, using automated equipment such as a rheofermentometer, or by manually monitoring the dough expansion (Palla et al. 2019; Korcari et al. 2021). However, the comparison of the results of this analysis must take into account, in addition to cell density, all other technological parameters of dough preparation, including flour composition, flour:sourdough ratio, humidity, and temperature, among others. Moreover, the sourdough leavening activity is also affected by other external factors such as acidity and fluctuations in temperature. The metabolic rate of cells decreases under cold conditions, reducing the production of CO_2 and, consequently, the air bubbles formation in the dough (Häggman and Salovaara 2008; Guerzoni et al. 2013). There seems to be no exception in stating that the yeasts with the highest fermentative and leavening activity belong to the species *Saccharomyces cerevisiae* (Palla et al. 2020a), which is also the most studied species.

5.3.1.2 Tolerance to High Acidity Levels

The selection of acid-tolerant yeast strains can also be interesting, especially for processes involving very acidic sourdoughs. This high acidity, which results from the production of lactic and acetic acids by LAB, can reduce the activity of yeasts, mainly through interactions of undissociated acids. These acids diffuse into cells and interfere with the leavening ability of yeasts (Guerzoni et al. 2013; Ogunsakin et al. 2017).

The ability of yeasts to tolerate acidic environments can be easily verified using culture media, usually YNB broth, with a pH standardized between 2.5 and 5, using hydrochloric acid. Microbial growth can be evaluated after 24 h at 30 °C using a spectrophotometer at 600 nm or by plating sample dilutions in a nutrient medium, such as YPD (Vrancken et al. 2010). Simultaneously, the tolerance of yeast strains to acetic acid can be verified using 1% glacial acetic acid in YNB broth (Vrancken et al. 2010). These simple assays can avoid the selection of very acid-sensitive strains and future incompatibility problems with the sourdough process.

5.3.1.3 Antioxidant Activity

Another parameter that has been evaluated in some studies is the antioxidant activity that some yeasts have shown in sourdough-based goods (Heitmann et al. 2018). In fact, some yeasts have already been shown to be capable of producing antioxidant substances, most of them belonging to the species *Saccharomyces cerevisiae* (Câmara et al. 2019). Although the antioxidant activity is also related to the sourdough composition, e.g. rye wheat has a higher content of antioxidants, the yeast strain selection directly influences it (Palla et al. 2020a).

The antioxidant activity can be measured by determining the free radical scavenger ability and the reaction with the DPPH (2,2-diphenyl-1-picrylhydrazyl) radical in a liquid medium (Coda et al. 2013). After yeasts inoculation in the broth, the antioxidant activity of the strains is assessed after 24 h of growth at 30 °C. The supernatant is recovered and the calculations are performed against a blank using a spectrophotometer at 517 nm (Palla et al. 2020a).

5.3.1.4 Antimicrobial Activity

Fermentative processes may require the prevalence of certain species of microorganisms over others or the inhibition of spoilage, while the use of antibiotics or chemical substances should be avoided for food applications. Thus, selecting strains with natural antimicrobial activities over other species is an advantage and a technological trend. For example, secondary metabolites produced by yeast, e.g. ethyl acetate, have antifungal activity (Coda et al. 2013; Arora et al. 2021).

The antimicrobial activity of a microorganism is generally determined by measuring the inhibition halo of one strain over another. Coda et al. (2013) used this method to evaluate the antifungal activity of yeasts belonging to the genera *Candida, Metschnikovia, Debaryomyces, Wickerhamomyces, Hanseniaspora, Pichia, Aureobasidium*, and *Kazachstania*, against the fungus *Penicillium roqueforti*, where the strain *Meyerozyma guilliermondii* LCF1353 showed the best inhibitory activity. *Wickerhamomyces anomalus* has already been evaluated for its antifungal potential (Coda et al. 2011; Mo and Sung 2014). These authors observed that it is possible to extend the bake-goods shelf-life by prioritizing or mixing these species during product formulation (Heitmann et al. 2018). In addition, the antimicrobial activity can also be evaluated and quantified in extracts from the yeast suspension or by inhibiting the germination of fungal spores.

5.3.1.5 Phytase Activity

Phytate is an anti-nutritional compound that interacts with the absorption of certain nutrients. Therefore, sourdough processes have been highlighted due to the phytase activity of certain bacteria and yeasts. The simplest way to measure this activity is using a screening-specific phytase activity medium, PSM (Palla et al. 2020a), where the halo around the colonies inoculated on the agar surface is proportional to the enzymatic activity of the strain. However, for phytase activity quantification purposes, the enzyme activity in intracellular and extracellular extracts can be determined, for example, as described by Nuobariene et al. (2012).

5.3.2 *Characterization of Sourdough Bacteria*

Most of the nutritional, sensory, and functional benefits found in sourdoughs are directly related to the metabolism of the complex diversity of LAB present during cereal-based fermentations. Several studies have reported the current methods applied to assess the potential of LAB in the development of single/mixed synthetic starter cultures designed for industrial commercialization (Bartkiene et al. 2022; Meruvu and Harsa 2022; Olojede et al. 2022; Shen et al. 2022). Among these traits, the following stand out: the acidification potential, adaptation to strict conditions, ability to inhibit the growth of fungi and spoilage bacteria, production of important enzymes for flavor and textural enhancement, and some functional properties aimed at the well-being of consumers. The objective of this session is to make a brief review of the most applied methods in recent studies and to summarize their importance for the development of healthier and more sustainable foods.

5.3.2.1 Antimicrobial Activity

Sourdough production usually involves intense raw materials handling and contamination with some pathogens (*Salmonella enterica*, *Staphylococcus aureus*, and *Listeria*) and spoilage strains (*Bacillus* sp. and fungi species). *Aspergillus*, *Fusarium*, and *Penicillium* sp. are recognized as major genera present in bakery products (Demirbas et al. 2017). These microorganisms can be detected throughout the whole production chain of cereal-based products. In addition, they present risks to the consumers' health and great economic losses due to the rejection of contaminated products (Petkova et al. 2021). The main source of mold contamination in the final product (after baking) is the air (containing about 10^3 fungal spores/m^3) (Luz et al. 2019). Moreover, fungi can produce mycotoxins and allergens with carcinogenic potential, representing a serious public health issue (Hassan et al. 2016).

The most commonly applied method for antibacterial and antifungal determination in LAB are the agar-spot- and well-diffusion-based methods. Antimicrobial assessment can be carried out in petri dish plates containing the indicator microorganism by the direct spot (spot on the lawn) of grown LAB cultures or their cell-free supernatant (CFS) (agar spot-method), or by doing wells (5–10 mm diameter) in the agar plate (well diffusion method) (Bartkiene et al. 2020; Fraberger et al. 2020a; Dentice Maidana et al. 2020; Petkova et al. 2021). Pure LAB culture can be applied after incubation under ideal conditions of time and temperature in a specific medium (generally MRS broth), while CFS is usually treated with enzymes (catalase, trypsin, proteinase K, etc.) and/or bases (mostly NaOH), subjected (or not) to high temperatures (> 100 °C). These treatments indicate the nature of the compound produced, such as organic acids (mostly lactic, acetic, and phenyllactic acids), bacteriocins or bacteriocin-like substances (BLIS), bioactive peptides, and hydrogen peroxide, among others (Arsoy et al. 2022; Bartkiene et al. 2022). Results are presented as the diameter (in mm) of halos (clear zones), indicating inhibition. However, this technique has some disadvantages. For example, the use of selective synthetic media containing glucose can stimulate the synthesis of antimicrobial metabolites, which does not represent a real process scenario (Mota-Gutierrez et al. 2021). Thus, some studies have developed an artificial culture media from the sourdough matrix, as liquid sterile chia flour extract (SCFE), and sterile flour extract (SFE) from chickpea or quinoa flour (Ruiz Rodríguez et al. 2016a; Dentice Maidana et al. 2020). For SCFE preparation, chia seeds powder is diluted in sterile tap water (1:3), and the obtained dough is left 10 min at room temperature. Then, 10 g of dough plus 190 mL of deionized water are mixed, and pH adjusted to 12 with 0.1 N NaOH. After this, the mixture is shaken at room temperature for 1 h and the slurry is centrifuged (10,000 g, 4 °C) and the supernatant is filtered using a cheese cloth (Dentice Maidana et al. 2020). For SFE broth preparation, chickpea flour is suspended at 40% (w/v) in distilled H_2O, stirred at 4 °C during 2 h and then sterilized at 121 °C for 20 min. Then, the flour from the suspension is removed by centrifugation (8000 rpm, 10 min at 4 °C) and the super- natant is used as culture medium (Sáez et al. 2018).

Studies concerning the application of sourdough autochthonous starter cultures in bread have demonstrated an increase in shelf life (up to 3–16 days), as well as technological and functional properties (such as crumb porosity and antioxidant capacity), revealing a natural alternative to synthetic preservatives (Luz et al. 2019; Hajinia et al. 2021; Müller et al. 2022). Another study indicates the anti-aflatoxigenic potential, mostly against aflatoxin B_1, of *Levilactobacillus brevis* (isolated from rice bran sourdough) (Sadeghi et al. 2019). Anti-bacilli activity is also an important feature attributed to sourdough LAB strains, as some *Bacillus* sp. (mostly *B. subtilis* and *B. licheniformis*) can cause the so-called "potato disease" spoilage in baked products due to the ropy phenotype, leading to stringy strands of mucilage, stickiness, and unpleasant flavor and taste (Fraberger et al. 2020a; Petkova et al. 2021).

5.3.2.2 Enzymatic Activity

The proteolytic system of LAB plays a key role in the catabolism of proteins and oligopeptides, releasing free amino acids and bioactive peptides. This feature enhances the flavor, antimicrobial, and general quality of sourdough (as organoleptic, digestibility, and nutritional value), including the extension of shelf life (Sáez et al. 2018; Reale et al. 2021). In addition, the proteolytic activity of LAB is important not only for their development in a cereal matrix but also for the potential to decrease gluten and other anti-nutritional factors, which also contribute to improve human health (Petkova et al. 2021).

The most used methods for evaluating the proteolytic activity of LAB are performed in protein-rich media by the spot of previously prepared LAB cultures (20–100 μL) onto Petri dish agar plates or in wells (5–10 mm size). After inoculation, plates are incubated under appropriate conditions, and the diameter of the halos is measured. The most commonly used media are: skim milk agar (casein 0.5%, yeast extract 0.25%, dextrose 0.1%, skimmed milk powder 2.5%, and agar 1.5%); gelatin or casein agar (sucrose 2%, yeast extract 0.5%, peptone 2%, agar 1.5%, and gelatin or casein 1%) (Mota-Gutierrez et al. 2021; Petkova et al. 2021). Another widely applied method is spectrophotometric using the o-phthaldialdehyde test (OPA) procedure, which measures the increase in absorbance at 340_{nm} relative to uninoculated wells (negative control). In this case, results are expressed as released α-amino acids, in mg/mL or mmoles/L of FAA (Ruiz Rodríguez et al. 2016b; Sáez et al. 2018; Dentice Maidana et al. 2020).

The amylolytic activity of LAB strains is also a good start attribute since it guarantees their survival under cereal-based sourdough fermentation by starch hydrolysis. LAB amylolytic activity has been reported to reduce bread staling by reducing starch retrogradation. α-Amylase activity is involved in starch degradation by catalyzing the hydrolysis of α-1,4-glycosidic linkages in starch (Zhang et al. 2019). The measurement of amylase production is also performed by qualitative agar-diffusion assays within solid media containing starch as the major carbon source. However, after inoculation and incubation, the plates are treated with an iodine solution to form a blue-colored starch-iodine complex that enables the appearance of a clear zone around the colonies (Sáez et al. 2018; Mota-Gutierrez et al. 2021; Petkova et al. 2021). Starch agar (10 g/L of wheat starch, 5 g/L of peptone, 5 g/L of yeast extract, 0.25 g/L of $MgSO_4 \cdot 7H_2O$, 0.01 g/L of $FeSO_4 \cdot 7H_2O$, and 15 g/L of agar, pH 6.8 ± 0.2) and MRS agar plates containing 1% starch instead of glucose are the most applied medium.

5.3.2.3 Exopolysaccharide (EPS) Formation

EPSs are biopolymers formed from repeating units of mono- or oligosaccharides. Depending on their composition, they can be classified as homo- or heteropolysaccharides. The most frequent EPS in sourdoughs are the HoPS: fructans and glucans, composed of either glucose or fructose units, respectively. They are synthesized from sucrose by the extracellular enzymes glucansucrases and fructansucrase by LAB of the genera *Weissella*, *Leuconostoc*, *Streptococcus*, *Pediococcus*, and lactobacilli species (Galli et al. 2020; Xu et al. 2020). These compounds improve the sensorial, texture, and staling properties of baked products, being recently employed as alternatives to boost gluten-free formulations (Dentice Maidana et al. 2020). The ability of *Weissella* strains to produce EPS has been investigated in different flours such as pseudocereal and legumes (Galli et al. 2020).

The most common method for evaluating EPS formation is by streaking pure cultures onto mMRS agar plates containing different sugars (glucose, fructose, sucrose, maltose, lactose, and starch as the only carbon sources at 1–15% w:v) and incubating them under appropriate conditions. LAB is recorded as positive-EPS by the presence of a ropy phenotype after touching the colony with a loop (Galle and Arendt 2014; Dentice Maidana et al. 2020; Mota-Gutierrez et al. 2021). According to sugar composition, EPS can be divided into homopolysaccharides – HoEPS and heteropolysaccharides – HeEPS. HoEPS is composed of only one type of monosaccharide, whereas HeEPS is composed of two to eight repeated types of monosaccharides (Sanalibaba and Cakmak 2016).

5.3.2.4 Acidification

The ability of LAB to acidify food matrices is, without a doubt, their utmost important characteristic, since it is directly related to dough's safety and organoleptic and nutritional attributes. According to Petkova et al. (2021), quality sourdough is achieved when the pH values range from 3.5 to 4.1. Low pH values can increase the solubilization of proteins due to the positive net charge and can also affect the viscoelastic behavior of the dough and the activity of cereal and bacterial enzymes (Shen et al. 2022). When the pH reduces, a serial phenomenon occurs in the dough, causing starch hydrolysis, gluten degradation, and the production of doughs with greater extensibility and internal humidity. It also favors yeast growth, increasing the volatile compound concentration, CO_2 production, and shelf-life (Siepmann et al. 2022).

The acid-producing capacity of LAB isolates can be assessed by pH measurements (using a ph meter) by spectrophotometric methods. In the latest method, grown LAB cultures are harvested, washed, and resuspended in a physiological solution (PBS or sterile saline at 0.85%) to eliminate carryover nutrients. Then, optical density (OD_{560nm} -OD_{600nm}) is adjusted (mostly at 0.5) to obtain an approximate level of 10^8 CFU/mL in the suspension. Subsequently, a liquid medium (MRS broth or synthetic flour extract broth) is inoculated with 2% (v:v) of LAB suspension and incubated at appropriate conditions. Samples for pH measurements are withdrawn over time (0, 4, 8, 24, 48, and 72 h) to build a kinetic acidification curve (Ruiz et al. 2016; Dentice Maidana et al. 2020; Petkova et al. 2021).

High throughput screening methods can also be applied. As described by Margalho et al. (2021), a large set of LAB isolated from cheeses were submitted to acidification tests in 96-well microplates containing milk added with 1 μM of -5-(6)-carboxyfluorescein. The acidification rates (drop in pH per hour [ΔpH/h]) were measured based on the decreasing fluorescence intensity of the pH-dependent indicator carboxy-fluoresceine. This methodology can be adapted to sourdough LAB strains by replacing the culture medium. Instead of milk, a flour-based broth can be used to obtain more realistic conditions. It shows a time and cost-effective alternative and can also be applied to other trait measurements, as online measurement such as ICINAC or Hydroplates.

5.3.2.5 Phytase Activity

Cereals and cereal-based products present high contents of vitamins and minerals. However, in most cases, they have reduced bioavailability due to the formation of a complex bond with phytic acid to produce phytates, which are anti-nutritional compounds. Thus, phytic acid degradation can increase the nutritional and functional traits of these products. (Çakır et al. 2020). Phytase is endogenously found in cereals, vegetables, and microorganisms, including lactic acid bacteria (Mohammadi-Kouchesfahani et al. 2019).

Phytase activity can be qualitatively evaluated by streaking grown LAB cultures onto Pikovskaya's medium containing calcium phytate or sodium phytate [D-glucose (20 g/L), sodium phytate (4 g/L), calcium chloride (2 g/L), ammonium nitrate (5 g/L), potassium chloride (0.5 g/L), magnesium sulfate (0.5 g/L), iron sulfate (0.01 g/L), manganese sulfate (0.01 g/L), and agar (15 g/L)] to characterize phytase positive species. After incubation in appropriate conditions, the colonies with a transparent zone around were accepted as phytase positive. Then, they were washed to determine if these zones were sourced by phytase or acid production (Karaman et al. 2018). Alternatively, modified Chalmers broth with 1% neutral red and 1% calcium phytate can be used, but Chalmer agar plates are flooded twice with 2% (w/v) aqueous cobalt chloride solution for 20 min before measurement of halos to eliminate false positive results caused by microbial acid production (Syrokou et al. 2021).

To determine the extracellular phytase activity of samples, grown LAB supernatants are submitted to enzymatic tests by spectrophotometric assays following standard methods (Rizzello et al. 2016; Karaman et al. 2018; Mohammadi-Kouchesfahani et al. 2019).

5.3.2.6 Probiotic Properties

Probiotics are living microorganisms that exert health benefits to the host when administered in proper doses. Beneficial effects of probiotics include the improvement of the immune system, anticancer activity, the reduction of blood cholesterol and pressure, the modification of lactose intolerance and diarrhea prevention, as well as the modulation of human intestinal microbiota (Sadeghi et al. 2019). One of the most important criteria to evaluate LAB probiotic potential is related to their resistance to gastric and pancreatic digestion and their ability to survive during the passage through the gastrointestinal tract (GIT) (Fekri et al. 2020). In addition, probiotics, which pass into the colon while maintaining their viability in the small intestine, must be able to attach to the epithelial and mucosal surfaces to colonize the intestinal surfaces (Doğan and Tekiner 2019).

Once pH values in the stomach range from 1–2 (during fasting) up to 4–5 (after meal) (Bao et al. 2010), most of the methods tested for acid resistance involve monitoring the growth of LAB suspensions in the simulated stomach, duodenal, and intestinal juices by plate count or optical density measurement after proper incubation condition (mostly at 37 °C for 3-6 h) (Sadeghi et al. 2019). Hafeez et al. (2014) and Adisa and Ifesan (2016) used mMRS broth containing 0.1–5% (standard = 0.3%) bile salts and/or pH adjusted to 1.5–3.0. Data values are compared with controls.

Adherence properties of LAB are usually assessed by measurements of cell surface hydrophobicity (H%), auto-aggregation (A%), and/or co-aggregation (CoA%) since these traits are correlated with the ability of strains to adhere to intestinal epithelial cells surfaces and the prevention of pathogen colonization. All methods are performed by spectrophotometry methods at 600 nm (Park et al. 2019; Sadeghi et al. 2019). Adhesion of LAB to human colon carcinoma cell-line Caco-2 is also applied; however, it requires special attention during the cultivation and maintenance of eukaryotic cells, which are very fastidious and susceptible to contamination (Galli et al. 2018; Park et al. 2019). Other characteristics, such as antimicrobial activity, antibiogram profile, hemolytic activity, and presence of virulent genes, are also detected before ensuring their safe application in food (Meruvu and Harsa 2022).

5.4 Biomolecular Methods and Approaches Used for Identification

5.4.1 Culture-Dependent Techniques

5.4.1.1 Random Amplified Polymorphic DNA (RAPD)

A PCR-based technique frequently used to characterize the DNA fingerprint of sourdough strains is RAPD. This technique is also applied as a pre-selection step for unknown strains and it is based on the use of small oligonucleotides to provide a genetic overview of the samples in a simple and fast way. Short-chain primers of up to 15 bp are used and some reported examples are: M13, RF2, OPL5, P1, and P2 (Paffetti et al. 1995; Andrighetto et al. 2001; Gutiérrez et al. 2011). The profile obtained with the hybridization of the primers can be revealed on an agarose gel. In general, at least two primers are used separately to obtain an accurate picture of the DNA fingerprint. Images can be combined into a composite data set using bioinformatics software and the similarities between the bands revealed on the agarose gel can be evaluated using Pearson or Dice coefficients and the unweighted pair group method with arithmetic averages (UPGMA) algorithm. The fingerprint can be also compared with the DNA of reference strains or serve as a basis clustering of representative candidates for DNA sequencing. For example, this method has already been used to detect intraspecific polymorphism among *S. cerevisiae* sourdough strains from Italy (Valmorri et al. 2010) and China (Yang et al. 2020), and LAB in Turkey (Boyaci Gunduz et al. 2022).

5.4.1.2 Amplified Ribosomal DNA Restriction Analysis (ARDRA)

The ARDRA is a technique used for the identification of LAB found in sourdoughs by fingerprint representation. This technique allows the exploration of the sequence variation present in amplicon regions, such as the 16S rRNA gene, by enzymatic reactions and PCR amplifications using specific primers (Yadav et al. 2019). Under this scope, each species has a specific profile. Therefore, a genetic similarity clustering can be generated to optimize the identification of all isolates from sourdough samples. In the alternative, double digestion with restriction enzymes, like *AluI* and *MspI* simultaneously, allows the discrimination of LAB species such as

Levilactobacillus brevis, *Lactiplantibacillus plantarum*, *Lacticaseibacillus casei*, *Furfurilactobacillus rossiae*, *Fructilactobacillus sanfranciscensis*, etc. In addition, facultatively heterofermentative Pediococci as *Pediococcus acidilactici* and *P. pentosaceus* can be differentiated (Venturi et al. 2012; Palla et al. 2017; Pino et al. 2022). Regarding ARDRA advantages, it is rapid and easy to use and interpret. Moreover, it presents high reproducibility and low or moderate costs. However, it is less discriminatory than RAPD and PFGE (El Sheikha and Hu 2020).

5.4.1.3 Amplified Fragment Length Polymorphism (AFLP)

The APLP technique occurs through the digestion of the DNA on restriction enzymes. The product of this digestion is used in a PCR-based technique that applies selective amplification of a subset of digested DNA fragments to generate and compare unique fingerprints for genomes of interest. This method does not require prior information regarding the targeted genome. Moreover, it has high reproducibility and sensitivity for detecting polymorphism at the level of DNA sequence (Paun and Schönswetter 2012). The APLP technique has been useful to support the description of *Lactobacillus acidifarinae* sp. and *Lactobacillus zymae* sp., as well as *Lactobacillus clustorum*, *Lactobacillus hammesii*, *Lactobacillus hammesii, Lactobacillus crustorum*, *Lactobacillus namurensis, Lactobacillus spicheri, Lactiplantibacillus plantarum,* and *Fructilactobacillus sanfranciscensis* (Scheirlinck et al. 2008; Valcheva et al. 2005), which were isolated from two traditional Belgian wheat sourdoughs (Ilse Scheirlinck et al. 2007; Valcheva et al. 2006; Vancanneyt et al. 2005). In addition, fluorescent amplified fragment length polymorphisms (FAFLP) can be used in studies in which the species can be the marker, such as *Levilactobacillus brevis* and *Weissella confuse*, from the design of the oligonucleotide primers for a species-specific PCR assay (Fusco et al. 2011; Fusco et al. 2016).

5.4.1.4 Restriction Fragment Length Polymorphism (RFLP)

Following the same target of the previous technique, RFLP is suitable to examine the general and sourdough microbiome (bacteria and yeasts) and to select unique strains before sequence analysis using the restriction enzymes (Slezak and Salzberg 2005; Valmorri et al. 2010; Palla et al. 2017). Bacteria species such as *Lactobacillus pentosus*, *Lacticaseibacillus casei*, *Lactobacillus kimchi*, *Lactobacillus alimentarius*, *Lactobacillus sanfranciscensis*, and *Lactiplantibacillus plantarum* were identified by this technique (Randazzo et al. 2005), as well as *Saccharomyces cerevisiae*, *Kazachstania humilis*, *Candida milleri, Candida krusei,* and *Torulaspora delbrueckii* (Valmorri et al. 2010; Palla et al. 2019). This method has advantages such as safety, efficiency, convenience, and good reproducibility.

5.4.1.5 Pulse Field Gel Electrophoresis (PFGE)

The most significant difference between PFGE and the previously described methods is the additional step of separating large DNA molecules (entire genomic DNA) in a gel matrix under an electric field that periodically changes direction. This step is performed after applying restriction enzymes and permits the analysis of DNA fragments in a superior order of magnitude compared to the conventional ones. PFGE provides a good representation of the entire bacterial chromosome in a single gel and with a highly reproducible restriction profile, providing distinct and well-resolved DNA fragments (Sharma-Kuinkel et al. 2016).

PFGE analysis is extensively used in studies involving sourdough, specifically for dominant bacteria species found in sourdoughs from a specific origin. *Lactiplantibacillus plantarum, Weissella confusa, Fructilactobacillus sanfranciscensis*, *Fructilactobacillus rossiae*, *Levilactobacillus brevis*, *Lactococcus lactis* subsp. *lactis*, *Pediococcus pentosaceus*, and *Lactobacillus* sp. were already identified by this method (Pepe et al. 2004; Sonya et al. 2009).

5.4.1.6 Repetitive Element Sequence-Based PCR (rep-PCR)

Rep-PCR is a typing method that differentiates microbes by using primers such as $(GTG)_5$ complementary to interspersed repetitive consensus sequences that enable amplification of diverse-sized DNA fragments consisting of sequences between the repetitive elements (Patrizia and Paola 2003). Clustering analysis can be generated following a similarity threshold of more than 50%. However, rep-PCR should be integrated with other techniques, such as 16S rRNA gene by ARDRA or RAPD, to properly select strains for sequencing and to avoid any misidentifications (Patrizia and Paola 2003).

When Rep-PCR is used together with 16S rRNA region, the diversity among bacterial isolates from sourdough can be demonstrated. Thus, heterogeneity in the band patterns revealed by rep-PCR must be subjected to 16S rRNA and the genetic relationship among isolated can be performed by clusters analysis (Dentice Maidana et al. 2020).

Moreover, rep-PCR is a useful method to study the performance of a starter culture directly in sourdoughs without cultivation on selective media. It was demonstrated by Dolci and Cocolin (2017) with *Lactiplantibacillus plantarum.* The separation can be achieved if other lactobacilli are present. However, the extraction of good quality microbial DNA suitable for amplification is required. *Fructilactobacillus sanfranciscensis*, *Kazachstania humilis*, and *Saccharomyces cerevisiae* were also identified by rep-PCR. The rep-PCR with $(GTG)_5$ primer method showed a high discriminatory index (up to 0.95), which was higher than those obtained with RAPD-PCR in combination with P4, P7, and M13 primers, considering the *Kazachstania humilis* at the strain level (Palla et al. 2020b).

For yeasts, the rep-PCR are designed using the ITS amplicons size and the restriction profiles obtained with different restriction enzymes allowed the identification at strain levels. Species such as *Wickerhamomyces anomalus*, *Saccharomyces cerevisiae*, *Torulaspora delbrueckii*, *Candida boidinii*, and *Candida ethanolica* were identified by this method (Pino et al. 2022). Thus, concerning the yeast community, the rep-PCR with $(GTG)_5$ primer has been a suitable method to discriminate biotypes in the yeast sourdough population.

5.4.1.7 Multilocus Sequence Typing (MLST)

Based on the housekeeping genes (loci) approach, MLST generates internal fragments of each loci used, as these can be accurately sequenced on both strands using an automated DNA sequencer (Urwin and Maiden 2003). Gene fragments of *gdh*, *gyrA*, *mapA*, *nox*, and *pta* are good candidates to be used in the MLST technique, mainly to identify LAB species in sourdoughs (Lhomme et al. 2016a; Yang et al. 2017). The advances of this technique are the differentiation of highly related genotypes and excellent discrimination. However, specialized equipment and trained personnel are required. The cost of the reagents can be highlighted as another disadvantage.

5.4.2 Culture-Independent Techniques

5.4.2.1 Denaturing Gradient Gel Electrophoresis (DGGE) and Temperature Gradient Gel Electrophoresis (TTGE)

These techniques allow rapid estimation of diversity, with simultaneous analysis of multiple samples using low-cost equipment. Besides, they represent the evolution and dynamics of microbial populations in fermentation and spoilage processes. On the other hand, these methods present similar migration characteristics of heterologous sequences and overestimation of microbial diversity because of heteroduplexes (Mantzourani et al. 2019; Baek et al. 2021).

DGGE has been used to separate mixtures of DNA fragments by a DNA strand separation at a given position in a gradient of chemical denaturant according to their melting point and to analyze microbial communities without cultivation. The principle of TGGE is based on temperature gradients that change the structure of DNA/RNA nucleic acids and proteins. Through electrophoresis, it is possible to visualize the separation on the agarose gel (Muyzer 1999; Levin et al. 2018). Moreover, even with some difficulties, this technique continues to be used (Mantzourani et al. 2019; Baek et al. 2021; Reale et al. 2019).

5.4.2.2 Real-Time PCR

In real-time PCR, DNA extracted from samples is quantified at each cycle using fluorescent dyes. The fluorescent signal detected is proportional to the number of amplicons generated. The fluorescent probes used can bind directly to the double-stranded DNA or can be associated with primers that hybridize to the PCR products in each amplification cycle (Wong and Medrano 2005). Real-time PCR enables some results that PCR cannot provide, such as calculation of DNA template concentration, including assessment of DNA or RNA copy number, allelic discrimination, and SNP detection (Scheirlinck et al. 2009; Michel et al. 2016; Palla et al. 2020b).

Several studies have used this technique to identify or quantify DNA or RNA copy numbers during sourdough fermentation (Scheirlinck et al. 2009; Michel et al. 2016; Palla et al. 2020b; Fraberger et al. 2020b; Bazalová et al. 2022).

Some characteristics of this method are high throughput, the requirement of a clone library for identification, robustness, reproducibility, efficiency, highly stable DNA biomarkers, high sensibility, and possible use with degraded samples. In this technique, the sequences must be known for enzyme selection and the same length of the fragment can be used for many species. In addition, it requires high expertise in molecular biology and expensive equipment.

5.5 Conclusion

The classical methods usually applied for the isolation and characterization of sourdough strains are the culture-dependent ones. They allow the selection of cultivable strains of technological interest and are of fundamental importance for fermentation processes involving sourdough. The characterization methods bring relevant physiological information about these microorganisms, such as resistance to high rates of acidity or fermentative and leavening activity, contributing to the quality of sourdough-based goods. Biomolecular methods, especially culture-independent have broadened our perspective on the composition of the sourdough microbiome. Thus, new culture media must be produced to allow their proper isolation. Moreover, the use of one or more biomolecular methods in combination to differentiate and identify the isolates obtained at the species level should be taken into account.

References

Adisa AM, Ifesan BO (2016) Probiotic potential of lactic acid bacteria (LAB) isolated from wholegrain millet sourdoughs. Ann Food Sci Technol 17:459–463

Andrighetto C, Psomas E, Tzanetakis N et al (2001) Randomly amplified polymorphic DNA (RAPD) PCR for the identification of yeasts isolated from dairy products. Lett Appl Microbiol 30:5–9. https://doi.org/10.1046/j.1472-765x.2000.00589.x

Anguluri K, Cassanelli S, Gullo M et al (2022) Candidate acetic acid bacteria strains for levan production. *Polymers* 14(10):2000

Aplevicz KS, Mazo JZ, Ilha EC et al (2014) Isolation and characterization of lactic acid bacteria and yeasts from the Brazilian grape sourdough. Brazilian J Pharm Sci 50:321–327. https://doi.org/10.1590/S1984-82502014000200011

Arena MP, Russo P, Spano G, Capozzi V (2019) Exploration of the microbial biodiversity associated with north Apulian sourdoughs and the effect of the increasing number of inoculated lactic acid bacteria strains on the biocontrol against fungal spoilage. Fermentation 5:97. https://doi.org/10.3390/fermentation5040097

Arici M, Ozulku G, Yildirim RM et al (2018) Biodiversity and technological properties of yeasts from Turkish sourdough. Food Sci Biotechnol 27:499–508. https://doi.org/10.1007/s10068-017-0282-0

Arora K, Ameur H, Polo A et al (2021) Thirty years of knowledge on sourdough fermentation: a systematic review. Trends Food Sci Technol 108:71–83. https://doi.org/10.1016/j.tifs.2020.12.008

Arsoy ES, Gül LB, Çon AH (2022) Characterization and selection of potential antifungal lactic acid bacteria isolated from Turkish spontaneous sourdough. Curr Microbiol 79:1–11. https://doi.org/10.1007/s00284-022-02839-z

Baek H-W, Bae J-H, Lee Y-G et al (2021) Dynamic interactions of lactic acid bacteria in Korean sourdough during back-slopping process. J Appl Microbiol 131:2325–2335. https://doi.org/10.1111/jam.15097

Bao Y, Zhang Y, Zhang Y et al (2010) Screening of potential probiotic properties of *Lactobacillus fermentum* isolated from traditional dairy products. Food Control 21:695–701. https://doi.org/10.1016/j.foodcont.2009.10.010

Bartkiene E, Lele V, Ruzauskas M et al (2020) Lactic acid bacteria isolation from spontaneous sourdough and their characterization including antimicrobial and antifungal properties evaluation. Microorganisms 8(1):64. https://doi.org/10.3390/microorganisms8010064

Bartkiene E, Özogul F, Rocha JM (2022) Bread sourdough lactic acid bacteria—technological, antimicrobial, toxin-degrading, immune system-, and faecal microbiota-modelling biological agents for the preparation of food, nutraceuticals and feed. Foods 11:1–20

Bazalová O, Cihlář JZ, Dlouhá Z et al (2022) Rapid sourdough yeast identification using panfungal PCR combined with high resolution melting analysis. J Microbiol Methods 199:106522. https://doi.org/10.1016/j.mimet.2022.106522

Boyaci-Gunduz CP, Erten H (2020) Predominant yeasts in the sourdoughs collected from some parts of Turkey. Yeast 37:449–466. https://doi.org/10.1002/yea.3500

Boyaci Gunduz CP, Agirman B, Gaglio R et al (2022) Evaluation of the variations in chemical and microbiological properties of the sourdoughs produced with selected lactic acid bacteria strains during fermentation. Food Chem X 14:100357. https://doi.org/10.1016/j.fochx.2022.100357

Cakir E, Arici M, Durak MZ (2021) Effect of starter culture sourdough prepared with Lactobacilli and Saccharomyces cerevisiae on the quality of hull-less barley-wheat bread. LWT 152:112230. https://doi.org/10.1016/j.lwt.2021.112230

Çakır E, Arıcı M, Durak MZ (2020) Biodiversity and techno-functional properties of lactic acid bacteria in fermented hull-less barley sourdough. J Biosci Bioeng 130:450–456. https://doi.org/10.1016/j.jbiosc.2020.05.002

Câmara AA, Maréchal PA, Tourdot-Maréchal R, Husson F (2019) Oxidative stress resistance during dehydration of three non-Saccharomyces wine yeast strains. Food Res Int 123:364–372. https://doi.org/10.1016/j.foodres.2019.04.059

Canesin MR, Cazarin CBB (2021) Nutritional quality and nutrient bioaccessibility in sourdough bread. Curr Opin Food Sci 40:81–86. https://doi.org/10.1016/j.cofs.2021.02.007

Carbonetto B, Nidelet T, Guezenec S et al (2020) Interactions between Kazachstania humilis yeast species and lactic acid bacteria in sourdough. Microorganisms 8:240. https://doi.org/10.3390/microorganisms8020240

Cocolin L, Ercolini D (eds) (2008) Molecular techniques in the microbial ecology of fermented foods. Springer New York, New York

Coda R, Cassone A, Rizzello CG et al (2011) Antifungal activity of Wickerhamomyces anomalus and Lactobacillus plantarum during sourdough fermentation: identification of novel compounds and long-term effect during storage of wheat bread. Appl Environ Microbiol 77:3484–3492. https://doi.org/10.1128/AEM.02669-10

Coda R, Rizzello CG, Di Cagno R et al (2013) Antifungal activity of Meyerozyma guilliermondii: identification of active compounds synthesized during dough fermentation and their effect on long-term storage of wheat bread. Food Microbiol 33:243–251. https://doi.org/10.1016/j.fm.2012.09.023

Comasio A, Verce M, Van Kerrebroeck S, De Vuyst L (2020) Diverse microbial composition of sourdoughs from different origins. Front Microbiol 11:1212. https://doi.org/10.3389/fmicb.2020.01212

Danilo E (2013) High-throughput sequencing and metagenomics: moving forward in the culture-independent analysis of food microbial ecology. Appl Environ Microbiol 79:3148–3155. https://doi.org/10.1128/AEM.00256-13

De Vuyst L, Van Kerrebroeck S, Harth H et al (2014) Microbial ecology of sourdough fermentations: diverse or uniform? Food Microbiol 37:11–29. https://doi.org/10.1016/j.fm.2013.06.002

De Vuyst L, Van Kerrebroeck S, Leroy F (2017) Microbial ecology and process technology of sourdough fermentation. Adv Appl Microbiol 100:49–160

Demirbas F, Ispirli H, Kurnaz AA et al (2017) Antimicrobial and functional properties of lactic acid bacteria isolated from sourdoughs. LWT – Food Sci Technol 79:361–366. https://doi.org/10.1016/j.lwt.2017.01.067

Dentice Maidana S, Aristimuño Ficoseco C, Bassi D et al (2020) Biodiversity and technological-functional potential of lactic acid bacteria isolated from spontaneously fermented chia sourdough. Int J Food Microbiol 316:108425. https://doi.org/10.1016/j.ijfoodmicro.2019.108425

Doğan M, Tekiner İH (2019) Assessment of probiotic properties of lactic acid bacteria from traditional sourdoughs for bread-making in Turkey against some gut conditions. J Appl Food Technol 6:34–40. https://doi.org/10.17728/jaft.5662

Dolci P, Cocolin L (2017) Direct application of Rep-PCR on Type I sourdough matrix to monitor the dominance and persistence of a lactobacillus plantarum starter throughout back-slopping. J Food Sci 82:1898–1901. https://doi.org/10.1111/1750-3841.13804

El Sheikha AF, Hu D-M (2020) Molecular techniques reveal more secrets of fermented foods. Crit Rev Food Sci Nutr 60:11–32. https://doi.org/10.1080/10408398.2018.1506906

El Sheikha AF, Levin R, Xu J (eds) (2018) Molecular techniques in food biology. Wiley, Chichester

Ewuoso MO, Animashaun OH, Adejumo AA (2020) Lactic acid bacteria and yeasts in spontaneously fermented sorghum sourdough. Am J Microbiol Res 8:63–72. https://doi.org/10.12691/ajmr-8-2-4

Fekri A, Torbati M, Yari Khosrowshahi A et al (2020) Functional effects of phytate-degrading, probiotic lactic acid bacteria and yeast strains isolated from Iranian traditional sourdough on the technological and nutritional properties of whole wheat bread. Food Chem 306:125620. https://doi.org/10.1016/j.foodchem.2019.125620

Ferchichi M, Valcheva R, Prevost H et al (2007) Molecular identification of the microbiota of French sourdough using temporal temperature gradient gel electrophoresis. Food Microbiol 24:678–686. https://doi.org/10.1016/j.fm.2007.04.001

Fernández-Peláez J, Paesani C, Gómez M (2020) Sourdough technology as a tool for the development of healthier grain-based products: an update. Agronomy 10:1962. https://doi.org/10.3390/agronomy10121962

Fraberger V, Ammer C, Domig KJ (2020a) Functional properties and sustainability improvement of sourdough bread by lactic acid bacteria. Microorganisms 8:1–17. https://doi.org/10.3390/microorganisms8121895

Fraberger V, Unger C, Kummer C, Domig KJ (2020b) Insights into microbial diversity of traditional Austrian sourdough. LWT 127:109358. https://doi.org/10.1016/j.lwt.2020.109358

Fujimoto A, Ito K, Narushima N, Miyamoto T (2019) Identification of lactic acid bacteria and yeasts, and characterization of food components of sourdoughs used in Japanese bakeries. J Biosci Bioeng 127:575–581. https://doi.org/10.1016/j.jbiosc.2018.10.014

Fusco V, Quero GM, Chieffi D, Franz CMAP (2016) Identification of lactobacillus brevis using a species-specific AFLP-derived marker. Int J Food Microbiol 232:90–94. https://doi.org/10.1016/j.ijfoodmicro.2016.06.002

Fusco V, Quero GM, Stea G et al (2011) Novel PCR-based identification of Weissella confusa using an AFLP-derived marker. Int J Food Microbiol 145:437–443. https://doi.org/10.1016/j.ijfoodmicro.2011.01.015

Galle S, Arendt EK (2014) Exopolysaccharides from sourdough lactic acid bacteria. Crit Rev Food Sci Nutr 54:891–901. https://doi.org/10.1080/10408398.2011.617474

Galli V, Mazzoli L, Luti S et al (2018) Effect of selected strains of lactobacilli on the antioxidant and anti-inflammatory properties of sourdough. Int J Food Microbiol 286:55–65. https://doi.org/10.1016/j.ijfoodmicro.2018.07.018

Galli V, Venturi M, Coda R et al (2020) Isolation and characterization of indigenous Weissella confusa for in situ bacterial exopolysaccharides (EPS) production in chickpea sourdough. Food Res Int 138:109785. https://doi.org/10.1016/j.foodres.2020.109785

Gänzle M, Ripari V (2016) Composition and function of sourdough microbiota: from ecological theory to bread quality. Int J Food Microbiol 239:19–25. https://doi.org/10.1016/j.ijfoodmicro.2016.05.004

Gélinas P, McKinnon C, Gagnon F (2016) Fructans, water-soluble fibre and fermentable sugars in bread and pasta made with ancient and modern wheat. Int J Food Sci Technol 51:555–564. https://doi.org/10.1111/ijfs.13022

Giraffa G, Carminati D (2008) Molecular techniques in food fermentation: principles and applications. In: Cocolin L, Ercolini D (eds) *Molecular techniques in the microbial ecology of fermented foods*. Springer New York, New York, NY, pp 1–30

Glen TC (2011) Field guide to next-generation DNA sequencers. Mol Ecol Resour 11:759–769. https://doi.org/10.1111/j.1755-0998.2011.03024.x

Graça C, Lima A, Raymundo A, Sousa I (2021) Sourdough fermentation as a tool to improve the nutritional and health-promoting properties of its derived-products. Fermentation 7:246. https://doi.org/10.3390/fermentation7040246

Guerzoni ME, Serrazanetti DI, Vernocchi P, Gianotti A (2013) Physiology and biochemistry of sourdough yeasts. In: Gobbetti M, Gänzle M (eds) Handbook on sourdough biotechnology. Springer US, Boston, pp 155–181

Guerzoni ME, Vernocchi P, Ndagijimana M et al (2007) Generation of aroma compounds in sourdough: effects of stress exposure and lactobacilli–yeasts interactions. Food Microbiol 24:139–148. https://doi.org/10.1016/j.fm.2006.07.007

Gutiérrez D, Martín-Platero AM, Rodríguez A et al (2011) Typing of bacteriophages by randomly amplified polymorphic DNA (RAPD)-PCR to assess genetic diversity. FEMS Microbiol Lett 322:90–97. https://doi.org/10.1111/j.1574-6968.2011.02342.x

Hafeez Z, Cakir-kiefer C, Roux E et al (2014) Strategies of producing bioactive peptides from milk proteins to functionalize fermented milk products. FRIN 63:71–80. https://doi.org/10.1016/j.foodres.2014.06.002

Häggman M, Salovaara H (2008) Effect of fermentation rate on endogenous leavening of Candida milleri in sour rye dough. Food Res Int 41:266–273. https://doi.org/10.1016/j.foodres.2007.12.002

Hajinia F, Sadeghi A, Sadeghi Mahoonak A (2021) The use of antifungal oat-sourdough lactic acid bacteria to improve safety and technological functionalities of the supplemented wheat bread. J Food Saf 41:e12873. https://doi.org/10.1111/jfs.12873

Hassan YI, Zhou T, Bullerman LB (2016) Sourdough lactic acid bacteria as antifungal and mycotoxin-controlling agents. Food Sci Technol Int 22:79–90. https://doi.org/10.1177/1082013214565722

Heitmann M, Zannini E, Arendt E (2018) Impact of Saccharomyces cerevisiae metabolites produced during fermentation on bread quality parameters: a review. Crit Rev Food Sci Nutr 58:1152–1164. https://doi.org/10.1080/10408398.2016.1244153

Holt BH, Buchan A, DeBruyn JM et al (2022) Breaking barriers with bread: using the sourdough starter microbiome to teach high-throughput sequencing techniques. J Microbiol Biol Educ 23(2):e00306-21. https://doi.org/10.1128/jmbe.00306-21

Ispirli H, Demirbaş F, Yüzer MO, Dertli E (2018) Identification of lactic acid bacteria from spontaneous Rye sourdough and determination of their functional characteristics. Food Biotechnol 32:222–235. https://doi.org/10.1080/08905436.2018.1507913

Jakob F, Steger S, Vogel RF (2012) Influence of novel fructans produced by selected acetic acid bacteria on the volume and texture of wheat breads. Eur Food Res Technol 234:493–499. https://doi.org/10.1007/s00217-011-1658-7

Karaman K, Sagdic O, Durak MZ (2018) Use of phytase active yeasts and lactic acid bacteria isolated from sourdough in the production of whole wheat bread. LWT – Food Sci Technol 91:557–567. https://doi.org/10.1016/j.lwt.2018.01.055

Kim D, Chon J, Kim H, Seo K (2019) Development of a novel selective medium for the isolation and enumeration of acetic acid bacteria from various foods. Food Control 106:106717. https://doi.org/10.1016/j.foodcont.2019.106717

Korcari D, Ricci G, Quattrini M, Fortina MG (2020) Microbial consortia involved in fermented spelt sourdoughs: dynamics and characterization of yeasts and lactic acid bacteria. Lett Appl Microbiol 70:48–54. https://doi.org/10.1111/lam.13241

Korcari D, Secchiero R, Laureati M et al (2021) Technological properties, shelf life and consumer preference of spelt-based sourdough bread using novel, selected starter cultures. LWT 151:112097. https://doi.org/10.1016/j.lwt.2021.112097

Kurtzman CP, Fell JW, Boekhout T, Robert V (2011) Methods for isolation, phenotypic characterization and maintenance of yeasts. In: Kurtzman CP, Fell JW, Teun B (eds) The Yeasts, 5th edn. Elsevier, London, pp 87–110

Landis EA, Oliverio AM, McKenney EA et al (2021) The diversity and function of sourdough starter microbiomes. elife 10:e61644. https://doi.org/10.7554/eLife.61644

Lau SW, Chong AQ, Chin NL et al (2021) Sourdough microbiome comparison and benefits. Microorganisms 9(7):1355. https://doi.org/10.3390/microorganisms9071355

Levin RE, Ekezie F-GC, Sun D-W (2018) Chapter 14 – DNA-based technique: Polymerase Chain Reaction (PCR). In: Sun D-W (ed) Modern techniques for food authentication, 2nd edn. Academic, pp 527–616

Lhomme E, Onno B, Chuat V et al (2016a) Genotypic diversity of Lactobacillus sanfranciscensis strains isolated from French organic sourdoughs. Int J Food Microbiol 226:13–19. https://doi.org/10.1016/j.ijfoodmicro.2016.03.008

Lhomme E, Urien C, Legrand J et al (2016b) Sourdough microbial community dynamics: an analysis during French organic bread-making processes. Food Microbiol 53:41–50. https://doi.org/10.1016/j.fm.2014.11.014

Li H, Fu J, Hu S et al (2021) Comparison of the effects of acetic acid bacteria and lactic acid bacteria on the microbial diversity of and the functional pathways in dough as revealed by high-throughput metagenomics sequencing. Int J Food Microbiol 346.:109168 Contents:109168

Liu L, Li Y, Li S et al (2012) Comparison of next-generation sequencing systems. J Biomed Biotechnol 2012:251364

Liu T, Li Y, Chen J et al (2016) Prevalence and diversity of lactic acid bacteria in Chinese traditional sourdough revealed by culture dependent and pyrosequencing approaches. LWT – Food Sci Technol 68:91–97. https://doi.org/10.1016/j.lwt.2015.12.025

Loman NJ, Misra RV, Dallman TJ et al (2012) Performance comparison of benchtop high-throughput sequencing platforms. Nat Biotechnol 30:434–439. https://doi.org/10.1038/nbt.2198

Luangsakul N, Keeratipibul S, Jindamorakot S, Tanasupawat S (2009) Lactic acid bacteria and yeasts isolated from the starter doughs for Chinese steamed buns in Thailand. LWT – Food Sci Technol 42:1404–1412. https://doi.org/10.1016/j.lwt.2009.03.007

Luz C, D'Opazo V, Mañes J, Meca G (2019) Antifungal activity and shelf life extension of loaf bread produced with sourdough fermented by Lactobacillus strains. J Food Process Preserv 43:1–8. https://doi.org/10.1111/jfpp.14126

Mantzourani I, Plessas S, Odatzidou M et al (2019) Effect of a novel Lactobacillus paracasei starter on sourdough bread quality. Food Chem 271:259–265. https://doi.org/10.1016/j.foodchem.2018.07.183

Margalho LP, Kamimura BA, Brex RP et al (2021) High throughput screening of technological and biopreservation traits of a large set of wild lactic acid bacteria from Brazilian artisanal cheeses. Food Microbiol 100.:103872 Contents. https://doi.org/10.1016/j.fm.2021.103872

Martorana A, Giuffrè AM, Capocasale M et al (2018) Sourdoughs as a source of lactic acid bacteria and yeasts with technological characteristics useful for improved bakery products. Eur Food Res Technol 244:1873–1885. https://doi.org/10.1007/s00217-018-3100-x

Masoomshahi T, Ehsani J, Ebrahimi M (2020) Isolation and identification of predominant lactic acid bacterial Flora From Oat Bran sourdough and their antifungal ability. Iran J Med Microbiol 14:408–424. https://doi.org/10.30699/ijmm.14.5.408

Melini V, Melini F (2018) Strategies to extend bread and GF bread shelf-life: from sourdough to antimicrobial active packaging and nanotechnology. Fermentation 4:9. https://doi.org/10.3390/fermentation4010009

Meruvu H, Harsa ST (2022) Lactic acid bacteria : isolation – characterization approaches and industrial applications. Crit Rev Food Sci Nutr 0:1–20. https://doi.org/10.1080/10408398.2022.2054936

Michel E, Monfort C, Deffrasnes M et al (2016) Characterization of relative abundance of lactic acid bacteria species in French organic sourdough by cultural, qPCR and MiSeq high-throughput sequencing methods. Int J Food Microbiol 239:35–43. https://doi.org/10.1016/j.ijfoodmicro.2016.07.034

Mo EK, Sung CK (2014) Production of white pan bread leavened by Pichia anomala SKM-T. Food Sci Biotechnol 23:431–437. https://doi.org/10.1007/s10068-014-0059-7

Mohammadi-Kouchesfahani M, Hamidi-Esfahani Z, Azizi MH (2019) Isolation and identification of lactic acid bacteria with phytase activity from sourdough. Food Sci Nutr 7:3700–3708. https://doi.org/10.1002/fsn3.1229

Montemurro M, Coda R, Rizzello C (2019) Recent advances in the use of sourdough biotechnology in pasta making. Foods 8:129. https://doi.org/10.3390/foods8040129

Mota-Gutierrez J, Franciosa I, Ruggirello M, Dolci P (2021) Technological, functional and safety properties of lactobacilli isolates from soft wheat sourdough and their potential use as antimould cultures. World J Microbiol Biotechnol 37:1–13. https://doi.org/10.1007/s11274-021-03114-2

Müller DC, Mischler S, Schönlechner R, Miescher Schwenninger S (2021) Multiple technofunctional characteristics of leuconostoc and their potential in sourdough fermentations. Microorganisms 9:1–15. https://doi.org/10.3390/microorganisms9081633

Müller DC, Schipali S, Näf P et al (2022) Potential of a techno-functional Sourdough and its application in sugar-reduced soft buns. Fermentation 8. https://doi.org/10.3390/fermentation8020042

Muyzer G (1999) DGGE/TGGE a method for identifying genes from natural ecosystems. Curr Opin Microbiol 2:317–322. https://doi.org/10.1016/S1369-5274(99)80055-1

Nuobariene L, Cizeikiene D, Gradzeviciute E et al (2015) Phytase-active lactic acid bacteria from sourdoughs: isolation and identification. Lwt 63:766–772. https://doi.org/10.1016/j.lwt.2015.03.018

Nuobariene L, Hansen ÅS, Arneborg N (2012) Isolation and identification of phytase-active yeasts from sourdoughs. LWT – Food Sci Technol 48:190–196. https://doi.org/10.1016/j.lwt.2012.03.011

Ogunsakin AO, Vanajakshi V, Anu-Appaiah KA et al (2017) Evaluation of functionally important lactic acid bacteria and yeasts from Nigerian sorghum as starter cultures for gluten-free sourdough preparation. LWT – Food Sci Technol 82:326–334. https://doi.org/10.1016/j.lwt.2017.04.048

Olojede AO, Sanni AI, Banwo K, Michael T (2022) Improvement of texture, nutritional qualities, and consumers' perceptions of Sorghum-based sourdough bread made with pediococcus pentosaceus and weissella confusa strains. Fermentation 8. https://doi.org/10.3390/fermentation8010032

Oshiro M, Zendo T, Nakayama J (2021) Diversity and dynamics of sourdough lactic acid bacteriota created by a slow food fermentation system. J Biosci Bioeng 131:333–340. https://doi.org/10.1016/j.jbiosc.2020.11.007

Paffetti D, Barberio C, Casalone E et al (1995) DNA fingerprinting by random amplified polymorphic DNA and restriction fragment length polymorphism is useful for yeast typing. Res Microbiol 146:587–594. https://doi.org/10.1016/0923-2508(96)80565-1
Palla M, Agnolucci M, Calzone A et al (2019) Exploitation of autochthonous Tuscan sourdough yeasts as potential starters. Int J Food Microbiol 302:59–68. https://doi.org/10.1016/j.ijfoodmicro.2018.08.004
Palla M, Blandino M, Grassi A et al (2020a) Characterization and selection of functional yeast strains during sourdough fermentation of different cereal wholegrain flours. Sci Rep 10:12856. https://doi.org/10.1038/s41598-020-69774-6
Palla M, Cristani C, Giovannetti M, Agnolucci M (2017) Identification and characterization of lactic acid bacteria and yeasts of PDO Tuscan bread sourdough by culture dependent and independent methods. Int J Food Microbiol 250:19–26. https://doi.org/10.1016/j.ijfoodmicro.2017.03.015
Palla M, Cristani C, Giovannetti M, Agnolucci M (2020b) Large genetic intraspecific diversity of autochthonous lactic acid bacteria and yeasts isolated from PDO tuscan bread sourdough. Appl Sci 10:1043. https://doi.org/10.3390/app10031043
Park DM, Bae JH, Kim MS et al (2019) Suitability of lactobacillus plantarum SPC-SNU 72-2 as a probiotic starter for sourdough fermentation. J Microbiol Biotechnol 29:1729–1738. https://doi.org/10.4014/jmb.1907.07004
Patrizia S, Paola M (2003) Evaluation of repetitive element sequence-based PCR as a molecular typing method for clostridium difficile. J Clin Microbiol 41:2454–2457. https://doi.org/10.1128/JCM.41.6.2454-2457.2003
Paun O, Schönswetter P (2012) Amplified fragment length polymorphism: an invaluable fingerprinting technique for genomic, transcriptomic, and epigenetic studies. In: Sucher NJ, Hennell JR, Carles MC (eds) Plant DNA fingerprinting and barcoding: methods and protocols. Humana Press, Totowa, pp 75–87
Pepe O, Blajotta G, Anastasio M et al (2004) Technological and molecular diversity of lactobacillus plantarum strains isolated from naturally fermented sourdoughs. Syst Appl Microbiol 27:443–453. https://doi.org/10.1078/0723202041438446
Petkova M, Stefanova P, Gotcheva V, Angelov A (2021) Isolation and characterization of lactic acid bacteria and yeasts from typical Bulgarian sourdoughs. Microorganisms 9:1346. https://doi.org/10.3390/microorganisms9071346
Pino A, Russo N, Solieri L et al (2022) Microbial consortia involved in traditional sicilian sourdough: characterization of lactic acid bacteria and yeast populations. Microorganisms 10:283. https://doi.org/10.3390/microorganisms10020283
Randazzo CL, Heilig H, Restuccia C et al (2005) Bacterial population in traditional sourdough evaluated by molecular methods. J Appl Microbiol 99:251–258. https://doi.org/10.1111/j.1365-2672.2005.02624.x
Reale A, Di Renzo T, Boscaino F et al (2019) Lactic acid bacteria biota and aroma profile of Italian traditional sourdoughs from the irpinian area in Italy. Front Microbiol 10. https://doi.org/10.3389/fmicb.2019.01621
Reale A, Di Stasio L, Di Renzo T et al (2021) Bacteria do it better! Proteomics suggests the molecular basis for improved digestibility of sourdough products. Food Chem 359:129955. https://doi.org/10.1016/j.foodchem.2021.129955
Rizzello CG, Lorusso A, Montemurro M, Gobbetti M (2016) Use of sourdough made with quinoa (Chenopodium quinoa) flour and autochthonous selected lactic acid bacteria for enhancing the nutritional, textural and sensory features of white bread. Food Microbiol 56:1–13. https://doi.org/10.1016/j.fm.2015.11.018
Ruiz L, Hidalgo C, Blanco-míguez A et al (2016) Tackling probiotic and gut microbiota functionality through proteomics. J Proteome 147:28–39. https://doi.org/10.1016/j.jprot.2016.03.023
Ruiz Rodríguez L, Vera Pingitore E, Rollan G et al (2016a) Biodiversity and technological-functional potential of lactic acid bacteria isolated from spontaneously fermented quinoa sourdoughs. J Appl Microbiol 120:1289–1301. https://doi.org/10.1111/jam.13104

Ruiz Rodríguez L, Vera Pingitore E, Rollan G et al (2016b) Biodiversity and technological potential of lactic acid bacteria isolated from spontaneously fermented amaranth sourdough. Lett Appl Microbiol 63:147–154. https://doi.org/10.1111/lam.12604

Sadeghi A, Ebrahimi M, Raeisi M, Nematollahi Z (2019) Biological control of foodborne pathogens and aflatoxins by selected probiotic LAB isolated from rice bran sourdough. Biol Control 130:70–79. https://doi.org/10.1016/j.biocontrol.2018.09.017

Sáez GD, Saavedra L, Hebert EM, Zárate G (2018) Identification and biotechnological characterization of lactic acid bacteria isolated from chickpea sourdough in northwestern Argentina. Lwt 93:249–256. https://doi.org/10.1016/j.lwt.2018.03.040

Sanalibaba P, Cakmak GA (2016) Exopolysaccharides production by lactic acid bacteria. Appl Microbiol open access 2:1–5. https://doi.org/10.4172/2471-9315.1000115

Scheirlinck I, Van Der Meulen R, Van Schoor A et al (2008) Taxonomic structure and stability of the bacterial community in Belgian sourdough ecosystems as assessed by culture and population fingerprinting. Appl Environ Microbiol 74(8):2414–2423. https://doi.org/10.1128/aem.02771-07

Scheirlinck I, Van der Meulen R, De Vuyst L et al (2009) Molecular source tracking of predominant lactic acid bacteria in traditional Belgian sourdoughs and their production environments. J Appl Microbiol 106:1081–1092. https://doi.org/10.1111/j.1365-2672.2008.04094.x

Scheirlinck I, Van der Meulen R, Van Schoor A et al (2007) Lactobacillus crustorum sp. nov., isolated from two traditional Belgian wheat sourdoughs. Int J Syst Evol Microbiol 57:1461–1467. https://doi.org/10.1099/ijs.0.64836-0

Schillinger U, Holzapfel WH (2003) Culture media for lactic acid bacteria. In: Corry JEL et al (eds) Handbook of culture media for food microbiology. Elsevier Science, pp 127–140

Sevgili A, Erkmen O, Koçaslan S (2021) Identification of lactic acid bacteria and yeasts from traditional sourdoughs and sourdough production by enrichment. Czech J Food Sci 39:312–318. https://doi.org/10.17221/56/2021-CJFS

Sharma-Kuinkel BK, Rude TH, Fowler VG (2016) Pulse field gel electrophoresis. In: Bose JL (ed) The genetic manipulation of staphylococci: methods and protocols. Springer New York, New York, NY, pp 117–130

Sharma A, Lee S, Park Y-S (2020) Molecular typing tools for identifying and characterizing lactic acid bacteria: a review. Food Sci Biotechnol 29:1301–1318. https://doi.org/10.1007/s10068-020-00802-x

Shen J, Shi K, Dong H et al (2022) Screening of sourdough starter strains and improvements in the quality of whole wheat steamed bread. Molecules 27:3510. https://doi.org/10.3390/molecules27113510

Shendure J, Ji H (2008) Next-generation DNA sequencing. Nat Biotechnol 26:1135–1145. https://doi.org/10.1038/nbt1486

Shevchenko A, Yang Y, Knaust A et al (2014) Proteomics identifies the composition and manufacturing recipe of the 2500-year old sourdough bread from Subeixi cemetery in China. J Proteome 105:363–371. https://doi.org/10.1016/j.jprot.2013.11.016

Siepmann FB, De Almeida BS, Gomes TA et al (2022) Biochemical, microbiological and technological characteristics of type II sourdoughs produced with lactic acid bacteria single-strains. Semin Agrar 43:693–712. https://doi.org/10.5433/1679-0359.2022v43n2p693

Slezak TR, Salzberg SL (2005) Bioinformatics methods for microbial detection and forensic diagnostic design. In: Breeze RG, Budowle B, SEBT-MF S (eds) Microbial forensics. Elsevier, Burlington, pp 313–353

Sonya S, Raffaella DC, Danilo E et al (2009) Taxonomic structure and monitoring of the dominant population of lactic acid bacteria during wheat flour sourdough Type I propagation using lactobacillus sanfranciscensis starters. Appl Environ Microbiol 75:1099–1109. https://doi.org/10.1128/AEM.01524-08

Suo B, Chen X, Wang Y (2021) Recent research advances of lactic acid bacteria in sourdough: origin, diversity, and function. Curr Opin Food Sci 37:66–75. https://doi.org/10.1016/j.cofs.2020.09.007

Syrokou MK, Tziompra S, Psychogiou E et al (2021) Technological and safety attributes of lactic acid bacteria and yeasts isolated from spontaneously fermented greek wheat sourdoughs. Microorganisms 9(4):671

Ua-arak T, Jakob F, Vogel RF (2016) International journal of food microbiology characterization of growth and exopolysaccharide production of selected acetic acid bacteria in buckwheat sourdoughs. Int J Food Microbiol 239:103–112. https://doi.org/10.1016/j.ijfoodmicro.2016.04.009

Ua-Arak T, Jakob F, Vogel RF (2017) Influence of levan-producing acetic acid bacteria on buckwheat-sourdough breads. Food Microbiol 65:95–104. https://doi.org/10.1016/j.fm.2017.02.002

Urien C, Legrand J, Montalent P et al (2019) Fungal species diversity in french bread sourdoughs made of organic wheat flour. Front Microbiol 10. https://doi.org/10.3389/fmicb.2019.00201

Urwin R, Maiden MCJ (2003) Multi-locus sequence typing: a tool for global epidemiology. Trends Microbiol 11:479–487. https://doi.org/10.1016/j.tim.2003.08.006

Valcheva R, Ferchichi MF, Korakli M et al (2006) Lactobacillus nantensis sp. nov., isolated from French wheat sourdough. Int J Syst Evol Microbiol 56:587–591. https://doi.org/10.1099/ijs.0.63619-0

Valcheva R, Korakli M, Onno B et al (2005) Lactobacillus hammesii sp. nov., isolated from French sourdough. Int J Syst Evol Microbiol 55:763–767. https://doi.org/10.1099/ijs.0.63311-0

Valmorri S, Tofalo R, Settanni L et al (2010) Yeast microbiota associated with spontaneous sourdough fermentations in the production of traditional wheat sourdough breads of the Abruzzo region (Italy). Antonie Van Leeuwenhoek 97:119–129. https://doi.org/10.1007/s10482-009-9392-x

Vancanneyt M, Neysens P, De Wachter M et al (2005) Lactobacillus acidifarinae sp. nov. and Lactobacillus zymae sp. nov., from wheat sourdoughs. Int J Syst Evol Microbiol 55:615–620. https://doi.org/10.1099/ijs.0.63274-0

Venturi M, Guerrini S, Vincenzini M (2012) Stable and non-competitive association of Saccharomyces cerevisiae, Candida milleri and Lactobacillus sanfranciscensis during manufacture of two traditional sourdough baked goods. Food Microbiol 31:107–115. https://doi.org/10.1016/j.fm.2012.02.011

Vrancken G, De Vuyst L, Van der Meulen R et al (2010) Yeast species composition differs between artisan bakery and spontaneous laboratory sourdoughs. FEMS Yeast Res 10:471–481. https://doi.org/10.1111/j.1567-1364.2010.00621.x

Wong ML, Medrano JF (2005) Real-time PCR for mRNA quantitation. BioTechniques 39:75–85. https://doi.org/10.2144/05391RV01

Xu D, Hu Y, Wu F et al (2020) Comparison of the functionality of exopolysaccharides produced by sourdough lactic acid bacteria in bread and steamed bread. J Agric Food Chem 68:8907–8914. https://doi.org/10.1021/acs.jafc.0c02703

Xu D, Zhang Y, Tang K et al (2019) Effect of mixed cultures of yeast and lactobacilli on the quality of wheat sourdough bread. Front Microbiol 10. https://doi.org/10.3389/fmicb.2019.02113

Yadav AN, Yadav N, Kour D et al (2019) Bacterial community composition in lakes. In: Bandh SA, Shafi S, Shameem NBT-FM (eds) Freshwater microbiology. Elsevier, pp 1–71

Yan B, Sadiq FA, Cai Y et al (2019) Microbial diversity in traditional type I sourdough and jiaozi and its influence on volatiles in Chinese steamed bread. LWT 101:764–773. https://doi.org/10.1016/j.lwt.2018.12.004

Yang H, Liu T, Zhang G et al (2017) Genotyping of Lactobacillus sanfranciscensis isolates from Chinese traditional sourdoughs by multilocus sequence typing and multiplex RAPD-PCR. Int J Food Microbiol 258:50–57. https://doi.org/10.1016/j.ijfoodmicro.2017.07.013

Yang H, Liu T, Zhang G, He G (2020) Intraspecific diversity and fermentative properties of Saccharomyces cerevisiae from Chinese traditional sourdough. LWT 124:109195. https://doi.org/10.1016/j.lwt.2020.109195

Yarrow D (1998) Methods for the isolation, maintenance and identification of yeasts. In: Kurtzman CP, Fell JWBT-TY (eds) The Yeasts, 4th edn. Elsevier, Amsterdam, pp 77–100

Yu Y, Wang L, Qian H et al (2019) Effect of selected strains on physical and organoleptic properties of breads. Food Chem 276:547–553. https://doi.org/10.1016/j.foodchem.2018.10.048

Zhang G, Tu J, Sadiq FA et al (2019) Prevalence, genetic diversity, and technological functions of the Lactobacillus sanfranciscensis in sourdough: a review. Compr Rev Food Sci Food Saf 18:1209–1226. https://doi.org/10.1111/1541-4337.12459

Zhou Y, She X, Zhu S, Zhou X (2022) The study of microbial diversity and volatile compounds in Tartary buckwheat sourdoughs. Food Chem X 14:100353. https://doi.org/10.1016/j.fochx.2022.100353

Chapter 6
Undesirable Microorganisms in Cereal Flours and Their Impact on the Stability of Sourdough Microbiome

Anna Otlewska, Katarzyna Dybka-Stępień, and Katarzyna Rajkowska

6.1 Introduction

Wheat flour (*Triticum aestivum* L. or *Triticum durum* L.) consists of carbohydrates, mainly starch (70–75%, m/m), free saccharides (1–2%), crude fiber (2–3%, fructans and arabinoxylan pentosans), proteins (8–15%, gliadins and glutenins) including enzymes (α-amylase, β-amylase, proteases, lipoxygenase, hydroperoxide lyase, and polyphenol oxidase), and lipides (1–2% glycolipids). It also contains other valuable compounds, such as phenols, minerals, and vitamins. The composition of rye flour (*Secale cereale* L.) is slightly different from that of wheat flour. Rye flour contains less starch (65%), but more fiber (6–8%), mainly arabinoxylan pentosans. The amount of protein is similar in both types of flour, but in rye flour gluten proteins (secalins and secalinins) predominate (De Vuyst et al. 2023).

The chemical composition of flour and the content of macronutrients contribute to the development of a specific microbiota of both the flour and flour-based products. Flour also provides various micronutrients, including ashes, which can also influence microbial composition and succession (Calvert et al. 2021). In addition, flour with an increased ash content, containing more bran and typically used in the production of sourdoughs, can be a significant source of microbial contamination.

Although the composition of flour, due to its nutritional value, is favorable for the development of microorganisms, it is also a product with a low water content (De Vuyst et al. 2023). Low moisture content is one of the key factors limiting the growth of microorganisms and therefore the flour is generally recognized as microbiologically safe.

A. Otlewska (✉) · K. Dybka-Stępień · K. Rajkowska
Institute of Fermentation Technology and Microbiology, Faculty of Biotechnology and Food Sciences, Lodz University of Technology, Lodz, Poland
e-mail: anna.otlewska@p.lodz.pl

E. B. Ceresino et al. (eds.), *Sourdough Microbiota and Starter Cultures for Industry*, https://doi.org/10.1007/978-3-031-48604-3_6

According to Codex Alimentarius Standard for Wheat Flour (No. 152), the moisture content must not exceed 15.5% m/m. Sometimes lower moisture limits are required, depending on the climate, duration of transport, and length of storage. Due to the low water content in flour, many groups of microorganisms are unable to grow, and the risk of their presence is limited. However, certain microorganisms can survive in flour for a long time, due to adaptation mechanisms (e.g., increased heat resistance). The risk of microbial contamination from such microorganisms is quite high, starting from the crop production chain through preharvest, harvest, storage, transport, and processing (Magallanes López and Simsek 2021; Stuper-Szablewska and Perkowski 2017). Sources of microbial contamination include air, water, dust, shipping containers, handling, and processing equipment (Thielecke and Nugent 2018).

The microbial community in flour (defined as the number and relative proportions of species) is a consequence of the autochthonic microbiota found in cereal grains. The population of microbes found in grains is affected by such factors as (i) pedoclimatic conditions (e.g., temperature, soil type, salinity level, geographical location linked to rainfall ratio); (ii) grain damage and related plant diseases; (iii) the crop cultivation strategy (use of chemicals such as insecticides and fungicides) (Corsetti et al. 2007; Otlewska et al. 2020; Pontonio et al. 2016).

6.2 Flour as a Source of Undesirable Microorganisms and Its Influence on the Sourdough Stability

Most studies in the literature focus on the presence in flour of aerobic mesophiles, yeasts, and molds, as well as spore-forming bacteria, both aerobic and anaerobic (Berghofer et al. 2003; Eglezos 2010; Czerwińska and Piotrowski 2009; Cardoso et al. 2019; Aydin et al. 2009; Ntuli et al. 2013; Ennadir et al. 2012). The numbers of these groups of microorganisms vary, depending on the flour samples and their geographical origin (Table 6.1). There are significant differences in the amounts of contamination with mesophilic aerobic microorganisms, with numbers ranging from less than 1 log CFU/g in flour from the United States (Manthey et al. 2004) to up to 7 log CFU/g for some flour samples from Australia and Turkey (Berghofer et al. 2003; Aydin et al. 2009). The highest level of aerobic mesophiles has been detected in flour samples originating from the USA. This was correlated with low kernel weight and heavily damaged kernel content (Manthey et al. 2004). Wheat with high microbial loads produces flour of low microbiological quality (Berghofer et al. 2003). In Nigeria, 20 commercially available wheat flour samples were characterized by low total microbial levels. However, *Salmonella* spp. was identified in 10% of the samples, and *E. coli* was found in one wheat semolina sample (Nosa-Obamwonyi and Antai 2018).

The presence of pathogenic bacteria, especially from the family *Enterobacteriaceae*, affects not only the microbiological quality of flour. Most importantly, it also affects the safety of both the flour and the products of which it is

Table 6.1 Groups of microorganisms detected in the wheat and rye flours, and the ranges of their counts

Microorganisms	Number of microorganisms in wheat/rye flour (CFU/g)	Country	References
Aerobic mesophiles	10^1–10^7	Australia	Berghofer et al. (2003)
Coliforms	10^0–10^3		
Escherichia coli	10^0 MPN[a]/g		
Bacillus spp.	10^2–10^5		
Bacillus cereus	10^{-1}–10^1 MPN[a]/g		
Mesophilic aerobic spores	10^0–10^3		
Aerobic thermophiles	10^1–10^2		
Yeasts	10^2–10^3		
Molds	10^2–10^3		
Yeasts	10^3	Australia	Eglezos (2010)
Molds	10^2–10^4		
Aerobic plate count	10^4–10^5		
Escherichia coli	<3–6 MPN[a]/g		
Bacillus cereus	<10		
Salmonella spp.	ND in 25 g		
Total fungal counts	10^3	Germany	Weidenbörner et al. (2000)
Lactic acid bacteria	<10–10^4	Italy	Alfonzo et al. (2013)
Bacillus subtilis spores	<10	Poland	Cichoń and Ptak (2005)
Aerobic mesophiles	<10^2		
Molds	<10^2		
Yeasts	<10^2		
Mesophilic bacteria	10^2/10^1–10^2	Poland	Czerwińska and Piotrowski (2009, 2010)
Thermophilic bacteria	10^1–10^2/10^1		
Spores	10^1/10^1		
Molds	10^1/10^1–10^2		
Mesophilic microorganisms	10^3–10^5	Poland	Warzecha et al. (1987)
Thermophilic microorganisms	10^1–10^3		
Amylolytic microorganisms	10^1–10^3		
Aerobic plate count	10^4/10^3–10^5	Portugal	Cardoso et al. (2019)
Coliforms	10^2–10^3/<10–10^2		
Yeasts	< 10–10^2/10^2–10^4		
Molds	10^2–10^5/10^1–10^3		
Escherichia coli	<10/<10		
Clostridium spp.	<10^2/<10^2		
Salmonella spp.	ND in 25 g/ ND in 25 g		
Molds	10^1–10^2	Serbia	Plavšić et al. (2017)

(continued)

Table 6.1 (continued)

Microorganisms	Number of microorganisms in wheat/rye flour (CFU/g)	Country	References
Mesophilic aerobic bacteria	10^1–10^7	Turkey	Aydin et al. (2009)
Escherichia coli	<3–>9 MPN[a]/g		
Bacillus cereus	<10–10^3		
Clostridium perfringens	<10–10^3		
Molds	<10–10^5		
Rope spore	<30–>4500 MPN[a]/g		
Salmonella spp.	ND in 25 g		
Aerobic plate count	10^2–10^6	Iran	Abedi et al. (2020)
Molds	<10–10^1		
Yeasts	<10–10^1	Pakistan	Shahzad et al. (2019)
Molds	<10–10^2		
Aerobic plate counts	<10–10^8	USA	Manthey et al. (2004)
Molds and yeasts	10^1–10^6		
Aerobic plate count	10^3–10^4	USA	Richter et al. (1993)
Molds	10^2–10^3		
Yeasts	10^2		
Aerobic plate count	$10^{4¥}$–10^6	USA	Sperber (2007)
Yeasts	$10^{1¥}$–10^5		
Molds	$10^{2¥}$–10^5		
Coliforms	$10^{3¥}$–10^5		
Escherichia coli	<$10^{¥}$–10^2		
Mesophilic aerobic bacteria	10^1–10^2	Nigeria	Nosa-Obamwonyi and Antai (2018)
Coliforms	10^1–10^2		
Escherichia coli	<10		
Bacillus cereus	<10		
Salmonella spp.	<10–10^1		
Clostridium perfringens	<10		
Total plate count	10^3–10^5	Lesotho	Ntuli et al. (2013)
Coliforms	10^3–10^4		
Bacillus cereus	10^4–10^5		
Salmonella spp.	<10		
Escherichia coli	10^3–10^4		
Yeasts and molds	10^1–10^3		

(continued)

Table 6.1 (continued)

Microorganisms	Number of microorganisms in wheat/rye flour (CFU/g)	Country	References
Aerobic mesophiles	10^2–10^5	Morocco	Ennadir et al. (2012)
Total coliforms	<1–10^4		
Faecal coliforms	<1–10^3		
Clostridium spp.	<10		
Salmonella spp.	ND in 25 g		
Staphylococcus aureus	<10		
Listeria monocytogenes	ND in 10 g		
Lactic acid bacteria	<1–10^2		
Yeasts	<1–10^3		
Molds	<1–10^4		
Total heterotrophic bacteria	10^4–10^7	Bangladesh	Khanom et al. (2016)
Total yeast and molds	<10–10^6		
Total *Enterobacteriaceae*	<10–10^6		
Coliforms	<10–10^6		
Staphylococcus aureus	<10–10^4		

[a]*MPN* Most Probable Number, ¥ geometric mean, *ND* not detected

an ingredient. Although flour does not support the growth of pathogenic bacteria, pathogens may survive in products with low water activity, and can exhibit enhanced heat resistance (Beuchat et al. 2013).

There are few reported cases of food poisoning resulting from flour contaminated with *Salmonella* spp. or *Escherichia coli*. The presence of pathogens in wheat flour has been reported to the RASFF only once in the last 20 years, in a case related to *Salmonella* contamination (RASFF notification 2021.0931). The RASFF notification was made by a flour producer in Belgium. In the same period, the CDC reported two outbreaks of *Escherichia coli* infections linked to flour. The first multistate outbreak started in December 2015 and ended in September 2016. A total of 63 people were infected with Shiga toxin-producing *E. coli* O121 or O26, with 17 people requiring hospitalization (CDC 2016). During the second outbreak, caused by *E. coli* O26, 21 people were infected and 3 people were hospitalized. The illnesses occurred in the period from December 2018 to July 2019 (CDC 2019).

Australian, European, African, and US studies have found that *Salmonella* spp. and *E. coli* were either not present in wheat flour or were present at low levels (Table 6.1). Analysis during the period 2003–2005 in the United Sates revealed that 0.14% of wheat flour samples were contaminated with *Salmonella* (Sperber 2007). Eglezos (2010) found *E. coli* in only 2 out of 300 flour samples (0.7%) originating from Australia. In another study, 1 in 71 flour samples was *E. coli* positive (Berghofer et al. 2003). On the other hand, in Thrace (Turkey) the number of *E. coli* exceed national legal limits in more than 50% of the flour samples tested (Aydin et al. 2009). High *E. coli* counts (3.11–4.45 log CFU/g) were detected in all samples analysed in Lesotho (Ntuli et al. 2013). Mäde et al. (2017) showed the presence of Shiga toxin-producing *E. coli* in 39% of wheat flour samples in Germany, indicating a potential risk of foodborne disease.

Another microorganism that can cause spoilage of food prepared with contaminated flour is *Bacillus cereus* (Table 6.1). The growth of this bacterium in food products can pose a health risk to consumers, contributing to the incidence of foodborne diseases: emetic syndrome and diarrheal syndrome (Jovanovic et al. 2021). Emetic syndrome is poisoning caused by the ingestion of the heat-stable emetic toxin cereulide, which is produced in food. Diarrheal syndrome is caused by enterotoxins such as non-hemolytic enterotoxin (Nhe), non-hemolytic enterotoxin (Hbl), and cytotoxin K (CytK) which form in the human intestine after ingestion of vegetative cells or spores of enteropathogenic *B. cereus sensu stricto.*

Kindle et al. (2019) detected bacteria belonging to *Bacillus cereus sensu lato* in 75 out of 89 flour samples tested in Switzerland. Berghofer et al. (2003) reported that of 71 flour samples analyzed in Australia, as many as 93% were positive for *B. cereus*. However, in more recent studies on 300 samples of Australian flour, none tested positive for this foodborne pathogen (Eglezos 2010). All investigated flours in Brazil were found to be contaminated with *B. cereus* bacteria, in amounts ranging from 0.25 to 1.57 log CFU/g (Rossi et al. 2021). In a study in Turkey (Aydin et al. 2009) of 142 flour samples, only 4.2% tested positive for *B. cereus*, with maximum counts not exceeding 10^3 CFU/g. A study in Nigeria did not detect *B. cereus* in the tested flours (Nosa-Obamwonyi and Antai 2018). However, in Lesotho the counts of this bacterium were surprisingly high in 90% of flour samples, at 10^4–10^5 CFU/g (Ntuli et al. 2013). Although *B. cereus* cannot grow in dry flour and therefore does not present a hazard, it can negatively affect the quality of end products and pose a health risk (ICMFS 1998). There are few reports of illnesses related to the consumption of food containing flour contaminated with *B. cereus*. These cases concern diarrheal diseases caused by a wheat flour dessert in Bulgaria and two incidents related to bakery products in the Netherlands (Kindle et al. 2019).

Products made from flour contaminated with aerobic spore-forming bacteria of the genus *Bacillus* can be spoiled by a phenomenon known as ropiness. Ropiness may occur even with low counts of *Bacillus* spores in the flour, such as 10^3 spores/g (Vaičiulytė-Funk et al. 2015). According to the literature, in some flour samples the numbers of *Bacillus* spp. or their spores may be higher (Table 6.1). For example, in one study 65% of flour samples in Australia were contaminated with *Bacillus* spp. in amounts ranging from 10^2 to 10^5 CFU/g (Berghofer et al. 2003). Bacterial spores have also been detected in flour from Poland and Turkey, although usually at lower levels (Czerwińska and Piotrowski 2009; Czerwińska and Piotrowski 2010; Aydin et al. 2009). Pereira et al. (2020) reported about 0.5–3.3 log spores/g of mesophilic bacteria and 1.0–3.4 log spores/g of thermophilic bacteria in flour samples obtained from a bakery and the baking industry in Brazil. Of the 327 strains isolated from the flours, 45 produced amylase and were able to cause spoilage (Pereira et al. 2020).

These isolates were identified as *B. licheniformis*, *B. sonorensis*, *B. cereus*, *B. pumilus*, and *Paenibacillus polymyxa. Bacillus* spp., whereas mainly *B. laterosporus* and *Bacillus lentus* were found in rye flours from Poland. Interestingly, these bacterial species were also detected in the air of the production hall, but the pathway of contamination was undetermined (Czerwińska and Piotrowski 2010).

Most analyses of microbiological contamination of flours have aimed to enumerate and identify specific groups of bacteria, namely mesophilic bacteria, *Enterobacteriaceae*, coliforms, *Bacillus cereus*, or/and *Clostridium* spp. (Table 6.1). The use of modern genetic methods (PCR-DGGE, high-throughput sequencing) can provide detailed information about contaminated and undesirable flour microbiota. Such approaches can also be used to monitor normal and abnormal microbial communities during sourdough fermentation (De Filippis et al. 2017). Based on DNA and RNA analysis, Ercolini et al. (2013) found that wheat flours from Italy were contaminated mainly with bacteria belonging to the phylum *Proteobacteria, Bacteroidetes, Cyanobacteria*, and *Firmicutes*, while rye flours were mainly contaminated by *Proteobacteria* and *Firmicutes*. Metabolically active bacteria, such as *Acinetobacter, Pantoea, Pseudomonas, Comamonas, Enterobacter, Erwinia, Sphingomonas, Staphylococcus, Chryseobacterium*, and *Luteibacter*, predominated. Wheat flours from Italy were also characterized by a high predominance of *Proteobacteria* (over 90%), comprising the *Enterobacteriaceae, Xanthomonadaceae*, and *Moraxellaceae* families (De Angelis et al. 2019). In both cases, the composition of the individual bacteria changed significantly during sourdough fermentation (see Sect. 6.4.1).

Molds represent another important health hazard in flours, especially due to the mycotoxins they produce. Molds may be present in flour at various levels (Table 6.1), in the range of 10^2–10^5 CFU/g (Australia, Germany, Portugal, Turkey, USA), but in some flour samples their counts were significantly lower, <10^2 CFU/g (Poland, Serbia, Iran, Pakistan). In flour samples analysed by Berghofer et al. (2003), the predominant molds were *Penicillium, Aspergillus, Cladosporium, Aureobasidium* and *Eurotium. Penicillium verrucosum* and *Aspergillus ochraceus* strains isolated from flour are able to produce mycotoxins on laboratory media (International Commission on Microbiological Specifications for Foods 1998). Similar results were obtained by Plavšić et al. (2017), who identified 7 fungal genera (*Alternaria, Aspergillus, Cladosporium, Chrysonilia, Fusarium, Penicillium, Rhizopus*) in flour, with about 66.7% of the isolated molds belonging to potentially toxigenic species. The mycobiota of whole wheat flour and white wheat flour in Germany was overwhelmingly dominated by *Aspergillus* spp., with as many as 93.3% of isolated strains belonging to 32 fungal species known for their mycotoxin producing potential (Weidenbörner et al. 2000). In 2020, the United Kingdom notified the RASFF of aflatoxin contamination in wheat flour, detected during border control (RASFF notification 2020.3307).

When Ghali et al. (2008) analyzed the mycotoxin content in wheat flour in Tunisia, more than 30% of the samples were found to be contaminated with aflatoxins, with concentrations higher than the limits established by the European Union. The European Commission has set maximum limits (MLs) of 2 μg/kg for aflatoxin

B1 and 4 μg/kg for aflatoxins B1, B2, G1, and G2 in all cereals and all products derived from cereals, including processed cereal products. The ML for ochratoxin A is 3 μg/kg (European Commission 2006). Aflatoxins were also found to be dominant in 69% of flour samples from Bulgaria. On the other hand, studies by Kara et al. (2015) revealed the presence of ochratoxin in 16 out of 60 wheat flour samples from Bulgaria. Apart from aflatoxins and ochratoxins, mycotoxins produced by *Fusarium* spp. (*F. graminearum, F. crookwellense, F. culmorum*), deoxynivalenol and nivalenol, have been detected in both rye and wheat flour. More than half of flour samples collected from Denmark between 1998 and 2000 contained deoxynivalenol at a concentration of 20–527 μg/kg, and 13–26.5% nivalenol at a concentration of 10–440 μg/kg (Rasmussen et al. 2003; Thielecke and Nugent 2018).

In the Codex Alimentarius Standard for Wheat Flour (No. 152), there are no microbiological criteria. The standard only recommends that "the product shall be free from microorganisms in amounts which may represent a hazard to health". The microbiological requirements for flour recommended by other authorities, which can be used as a reference, are listed in Table 6.2. Based on the results of microbiological analyses, some authors have also proposed microbiological limits for wheat flour, intended to ensure high quality and consumer safety (Berghofer et al. 2003; Stannard 1997).

Table 6.2 Microbiological limits for wheat flour

Microorganism	n	c	m (CFU/g)	M (CFU/g)	References
Escherichia coli	5	2	10	10^2	Official Journal of Republic of Algeria No 39, 2017
Coagulase-positive staphylococci	5	2	10^2	10^3	
Bacillus cereus	5	2	10^3	10^4	
Molds	5	2	10^3	10^4	
Sulfite-reducing anaerobes	5	2	10^2	10^3	
Molds	5	2	10^2	10^4	GSO 1016:2015, Saudi Arabia
Bacillus cereus	5	2	10^3	10^4	
Clostridium perfringens	5	0	10^2	–	
Molds	5	2	10^2	10^4	FDA Philippines (2013)
Yeasts	5	2	10	10^2	
Coliform	5	2	10	10^2	
Bacillus subtilis "rope spores"	5	2	10	10^2	
Coliform bacteria	5	2	10^3	10^4	Turkish Food Codex (2011)
Molds	5	2	10^4	10^5	
	Maximum limit				
Total aerobic count	10^5 CFU/g				EAS 1:2018, East Africa
Escherichia coli	<1 × 10^2 CFU/g				
Salmonella spp.	absent in 25 g				
Yeast and moulds	10^4 CFU/g				
Staphylococcus aureus	10^2 CFU/25 g				

(continued)

Table 6.2 (continued)

Microorganism	n	c	m (CFU/g)	M (CFU/g)	References
Total aerobic plate count	10^5 CFU/g				CRS 64:2019, Barbados
Escherichia coli	not detectable in 1 g				
Salmonella spp.	not detectable in 1 g				
Yeast and molds	10^4 CFU/g				
Staphylococcus aureus	not detectable in 1 g				
Total plate count	10^3 CFU/g				ES 1052:2017, Ethiopia
Staphylococcus aureus	10^2 CFU/g				
Escherichia coli	absent in 1 g				
Salmonella spp.	absent in 25 g				
Coliforms	absent in 100 g				
Yeast and moulds	10^3 CFU/g				
Total viable counts	10^4 CFU/g				KS 169:2009, Kenya
Coliforms	absent in 1 g				
Escherichia coli	absent in 1 g				
Staphylococcus aureus	absent in 1 g				
Salmonella spp.	absent in 25 g				
Yeasts and moulds	10^4 CFU/g				
Bacterial rope spore count	10^2 CFU/g				
Mesophilic aerobes	$<10^4$ CFU/g				Berghofer et al. (2003)
Coliforms	$<10^2$ CFU/g				
Escherichia coli	<10 CFU/g				
Bacillus cereus	<10 CFU/g				
Mesophilic aerobic spores	$<10^2$ CFU/g				
Yeasts and moulds	$<10^3$ CFU/g				
	GMP (CFU/g)		**Maximum (CFU/g)**		
Bacillus cereus	$<10^2$		10^4		Stannard (1997)
Clostridium perfringens	$<10^2$		10^3		
Escherichia coli	$<10^2$		10^3		
Molds	$<10^3$		10^5		

Legend: *n* number of sample units to be examined, *m* value or level of microbiological criterion to be met in the food product, *c* the maximum number of sample units allowed to have a microbiological criterion value greater than "m" and below the value of "M", *M* the maximum criterion value that should not achieved or exceeded any of "n" units, *GMP* values expected following production of the food under good manufacturing conditions (occasional samples may exceed these limits and such results should not be used as rejection criteria without consideration of all the factors important in an entire sampling plan), *Maximum* the maximum acceptable values at any point in the shelf life of the product

Although various parameters of flour, and above all moisture content, prevent the active growth of microorganisms, it may become an important source of microbial contamination in the production chain of bread. Many microorganisms have developed mechanisms of adaptation to unfavourable environmental conditions. The link between flour and the final product is sourdough. The survival of microorganisms transferred from the flour to the fermenting flour-water mixture depends on the competitiveness of the particular microbial species present under the harsh conditions of the sourdough ecosystem. Their survival and growth are also determined by the presence of appropriate substrates, whether or not provided by the ingredients.

6.3 Key Players Determining the Composition and Contamination of Sourdough

Sourdough is an intermediate product for dough and bread preparations. It contains metabolically active microorganisms, which impact the nutritional and sensory properties of bakery products. The sourdough in each geographical region contains its own unique cocktail of microbes, which varies in composition and number depending on the artisanal tradition and region-dependent processes (De Vuyst and Neysens 2005). Sourdoughs are an enormous source of diverse microorganisms, mostly lactic acid bacteria (LAB) and yeast species. Some undesirable microbial strains may also occur, as sourdough contaminants (Fig. 6.1). Microbiological spoilage has been recognized as a key limiting factor for the shelf life of bakery products, causing considerable economic losses.

The sourdough research to date suggests that the connections between the sourdough microbiome and physical, chemical, and biological factors driving its composition and changes cause the selection of characteristic LAB and yeast microflora during the propagation of sourdough, and prevent the growth of other unwanted microorganisms originating from the raw materials or the bakery environment. Yeasts present in sourdough act as leavening agents in bakery production, mostly due to sugar fermentation resulting in the release of carbon dioxide, which is a by-product of their metabolism. Lactic acid bacteria influence the ecosystem of sourdough and its properties. Because they serve as acidifying agents and produce volatile compounds, and additionally heterofermentative LAB species are involved in sourdough leavening (Calvert et al. 2021).

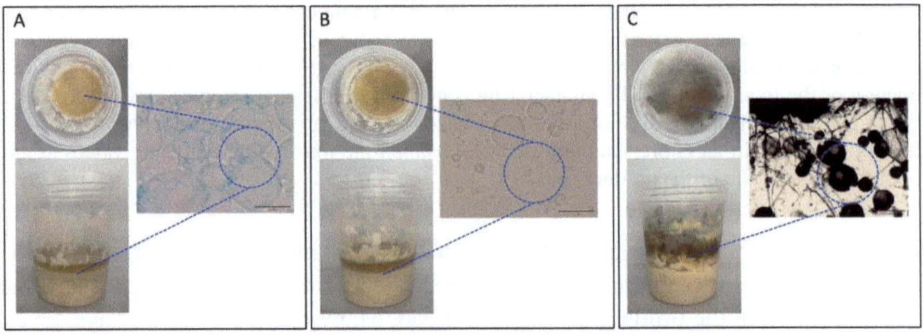

Fig. 6.1 Undesirable microorganisms in rye sourdough: (**a**) bacteria; (**b**) yeast; (**c**) molds

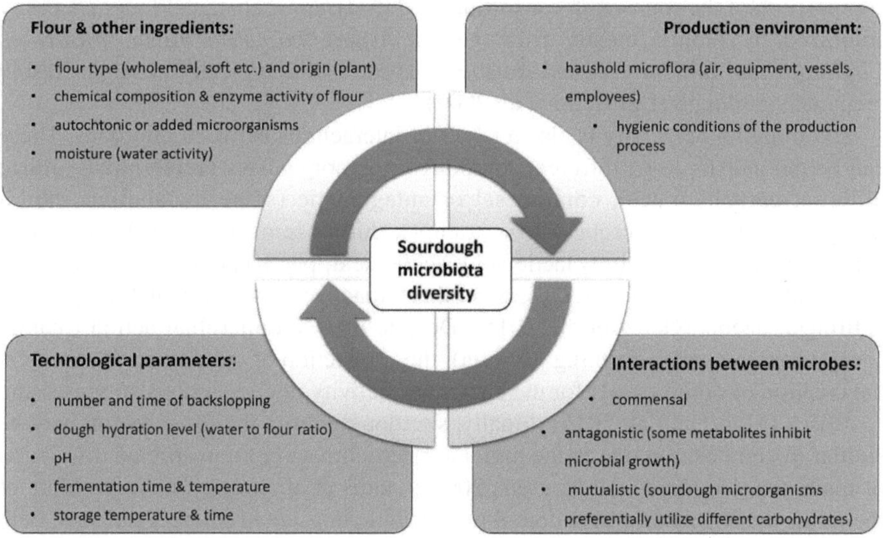

Fig. 6.2 Factors determining the composition and contamination of sourdough

The microbial community in sourdough impacts the quality of bakery products and is formed in connection to many factors of chemical, physical and biological origin (Fig. 6.2). Generally, factors driving sourdough microbial community divergence can be categorized as intrinsic or exogenous. Sourdough formation processes are influenced by conditions intrinsic to the sourdough itself, including the ingredients and the recipe, as well as the microbial consortium and its dynamic change. Extrinsic factors cover propagation and baking practices, the temperature of sourdough storage and propagation, the time and number of propagation steps, the presence of oxygen, the initial pH, water activity (a_w), as well as the flour to water ratio (also known as dough yield). Endogenous factors are strongly affected by the chemical and microbiological composition of the sourdough ecosystem, whereas exogenous factors are mainly affected by temperature and redox potential (Calvert et al. 2021).

The ecology of sourdough microbiota depends on the ingredients and recipe used, how the sourdough is maintained, and the type of the sourdough. Four types of sourdough can be prepared. They are distinguished on the basis of the inoculum and the desired properties of the final bakery products, such as texture, flavour, smell, stability, and nutritional value. An in-depth comparison of sourdough characteristics and re-feeding practices (including inoculum dosage, temperature, and the time of each step during sourdough maintenance) was conducted by Pérez-Alvarado et al. (2022). Sourdough ingredients include cereal flour with the possible addition of pseudocereals and/or legumes, water, salt, and other ingredients such as sugar, and a leavening agent (Pérez-Alvarado et al. 2022). The type and origin of the flour seem to be important factors for the sourdough microbiome. The numbers of autochthonic microflora from raw materials, especially different cereal flours, affect the composition of the sourdough consortium and its dynamic change. Flours are easily colonized by molds, mostly from genera *Aspergillus, Penicillium, Fusarium, Cladosporium, Endomyces*, and *Rhizopus*. These molds typically occur in freshly prepared sourdough (Dal Bello et al. 2007).

Sourdough biodiversity is also a result of interactions between microbes. These can be mutualistic (e.g., different sourdough microorganisms preferentially utilize different carbohydrates), commensal or antagonistic (some metabolites inhibit the growth of other microbes). For example, during fermentation LAB produces various metabolites, such as lactic acid, acetic acid, phenyllactic acid, short-chain fatty acids (SCFAs), bacteriocins, and other bioactive peptides, which may act as antifungal agents (Gerez et al., 2015). Despite their antimicrobial action (mainly against ropiness spore-forming bacteria), the production of organic acids results in the creation of optimum pH for the enzymatic activity of endogenous amylases and proteases (Mohsen et al. 2016). Finally, sourdoughs from the same region may be similar in composition due to the regional microclimate or the restricted dispersion of microbes (De Vuyst and Neysens 2005; Landis et al. 2021). The microbiota of the air in the bakery has been found to distribute household microorganisms, both bacteria and fungi, characteristic for the specific facility, including spores (Wlazło et al. 2020). Microbes from vessels or other equipment, or even from the bodies of bakers, may also contaminate sourdough. The sourdough consortium can also by affected by random contamination with lytic bacteriophages (Foschino et al. 2005). Both the temperature of sourdough storage and the temperature of fermentation during the feeding protocol have an impact on the microbial biodiversity of sourdough, affecting the organoleptic and technological quality of sourdough-based products (Urien et al. 2019; Katsi et al. 2021; Oshiro et al. 2021; Reese et al. 2020).

6.4 Microbial Contaminants of Sourdough: A Main Risk Factor in Bread Spoilage

6.4.1 Bacterial Contamination in the Process of Bakery Production

There is little data on the direct occurrence of different microorganisms in sourdoughs because the acidic pH of sourdough inhibits their proliferation. Nonetheless, the presence of bacteria in the bread production chain should not be ignored. De Angelis et al. (2019) compared the microbiota of flours (soft wheat flour, wholemeal flour, and durum wheat flour) and sourdoughs based on them, employing high-throughput sequencing. Soft and durum wheat flours were characterized by the highest relative abundance of *Enterobacteriaceae*. Wholemeal wheat flours contained the lowest amounts. In the wholemeal flours, *Xanthomonadaceae* (mainly *Xanthomonas albilineans* and *X. oryzae*) dominated, followed by *Staphylococcaceae* (*Staphylococcus gallinarum* and *S. kloosii*) and *Leuconostocaceae*. After 8 h of fermentation, the sourdough microbiota was close to that of wholemeal flour. This was in contrast to sourdoughs obtained from soft wheat flour, where a sharp increase was observed in the amounts of bacteria from the *Enterobacteriaceae* family (*Enterobacter, Serratia, Erwinia*), as well as *Clostridium* and *Enterococcus* (Table 6.3). This was explained by the chemical composition of the flour and the higher buffer capacity of the sourdough, which did not allow the pH of the dough to drop quickly. On the other hand, Ercolini et al. (2013) reported that after 8 h of propagation at 25 °C the undesired bacteria from flour were completely eliminated, except for the *Enterobacteriaceae*. The structure of microbial community depends on the type and origin of the flour and number and time of propagation.

Contamination by *Bacillus* spp. is unavoidable at every stage of bread production because these bacteria originate from the raw ingredients and are part of the endophytic microbiota of cereals. Moreover, they are able to form endospores that can survive processing, baking, and storage (Thompson et al. 1998). Their endospores are not only thermally stable, but also highly resistant to chemical substances and radiation, as well as desiccation. *Bacillus* spp. can growth at temperatures from 28 to 40 °C. The maximum growth temperature may be as high as 55 °C. Moreover, these bacteria can grow in the pH range of 4.9–9.3, as well as in the presence of NaCl at concentrations of <5–10% (Khanna et al. 2020; Pacher et al. 2022). The most favourable conditions for endospore germination are temperatures above 25 °C in combination with aw ≥0.95 and pH > 5. Decimal reduction times at 100 °C (D_{100}) reflect the heat resistance of endospores. For *Bacillus* spp., the D_{100} is from 10 to 56 min (Leuschner et al. 1998).

Traditional sourdough bread production can limit the growth and germination of endospores. It has been proven that with increasing sourdough content in the process, the thermal resistance of endospores decreases (Pepe et al. 2003). However, the process of making sourdough bread is time consuming compared to other simpler technologies and the spores are not always fully eliminated. Czerwińska and Piotrowski (2010) showed the presence of a higher number of mesophilic bacteria in rye sourdough (3.0×10^2–4.4×10^2 CFU/g) than in rye flour (8.0×10–7.8×10^2 CFU/g). A spore-forming bacteria in the studied sourdough (3.3×10 CFU/mL) was identified as *Bacillus subtilis. Bacillus subtilis* also accounted for 70% of microbial contaminants in bakery ingredients analyzed by Rosenkvist and Hansen (1995). *Bacillus licheniformis, B. pumilus*, and the *B. cereus* group were also detected, but only *B. subtilis* isolates were associated with ropiness. These results are consistent with a study by Sorokulova et al. (2003), in which 30 bacterial strains were isolated from flour and laboratory-baked bread. The percentage of *B. subtilis* was 80%, the other 20% being *B. licheniformis.* Almost half of the isolates showed the ability to cause ropiness in bread. Pepe et al. (2003) combined the randomly amplified polymorphic DNA PCR assay (RAPD-PCR) and denaturing gradient gel electrophoresis (DGGE) to detect bacteria in white wheat bread. All isolated strains belonging to *B. subtilis, B. licheniformis, B. cereus, B. clausii* and *B. firmus* species were characterized by amylase activity, but just over 30% were capable of producing ropiness.

Table 6.3 Bacterial microbiota predominated in traditional or artisan bakeries sourdoughs in different regions of Europe

	Denmark	Italy	United Kingdom	Ukraine	Poland	Switzerland	Estonia
***Acinetobacter* spp.**		Ercolini et al. (2013)					
Bacillus amyloliquefaciens		Valerio et al. (2012)					
Bacillus atrophaeus		Thompson et al. (1998)					
Bacillus cereus	Rosenkvist and Hansen (1995)	Pepe et al. (2003) and Valerio et al. (2012)				Kindle et al. (2019)	
Bacillus clausii		Pepe et al. (2003)					
Bacillus licheniformis	Rosenkvist and Hansen (1995)	Pepe et al. (2003)	Thompson et al. (1998)	Sorokulova et al. (2003)			
Bacillus pumilus	Rosenkvist and Hansen (1995)		Thompson et al. (1998)				
Bacillus subtilis		Pepe et al. (2003) and Valerio et al. (2012)	Thompson et al. (1998)	Sorokulova et al. (2003)	Czerwińska and Piotrowski (2010)		
Bacillus thuringiensis						Kindle et al. (2019)	
***Comamonas* spp.**		Ercolini et al. (2013)					
Chryseobacterium		De Angelis et al. (2019) and Ercolini et al. (2013)					
Clostridium aurantibutyricum		De Angelis et al. (2019)					
Clostridium saccharobutylicum		De Angelis et al. (2019)					

Cytobacillus firmus		Pepe et al. (2003)					
Enterobacter amnigenus		De Angelis et al. (2019)					
Enterobacter asburiae		De Angelis et al. (2019)					
Enterobacter hormaechei		De Angelis et al. (2019)					
Enterococcus hermanniensis		De Angelis et al. (2019)					
***Enterobacter* spp.**		Ercolini et al. (2013)					Bessmeltseva et al. (2014)
Erwinia tasmaniensis		De Angelis et al. (2019)					
***Erwinia* spp.**		Ercolini et al. (2013)					
***Pantoea* spp.**		Ercolini et al. (2013)					Bessmeltseva et al. (2014)
Paenibacillus polymyxa		Valerio et al. (2012)					
Priestia megaterium		Valerio et al. (2012)	Thompson et al. (1998)				
***Sphingomonas* spp.**		Ercolini et al. (2013)					
Staphylococcus gallinarum		De Angelis et al. (2019)					
Staphylococcus kloosii		De Angelis et al. (2019)					
Xanthomonas albilineans		De Angelis et al. (2019)					
Xanthomonas oryzae		De Angelis et al. (2019)					

6.4.2 Yeast Contamination in the Process of Bakery Production

Flour may be an important source of microorganisms in sourdough. However, there have been only a few scientific attempts to identify yeasts in flour, and the species classification of mycobiota mainly concerns molds. This is likely due to the fact that wild-type yeasts have been mostly replaced by commercial *Saccharomyces cerevisiae* strains, which are commonly used as starter cultures. Weidenbörner et al. (2000) did not list any yeasts in the mycobiota of wheat flour, although 45 fungal species were isolated in the whole wheat flour and 49 species were found in the white flour. Similarly, no yeast species were identified among fungi detected in 160 flour and dough samples originating from Iran (Abedi et al. 2020). These results may suggest that yeast is present at only a very low level in flour. However, in 650 samples from the milling process and final products, yeasts were detected frequently in amounts above 10^2 CFU/g. No information was provided on the yeasts among the identified fungi (Berghofer et al. 2003). In wheat flour originating from Lesotho, yeasts belonging to the genera *Candida*, *Saccharomyces*, and *Rhodotorula* were found (Ntuli et al. 2013).

It seems that the processing environment may also be an important source of microorganisms in sourdough (Reese et al. 2020). A study of 18 bakers who used the same flour to make a single starter each showed that about 26% of bacterial and yeast taxa were shared between the skin microbiome on the bakers' hands and the sourdough. Moreover, the starter communities were more varied than expected and predictably were most similar to those in flour. In terms of yeasts, strains belonging to *Saccharomycetales* predominated in almost 90% of the starters (Reese et al. 2020). In the starter mycobiota, in addition to *Saccharomyces cerevisiae* and *Saccharomyces bayanus*, the yeasts *Kazachstania humilis*, *Kazachstania unispora*, *Kazachstania castelli*, and *Debaryomyces hansenii* were also identified.

This relationship is even more strongly emphasized by other published results (Minervini et al. 2015; Comasio et al. 2020), which indicate that the microbial diversity of sourdoughs is influenced more by the bakery environment than the cereal flour. Moreover, flour does not appear to function as a vector for yeasts, since any yeast species found in sourdough were detected in the flours (von Gastrow et al. 2021). Even when the baker uses different flours, the sourdoughs contain similar microbial communities (Comasio et al. 2020). On the other hand, comparative studies of mycobiota in bakery sourdoughs and sourdoughs made in the laboratory from the same flours showed differences in species composition (Vrancken et al. 2010). The dominant yeasts in the bakery sourdoughs were *Saccharomyces cerevisiae* and *Wickerhamomyces anomalus*. In the laboratory sourdough, *Wickerhamomyces anomalus* and *Candida glabrata* dominated. The presence of *S. cerevisiae* may be caused by contamination with commercial yeast. However, it appears that yeasts in laboratory sourdough obtained under aseptic conditions originated from the flour, as it was the only non-sterile ingredient.

Generally, yeast diversity in sourdough is lower than the diversity of lactic acid bacteria, as are their counts, which are usually around 10^7 CFU/g vs. 10^8–10^9 CFU/g (von Gastrow et al. 2021; De Vuyst et al. 2014; Lhomme et al. 2015). Typical sourdough yeast species in Europe include *Saccharomyces cerevisiae* and *Candida humilis* (current name *Kazachstania humilis*, synonyms *Candida milleri*, *Torulopsis acidi-lactici*, *Torulopsis humilis*) (Table 6.4). Other frequently reported species in wheat sourdoughs are *Wickerhamomyces anomalus* (anamorph *Candida pelliculosa*, synonyms inter alia *Pichia anomala*, *Hansenula anomala*, *Saccharomyces anomalus*, *Saccharomyces sphaericus*, *Hansenula sphaerica*, *Candida javanica*), and *Torulaspora delbrueckii* (anamorph *Candida colliculosa*, synonyms *Saccharomyces delbrueckii*, *Zygosaccharomyces delbrueckii*, *Debaryomyces delbrueckii*). Less frequently, yeasts belonging to the species *Kazachstania exigua* (formerly *Saccharomyces exiguus*, synonyms *Torulaspora exigua*, *Torula holmii*, *Torulopsis holmii*, *Cryptococcus holmii*, *Candida holmii*), *Kazachstania bulderi* (synonym *Saccharomyces bulderi*), *Kazachstania servazzii* (synonym *Saccharomyces servazzii*), and *Pichia membranifaciens* (anamorph *Candida valida*, synonyms inter alia *Saccharomyces membranifaciens*, *Saccharomyces piriformis*, *Pichia hyalospora*, *Saccharomyces anomalus*, *Hansenula belgica*, *Pichia alcoholophila*) have been reported (www.indexfungorum.org; Kurtzman et al. 2011).

Although different yeast species can be expected in sourdough depending on geographical region and processing conditions, a single sourdough usually contains only one or two yeast species at a time (De Vuyst et al. 2014; Vrancken et al. 2010). There is no information available about yeasts that contaminate sourdough and lead to spoilage. However, since most yeast species are also found in other ecosystems and usually come from the production environment, not from flour (Minervini et al. 2015; Comasio et al. 2020; von Gastrow et al. 2021), it cannot be excluded that some of the detected yeasts are not natural and desirable sourdough microbiota. Some yeasts, such as *Pichia fermentans* (anamorph *Candida lambica*) and *Pichia membranifaciens* (anamorph *Candida valida*), are typically found in fermented food, such as sausages, cheeses, and wine (Succi et al. 2003). However, other unusual strains found in sourdough including *Candida boidinii*, *Candida boleticola*, *Candida carpophila*, *Candida diddensiae*, *Debaryomyces udenii*, *Hansenula subpelliculosa*, and *Pichia ohmeri* may raise greater concern. Sources of *Candida boleticola* (mushrooms, soil, rotten leaves, and trunks) indicate a connection with fungi and an association with tree decay (Kurtzman et al. 2011). Likewise, the presence of *Candida carpophila* in sourdough, associated with insect-plant interactions, is also atypical (Kurtzman et al. 2011). *Candida boidinii* has been isolated from a variety of sources, sometimes related to human activity, and is often spread by drosophilid flies, which may explain its presence in sourdough (Kurtzman et al. 2011; Pino et al. 2022). Probably, insects are also the source of *Candida diddensiae*, whereas *Debaryomyces udenii* has hitherto been isolated only from soil samples (Kurtzman et al. 2011).

Hansenula anomala can be a microbiological spoilage problem in refrigerated dough products with high sugar content. These osmophilic yeasts metabolize sugars to carbon dioxide and alcohol, producing an undesirable odour and flavour during baking (Cook and Johnson 2009). Another species that grows on high osmotic food

Table 6.4 Yeasts microbiota predominated in traditional or artisan bakeries sourdoughs in different regions of Europe

	France	Bulgaria	Italy	Belgium	Greece	Germany	Ireland	Portugal	Spain
Candida boidinii			Pino et al. (2022)						
Candida boleticola	Urien et al. (2019)								
Candida carpophila	Urien et al. (2019)								
Candida diddensiae			Pino et al. (2022)						
Candida glabrata			Succi et al. (2003)	Vrancken et al. (2010)					
Debaryomyces udenii	Urien et al. (2019)								
Dekkera anomala				Comasio et al. (2020)					
Hansenula subpelliculosa									Barber et al. (1983)
Issatchenkia orientalis			Pulvirenti et al. (2004) and Valmorri et al. (2010)						
Kazachstania barnettii	Urien et al. (2019)	Petkova et al. (2021)					Moroni et al. (2011)		
Kazachstania bulderi	Urien et al. (2019) and von Gastrow et al. (2021)			Comasio et al. (2020)					
Kazachstania exigua	Urien et al. (2019)		Pulvirenti et al. (2004) and Minervini et al. (2015)						

Kazachstania humilis	Urien et al. (2019) and von Gastrow et al. (2021)	Petkova et al. (2021)	Pulvirenti et al. (2004), Di Cagno et al. (2014), Lattanzi et al. (2013), Foschino et al. (2004), Valmorri et al. (2010), Lombardi et al. (2007), and Venturi et al. (2012)		Syrokou et al. (2020)	Vogelmann et al. (2009)			
Kazachstania saulgeensis	Urien et al. (2019)								
Kazachstania servazzii	Urien et al. (2019) and Lhomme et al. (2015)		Di Cagno et al. (2014)						
Kazachstania unispora	Urien et al. (2019)		Ripari et al. (2016)						
Kluyveromyces marxianus		Petkova et al. (2021)							
Metschnikowia pulcherrima			Palomba et al. (2011)						
Pichia fermentans		Petkova et al. (2021)	Succi et al. (2003)		Syrokou et al. (2020)				
Pichia kluyveri			Pino et al. (2022)						
Pichia membranifaciens			Succi et al. (2003)		Syrokou et al. (2020) and Paramithiotis et al. (2000)				

(continued)

Table 6.4 (continued)

	France	Bulgaria	Italy	Belgium	Greece	Germany	Ireland	Portugal	Spain
Pichia ohmeri								Rocha and Malcata (1999)	
Pichia polymorpha									Barber et al. (1983)
Saccharomyces barnettii			Ripari et al. (2016)						
Saccharomyces bayanus			Di Cagno et al. (2014)						
Saccharomyces cerevisiae	Urien et al. (2019), Comasio et al. (2020), Lhomme et al. (2015), and von Gastrow et al. (2021)	Petkova et al. (2021)	Ripari et al. (2016), Pino et al. (2022), Pulvirenti et al. (2004), Di Cagno et al. (2014), Lattanzi et al. (2013), Minervini et al. (2015), Succi et al. (2003), Foschino et al. (2004), Valmorri et al. (2010), Lombardi et al. (2007), Palomba et al. (2011), and Venturi et al. (2012)	Comasio et al. (2020) and Vrancken et al. (2010)	Syrokou et al. (2020), Katsi et al. (2021), and Paramithiotis et al. (2000)	Vogelmann et al. (2009)		Rocha and Malcata (1999)	Barber et al. (1983)
Torulaspora delbrueckii			Pino et al. (2022), Di Cagno et al. (2014), Valmorri et al. (2010), and Succi et al. (2003)					Rocha and Malcata (1999)	

Trichosporon margaritiferum									Barber et al. (1983)
Wickerhamomyces anomalus			Ripari et al. (2016) and Pino et al. (2022)	Vrancken et al. (2010)	Syrokou et al. (2020)			Rocha and Malcata (1999)	
Yarrowia lipolytica		Petkova et al. (2021)			Paramithiotis et al. (2000)				

and causes a loss of quality is *Hansenula subpelliculosa*, which is also found in sourdough in Spain (Barber et al. 1983; Kurtzman et al. 2011).

The most studied alterations in the physical and sensorial properties of unbaked bread dough caused by yeasts spoilage are surface growth, gas production, discoloration, and off-flavours (Loureiro and Querol 1999). In sourdoughs, it is difficult to distinguish between the beneficial activity of yeasts, causing desirable organoleptic characteristics in the product, and activity causing spoilage (Fleet 1992). Sourdough spoilage cannot be also determined by the number of yeasts, unlike other foods where yeast counts above 10^5 cells/g can be an indicator of product deterioration.

Some yeast species present in sourdough appear on the list of emerging opportunistic pathogens i.e., *Kodamaea ohmeri* (synonym *Pichia ohmeri*) (Zhou et al. 2021), *Candida diddensiae* (Kim et al. 2020), *Candida colliculosa* (Kaygusuz et al. 2003), *Issatchenkia orientalis* (syn. *Candida krusei*) (Opulente et al. 2019). It should be emphasized that yeast spoilage is the least common of all types of microbial spoilage of sourdough. Moreover, yeasts are unlikely to be a risk because they do not survive the baking process. However, yeast contamination of final products may occur later during cooling (Melini and Melini 2018).

6.4.3 Mold Contamination in the Process of Bakery Production

The role of molds in spoilage of sourdough-based bakery products has been extensively examined and demonstrated in the literature. Molds are considered the dominant group of microorganisms responsible for sourdough spoilage. Bakery products are an excellent substrate for the development of mold spoilage. The typical chemical composition, pH (5.5–6.0), and water activity (0.95–0.98) of bread make it susceptible to mold development (Garcia et al. 2019b; Magan et al. 2003). Garcia et al. (2019a) reviewed the literature on the fungal spoilage of cereal-based products, its economic impact and causes. In Europe, economic losses related just to the mold contamination of bread (assessed in the range of 1–5%) have been estimated at more than £200 million per year. In some other regions of the world (in countries with higher temperatures and humidity), losses related to fungal deterioration in baked goods are reach more than 10%. Freshly prepared bakery products do not possess mold mycelium (vegetative cells) or spores but are susceptible to microbial deterioration. Spore contamination may occur during cooling, slicing, and packaging of baked products. Baked products may be contaminated with spores from the air and equipment, or transferred by bakery personnel. Therefore, bakery products often have a short shelf life (a few days when stored at room temperature) (Axel et al. 2016). Biodeterioration caused by molds starts after the appearance of visible mycelium from germinated spores that develop on the product surface. After spore germination and mycelium growth, the activity of exoenzymes in mold, such as lipases, proteases, and carbohydrases, causes changes in the sensory properties of bakery products (mainly off-flavours and odour, discolouring), as well as their texture (firming) (Baert et al. 2007; Dagnas and Membré 2013; Filtenborg et al. 1996), Garcia and Copetti 2019; Nielsen and Rios 2000).

Molds belonging to diverse genera including *Aspergillus, Penicillium, Rhizopus, Cladosporium, Fusarium, Wallemia, Mucor, Hyphopichia*, *Chrysonilia*, and *Endomyces* are known to colonize bakery products (Dal Bello et al. 2007; Garcia and Copetti 2019; Legan 1993; Quattrini et al. 2019). Molds responsible for spoilage in the bakery industry may occur in large numbers and varieties (see Tables 6.2 and 6.5). In a study by Demirel and Sariozlu (2013), the total numbers of molds found in samples of flour from different cereals were between 2.0×10^4 and 8.5×10^4 CFU/g. Alborch et al. (2012) found that the numbers of molds in corn flour were from <10 to 8.8×10^4 CFU/g. Plavšić et al. (2017) showed that the total number of molds in samples of whole grain wheat flour exceeded 1.0×10^4 CFU/g. This is in agreement with a study by Asadzadeh et al. (2014), who found the presence of molds in the samples of wheat flour in the range from 10^2 to 10^4 CFU/g. Rezazadeh et al. (2013) determined the presence of molds in flour in amounts greater than 10^4 CFU/g.

The diversity of mold species depends mostly on the geographical source and the nature of cereals. There is a comprehensive literature data on the quality characteristics of molds contaminating wheat flour and derivative products. According to Legan and Voysey (1991), wheat products are mostly spoiled by *Penicillium* spp. (ca. 90%). Gerez et al. (2009) and Legan (1993) identified molds including *Penicillium* spp. and *Aspergillus* spp. as the main causes of bread spoilage. In studies by Demirel and Sariozlu (2013), the most commonly found molds belonged to the genera *Aspergillus* and *Penicillium*. According to a study by Plavšić et al. (2017), the most frequently observed molds were *Aspergillus, Penicillium, Eurotium, Paecilomyces*, and *Rhizopus*. Apart from *Aspergillus* and *Penicillium*, representatives of the genera *Acremonium, Alternaria, Cladosporium, Eurotium, Fennellia, Fusarium, Mucor, Paecilomyces, Rhizopus, Scopulariopsis, Trichoderma,* and *Ulocladium* were detected in bread by Demirel and Sariozlu (2013). Similarly, Weidenbörner et al. (2000) identified *Aspergillus* spp. followed by *Penicillium* spp. as the dominant molds in German samples. Bread samples from Serbia were contaminated by molds classified as *Alternaria, Aspergillus, Cladosporium, Chrysonilia, Fusarium, Penicillium, Rhizopus,* and *Scopulariopsis,* with the highest number of species from the genus *Aspergillus* (Plavšić et al. 2017). In research by Čonková et al. (2006), *Fusarium* spp. contributed to wheat spoilage in Poland and Slovakia (95.5% and 70.5% of samples, respectively). Analyzed Polish samples also contained *Aspergillus* spp. (81.8%) and *Penicilium* spp. (72.3%), whereas *Penicillium* spp. (68.2% of samples), *Aspergillus* spp. (61.4%), and *Cladosporium* spp. (45.5%) were found in Slovak samples. In the same studies, barley samples were found to contain *Penicillium* spp., *Alternaria* spp. and *Cladosporium* spp. Minutillo et al. (2022) characterized the mycobiome in different types of Italian wheat flour, using culture-dependent and culture-independent techniques. They detected molds in amounts from 53 CFU/g to 1.8×10^3 CFU/g. Species of *Penicillium, Alternaria,*

Table 6.5 Molds microbiota involved in production of sourdough-based products in Europe (nonexhaustive list)

	France	Germany	Greece	Italy	Poland	Serbia	Spain	Slovakia	Turkey	United Kingdom
Absidia corymbifera		Weidenbörner et al. (2000)								
***Acremonium* spp.**[a]		Weidenbörner et al. (2000)							Demirel and Sariozlu (2013)	
***Alternaria* spp.**[a]		Legan (1993) and references therein	Katsi et al. (2021)	Minutillo et al. (2022)	Čonková et al. (2006)			Čonková et al. (2006)	Demirel and Sariozlu (2013)	
Alternaria alternata		Weidenbörner et al. (2000)				Plavšić et al. (2017)				
Alternaria infectoria				Minutillo et al. (2022)						
***Arthrinium* spp.**[a]	Urien et al. (2019)			Minutillo et al. (2022)						
***Aspergillus* spp.**[a]	Dagnas and Membré (2013)	Weidenbörner et al. (2000) and Legan (1993) and references therein		Minutillo et al. (2022)	Čonková et al. (2006)		Alborch et al. (2012)	Čonková et al. (2006)		Legan (1993) and references therein
A. alliaceus									Demirel and Sariozlu (2013)	
A. caesiellus	Urien et al. (2019)									
A. cibarius	Urien et al. (2019)									

A. flavus	Urien et al. (2019) and Dagnas and Membré (2013)	Weidenbörner et al. (2000)				Plavšić et al. (2017)	Alborch et al. (2012)		Demirel and Sariozlu (2013)	
A. fumigatus		Weidenbörner et al. (2000)					Alborch et al. (2012)			
A. japonicus									Demirel and Sariozlu (2013)	
A. niger	Dagnas and Membré (2013)					Plavšić et al. (2017)	Alborch et al. (2012)		Demirel and Sariozlu (2013)	Legan (1993) and references therein
A. ochraceus	Dagnas and Membré (2013)	Legan (1993) and references therein					Alborch et al. (2012)		Demirel and Sariozlu (2013)	
A. penicillioides	Urien et al. (2019)						Alborch et al. (2012)			
A. subversicolor	Urien et al. (2019)									
A. tubingensis							Alborch et al. (2012)		Demirel and Sariozlu (2013)	

(continued)

Table 6.5 (continued)

	France	Germany	Greece	Italy	Poland	Serbia	Spain	Slovakia	Turkey	United Kingdom
A. versicolor		Weidenbörner et al. (2000)				Plavšić et al. (2017)	Alborch et al. (2012)			
Botrytis caroliniana	Urien et al. (2019)									
***Chrysonilia* spp.**[a]						Plavšić et al. (2017)				
***Cladosporium* spp.**[a]			Katsi et al. (2021)	Minutillo et al. (2022)	Čonková et al. (2006)			Čonková et al. (2006)	Demirel and Sariozlu (2013)	Legan (1993) and references therein
Cladosporium cladosporioides		Weidenbörner et al. (2000)								
Claviceps purpurea	Urien et al. (2019)		Katsi et al. (2021)							
Epicoccum nigrum		Weidenbörner et al. (2000)		Minutillo et al. (2022)						
***Eurotium* spp.**[a]		Weidenbörner et al. (2000)	Katsi et al. (2021)				Alborch et al. (2012)		Demirel and Sariozlu (2013)	
E. amstelodami		Weidenbörner et al. (2000)					Alborch et al. (2012)			
E. herbariorum		Weidenbörner et al. (2000)								

***Fennellia* spp.**[a]									Demirel and Sariozlu (2013)	
***Fusarium* spp.**[a]	Urien et al. (2019) and Dagnas and Membré (2013)	Weidenbörner et al. (2000)			Čonková et al. (2006)	Plavšić et al. (2017)	Alborch et al. (2012)	Čonková et al. (2006)	Demirel and Sariozlu (2013)	
F. oxysporum				Minutillo et al. (2022)						
F. poae			Katsi et al. (2021)							
Lichtheimia corymbifera				Minutillo et al. (2022)						
***Mucor* spp.**[a]				Minutillo et al. (2022)	Čonková et al. (2006)			Čonková et al. (2006)	Demirel and Sariozlu (2013)	
***Paecilomyces* spp.**[a]			Katsi et al. (2021)						Demirel and Sariozlu (2013)	
***Penicillium* spp.**[a]	Dagnas and Membré (2013)	Legan (1993) and references therein	Katsi et al. (2021)	Minutillo et al. (2022)	Čonková et al. (2006)	Plavšić et al. (2017)	Alborch et al. (2012)	Čonková et al. (2006)		Legan (1993) and references therein

(continued)

Table 6.5 (continued)

	France	Germany	Greece	Italy	Poland	Serbia	Spain	Slovakia	Turkey	United Kingdom
P. aurantiogriseum		Weidenbörner et al. (2000)		Minutillo et al. (2022)						
P. brevicompactum		Weidenbörner et al. (2000)		Minutillo et al. (2022)						
P. griseofulvum		Weidenbörner et al. (2000)		Minutillo et al. (2022)			Alborch et al. (2012)			
P. implicatum		Weidenbörner et al. (2000)								
P. janczewskii		Weidenbörner et al. (2000)								
P. polonicum	Urien et al. (2019) and Dagnas and Membré (2013)			Minutillo et al. (2022)						
P. verrucosum		Weidenbörner et al. (2000)		Minutillo et al. (2022)			Alborch et al. (2012)			
P. viridicatum		Weidenbörner et al. (2000)		Minutillo et al. (2022)					Demirel and Sariozlu (2013)	Legan (1993) and references therein
***Rhizopus* spp.**[a]			Katsi et al. (2021)						Demirel and Sariozlu (2013)	
R. oryzae				Minutillo et al. (2022)						

R. stolonifer		Weidenbörner et al. (2000)				Plavšić et al. (2017)				
***Scopulariopsis* spp.**[a]									Demirel and Sariozlu (2013)	
S. brevicaulis		Weidenbörner et al. (2000)				Plavšić et al. (2017)				
Sporisorium panici	Urien et al. (2019)									
***Trichoderma* spp.**[a]									Demirel and Sariozlu (2013)	
***Ulocladium* spp.**[a]									Demirel and Sariozlu (2013)	
Wallemia sebi	Urien et al. (2019)									

[a] Other or non-identified species belonging to the genera

Cladosporium, Mucor, Epicoccum, Arthrinium, Lichtheimia, and *Rhizopus* were isolated and identified as representing the Italian mycobiota.

In comparison to the extensive research on the fungal microbiome of flours and grains, even when we look only at literature data related strictly to Europe, there is little information about the presence and diversity of molds in sourdough. Urien et al. (2019) analyzed the fungal diversity in 14 sourdoughs collected from French bakeries made from organic wheat flour using different bread-making practices. The presence was found of *Aspergillus caesiellus, A. cibarius, A. flavus, A. penicillioides, A. subversicolor, Botrytis coroliniana, Penicillium polonicum, Sporisorium panici,* and *Wallemia sebi.* Known plant fungal pathogens *Claviceps purpurea, Fusarium poae*, and *Monographella nivalis* were found to colonize the sourdoughs. Katsi et al. (2021) analysed microbiota in sourdough for bread production in Greece by sequencing the V7–V8 hypervariable regions of the 18S rRNA gene and the internal transcribed spacer 1 (ITS1) gene. Eucaryotic profiles obtained in the research also contained microscopic molds belonging to the genera *Alternaria, Fusarium,* and *Cladosporium.*

In addition to the spoilage of bakery products by mold mycelium or spores, mycotoxin production is also of great concern for biological safety. There are more than 300 known mycotoxins, which are considered one of the most important risks associated with the consumption of cereal-based products (Codex Alimentarius Commission 2000). Commission Regulation (EC) No. 1881/2006 on maximum levels for certain contaminants in foodstuffs includes information about mycotoxins produced by crop-related molds. It defines maximum levels for aflatoxin (4 μg/kg) and ochratoxin A (3 μg/kg) in all cereal-based products. Bread and other bakery products should not contain more than 500 μg/kg of deoxynivalenol and 50 μg/kg of zearalenone. The concentration of deoxynivalenol in unprocessed cereals like durum wheat, maize, and oats, may be up to 1750 μg/kg. A maximum zearalenone concentration of 50 μg/kg is accepted in bread and other bakery products (pastries, biscuits, cereal snacks, and breakfast cereals, excluding maize snacks and maize-based breakfast cereals). Cereals intended for direct human consumption, cereal flour, or bran as end product marketed for direct human consumption, cannot contain more than 75 μg/kg of zearalenone. The maximum levels for fumonisins and other mycotoxins produced by *Fusarium* (T-2 and HT-2 toxin) in cereals and maize-based products vary from 200 μg/kg to 2000 μg/kg. In future, this list will probably be extended to include limits for other mycotoxins.

Mycotoxins are secondary metabolites of some molds. As toxic compounds, their consumption may result in a wide variety of short or long-term effects, from immune deficiencies and cancer to acute poisoning. Most fungi are able to produce several mycotoxins simultaneously in the same food or feed. Their co-occurrence can change the nature of their harmfulness to humans and animals, due to possible additive, antagonistic, or synergistic effects (Alassane-Kpembi et al. 2017).

It is well known that fungi associated with grain and flour are able to form mycotoxins. These toxic compounds may be produced by molds either immediately before, or after harvest (during grain storage or before further processing) (Minutillo et al. 2022; Amirahmadi et al. 2018; Škrbic et al. 2012; Dos Santos et al. 2021).

Commercially available wheat flour has been found to host a wide variety of potentially harmful fungal species, which are known to produce mycotoxins. Flour quality depends on the quality of the grains. Mesterházy et al. (2020) highlighted that grain losses cost several millions of dollars annually. Total grain losses before and after harvest or processing are estimated as follows: about one third of the possible yield is lost before harvest; another 20% is wasted during the grain storage and related with mycotoxin contamination; almost 10% is wasted by consumers; only about 36% of the total grain production is really consumed (Mesterházy et al. 2020). Table 6.6 summarize the distribution of mycotoxins related to the bakery industry in Europe.

Of the many fungal genera, *Alternaria, Fusarium, Aspergillus,* and *Penicillium* are frequently identified as forming mycotoxins as pre- or postharvest pathogens in grain (Plavšić et al. 2017). Since grinding does not eliminate the fungi present in the grain, if mycotoxin-producing fungi are present they will also be contained in the

Table 6.6 Non exhaustive list of the distribution of commonly occurring mycotoxins related to the bakery industry in Europe

Mycotoxin	Mycotoxin producers[a]	Infected material	Countries of mycotoxin detection	References
Aflatoxins ($AFB_1B_2G_1G_2$)	*Aspergillus flavus, A. parasiticus*	cereal based products, kernels, maize flour	Spain, Slovenia	Alborch et al. (2012), Kirinčič et al. (2014), and Oliveira et al. (2014) and references therein
Citrinin	*Penicillium* spp.	Kernels, maize flour	Spain	Alborch et al. (2012)
Deoxynivalenol	*Fusarium* spp.	Barley, bread and bakery products, cereal based products, durum wheat, maize, maize flour, oat, rye flour, wheat, wheat flour	Croatia, Denmark, Finland, Italy, Poland, Serbia, Slovenia, Spain, Sweden, United Kingdom	Błajet-Kosicka et al. (2014), Covarelli et al. (2015), Čonková et al. (2006), Ferrigo et al. (2016) and references therein, Kirinčič et al. (2014), Magan et al. (2003), Oliveira et al. (2014) and references therein, Pokrzywa and Surma (2022), Rasmussen et al. (2003), and Vidal et al. (2015)
Fumonisins (B_1B_2)	*Fusarium* spp., *F. verticillioides*	Maize, maize flour, maize-based products, wheat	Croatia, France, Germany, Italy, Poland, Slovenia, Spain, Sweden	Ferrigo et al. (2016) and references therein, Kirinčič et al. (2014), Oliveira et al. (2014) and references therein, and Pokrzywa and Surma (2022)

(continued)

Table 6.6 (continued)

Mycotoxin	Mycotoxin producers[a]	Infected material	Countries of mycotoxin detection	References
HT-2 and T-2 toxins	*Fusarium* spp., *F. langsethiae, F. poae, F. sporotrichioides*	Barley, durum wheat, durum wheat flour, maize, oat, oat flour, oat meals and other oat-based products from mills, rye flour, wheat, wheat flour	Croatia, Denmark, Finland, Germany, Italy, Ireland, Lithuania, United Kingdom, Poland, Slovenia, Sweden	Błajet-Kosicka et al. (2014), Ferrigo et al. (2016) and references therein, Kirinčič et al. (2014), Rasmussen et al. (2003), de Oliveira et al. (2014) and references therein
Nivalenol	*Fusarium* spp.	Bread, durum maize-based products, wheat flour, rye flour, wheat flour,	Czech Republic, Denmark, Germany, Spain, United Kingdom	Magan et al. (2003), Rasmussen et al. (2003), Oliveira et al. (2014) and references therein
Ochratoxin A	*A. niger, A. ochraceus*	Kernels, maize, maize flour, rye, wheat	Poland, Slovenia, Spain, United Kingdom	Alborch et al. (2012), Čonková et al. (2006), Kirinčič et al. (2014), and Magan et al. (2003)
Patulin	*Penicillium* spp., *A. clavatus*	Kernels, maize flour	Spain	Alborch et al. (2012)
Zearalenone	*Fusarium* spp.	Barley, bread and bakery products, cereal based products, durum wheat flour, maize, maize flour, oat, rye flour, wheat, wheat flour	Croatia, Denmark, Finland, Italy, Poland, Slovenia, Spain, Sweden	Błajet-Kosicka et al. (2014), Ferrigo et al. (2016) and references therein, Kirinčič et al. (2014), Oliveira et al. (2014) and references therein, Pokrzywa and Surma (2022), and Rasmussen et al. (2003)

[a] Isolates identification to the genera or species made by classic or by molecular methods; the way of the mold affiliation depending on the Author

flour obtained. Some mycotoxins (e.g. fumonisin B1 and B2, deoxynivalenol, ochratoxin A, and zearalenone) are highly stable during thermal treatment and pose a threat to the health of consumers due to the impossibility of eliminating mycotoxins at later stages of processing (such as baking, cooking, frying, or steaming) (Cheli et al. 2013; Boudra et al. 1995; Ryu et al. 2003; Vidal et al. 2015). Tralamazza et al. (2016) reported that freshly harvested wheat grains were contaminated mostly with *Alternaria, Epicoccum,* and *Fusarium.* Covarelli et al. (2015) found *Fusarium* to be the most abundant fungal genus in grains of durum wheat. Cereal fusariosis strongly affects crop production and causes problems due to toxin production. In general,

Fusarium spp. produce trichothecenes (TCTC), zearalenones (ZEA), fumonisins (FUM), enniatins (ENs), moniliformin (MON), beauvericin (BEA), and fusaproliferin (FP). Mycotoxins produced by *Fusarium* are described in detail by Ferrigo et al. (2016). Some are highly cytotoxic and extremely dangerous to consumers. Aflatoxins (AFs) are considered the most toxic, but research is still being conducted on many other mycotoxins, including "newly emerging" mycotoxins. Alshannaq and Yu (2017) provides an overview of research on the occurrence and toxicity of mycotoxins, along with methods used to investigate mycotoxins in food. In 2020, the European Food Safety Authority (EFSA) published a report on ochratoxin A, which can be found in a variety of foodstuffs including cereals, preserved meats, fresh and dried fruit, and cheese. Pinotti et al. (2016) described mycotoxin contamination in the EU feed supply chain, with focus on cereal by-products.

Although mycotoxin monitoring is implemented for specific grains in most European countries, mycotoxins are still detected in cereal-based products (Błajet-Kosicka et al. 2014; Škrbic et al. 2012). Pokrzywa and Surma (2022) analyzed flours, grits, pastas, snacks based on cereal, and bakery products available on the Polish market. They confirmed the presence of deoxynivalenol (DON), zearalenone, and fumonisins in the analyzed products (including products intended for infants) at different levels. In two cases, their concentrations exceeded the acceptable level of contamination.

6.5 Future Perspectives for Solving the Problem of Undesirable Microorganisms in Bakery Products

Bacteria and yeasts, as well as molds and their metabolites, strongly impact the shelf life and safety of bakery products. The presence of these microorganisms in cereal-based products has led to widespread research efforts to inhibit their growth. Multiple strategies have been developed to protect grains, by sanitizing them during storage. However, it is impossible to eliminate microorganisms and their harmful metabolites entirely from bakery goods. Therefore, over the years in parallel researchers have developed methods of chemical preservation. Conventionally, chemical preservatives, such as propionic, sorbic, and acetic acids, as well as their salts, have been used to inhibit microbial growth. Now, new alternative methods of biopreservation (especially against spoiling fungi) are under investigation. Such alternatives include the possible use of LAB and their metabolites as antibacterial and antifungal agents. Many LAB strains are known to produce different antimicrobial compounds (including organic acids mainly acetic acid, phenyllactic acid, bacteriocins, cyclic dipeptides, and fatty acids), which are able to control bacteria and fungi (Axel et al. 2016, 2017; Debonne et al. 2020; Quattrini et al. 2019).

The antifungal activity of LAB is strain-specific, as described by Iosca et al. (2022) against *Fusarium graminearum, Aspergillus flavus, Penicillium paneum,* and *Aspergillus niger.* On the other hand, the antifungal effect of LAB depends on the sensitivity of the species (particularly to acetic acid and lactic acid) (Debonne et al.

2020). Eighty-eight strains of *Lactiplantibacillus plantarum* were screened by Russo et al. (2017) for their antifungal activity against *Aspergillus niger, Aspergillus flavus, Fusarium culmorum, Penicillium roqueforti, Penicillium expansum, Penicillium chrysogenum,* and *Cladosporium* spp. An interesting concept for acrylamide reduction in wheat bread to prevent mold spoilage was proposed by Bartkiene et al. (2018). These authors investigated strains belonging to *Pediococcus, Lactobacillus* and *Leuconostoc* as antifungal agents against *Aspergillus nidulans, Penicillium funiculosum* and *Fusarium poae*.

The problem of undesirable microorganisms in sourdough starts with the grain quality and the hygienic conditions during the production process. Bakery products start to deteriorate as soon as baking finishes. Mycotoxin contamination of bakery products is extremely dangerous for consumers and is associated with the grain quality. Many studies have reported the occurrence of mycotoxins in grain and flour-based products. Sourdough-contaminating molds are less well understood. Both raw materials and cereal-based products should be routinely examined for the occurrence of mycotoxins and pathogenic bacteria, especially of the *Enterobacteriaceae* family and *Bacillus cereus sensu lato*. It would also be desirable to establish microbiological limits for different types of flour, primarily to ensure consumer safety, but also to improve the quality of end products. Additional studies, including on the use of culture and culture-independent methods along with wider collection of sourdoughs obtained in different bakery practices, are needed to better understand the origin and number of microorganisms species along with their diversity. The possible risks to consumer health should be minimized by the development and implementation of innovative methods based on bio-preservation of bakery products or their combination with physical or chemical factors used for food protection, and by employing efficient packaging and storage techniques.

References

Abedi R, Sami M, Mohammadi R, Mirlohi M (2020) Microbial and fungal contamination of wheat flour, dough, and bread samples collected from Isfahan, Iran. Egypt J Vet Sci 51(2):163–170

Alassane-Kpembi I, Schatzmayr G, Taranu I, Marin D, Puel O, Oswald IP (2017) Mycotoxins co-contamination: methodological aspects and biological relevance of combined toxicity studies. Crit Rev Food Sci Nutr 57:3489–3507

Alborch L, Bragulat MR, Castellá G, Abarca ML, Cabañes FJ (2012) Mycobiota and mycotoxin contamination of maize flours and popcorn kernels for human consumption commercialized in Spain. Food Microbiol 32(1):97–103

Alfonzo A, Ventimiglia G, Corona O, Di Gerlando R, Gaglio R, Francesca N, Moschetti G, Settanni L (2013) Diversity and technological potential of lactic acid bacteria of wheat flours. Food Microbiol 36(2):343–354

Alshannaq A, Yu J-H (2017) Occurrence, toxicity, and analysis of major mycotoxins in food. Int J Environ Res Public Health 14(6):632

Amirahmadi M, Shoeibi S, Rastegar H, Elmi M, Mousavi Khaneghah A (2018) Simultaneous analysis of mycotoxins in corn flour using LC/MS-MS combined with a modified QuEChERS procedure. Toxin Rev 37:187–195

Asadzadeh J, Teymori R, Ghazanfarirad N, Fakhernia M, Haghighat-Afshar N, Blouki M, Kheiri A, Hassanzadazar H, Bahmani M (2014) Fungal contamination of produced wheat flour in West Azerbaijan, northwest of Iran. Asian Pac J Trop Dis 4:S836–S839

Axel C, Brosnan B, Zannini E, Peyer LC, Furey A, Coffey A, Arendt EK (2016) Antifungal activities of three different *Lactobacillus* species and their production of antifungal carboxylic acids in wheat sourdough. Appl Microbiol Biotechnol 100:1701–1711

Axel C, Zannini E, Arendt EK (2017) Mold spoilage of bread and its biopreservation: a review of current strategies for bread shelf life extension. Crit Rev Food Sci Nutr 57:3528–3542

Aydin A, Paulsen P, Smulders FJM (2009) The physico-chemical and microbiological properties of wheat flour in Thrace. Turk J Agric For 33:445–454

Baert K, Valero A, De Meulenaer B, Samapundo S, Ahmed MM, Bo L, Debevere J, Devlieghere F (2007) Modeling the effect of temperature on the growth rate and lag phase of *Penicillium expansum* in apples. Int J Food Microbiol 118(2):139–150

Barber S, Báguena R, Martínez-Anaya MA, Torner MJ (1983) Microflora of the sour dough of wheat flour bread. I. Identification and functional properties of microorganisms of industrial sour doughs. Rev Agroquim Tecnol Alimen 23:552–562

Bartkiene E, Bartkevics V, Lele V, Pugajeva I, Zavistanaviciute P, Mickiene R, Zadeike D, Juodeikiene G (2018) A concept of mould spoilage prevention and acrylamide reduction in wheat bread: application of lactobacilli in combination with a cranberry coating. Food Control 91:284–293

Berghofer LK, Hocking AD, Miskelly D, Jansson E (2003) Microbiology of wheat and flour milling in Australia. Int J Food Microbiol 85(1–2):137–149

Bessmeltseva M, Viirad E, Simm J, Paalme T, Sarad I (2014) Evolution of Bacterial Consortia in spontaneously started rye sourdoughs during two months of daily propagation. PLoS One 9(4):e95449. https://doi.org/10.1371/journal.pone.0095449

Beuchat LR, Komitopoulou E, Beckers H, Betts RP, Bourdichon F, Fanning S, Joosten HM, Ter Kuile BH (2013) Low-water activity foods: increased concern as vehicles of foodborne pathogens. J Food Protect 76:150–172

Błajet-Kosicka A, Twarużek M, Kosicki R, Sibiorowska E, Grajewski J (2014) Co-occurrence and evaluation of mycotoxins in organic and conventional rye grain and products. Food Contr 38:61–66

Boudra H, Le Bars P, Le Bars J (1995) Thermostability of ochratoxin A in wheat under two moisture conditions. Appl Environ Microbiol 61:1156–1158

Calvert MD, Madden AA, Nichols LM, Haddad NM, Lahne J, Dunn RR, McKenney EA (2021) A review of sourdough starters: ecology, practices, and sensory quality with applications for baking and recommendations for future research. PeerJ 9:e11389

Cardoso RVC, Fernandes Â, Heleno SA, Rodrigues P, Gonzaléz-Paramás AM, Barros L, Ferreira ICFR (2019) Physicochemical characterization and microbiology of wheat and rye flours. Food Chem 280:123–129

Centers for disease controlling and prevention CDC (2016) Multistate outbreak of Shiga toxin-producing *Escherichia coli* infections linked to flour. https://www.cdc.gov/ecoli/2016/o121-06-16/index.html. Accessed 20 May 2022

Centers for disease controlling and prevention CDC (2019) Outbreak of *E. coli* infections linked to flour. https://www.cdc.gov/ecoli/2019/flour-05-19/index.html. Accessed 20 May 2022

Cheli F, Pinotti L, Rossi L, Dell'Orto V (2013) Effect of milling procedures on mycotoxin distribution in wheat fractions: a review. LWT – Food Sci Technol 54:307–314

Cichoń Z, Ptak M (2005) Analysis of quality of selected wheat flour types. Zesz Nauk AEK 678:89–102. (in Polish)

Codex Alimentarius Commission (2000) Codex General Standard for Contaminants and Toxins in Food and Feed; CODEX STAN 193-1995 (Rev. 1-1997); In Joint FAO/WHO Food Standards Programme

Codex Stan 152-1985 (Rev. 1-1995). Codex Alimentarius Standard for Wheat Flour

Comasio A, Verce M, Van Kerrebroeck S, De Vuyst L (2020) Diverse microbial composition of sourdoughs from different origins. Front Microbiol 11:1212

COMMISSION REGULATION (EC) No 1881/2006 of 19 December 2006 setting maximum levels for certain contaminants in foodstuffs

Conková E, Laciaková A, Štyriak I, Czerwiecki L, Wilczyńska G (2006) Fungal contamination and the levels of mycotoxins (DON and OTA) in cereal samples from Poland and East Slovakia. Czech J Food Sci 24: 33–40

Cook FK, Johnson BL Microbiological Spoilage of Cereal Products. In: Sperber W, Doyle M, editoes. Compendium of the microbiological spoilage of foods and beverages. Springer; 2009, p. 223–244

Corsetti A, Settanni L, Lopez CC, Felis GE, Mstrangelo M, Suzz G (2007) A taxonomic survey of lactic acid bacteria isolated from wheat (*Triticum durum*) kernels and nonconventional flours. Syst Appl Microbiol 30:561–571

Covarelli L, Beccari G, Prodi A, Generotti S, Etruschi F, Juan C, Ferrer E, Mañes J (2015) *Fusarium* species, chemotype characterisation and trichothecene contamination of durum and soft wheat in an area of central Italy. J Sci Food Agric 95:40–551

CRS 64:2019, CARICOM Regional Organisation for Standards and Quality (CROSQ), Barbados

Czerwińska E, Piotrowski W (2009) Technological aspects of bread backing with assessment of critical points in microbiological contamination. Rocz Ochr Srodowiska 11:449–464. (in Polish)

Czerwińska E, Piotrowski W (2010) Ocena ryzyka zanieczyszczenia mikrobiologicznego w piekarni z uwzględnieniem procesu wytwarzania pieczywa żytniego. Nauka Przyr Techn 4(2):1–15. (in Polish)

Dagnas S, Membré JM (2013) Predicting and preventing mold spoilage of food products. J Food Protect 76(3):538–551

Dal Bello F, Clarke CI, Ryan LAM, Ulmer H, Schober TJ, Ström K, Sjögren J, van Sinderen D, Schnürer J, Arendt EK (2007) Improvement of the quality and shelf life of wheat bread by fermentation with the antifungal strain *Lactobacillus plantarum* FST 1.7. J Cereal Sci 45:309–318

De Angelis M, Minervini F, Siragusa S, Rizzello CG (2019) Wholemeal wheat flours drive the microbiome and functional features of wheat sourdoughs. Int J Food Microbiol 302:35–46

De Filippis F, Parente E, Ercolini D (2017) Metagenomics insights into food fermentations. Microbial Biotechnol 10(1):91–102

De Vuyst L, Neysens P (2005) The sourdough microflora: biodiversity and metabolic interactions. Trends Food Sci Technol 16(1–3):43–56

De Vuyst L, Van Kerrebroeck S, Harth H, Huys G, Daniel HM, Weckx S (2014) Microbial ecology of sourdough fermentations: diverse or uniform? Food Microbiol 37:11–29

De Vuyst L, Comasio A, Kerrebroeck SV (2023) Sourdough production: fermentation strategies, microbial ecology, and use of non-flour ingredients. Crit Rev Food Sci Nutr 63(15):2447–2479

Debonne E, Maene P, Vermeulen A, Van Bockstaele F, Depredomme L, Vermeir P, Eeckhout M, Devlieghere F (2020) Validation of in-vitro antifungal activity of the fermentation quotient on bread spoilage moulds through growth/no-growth modelling and bread baking trials. LWT 117:108636. https://doi.org/10.1016/j.lwt.2019.108636

Demirel R, Sariozlu NY (2013). Mycotoxigenic moulds and mycotoxins in flours consumend in Turkey. J Sci Food Agri 94:1577–1584

Di Cagno R, Pontonio E, Buchin S, De Angelis M, Lattanzi A, Valerio F, Gobbetti M, Calasso M (2014) Diversity of the lactic acid bacterium and yeast microbiota in the switch from firm- to liquid-sourdough fermentation. Appl Environ Microbiol 80(10):3161–3172

Dos Santos ID, Pizzutti IR, Dias JV, Fontana MEZ, Souza DM, Cardoso CD (2021) Mycotoxins in wheat flour: Occurrence and co-occurrence assessment in samples from Southern Brazil. Food Addit Contam Part B 14:151–161

East African Standard EAS 1:2018. Wheat flour – specification. https://members.wto.org/crnattachments/2018/SPS/KEN/18_6256_00_e.pdf. Accessed 18 May 2022

EFSA Panel on Contaminants in the Food Chain (CONTAM), Schrenk S, Bodin L, Chipman JK, Del Mazo J, Grasl-Kraupp B, Hogstrand C, Hoogenboom LR, Leblanc JC, Nebbia CS, Nielsen E, Ntzani E, Petersen A, Sand S, Schwerdtle T, Vleminckx C, Wallace H, Alexander J, Dall'Asta C, Bignami M (2020) Risk assessment of ochratoxin A in food. EFSA J 18(5):e06113. https://doi.org/10.2903/j.efsa.2020.6113

Eglezos S (2010) Microbiological quality of wheat grain and flour from two mills in Queensland. Australia J Food Protect 73(8):1533–1536

Ennadir J, Hassikou R, Ohmani F, Hammamouchi J, Bouazza F, Qasmaoui A, Mennane Z, Touhami AO, Charof R, Khedid K (2012) Qualité microbiologique des farines de blé consommées au Maroc. Can J Microbiol 58:145–150

Ercolini D, Pontonio E, De Filippis F, Minervini F, La Storia A, Gobetti M, Di Cagno R (2013) Microbial ecology dynamics during rye and wheat sourdough preparation. Appl Environ Microbiol 79(24):7827–7836

Ethiopian Standard ES 1052:2017. Wheat flour – specification

European Commission (2006) Commission Regulation (EC) No 1881/2006 of 19 December 2006 setting maximum levels for certain contaminants in foodstuffs. Off J Eur Union 364:5–24

Ferrigo D, Raiola A, Causin R (2016) *Fusarium* toxins in cereals: occurrence, legislation, factors promoting the appearance and their management. Molecules 21(5):627. https://doi.org/10.3390/molecules21050627

Filtenborg O, Frisvad JC, Thrane U (1996) Molds in food spoilage. Int J Food Microbiol 33(1):85–102

Fleet GH (1992) Spoilage yeasts. Crit Rev Biotechnol 12:1–44

Food and Drug Administration (2013) FDA Revised Guidelines for the Assessment of Microbiological Quality of Processed Foods. Muntinlupa, Philippines

Foschino R, Gallina S, Andrighetto C, Rossetti L, Galli A (2004) Comparison of cultural methods for the identification and molecular investigation of yeasts from sourdoughs for Italian sweet baked products. FEMS Yeast Res 4(6):609–618

Foschino R, Venturelli E, Picozzi C (2005) Isolation and characterization of a virulent *Lactobacillus sanfranciscensis* bacteriophage and its impact on microbial population in sourdough. Curr Microbiol 51(6): 413–418

Garcia MV, Copetti M (2019) Alternative methods for mold spoilage control in bread and bakery products. Int Food Res J 26(3):737–749

Garcia MV, Sonnenstrahl Bregao A, Parussolo G, Bernardi OA, Stefanello A, Venturini Copetti M (2019a) Incidence of spoilage fungi in the air of bakeries with different hygienic status. Int J Food Microbiol 290:254–261

Garcia MV, Bernardi OA, Venturini Copetti M (2019b) The fungal problem in bread production: insights of causes, consequences, and control methods. Curr Opin Food Sci 29:1–6

GCC Standardization Organization GSO 1016:2015. Microbiological Criteria for Foodstuffs. https://www.gso.org.sa/store/standards/GSO:693280/GSO%201016:2015?lang=en. Accessed 24 Apr 2022

Gerez CL, Torino MI, Rollán G, de Valdez GF (2009) Prevention of bread mould spoilage by using lactic acid bacteria with antifungal properties. Food Contr 20(2):144–148

Gerez CL, Fornaguera M, Obregozo MD, de Valdez G, Torino MI (2015) Antifungal starter culture for packed bread: influence of two storage conditions. Rev Argent Microbiol 47(2):118–124

Ghali R, Hmaissia-Khlifa K, Ghorbel H, Maaroufi K, Hedili A (2008) Incidence of aflatoxins, ochratoxin A and zearalenone in Tunisian foods. Food Contr 19:921–924

Index Funforum (2023) http://www.indexfungorum.org. Accessed 24 Apr 2022

International Commission on Microbiological Specifications for Foods ICMFS (1998) Cereals and cereal products. In: Roberts TA, Pitt JI, Farkas J, Grau FH (eds) Microorganisms in foods: microbial ecology of food commodities, vol 6. Blackie Academic & Professional, London, pp 313–355

Iosca G, De Vero L, Di Rocco G, Perrone G, Gullo M, Pulvirenti A (2022) Anti-spoilage activity and exopolysaccharides production by selected lactic acid bacteria. Foods 11(13):1914. https://doi.org/10.3390/foods11131914

Jovanovic J, Ornelis VFM, Madder A, Rajkovic A (2021) *Bacillus cereus* food intoxication and toxicoinfection. Compr Rev Food Sci Food Saf 20:3719–3761

Kara GN, Ozbey F, Kabak B (2015) Co-occurrence of aflatoxins and ochratoxin A in cereal flours commercialised in Turkey. Food Control 54:275–281

Katsi P, Kosma IS, Michailidou S, Argiriou A, Badeka AV, Kontominas MG (2021) Characterization of artisanal spontaneous sourdough wheat bread from central Greece: evaluation of physico-chemical, microbiological, and sensory properties in relation to conventional yeast leavened wheat bread. Foods 10(3):635. https://doi.org/10.3390/foods10030635

Kaygusuz I, Mulazimoglu L, Cerikcioglu N, Toprak A, Oktay A, Korten V (2003) An unusual native tricuspid valve endocarditis caused by *Candida colliculosa*. Clin Microbiol Infect 9(4):319–322

Kenya Standard KS 169:2009. Wheat flour – specification. https://archive.org/details/ke.169.ds.2009. Accessed 5 May 2022

Khanna K, Lopez-Garrido J, Pogliano K (2020) Shaping an endospore: architectural transformations during *Bacillus subtilis* sporulation. Annu Rev Microbiol 74:361–386

Khanom A, Shammi T, Kabir MS (2016) Determination of microbiological quality of packed and unpacked bread. Stamford J Microbiol 6:24–29

Kim SE, Jung SI, Park KH, Choi YJ, Won EJ, Shin JH (2020) Case report: nosocomial fungemia caused by *Candida diddensiae*. BMC Infect Dis 20(1):377. https://doi.org/10.1186/s12879-020-05095-3

Kindle P, Etter D, Stephan R, Johler S (2019) Population structure and toxin gene profiles of *Bacillus cereus sensu lato* isolated from flour products. FEMS Microbiol Lett 366(20):fnz240. https://doi.org/10.1093/femsle/fnz240

Kirinčič S, Škrjanc B, Kos N, Kozolc B, Pirnat N, Tavčar-Kalcher G (2014) Mycotoxins in cereals and cereal products in Slovenia official control of foods in the years 2008–2012. Food Contr 50:157–165

Kurtzman CP, Fell JW, Boekhout T (2011) The Yeasts: A Taxonomic Study. Elsevier Science

Landis EA, Oliverio AM, McKenney EA, Nichols LM, Kfoury N, Biango-Daniels M, Shell LK, Madden A, Shapiro L, Sakunala S, Drake K, Robbat A, Booker M, Dunn RR, Fierer N, Wolfe BE (2021) The diversity and function of sourdough starter microbiomes. elife 10:1–24

Lattanzi A, Minervini F, Di Cagno R, Diviccaro A, Antonielli L, Cardinali G, Cappelle S, De Angelis M, Gobbetti M (2013) The lactic acid bacteria and yeast microbiota of eighteen sourdoughs used for the manufacture of traditional Italian sweet leavened baked goods. Int J Food Microbiol 163(2–3):71–79

Legan JD (1993) Mould spoilage of bread: The problem and some solutions. Int Biodeter Biodegrad 32(1):33–53

Legan JD, Voysey PA (1991) Yeast spoilage of bakery products and ingredients. J Appl Bacteriol 70(5):361–371

Leuschner R, O'Callaghan M, Arendt E (1998) Bacilli Spoilage in Part-baked and Rebaked Brown Soda Bread. J Food Sci 63:915–918

Lhomme E, Lattanzi A, Dousset X, Minervini F, De Angelis M, Lacaze G, Onno B, Gobbetti M (2015) Lactic acid bacterium and yeast microbiotas of sixteen French traditional sourdoughs. Int J Food Microbiol 215:161–170

Lombardi A, Zilio F, Andrighetto C, Zampese L, Loddo A (2007) Micobiological characterization of sourdoughs of Veneto region. Ind Aliment XLVI:147–151

Loureiro V, Querol A (1999) The prevalence and control of spoilage yeasts in foods and beverages. Trends Food Sci Technol 10(11):56–365

Mäde D, Geuthner AC, Imming R, Wicke A (2017) Detection and isolation of Shiga-Toxin producing *Escherichia coli* in flour in Germany between 2014 and 2017. J Cons Protect Food Saf 12:245–253

Magallanes López AM, Simsek S (2021) Pathogens control on wheat and wheat flour: A review. Cereal Chem 98:17–30

Magan N, Hope R, Cairns V, Aldred DCN (2003) Post-harvest fungal ecology: impact of fungal growth and mycotoxin accumulation in stored grain. Eur J Plant Pathol 109:723–730

Manthey FA, Wolf-Hall CE, Yalla S, Vijayakumar C, Carlson D (2004) Microbial loads, mycotoxins, and quality of durum wheat from the 2001 harvest of the northern plains region of the United States. J Food Protect 67(4):772–780

Melini V, Melini F (2018) Strategies to extend bread and gf bread shelf-life: from sourdough to antimicrobial active packaging and nanotechnology. Fermentation 4(1):9. https://doi.org/10.3390/fermentation4010009

Mesterházy Á, Oláh J, Popp J (2020) Losses in the grain supply chain: causes and solutions. Sustainability 12:2342. https://doi.org/10.3390/su12062342

Minervini F, Lattanzi A, De Angelis M, Celano G, Gobbetti M (2015) House microbiotas as sources of lactic acid bacteria and yeasts in traditional Italian sourdoughs. Food Microbiol 52:66–76

Minutillo SA, Ruano-Rosa D, Abdelfattah A, Schena L, Malacrinò A (2022) The fungal microbiome of wheat flour includes potential mycotoxin producers. Foods 11(5):676. https://doi.org/10.3390/foods11050676

Mohsen SM, Aly MH, Attia AA, Osman DB (2016) Effect of sourdough on shelf sife, sreshness and sensory characteristics of Egyptian Balady bread. J App Environ Microbiol 4:39–45

Moroni AV, Arendt EK, Dal Bello F (2011) Biodiversity of lactic acid bacteria and yeasts in spontaneously-fermented buckwheat and teff sourdoughs. Food Microbiol 28:497–502

Nielsen PV, Rios R (2000) Inhibition of fungal growth on bread by volatile components from spices and herbs, and the possible application in active packaging, with special emphasis on mustard essential oil. Int J Food Microbiol 60(2–3):219–229

Nosa-Obamwonyi JA, Antai SP (2018) Physicochemical properties and bacteriological profile of wheat flours produced and (or) sold in Calabar, Nigeria. Asian Res J Agric 10(2):1–11

Ntuli V, Mekbib SB, Asita AO, Molebatsi N, Makotoko M, Chatanga P (2013) Microbial and physicochemical characterization of maize and wheat flour from a milling company, Lesotho. Internet J Food Saf 15:11–19

Official Journal of Republic of Algeria No 39, 2017, Inter-ministerial regulation of October 4, 2016 regulating microbiological criteria for food products

Oliveira PM, Zannini E, Arendt EK (2014) Cereal fungal infection, mycotoxins, and lactic acid bacteria mediated bioprotection: From crop farming to cereal products. Food Microbiol 37:78–95

Opulente DA, Langdon QK, Buh KV, Haase MAB, Sylvester K, Moriarty RV, Jarzyna M, Considine SL, Schneider RM, Hittinger CT (2019) Pathogenic budding yeasts isolated outside of clinical settings. FEMS Yeast Res 19(3):foz032. https://doi.org/10.1093/femsyr/foz032

Oshiro M, Zendo T, Nakayama J (2021) Diversity and dynamics of sourdough lactic acid bacteriota created by a slow food fermentation system. J Biosci Bioeng 131:333–340

Otlewska A, Migliore M, Dybka-Stępień K, Manfredini A, Struszczyk-Świta K, Napoli R, Białkowska A, Canfora L, Pinzari F (2020) When salt meddles between plant, soil, and microorganisms. Front Plant Sci 16:11553087. https://doi.org/10.3389/fpls.2020.553087

Pacher N, Burtscher J, Johler S, Etter D, Bender D, Fieseler L, Domig KJ (2022) Ropiness in bread—A reemerging spoilage phenomenon. Foods 11:3021. https://doi.org/10.3390/foods11193021

Palomba S, Blaiotta G, Ventorino V, Saccone A, Pepe O (2011) Microbial characterization of sourdough for sweet baked products in the Campania region (southern Italy) by a polyphasic approach. Ann Microbiol 61:307–314

Paramithiotis S, Müller MRA, Ehrmann MA, Tsakalidou E, Seiler H, Vogel R, Kalantzopoulos G (2000) Polyphasic identification of wild yeast strains isolated from Greek sourdoughs. Syst Appl Microbiol 23(1):156–164

Pepe O, Blaiotta G, Moschetti G, Greco T, Villani F (2003) Rope-producing strains of *Bacillus* spp. from wheat bread and strategy for their control by lactic acid bacteria. Appl Environ Microbiol 69(4):2321–2329

Pereira APM, Freire L, Alvarenga VO, Crucello A, Morassi LLP, Silva FP, Stradiotto GC, Sant'Ana AS (2020) Occurrence and enumeration of rope-producing spore forming bacteria in flour and their spoilage potential in different bread formulations. LWT – Food Sci Technol 133:110108. https://doi.org/10.1016/j.lwt.2020.110108

Pérez-Alvarado O, Zepeda-Hernández A, Garcia-Amezquita LE, Requena T, Vinderola G, García-Cayuela T (2022) Role of lactic acid bacteria and yeasts in sourdough fermentation during breadmaking: Evaluation of postbiotic-like components and health benefits. Front Microbiol 13:969460. https://doi.org/10.3389/fmicb.2022.969460

Petkova M, Stefanova P, Gotcheva V, Angelov A (2021) Isolation and characterization of lactic acid bacteria and yeasts from typical Bulgarian sourdoughs. Microorganisms 9(7):1346

Pino A, Russo N, Solieri L, Sola L, Caggia C, Randazzo CL (2022) Microbial consortia involved in traditional Sicilian sourdough: characterization of lactic acid bacteria and yeast populations. Microorganisms 10(2):283

Pinotti L, Ottoboni M, Giromini C, Dell'Orto V, Cheli F (2016) Mycotoxin Contamination in the EU feed supply chain: a focus on cereal byproducts. Toxins 8(2):45

Plavšić DV, Škrinjar MM, Psodorov DB, Šarić LĆ, Psodorov DD, Varga AO, Mandić AI (2017) Mycopopulations of grain and flour of wheat, corn and buckwheat. Food Feed Res 44(1):39–45

Pokrzywa P, Surma M (2022) Level of contamination with deoxynivalenol, zearalenone and fumonisins in cereal products – assessment of consumer exposure. Ann Agric Environ Med 29(1):2–79

Pontonio E, Rizzello CG, Di Cagno R, Dousset X, Clement H, Onno B, Gobbetti M, Filannino P (2016) How organic farming of wheat may affect the sourdough and the nutritional and technological features of leavened baked goods. Int J Food Microbiol 239:44–53

Pulvirenti A, Solieri L, Gullo M, De Vero L, Giudici P (2004) Occurrence and dominance of yeast species in sourdough. Lett Appl Microbiol 38(2):113–117

Quattrini M, Lang N, Fortina MG, Xiang S, Curtis JM, Gänzle M (2019) Exploiting synergies of sourdough and antifungal organic acids to delay fungal spoilage of bread. Int J Food Microbiol 302:8–14

RASFF Notification 2020.3307 Chakki Atta Wheat Flour – Aflatoxins. https://webgate.ec.europa.eu/rasff-window/screen/notification/438005. Accessed 6 May 2022

RASFF Notification 2021.0931 *Salmonella* in flour. https://webgate.ec.europa.eu/rasff-window/screen/notification/46682. Accessed 6 May 2022

Rasmussen PH, Ghorbani F, Berg T (2003) Deoxynivalenol and other Fusarium toxins in wheat and rye flours on the Danish market. Food Addit Contam 20(4):396–404

Reese AT, Madden AA, Joossens M, Lacaze G, Dunn RR (2020) Influences of ingredients and bakers on the bacteria and fungi in sourdough starters and bread. mSphere 5(1):e00950–e00919

Rezazadeh A, Pirzeh L, Hosseini M, Razavich SV (2013) Evaluation of fungal contaminations and humidity percent of consumed flour in the bakeries of Tabritz city. J Paramed Sci 4(4):83–87

Richter KS, Dorneanu E, Eskridge KM, Rao CS (1993) Microbiological quality of flour. Cereals Foods World 38:367–369

Ripari V, Gänzle MG, Berardi E (2016) Evolution of sourdough microbiota in spontaneous sourdoughs started with different plant materials. Int J Food Microbiol 232:35–42

Rocha JM, Malcata FX (1999) On the microbiological profile of traditional Portuguese sourdough. J Food Protect 62(12):1416–1429

Rosenkvist H, Hansen A (1995) Contamination profiles and characterisation of *Bacillus* species in wheat bread and raw materials for bread production. Int J Food Microbiol 26:353–363

Rossi EM, Mahl SC, Spaniol AC, Honorato JFB, Rocha T (2021) Evaluating the thermoresistance of *Bacillus cereus* strains isolated from wheat flour. Res Soc Dev 10(6):e2510615268. https://doi.org/10.33448/rsd-v10i6.15268

Russo P, Arena MP, Fiocco D, Capozzi V, Drider D, Spano G (2017) *Lactobacillus plantarum* with broad antifungal activity: a promising approach to increase safety and shelf-life of cereal-based products. Int J Food Microbiol 247:48–54

Ryu D, Hanna MA, Eskridge KM, Bullerman LB (2003) Heat stability of zearalenone in an aqueous buffered model system. J Agric Food Chem 51:1746–1748

Shahzad K, Nawaz S, Zahid S, Saeed S, Saleem Y, Rashid AA, Hassan A, Husain S (2019) Comparative physicochemical and microbiological assessment of branded and unbranded whole wheat flours in Pakistan. Pak J Sci Ind Re B Biol Sci 62B(1):24–32

Škrbic B, Živancev J, Durišic-Mladenovic N, Godula M (2012) Principal mycotoxins in wheat flour from the Serbian market: levels and assessment of the exposure by wheat-based products. Food Contr 25:89–396

Sorokulova I, Reva O, Smirnov V, Pinchuk I, Lapa S, Urdaci M (2003) Genetic diversity and involvement in bread spoilage of *Bacillus* strains isolated from flour and ropy bread. Lett Appl Microbiol 37:169–173

Sperber WH (2007) The North American Millers' Association Microbiology Working Group. Role of microbiological guidelines in the production and commercial use of milled cereal grains: a practical approach for the 21st century. J Food Prot 70:1041–1053

Stannard C (1997) Development and use of microbiological criteria for foods. Food Sci Technol Today 11:137–177

Stuper-Szablewska K, Perkowski J (2017) Level of contamination with mycobiota and contents of mycotoxins from the group of trichothecenes in grain of wheat, oats, barley, rye and triticale harvested in Poland in 2006–2008. Ann Agric Environ Med 24(1):49–55

Succi M, Reale A, Andrighetto C, Lombardi A, Sorrentino E, Coppola R (2003) Presence of yeasts in southern Italian sourdoughs from *Triticum aestivum* flour. FEMS Microbiol Lett 225:143–148

Syrokou MK, Themeli C, Paramithiotis S, Mataragas M, Bosnea L, Argyri AA, Chorianopoulos NG, Skandamis PN, Drosinos EH (2020) Microbial ecology of Greek wheat sourdoughs, identified by a culture-dependent and a culture-independent approach. Foods 9(11):1603

Thielecke F, Nugent AP (2018) Contaminants in Grain – a major risk for whole grain safety? Nutrients, 10(9):1213. https://doi.org/10.3390/nu10091213

Thompson J, Waites W, Dodd C (1998) Detection of rope spoilage in bread caused by *Bacillus* species. J Appl Microbiol 85: 481–486

Tralamazza SM, Bemvenuti RH, Zorzete P, de Souza GF, Corrêa B (2016) Fungal diversity and natural occurrence of deoxynivalenol and zearalenone in freshly harvested wheat grains from Brazil. Food Chem 196:445–450

Turkish Food Codex (2011) Regulation on Microbiological Criteria; Annex-1 Food Safety Criteria; No. 28157

Urien C, Legrand J, Montalent P, Casaregola S, Sicard D (2019) Fungal species diversity in French bread sourdoughs made of organic wheat flour. Front Microbiol 10(18):201. https://doi.org/10.3389/fmicb.2019.00201

Vaičiulytė-Funk L, Žvirdauskienė R, Šalomskienė J, Šarkinas A (2015) The effect of wheat bread contamination by the *Bacillus* genus bacteria on the quality and safety of bread. Zemdirbyste-Agriculture 102:e3. https://doi.org/10.13080/z-a.2015.102.045

Valerio F, De Bellis P, Di Biase M, Lonigro SL, Giussani B, Visconti A, Lavermicocca P, Sisto A (2012) Diversity of spore-forming bacteria and identification of Bacillus amyloliquefaciens as a species frequently associated with the ropy spoilage of bread. Int J Food Microbiol 156(3):278–285

Valmorri S, Tofalo R, Settanni L, Corsetti A, Suzzi G (2010) Yeast microbiota associated with spontaneous sourdough fermentations in the production of traditional wheat sourdough breads of the Abruzzo region (Italy). Antonie Van Leeuwenhoek 97(2):119–129

Venturi M, Guerrini S, Vincenzini M (2012) Stable and non-competitive association of *Saccharomyces cerevisiae*, *Candida milleri* and *Lactobacillus sanfranciscensis* during manufacture of two traditional sourdough baked goods. Food Microbiol 31:107–115

Vidal A, Sanchis V, Ramos AJ, Marín S (2015) Thermal stability and kinetics of degradation of deoxynivalenol, deoxynivalenol conjugates and ochratoxin A during baking of wheat bakery products. Food Chem 1(178):276–286

Vogelmann SA, Seitter M, Singer U, Brandt MJ, Hertel C (2009) Adaptability of lactic acid bacteria and yeasts to sourdoughs prepared from cereals, pseudocereals and cassava and use of competitive strains as starters. Int J Food Microbiol 130:205–212

von Gastrow L, Amelot R, Segond D, Guézennec S, Valence F, Sicard D (2021) Microbial community dispersal in sourdough. A contribution of participatory research. Mol Ecol. https://doi.org/10.1111/mec.16630

Vrancken G, De Vuyst L, Van der Meulen R, Huys G, Vandamme P, Daniel HM (2010) Yeast species composition differs between artisan bakery and spontaneous laboratory sourdoughs. FEMS Yeast Res 10:471–481

Warzecha A, Jakubowska B, Siwicka H (1987) Microbiologic contamination of wheat and rye flour used for bread baking in the Szczecin Bakery and Confectionery Plant. Rocz Panstw Zakl Hig 38(6):498–505. (in Polish)

Weidenbörner M, Wieczorek C, Appel S, Kunz B (2000) Whole wheat and white wheat flour-the mycobiota and potential mycotoxins. Food Microbiol 17:103–107

Wlazło Ł, Nowakowicz-Dębek B, Chmielowiec-Korzeniowska A, Maksym P, Pawlak H, Kapica J (2020) Assessment of the level of organic dust and mould spores in the work environment of baker. Indian J Occup Environ Med 24(3):137–141

Zhou M, Li Y, Kudinha T, Xu Y, Liu Z (2021) *Kodamaea ohmeri* as an emerging human pathogen: a review and update. Front Microbiol 12:736582. https://doi.org/10.3389/fmicb.2021.736582

Chapter 7
Metagenomics Analysis of Sourdough Microbiota

Wilson J. F. Lemos Jr, Antonio A. Câmara Jr, Larissa P. Margalho, Vanessa S. de Oliveira, André F. Guerra, and Anderson S. Sant'Ana

7.1 Introduction

Sourdough ecosystems are characterized as microbial communities composed of yeasts and lactic acid bacteria (LAB), mainly heterofermentative coexisting in a matrix composed of cereal flour and water (Gobbetti et al. 2019; Comasio et al. 2020).

Yeasts metabolize carbohydrates to synthesize organic acids and produce ethanol and CO_2, contributing to dough expansion, while LAB metabolizes hexoses into lactic acid and reduces the dough's pH (Gänzle and Ripari 2016). Besides, the stress response due to the fermentation process directly affects the evolutionary mechanisms of these microorganisms. The prevalence of specific sourdough species is also due to multiple technological factors (Fig. 7.1), such as temperature, backslopping method, dough yield, and pH (Martín-Garcia et al. 2023). In addition, several interaction mechanisms can influence the structure and composition of microbial communities in sourdough (Carbonetto et al. 2018). For example: (i) the type of

W. J. F. Lemos Jr
Bioresources Unit, Center for Health & Bioresources, AIT Austrian Institute of Technology GmbH, Tulln, Austria

A. A. Câmara Jr · L. P. Margalho · A. S. Sant'Ana (✉)
Department of Food Science and Nutrition, Faculty of Food Engineering, University of Campinas, Campinas, SP, Brazil
e-mail: and@unicamp.br

V. S. de Oliveira
Department of Food Technology, Institute of Technology, University Federal Rural of Rio de Janeiro, Seropédica, RJ, Brazil

A. F. Guerra
Department of Food Engineering, Federal Center of Technological Education Celso Suckow da Fonseca, Valença, RJ, Brazil

E. B. Ceresino et al. (eds.), *Sourdough Microbiota and Starter Cultures for Industry*, https://doi.org/10.1007/978-3-031-48604-3_7

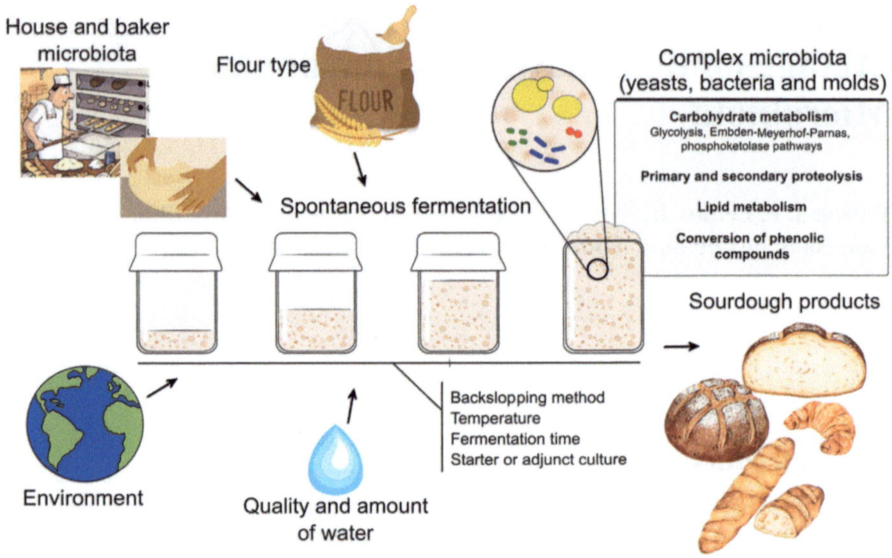

Fig. 7.1 Main factors that influence the sourdough microbiota

flour used reflects the amount of nutrients, minerals, carbohydrates, and proteins available and may favor the development of certain species over others; (ii) fermentation parameters, such as temperature, pH, the hydration level of the dough, etc. causes certain microorganisms to thrive under specific environmental conditions, while others can be inhibited or overcome; (iii) the use of starter or adjunct cultures can alter or influence the community structure and fermentation dynamics; (iv) interactions between species, such as competition and metabolic exchanges, can modify the growth and survival of different species; (v) the adaptation of certain species to the sourdough environment over time can increase its resistance and favor microbial growth. It is important to note that these and other mechanisms are interconnected and can influence each other. Thus, sourdough is a matrix in constant evolution where the microbial dynamics involved in fermentation processes are still poorly understood.

Despite providing useful information about cultivable species and allowing the selection of phenotypes of interest, the basic culture-dependent methods cannot be used for monitoring the sourdough metacommunity. Thus, a broader and deeper approach is crucial to profile these microbiomes. Indeed, monitoring sourdough metacommunities provides essential information about the dynamics and composition of microorganisms involved in fermentation. This knowledge can help to maintain the quality and uniformity of sourdough-based products, ensuring the desirable flavor profiles, texture, and overall product characteristics. In addition, this monitoring of microbial diversity allows the optimization of fermentation processes, identifying the predominant microbial species and possible interactions between species that contribute to maintaining the quality of the final product.

The advent of "omics" technologies, including metagenomics, metabolomics, and metatranscriptomics, stems from the desire to understand the intricate biodiversity in complex ecosystems like fermented food systems, such as sourdough. Hence, metagenomics analysis serves as a valuable tool to mitigate the constraints posed by culture-dependent methods, enabling the detection of yeast and bacterial species that may go unnoticed using traditional methodologies.

Metagenomics entails thoroughly examining genetic material directly extracted from environmental samples, including soil, water, and fermented products. It offers a comprehensive perspective on the genetic information within a specific ecosystem, enabling the identification and characterization of microorganisms without requiring isolation or individual cultivation. This approach has brought about a paradigm shift in our comprehension of microbial communities, unveiling undiscovered biodiversity and shedding light on these microorganisms' pivotal roles in various biological processes (De Filippis et al. 2017).

However, even metagenomics methods have limitations, such as insufficient detection capacity of subdominant species and low taxonomic differentiation of phylogenetically close related species. Thus, biomolecular and metagenomic methods are often combined to precisely identify these ecosystems' members (Huys et al. 2013).

While starter or adjunct cultures are frequently employed to achieve specific characteristics or expedite formulation time in fermented products, it is crucial to recognize that the success of traditional fermented products stems from biodiversity (De Vuyst et al. 2014; Arici et al. 2018; Arena et al. 2019; Dentice Maidana et al. 2020). For example, microbial communities of greater diversity may be more resistant to environmental fluctuations (Brenner et al. 2008; Antoniewicz 2020), and their interactions favor efficient substrate conversion (Sabra et al. 2010). In other words, the complexity and quality of fermented food systems are based on their rich biodiversity (De Vuyst and Neysens 2005; Gänzle and Ripari 2016).

Recent studies involving sourdough have detected the presence of specific species of yeasts and bacteria, and some niches are already considered typical of sourdough. About 50 species of LAB and 30 species of yeasts have been identified in sourdoughs (Lhomme et al. 2014, 2016; Carbonetto et al. 2018, 2020; Gänzle and Zheng 2019). For example, the yeasts *Kazachstania humilis* (synonym *Candida humilis* and *C. milleri*), *Torulaspora delbrueckii*, *Wickerhamomyces anomalus* (*Pichia anomala*), *K. exigua* (*Saccharomyces exiguus*) and *Pichia kudriavzevii* (*Issatchenkia orientalis*) and the bacteria *Fructilactobacillus sanfranciscensis* (*F. sanfranciscensis*), *Lactiplantibacillus plantarum* (*L. plantarum*), *Levilactobacillus brevis* (*Lv. brevis*), *P. pentosaceus*, *Limosilactobacillus fermentum* (*L. fermentum*), and *Companilactobacillus paralimentarius* (*C. paralimentarius*) have been the most frequent microorganisms that have been detected and identified in sourdoughs (De Vuyst et al. 2017). Other studies demonstrate that cultural practices related to production methods contribute more significantly to the composition of the sourdough microbiome than geographic factors (De Vuyst and Neysens 2005; De Vuyst et al. 2014, 2017).

Therefore, this chapter discusses recent aspects of the metagenomics of bacteria and yeasts in sourdough microbiomes of different origins.

7.2 "Omics" Methods Used for Sourdough Analysis

The sourdough fermentation process is composed of a diverse community of microorganisms. Dependent-culture methods may be limited to cultivating a small portion of these microorganisms, limiting an in-depth study of the microbiota. Microorganisms live in multi-species communities allowing microbial interactions in microbiomes. However, these interactions are lost upon establishing a pure culture, increasing the metabolic burden and limiting the metabolic potential of the isolated microbe (Diender et al. 2021). Therefore, "omics" techniques are suitable for determining microbial communities' presence and interaction. Even keeping in mind that this chapter intends not to review the already well-defined knowledge about "omics" techniques, i.e., metagenomics, metatranscriptomics, and metabolomics…, some points are essential to be raised for better understanding.

Metagenomics based on high-throughput shotgun sequencing of whole-community DNA samples obtained from sourdough fermentation processes allows the elucidation of taxonomic composition and the functional potential of the microbial ecosystem. Metagenomics is the most suitable method to reveal fermented-related agents of sourdoughs, allowing the elucidation of their microbial structure. It is based on the sequencing of a tiny part of whole-community DNA. **Metabolomics** often deals with metabolite target analysis of sourdough fermentation to determine the actual concentrations of residual substrates and metabolites produced and list their volatile solely. **In contrast, metatranscriptomics** examines RNA molecules transcribed from genes within a microbial community. By capturing gene expression patterns, it unveils the active transcription of specific genes and sheds light on the functional activities of microorganisms at a particular moment (Weckx et al. 2019).

7.3 Bacteria in Sourdough

LAB identified by metagenomics may also be identified by other omics methods, such as metatranscriptomics; the results could agree with the shotgun technique if the taxonomy database is the same and updated. For example, at the genera level, *Lactococcus* and *Enterococcus* are generally present at the beginning of sourdough fermentation. These species identified by metagenomics may be culturable but not transcriptionally active. In other words, metagenomics can identify bacterial species in a microbial community. However, it does not provide direct information about whether those bacteria are transcriptionally active or which genes they are actively expressing.

Regarding *Leuconostoc*, *Oneococcus,* and *Pediococcus* species identified by metagenomics during isolation, several cannot be cultured using selected media, or the growth conditions are optimal for colony visualization. In addition, the morphology of colonies from the same genera is very similar, resulting in a lower diversity of isolated strains (Calabrese et al. 2022).

The synergetic presence of LAB and acetic acid bacteria (AAB) increases during the leavening storage time of sourdough. *Pediococcus* spp. and *Glutanobacter* spp. are dominant in organic sourdoughs, while *Acetobacter* spp. and *Limosilactobacillus* spp. in conventional sourdoughs (Calabrese et al. 2022; Baev et al. 2023).

In the population dynamics that involve the transformation of flour into mature sourdough, LAB (several species of *Lactobacillus* spp., *Leuconostoc* spp., and *Weissella* spp.) and yeasts (mainly *S. cerevisiae* and *Candida* spp.) outcompete other microbial groups contaminating flour and interact with each other at different levels (Diender et al. 2021). Among prokaryotes, sourdoughs harbored 202 genera and 275 species. Eukaryotes accounted for 30 genera and 31 species of yeasts. Within the core microbiome, *Lactiplantibacillus plantarum*, *Furfurilactobacillus sanfranciscensis*, *Limosilactobacillus fermentum*, *Leuc. citreum* and *Weissella confusa* were classified as dominant (Calabrese et al. 2022). However, several genera contribute indirectly during fermentation. Geographical position, microclimate, and type of flour are significant factors in determining microbial community composition during sourdough preparation (Landis et al. 2021). The propagation temperature also exerts a remarkable influence on the sourdough community. Propagation of sourdough at 30 °C favors the persistence of enterobacteria; otherwise, lower propagation temperatures selectively suppress enterobacteria, favoring LAB.

The dominant microorganisms in a stable community can be influenced by the subculture steps (Calabrese et al. 2022). In wheat flour, the only phylum detected was proteobacteria. Metagenomics detected LAB only after the first step of propagation; they were possibly present in the flour but below the detection limits. *Clostridium* and *Bacillus cereus* groups became predominant after one step of sourdough propagation, and LAB became dominant after successive steps of propagation. *Bacillus*, *Clostridium*, *Pseudomonas*, and Enterobacteriaceae may be gradually suppressed at temperatures below 21 °C (Menezes et al. 2020). However, almost all competitions resulted in a unique and stable community whose composition was independent of the initial species fractions (Friedman et al. 2017).

The 16S metagenetic analysis showed that the succession of strains in traditional sourdough might occur even during 2–3 months of daily backslopping. The content of residual carbohydrates, lactic acid, ethanol, or free amino acids may vary depending on the bakery considered. In some cases, variations of these biochemical features were also found within the same sourdough, possibly affected by strains' successions. However, the fermentation temperature, by influencing metabolic activities (acid production and release of free amino acids) and LAB growth, also changes some characteristics, as well as the sourdough pH and the activity of flour endogenous enzymes (Minervini et al. 2018).

Based on these several influencers on the fermentation process, proper community screening may contribute to fingerprinting its main characteristics, allowing a certain level of reproducibility. Fermentation-relevant bacteria or yeast may be enriched during preparation to mimic that specific community (Landis et al. 2021). Homofermentative species produce more aldehydes and ketones with more than six carbon atoms, while heterofermentative species produce predominantly ethanol and esters. On the other hand, *L. plantarum* and *L. sanfranciscensis* are the leading volatile compounds producers (Minervini et al. 2014).

Presumptive coliforms seem limited to a few cases (*ca.* 2.5–4.5 log CFU/g). Frequently, the concepts of dominant or sub-dominant are defined based on the cell density, considering if it is higher or lower than 6 log CFU/g. Therefore, *Lactiplantibacillus plantarum* and *Leuconostoc citreum* are dominant or co-dominant in sourdoughs. Sub-dominant phylum may be denoted by the presence of enterococci, staphylococcal, and coliforms (Calabrese et al. 2022).

Satellite microorganisms are considered metabolically active, and their presence in sourdough depends on metabolites produced by other dominant or subdominant LAB. Satellite members are equipped with unique enzymes and genes essential for metacommunity stability. In general, non-lactic acid bacteria that occasionally occur in sourdough, such as those of the genera staphylococci, e.g., *Staphylococcus aureus*, and a few enterobacteria, are considered satellites (Calabrese et al. 2022). Different microbial assemblies in flour may influence traditional sourdoughs' microbiological and biochemical characteristics. *Pediococcus pentosaceus* was the dominant LAB species in sourdoughs when considering *Pediococcus pentosaceus, Saccharomyces cerevisiae, Pantoea agglomerans*, and *Escherichia hermannii*. In addition, quantitative and qualitative differences in free amino acid concentration may be due to different microbial communities (Calabrese et al. 2022).

7.4 Yeasts in Sourdough

Yeasts are mainly responsible for sourdoughs' leavening process and extensive aroma formation (Van Kerrebroeck et al. 2017). Nevertheless, different species and strains may have different metabolic activities, conferring specific properties to each sourdough final product (Gobbetti et al. 2014; Palla et al. 2020). Therefore, studying yeast biodiversity in sourdough, influenced by factors such as the degree of yeast tolerance to acidity, temperature, time of fermentation, and available carbon sources, is pivotal for ecological knowledge and its related practical applications. By gaining insights into yeast diversity, researchers and practitioners can enhance sourdough products' quality, flavor, and stability. Furthermore, this knowledge can help develop tailored fermentation strategies, select suitable starter cultures, and preserve traditional sourdough ecosystems (Pulvirenti et al. 2004; Palla et al. 2020). However, there need to be more studies concerning the use of metagenomics to elucidate the yeast taxonomic composition in sourdoughs.

Four artisan bakery-scale backslopped sourdough samples from Belgium were analyzed by their yeast microbiota through culture-dependent and culture-independent methods. The later techniques include shotgun metagenomic sequencing, which, according to the authors, was applied for the first time in sourdough (Comasio et al. 2020). Regarding the metagenomic sequence reads obtained with the taxonomic analysis, less than 1% of the total reads were assigned to the yeast genera. Although representing low read numbers (from 0.1 to 0.6% of the total reads), *Kazachstania turicensis* was determined in all sourdough samples, which include samples obtained from wheat, multigrain, rye, and spelled flour. Culture-dependent analysis revealed *Kazachstania bulderi* as the predominant yeast species. This was not included during the metagenomic recruitment plotting analysis due to the need for more information concerning the complete genome in the National Center for Biotechnology Information (NCBI) database. *Saccharomyces cerevisiae,* with values ranging from 0.01 to 0.02% of the total reads, was present in all sourdough samples except when rye flour was used. These findings agree with the identifications obtained by culture-dependent and metagenetic analyses. In addition, reads belonging to species related to *Rhizophagus irregularis* and *Rhizophagus clarus* were determined in sourdough samples (Comasio et al. 2020).

Later, Li et al. (2021) used high-throughput metagenomic sequencing to determine the microbial diversity of sourdoughs fermented by different starters to compare the effects of AAB and LAB. Moreover, the functional capacities of the starters were also investigated, showing remarkable differences in microbial diversity and metabolism among sourdoughs started by different cultures. *Saccharomyces cerevisiae* Y5, *Levilactobacillus brevis* B1, and *Acetobacter tropicalis were the starters applied.* Therefore, different treatments were obtained: only yeast, yeast + LAB, yeast + AAB, and yeast + LAB + AAB. The results showed that *Saccharomyces* corresponded to 6.13% when the inoculated species were yeast and LAB, while for the treatment using yeast and AAB as starters, *Saccharomyces* represented 28.88%. When both AAB and LAB were employed, *Saccharomyces* were the least representative (8.31%) of the starter groups.

Landis et al. (2021) worked with a global community-scientist network of bread bakers to perform a study to determine the microbial diversity of household sourdough starters. The authors evaluated 500 sourdough starters from different places around the world: the United States (n = 429), Canada (n = 29), Europe (n = 24), Australia and New Zealand (n = 17), and Thailand (n = 1). Amplicon ITS rRNA gene sequencing was used to characterize the fungal community at the species level. In addition, 40 sourdough starters were used for functional analyses.

Thus, shotgun metagenomic sequencing was performed to confirm that the experimental communities revived from glycerol stocks were similar to the original samples. The results showed that each starter sample contained a median of 35 fungal amplicon sequence variants, with yeasts (order: Saccharomycetales) encompassing over 70% of fungal reads. The other fungi detected (e.g., molds, plant pathogens) were not considered due to their restricted roles in sourdough fermentation. Many communities were dominated by a single yeast and bacterial species per starter, where *Saccharomyces cerevisiae* accounted for >50% of fungal ITS reads in 77% of samples (Landis et al. 2021).

In addition to *Saccharomyces cerevisiae,* other yeasts were identified with relative abundance (expressed as a fraction of one) ranging from 0.032 to 0.014:

Kazachstania humilis, Kazachstania servazzii, Wickerhamomyces anomalus, Pichia membranifaciens, Kazachstania unispora, Naumovozyma castellii, Candida sake and *Saccharomyces bayanus*. Moreover, *Kazachstania barnettii, Issatchenkia orientalis, Kazachstania exigua, Candida quercitrusa, Nakaseomyces* spp., *Pichia mandshurica, Pichia kluyveri, Vanderwaltozyma yarrowii, Kazachstania naganishii, Candida solani, Pichia norvegensis, Hanseniaspora uvarum, Candida santamariae, Candida tropicalis, Cyberlindnera fabianii, T. delbrueckii, Candida parapsilosis, Meyerozyma guilliermo, Trichomonascus ciferrii,* and *Dipodascaceae* spp. could also be determined.

This study demonstrated that geographical location does not determine the yeast taxonomic composition of sourdough at a global scale, while microbial interactions shape sourdough community structure. Although the authors were interested in species-level identification, it is well established that intraspecific strain diversity plays a vital role in sourdough, and amplicon sequencing only partially captures strain-level diversity. Therefore, further research using shotgun metagenomic or whole-genome sequencing should be developed to obtain more nuanced biogeographic patterns in sourdough biodiversity (Arora et al. 2021).

7.5 Functional Annotation of Metagenomic Data

The potential functional roles of microbial communities in diverse sourdoughs can be determined by shotgun analysis following two kinds of annotation using two different databases such as KEGG and COG. The difference between both databases is that KEGG assigns molecular functions based on KEGG pathways. In contrast, COG function is generated by protein sequences of microbial genomes with known and predicted proteins. It is essential to highlight that studies on microbial functions should be performed considering the specific critical pathways involved in the production or degradation of carbohydrates, amino acids, secondary metabolite biosynthesis, transport, and metabolism, as well as the transport and metabolism of inorganic iron and energy production and conversion (Tatusov et al. 2003; Tanabe and Kanehisa 2012).

Moreover, it is vital to determine the dominant players based on the occurrence of the primary genes annotated from the species identified by the shotgun. Afterward, the classification correlated to the bioinformatic estimative regarding the dominance and sub-dominance can be carried out. Studies evaluating spontaneous fermentation present potential functionality concerning player interactions, including bacteria and yeasts. Shotgun analyses in sourdoughs share many genes (Calabrese et al. 2022).

Thus, regarding the particularity of this method and sourdough, it is possible to find unique genes assigned to unique sourdoughs (Calabrese et al. 2022). For this purpose, the metabolisms of carbohydrates, pyruvate, energy, and nitrogen should be the focus of these studies, allowing the identification of species with genes annotated with high abundance.

To be considered a suitable and valid study, all functions should be present in reconstructed pathways, which are fundamental to understanding the resilience and performance of each sourdough. In addition, this strategy also confirms the presence of many players within the metacommunity since they may be originated from other ecosystems such as the human gut, animals, foods, and plants (Vitulo et al. 2019; Magro et al. 2019; Arora et al. 2021). The interactions between LAB and *S. cerevisiae* are observed in most sourdoughs during spontaneous fermentation, forming a complex metacommunity. Despite their differences, LAB uses homofermentative or heterofermentative pathways, while *S. cerevisiae* performs alcoholic fermentation. Overall, the highest number of annotated genes can help distinguish the key enzymes considering LAB competitiveness. Besides, all sourdoughs can harbor the gene carbohydrate metabolism (Calabrese et al. 2022).

The KEGG database can predict the essential enzymes and pathways in the different sourdoughs. L-lactate dehydrogenase (cytochrome) (EC 1.1.2.3), alcohol dehydrogenase (cytochrome c) (EC 1.1.2.8), and D-lactate dehydrogenase (cytochrome) (EC 1.1.2.4) are considered the genes with high abundance (Calabrese et al. 2022).

More studies must be performed to comprehensively understand how subdominants may cooperate synergistically for the dominant players to express essential genes, as well as the satellite members equipped with unique genes, which are also crucial for the stability of the metacommunity.

7.6 Conclusion

Numerous research articles have employed microbiological and biochemical approaches to characterize sourdough. Consequently, compelling evidence has emerged, highlighting the influence of the flour type, sourdough environment, and the type of fermentation (controlled or spontaneous) on the composition and dynamics of the sourdough microbial metacommunity (Arora et al. 2021). To understand the impact of overlooked microorganisms, which often go unnoticed due to the limitation of culture-dependent methods, novel design strategies should be implemented by employing innovative approaches such as metagenomics, non-culturable or challenging-to-culture microorganisms can be detected. Metagenomics can amplify the sourdough biobank. Moreover, new species can be obtained, improving the quality of sourdough fermentation. Studies must be carried out since the role of the satellite members needs to be sufficiently elucidated. It may be attributed to the lower abundance of these species compared to the other players, making it difficult to annotate and assemble the whole genome sequencing during the shotgun analysis. This finding expanded knowledge and holds significant potential for various applications, ranging from the food industry to scientists. It provides valuable insights to comprehensively understand microbial diversity and their roles in complex interactions and biological processes.

References

Antoniewicz MR (2020) A guide to deciphering microbial interactions and metabolic fluxes in microbiome communities. Curr Opin Biotechnol 64:230–237. https://doi.org/10.1016/j.copbio.2020.07.001

Arena MP, Russo P, Spano G, Capozzi V (2019) Exploration of the microbial biodiversity associated with north Apulian sourdoughs and the effect of the increasing number of inoculated lactic acid bacteria strains on the biocontrol against fungal spoilage. Fermentation 5:97. https://doi.org/10.3390/fermentation5040097

Arici M, Ozulku G, Yildirim RM et al (2018) Biodiversity and technological properties of yeasts from Turkish sourdough. Food Sci Biotechnol 27:499–508. https://doi.org/10.1007/s10068-017-0282-0

Arora K, Ameur H, Polo A et al (2021) Thirty years of knowledge on sourdough fermentation: a systematic review. Trends Food Sci Technol 108:71–83. https://doi.org/10.1016/j.tifs.2020.12.008

Baev V, Apostolova E, Gotcheva V et al (2023) 16S-rRNA-based metagenomic profiling of the bacterial communities in traditional Bulgarian sourdoughs. Microorganisms 11:803. https://doi.org/10.3390/microorganisms11030803

Brenner K, You L, Arnold FH (2008) Engineering microbial consortia: a new frontier in synthetic biology. Trends Biotechnol 26:483–489. https://doi.org/10.1016/j.tibtech.2008.05.004

Calabrese FM, Ameur H, Nikoloudaki O et al (2022) Metabolic framework of spontaneous and synthetic sourdough metacommunities to reveal microbial players responsible for resilience and performance. Microbiome 10:148. https://doi.org/10.1186/s40168-022-01301-3

Carbonetto B, Ramsayer J, Nidelet T et al (2018) Bakery yeasts, a new model for studies in ecology and evolution. Yeast 35:591–603. https://doi.org/10.1002/yea.3350

Carbonetto B, Nidelet T, Guezenec S et al (2020) Interactions between Kazachstania humilis yeast species and lactic acid bacteria in sourdough. Microorganisms 8:240. https://doi.org/10.3390/microorganisms8020240

Comasio A, Verce M, Van Kerrebroeck S, De Vuyst L (2020) Diverse microbial composition of sourdoughs from different origins. Front Microbiol 11:1212. https://doi.org/10.3389/fmicb.2020.01212

De Filippis F, Parente E, Ercolini D (2017) Metagenomics insights into food fermentations. Microb Biotechnol 10:91–102. https://doi.org/10.1111/1751-7915.12421

De Vuyst L, Neysens P (2005) The sourdough microflora: biodiversity and metabolic interactions. Trends Food Sci Technol 16:43–56. https://doi.org/10.1016/j.tifs.2004.02.012

De Vuyst L, Van Kerrebroeck S, Harth H et al (2014) Microbial ecology of sourdough fermentations: diverse or uniform? Food Microbiol 37:11–29. https://doi.org/10.1016/j.fm.2013.06.002

De Vuyst L, Van Kerrebroeck S, Leroy F (2017) Microbial ecology and process Technology of Sourdough Fermentation. Adv Appl Microbiol 100:49–160

Dentice Maidana S, Aristimuño Ficoseco C, Bassi D et al (2020) Biodiversity and technological-functional potential of lactic acid bacteria isolated from spontaneously fermented chia sourdough. Int J Food Microbiol 316:108425. https://doi.org/10.1016/j.ijfoodmicro.2019.108425

Diender M, Parera Olm I, Sousa DZ (2021) Synthetic co-cultures: novel avenues for bio-based processes. Curr Opin Biotechnol 67:72–79. https://doi.org/10.1016/j.copbio.2021.01.006

Friedman J, Higgins LM, Gore J (2017) Community structure follows simple assembly rules in microbial microcosms. Nat Ecol Evol 1:0109. https://doi.org/10.1038/s41559-017-0109

Gänzle M, Ripari V (2016) Composition and function of sourdough microbiota: from ecological theory to bread quality. Int J Food Microbiol 239:19–25. https://doi.org/10.1016/j.ijfoodmicro.2016.05.004

Gänzle MG, Zheng J (2019) Lifestyles of sourdough lactobacilli – do they matter for microbial ecology and bread quality? Int J Food Microbiol 302:15–23. https://doi.org/10.1016/j.ijfoodmicro.2018.08.019

Gobbetti M, Rizzello CG, Di Cagno R, De Angelis M (2014) How the sourdough may affect the functional features of leavened baked goods. Food Microbiol 37:30–40. https://doi.org/10.1016/j.fm.2013.04.012

Gobbetti M, De Angelis M, Di Cagno R et al (2019) Novel insights on the functional/nutritional features of the sourdough fermentation. Int J Food Microbiol 302:103–113. https://doi.org/10.1016/j.ijfoodmicro.2018.05.018

Huys G, Daniel H-M, De Vuyst L (2013) Taxonomy and biodiversity of sourdough yeasts and lactic acid bacteria. In: Gobbetti M, Gänzle M (eds) Handbook on sourdough biotechnology. Springer US, Boston, pp 105–154

Landis EA, Oliverio AM, McKenney EA et al (2021) The diversity and function of sourdough starter microbiomes. elife 10:e61644. https://doi.org/10.7554/eLife.61644

Lhomme E, Mezaire S, Ducasse MB et al (2014) A polyphasic approach to study the dynamics of the microbial population of an organic wheat sourdough during its conversion to gluten-free sourdough. Int Microbiol 17:1–9. https://doi.org/10.2436/20.1501.01.202

Lhomme E, Urien C, Legrand J et al (2016) Sourdough microbial community dynamics: an analysis during French organic bread-making processes. Food Microbiol 53:41–50. https://doi.org/10.1016/j.fm.2014.11.014

Li H, Fu J, Hu S et al (2021) Comparison of the effects of acetic acid bacteria and lactic acid bacteria on the microbial diversity of and the functional pathways in the dough as revealed by high-throughput metagenomics sequencing. Int J Food Microbiol 346.:109168 Contents:109168

Magro DO, Santos A, Guadagnini D et al (2019) Remission in Crohn's disease is accompanied by alterations in the gut microbiota and mucin production. Sci Rep 9:13263. https://doi.org/10.1038/s41598-019-49893-5

Martín-Garcia A, Riu-Aumatell M, López-Tamames E (2023) Influence of process parameters on sourdough microbiota, physical properties and sensory profile. Food Rev Int 39:334–348. https://doi.org/10.1080/87559129.2021.1906698

Menezes LAA, Sardaro MLS, Duarte RTD et al (2020) Sourdough bacterial dynamics revealed by metagenomic analysis in Brazil. Food Microbiol 85:103302. https://doi.org/10.1016/j.fm.2019.103302

Minervini F, De Angelis M, Di Cagno R, Gobbetti M (2014) Ecological parameters influencing microbial diversity and stability of traditional sourdough. Int J Food Microbiol 171:136–146. https://doi.org/10.1016/j.ijfoodmicro.2013.11.021

Minervini F, Dinardo FR, Celano G et al (2018) Lactic acid bacterium population dynamics in artisan sourdoughs over one year of daily propagations is mainly driven by flour microbiota and nutrients. Front Microbiol 9:1984. https://doi.org/10.3389/fmicb.2018.01984

Palla M, Blandino M, Grassi A et al (2020) Characterization and selection of functional yeast strains during sourdough fermentation of different cereal whole grain flours. Sci Rep 10:12856. https://doi.org/10.1038/s41598-020-69774-6

Pulvirenti A, Solieri L, Gullo M et al (2004) Occurrence and dominance of yeast species in sourdough. Lett Appl Microbiol 38:113–117. https://doi.org/10.1111/j.1472-765X.2003.01454.x

Sabra W, Dietz D, Tjahjasari D, Zeng A-P (2010) Biosystems analysis and engineering of microbial consortia for industrial biotechnology. Eng Life Sci 10:407–421. https://doi.org/10.1002/elsc.201000111

Tanabe M, Kanehisa M (2012) Using the KEGG database resource. Curr Protoc Bioinforma 1(12):38. https://doi.org/10.1002/0471250953.bi0112s38

Tatusov RL, Fedorova ND, Jackson JD et al (2003) The COG database: an updated version includes eukaryotes. BMC Bioinform 4:1–14. https://doi.org/10.1186/1471-2105-4-41

Van Kerrebroeck S, Maes D, De Vuyst L (2017) Sourdoughs as a function of their species diversity and process conditions, a meta-analysis. Trends Food Sci Technol 68:152–159. https://doi.org/10.1016/j.tifs.2017.08.016

Vitulo N, Lemos WJF, Calgaro M et al (2019) Bark and grape microbiome of Vitis vinifera: influence of geographic patterns and agronomic management on bacterial diversity. Front Microbiol 9:3203. https://doi.org/10.3389/fmicb.2018.03203

Weckx S, Van Kerrebroeck S, De Vuyst L (2019) Omics approaches to understand sourdough fermentation processes. Int J Food Microbiol 302:90–102. https://doi.org/10.1016/j.ijfoodmicro.2018.05.029

Chapter 8
Metabolic Engineering of Lactic Acid Bacteria and Yeasts for the Production of Compounds with Industrial Applications

Birsen Yılmaz, Hande Mortaş, Semra Navruz Varlı, and Duygu Ağagündüz

8.1 Introduction

Known short as LAB, lactic acid bacteria are classified as gram-positive bacteria which are able to tolerate low pH (Mokoena 2017). They convert carbohydrate substrates to organic acids (mostly lactic acid), and are capable of generating a variety of metabolites. LAB are commonly employed as probiotics, starter cultures, and microbial cell factories due to their desirable properties (Ruiz Rodríguez et al. 2019). There are more than 60 genera of LAB, but the most common genera found in food fermentation include *Enterococcus, Lactococcus, Pediococcus, Lactobacillus,* and *Leuconostoc,* among others. *Lactobacillaceae* has recently been proposed as a new genus through the merging of *Lactobacillaceae* and *Leuconostocaceae* into a single family. Also worth noting, *Lactobacillus* has received reclassification into 25 different genera (Zheng et al. 2020). Zheng et al. (2020) recently published the novel taxonomy on the genus *Lactobacillus* (Zheng et al. 2020), and throughout this chapter, this novel taxonomy for LAB was employed.

Yeasts are eukaryotic microorganisms that represent the Ascomycetes and Basidiomycetes phylum. They can withstand conditions low pH conditions along with low a_w, and salt of a high concentration. Humans have used them over a number of millennia as they possess a number of potential benefits to food production, medicine, agriculture, and bioengineering disciplines (Perricone et al. 2017a; Johnson and Echavarri-Erasun 2011a; Alvarez-Martin et al. 2008). In the food industry, LAB and yeasts are mainly used as a starter culture in fermenting foodstuffs which include grain-based products, dairy products, alcoholic

B. Yılmaz · H. Mortaş · S. N. Varlı · D. Ağagündüz (✉)
Faculty of Health Sciences, Department of Nutrition and Dietetics, Gazi University, Ankara, Türkiye

Department of Nutrition and Dietetics, Çukurova University, Adana, Türkiye
e-mail: duyguturkozu@gazi.edu.tr

E. B. Ceresino et al. (eds.), *Sourdough Microbiota and Starter Cultures for Industry*, https://doi.org/10.1007/978-3-031-48604-3_8

beverages, fermented meat products, fermented vegetables, fruits, and fish (De Vuyst and Leroy 2020). In addition to the beneficial uses of LAB and yeasts, which have been known for many years, they have recently become more popular with biotechnological applications (Nes et al. 2015). Even though these applications have a long history, as the techniques improve, the possibility of using these microorganisms in the industry is increasing (McKay and Baldwin 1990).

The utilization of both synthetic and systems biology technologies to enhance the use of present metabolic pathways or to build new ones is defined as metabolic engineering (Wang et al. 2021). The metabolic engineering methods of LAB and yeasts aim to enhance the synthesis of numerous metabolites and bioactive compounds. Various metabolic engineering strategies, including CRISPR-Cas9 and other gene-editing systems which will be mentioned in the following parts, are being developed to respond to the increasing demand for microbial metabolites in industrial applications (Börner et al. 2018).

Recently, metabolic engineering applications in LAB and yeasts have been even more popular (Nes et al. 2015). The applications of LAB include the production of nutraceuticals and high-value metabolites including low-calorie sugars, vitamins, polysaccharides; food ingredients such as alanine, diacetyl, acetaldehyde; and commodity chemicals such as lactic acid, ethanol, butanol (McKay and Baldwin 1990). On the other hand, yeasts have numerous applications and metabolic engineering of yeasts has shown to be a useful approach for understanding enzyme processes and producing high-value bioactive compounds (Johnson and Echavarri-Erasun 2011a). So, researchers have taken advantage of metabolic engineering in order to procure yeast strains that produce beneficial substances such as carotenoids, polyphenols, oligosaccharides, and folate. In this chapter, the metabolic engineering process upon LAB and yeast are covered in detail.

8.2 Genetic Tools/Applications Acting on Primary Metabolism of LAB

The use of genetic tools/applications that are effective on the primary metabolism of LAB has become widespread with the developments revealed by studies conducted for more than 40 years. Thus, new industrial initiatives gained momentum. After the use of genetic tools, cellular pathways were better understood and some desired changes in primary metabolism in LAB could be achieved. Genetic applications have a great impact on the creation of cell factories for the production of high-value products (Gaspar et al. 2013).

In genetic studies using LAB, various molecular techniques and tools have been developed by which strain-specific manipulation can be achieved (Zhu et al. 2009). Today, it is possible to access information on chromosome integrations, transformation procedures for gene expression, vectors, synthetic promoter libraries and systems for almost all LAB, especially Lactococci and Lactobacilli. The nisin-controlled gene expression (NICE) system, which is among the oldest and most basic chromosomal integration and gene expression systems developed and still widely used, has been successfully applied in regulated gene expression in Gram-positive bacteria (Goh et al. 2009; Mierau and Kleerebezem 2005). In metabolic engineering studies, the NICE system has been used for overexpression of homologous and heterologous genes and integral membrane proteins (Gaspar et al. 2013; Neves et al. 2006; Kunji et al. 2003). Although the NICE system was originally used only for Lactococci for regulated gene expression, its use in Lactobacilli and other Gram-positive bacteria was also promptly achieved (Gaspar et al. 2013). Some food-related tools in the NICE system support use in steps where antibiotic resistance is not present (Lu et al. 2013). In addition, this system is also used to increase cloning efficiency through vectors with various properties (Berlec and Štrukelj 2012).

Although it is known that inducible expression systems provide convenience in practice, it is a fact that inducer molecules are not suitable for cost balance. For this reason, constitutive promoters are preferably used in industrial applications. In metabolic engineering studies, it is recommended to use synthetic promoter libraries to adjust gene expression at the desired level. Promoter libraries consist of a spacer sequence that separates the consensus sequences of the promoter. Although this approach was first developed for *L. lactis*, it is stated that it can theoretically be applied to all LABs (Gaspar et al. 2013; Solem and Jensen 2002; Jensen and Hammer 1998).

The pORI system is a widely used gene exchange system for gene knockouts or substitutions. In this system, a cross homologous recombination process consisting of plasmid integration and excision is performed (Leenhouts et al. 1996). In this system, conditional replication of lactococcal plasmid pWV01 derivatives lacking the repA gene is provided. Various helper strains that provide RepA are readily available for the generation and isolation of interchangeable vectors. It has been stated that mutants produced by the method used in this system will not carry antibiotic resistance and can be safely used for gene inactivation or overexpression in *L. lactis* and Lactobacillus species (Goh et al. 2009; Douglas and Klaenhammer 2011; Gaspar et al. 2011). In a study, sequence-specific integration based on homologous recombination and integration into the bacteriophage binding site were combined with a developed vector (pCS1966). This vector has been used to rapidly introduce changes in specific site-directed nucleotide sequences in various genes encoding different subunits of glucose-PTS. A derivative of pCS1966 is currently available for integrations in a neutral region of the *L. lactis* chromosome. The viability of this derivative has been demonstrated in model strains MG1363 and IL1403 through the integration of a two-component nisin sensor (Gaspar et al. 2013; Pinto et al. 2011).

Site-specific recombination systems have also been developed, such as the Cre–loxP system. An important aspect of this system is the excision of selectable markers in gene replacements. However, this system is unsuitable for sequential gene replacement cycles and selectable marker excision, as it affects genome stability. This incompatibility was also demonstrated in *L. lactis* by the recombinase Cre inverting large DNA fragments flanked by loxP. In a study, a Cre-lox-based toolbox was developed for multiple gene deletions in a single genetic background for *Lb. plantarum* (Lambert et al. 2007). More specifically, a gene-specific mutagenesis vector is developed and used using the double cross-recombination method to replace the target with a lox66–P32-cat–lox71 cassette. It has been reported that the system can be adapted to other LABs, especially *L. lactis* and *S. thermophilus*, as well as *Lb. plantarum* (Gaspar et al. 2013).

Oligonucleotide-mediated recombination, also known as single-stranded (ssDNA) recombination, has been in use for more than a decade in LAB such as *L. lactis* and *Lb. reuteri* (Van Pijkeren and Britton 2012). With this technique, targeted point mutations in the chromosome can be performed effectively without the need for any selection. With the successful implementation of ssDNA recombination in LAB, methods such as multiplexed automated genome engineering (MAGE) can be applied more rapidly in genome-wide genetic engineering studies. It has been noted that ssDNA-directed LAB modification facilitates strain engineering by reducing the time required for administration and that the new strains created are as safe as existing natural strains. Additionally, it has been found safe to use in both industrial and medical applications (Gaspar et al. 2013; Wang et al. 2009; Van Pijkeren et al. 2012).

8.3 Metabolic Engineering of LAB for the Production of Compounds with Industrial Applications

"Metabolic engineering" is characterized by its use of both systems and synthetic biology processes to efficiently make use of existing metabolic pathways or establish new ones (Wang et al. 2021). Industry implements methods involving metabolic engineering in LAB to boost the production of various metabolites that lactic acid and its derivatives can produce (Fig. 8.1). Since the pathways and processes that enable the production of metabolites obtained through metabolic engineering have interfered with each other, it is quite difficult to make a systematic classification in metabolites in connection with metabolic transformations, and there is no consensus on this issue.

In response to the increase in requests for microbial metabolites in various industries, several cutting-edge metabolic engineering techniques are useful for implementation with the advancing impact of the developed CRISPR-Cas9 along with other noteworthy gene-editing systems (Wang et al. 2021; Hidalgo et al. 2019). It is possible to evaluate metabolic engineering approaches applied in LAB basically under four subtitles: (i) overexpression of the gene encoding the rate-limiting enzyme of the relevant biosynthetic pathway, (ii) inhibition of competing for metabolic reactions, (iii) engineering of existing biosynthetic pathways, (iv) directing the metabolic flow to the biosynthesis of the desired chemical (Hugenholtz and Smid 2002; Sharma et al. 2020).

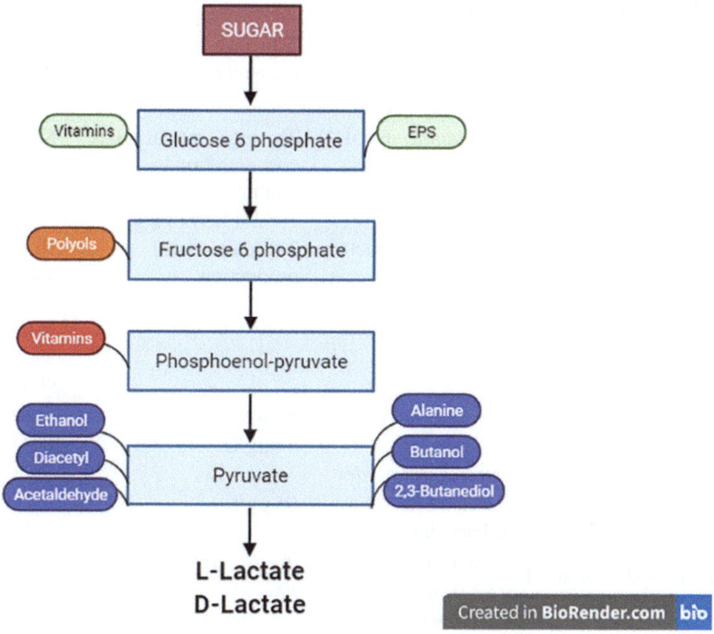

Fig. 8.1 Various metabolites associated with metabolic engineering of LAB

In this section, metabolic engineering strategies applied to LAB strains involved in the production of some industrially valuable metabolites about nutraceuticals and high-value metabolites such as low-calorie sugars, vitamins, polysaccharides; food ingredients such as alanine, diacetyl, acetaldehyde; and commodity chemicals such as lactic acid, ethanol, and butanol are presented under separate headings below (Table 8.1).

8.3.1 Lactic Acid

The homofermentative pathway, where lactic acid production takes place, is the Embden-Meyerhof pathway, while the heterofermentative pathway is the phosphoketolase pathway (Kandler 1983). In the first of these, the main end product is lactic acid, while in the second, the compounds acetic acid, propionic acid, ethanol, as well as lactic acid come into being. While the first of the L- and D-isomers, which are different forms of lactic acid produced by LAB, is popular within the food service industry, metabolic engineering strategies are mostly carried out through applications to increase the first form, since the second is toxic to humans (Papagianni and Avramidis 2011). The United States Food and Drug Administration generally recognized as safe (GRAS) approved that L-lactic acid is safe (Wee et al. 2006). LAB, some homofermentative and some heterofermentative, were used to produce

Table 8.1 Metabolically engineered LAB strains

Metabolic engineering applications	The parent or engineered strain	Effect on the biosynthesis	References
Folic acid			
When stored at 4 °C over 7 days of yoghurt, the highest levels of folic acid were produced.	*Lb. plantarum* 15HN	The highest total folate concentration in yoghurt: 1487 ± 96.42 μg/L	Khalili et al. (2020)
Overexpression of gchI gene	*Lc. lactis* NZ9000	3-fold enhanced production	Hugenholtz and Smid (2002), Sharma et al. (2020) and Wegkamp et al. (2007)
Elevated external pH boosted the manufacturing of folate and the supplementation of p-aminobenzoic acid also stimulated folate production, while high tyrosine concentrations led to reduced folate biosynthesis.	*Leuconostoc lactis*	45 μg/L	Sybesma et al. (2003)
	Leuconostoc paramesenteroides	44 μg/L	
	Lactococcus lactis subspecies	95 μg/L	
	S. thermophilus	29–202 μg/L	
	Lb. delbrueckii subsp. *bulgaricus*	54 μg/L	
	Lb. plantarum	45 μg/L	
	Lb. helveticus	2–89 μg/L	
Overexpression of folKE and gchI gene	*Lc. lactis* subsp. *cremoris* MG1363	3-fold enhanced total production	Sharma et al. (2020) and Sybesma et al. (2003)
Overexpression of pab and folB, folKE, folP, folQ, folC genes	*Lc. lactis*	High folate levels	Sharma et al. (2020) and Wegkamp et al. (2007)
Directed mutagenesis and strain selection	*Lc. lactis*	Overproduction of folic acid	Sharma et al. (2020) and Sybesma et al. (2004)
Riboflavin			
Overexpression of ribG, ribH, ribB and ribA genes	*Lc. lactis*	Overproduction of riboflavin	Burgess et al. (2004)
Directed mutagenesis and strain selection	*Lc. lactis*	Overproduction of riboflavin	Sharma et al. (2020) and Sybesma et al. (2004)

(continued)

Table 8.1 (continued)

Metabolic engineering applications	The parent or engineered strain	Effect on the biosynthesis	References
Roseoflavin may encourage an insertion of DNA fragments in LAB to increase riboflavin content. Insert a 1059-bp DNA fragment into the upstream regulatory region of the rib operon	*Lb. plantarum* RYG-YYG-9049-M10	1.013 to 2.332 mg/L	Ge et al. (2020)
Purine biosynthesis is capable of initiating riboflavin secretion more effectively considering that the purine nucleotide GTP is a precursor to riboflavin. Purine starvation in the mutants induced overexpression of the riboflavin biosynthesis cluster ribABGH.	*Lc.lactis* JC017	Riboflavin content of milk was improved to 2.81 mg/L from 0.99 mg/L	Chen et al. (2017)
Gene-specific genetic engineering Nucleotide replacements and deletions in the regulatory region of the rib operon	*Lc. lactis* subsp. *cremoris* NZ9000, (pNZGBAH)	24 mg/L	Burgess et al. (2004)
Production of riboflavin in MRS medium after 24 hrs of growth	*Lb. fermentum* KTLF1 *Lb. fermentum* KTLP13	2.13 mg/L 2.36 mg/L	Thakur and Tomar (2015)
Xylitol			
Genetic modifications Expressed P. stipitis XR	*Lc. lactis*	2.5 mol (mol glucose)$^{-1}$ 2.7 g l^{-1} h^{-1}	Nyyssölä et al. (2005)
Mannitol			
Genetic modifications LDH-deficient, disrupted PTSMtl gene mtlA or mtlF	*Lc. lactis*	0.33 mol (mol glucose)$^{-1}$	Gaspar et al. (2004)
Genetic modifications Expressed Mtl1PDH and mannitol-1-phosphatase	*Lc. lactis*	0.50 mol (mol glucose)$^{-1}$	Wisselink et al. (2005)
Sorbitol			
Genetic modifications Overexpressed Stl6PDH, LDH-deficient	*Lb.plantarum*	0.65 mol (mol glucose)$^{-1}$	Ladero et al. (2007)
Genetic modifications Expressed Stl6PDH, deleted ldhL	*Lacticaseibacillus casei*	0.04 mol (mol glucose)$^{-1}$	Nissen et al. (2005)

(continued)

Table 8.1 (continued)

Metabolic engineering applications	The parent or engineered strain	Effect on the biosynthesis	References
Lactic acid			
pH-stat strategy	*Lacticaseibacillus rhamnosus* B103	143.7 g/L	Bernardo et al. (2016)
Heterologous expression of gene encoding short-chain dehydrogenase	*Lacticaseibacillus casei*	Enhanced production	Shirai et al. (2001)
	Lactiplantibacillus pentosus		
	Lactobacillus sp.		
The addition of diverse nutrients including the substrate glucose or vitamin B compounds or taking on pH control strategies	*Lacticaseibacillus rhamnosus* HN001	Enhanced production	Wang et al. (2019)
Disruption of ldhD gene	*Lb. helveticus* CNRZ32	Product yield of 76.2%	Kylä-Nikkilä et al. (2000)
Disruption of phosphokeolase gene (ptk)	*Lactococcus lactis*	Purity – 99.6%, Titer – 50.1 g/l, Yield – 94.6%	Shinkawa et al. (2011)
Converting cellobiose to D(S)-lactic acid by gene mutation	*Lc.lactis*	1.52 g/l/h (cellulose)	Singhvi et al. (2010)
UV mutagenesis to enhance tolerance to glucose and lactate repression	*Lc. lactis* BME5–18	1.76 g/L.h	Bai et al. (2004b)
Construction of l-lactate dehydrogenase gene (ldh 1) deficiency	*Lb. plantarum*	0.85 g per g of consumed sugar	Okano et al. (2009)
Introducing the xylose-assimilating xylAB operon	*Lb. plantarum*	0.78 g per g of consumed sugar	Yoshida et al. (2011)
Disruption of L-LDH gene	*Lb. plantarum* NCIMB8826	2.18 g/L.h	Okano et al. (2017)
Introducing xylanase gene to *Levilactobacillus brevis*	*Levilactobacillus brevis*	Titer – 1.7 g/l (Xylan fermentation)	Hu et al. (2011)
Evolutionary of acid-resistant mutant	*Lacticaseibacillus casei*	Increased by 13.6%	Zhang et al. (2012)
Ultraviolet mutagenesis to foster strains of *Lactobacillus delbrueckii* (NCIM 2365) that produced amplified lactic acid concentrations	*Lactobacillus delbrueckii*	Titer – 135 g/l, Yield – 90.0%	Kadam et al. (2006)
Classical mutation using nitrous acid to improve Lactobacillus delbrueckii NCIM 2025 strain	*Lb. delbrueckii*	Titer – 54 g/l	Rajan and Madhavan (2008)
Interruption of the ldhD gene	*Lacticaseibacillus paracasei* 7BL	2.25–3.23 g/L.h	Kuo et al. (2015)

(continued)

Table 8.1 (continued)

Metabolic engineering applications	The parent or engineered strain	Effect on the biosynthesis	References
ldhD gene disruption in the wild-type *Pediococcus acidilactici* DQ2 strain	*Pediococcus acidilactici* TY112	77.66 g/L of L-lactic acid acquired at 25% (w/w) solids content of dry dilute acid pretreated and biodetoxified corn stover feedstock	Yi et al. (2016)
Overexpression of ldhL gene and interference with pfl gene	*Enterococcus faecalis*	a yield of >99% within 48 h	Doi (2015)
Acetaldehyde			
Overexpression of glyA gene	*S. thermophillus*	80–90% production yield	Chaves et al. (2002)
Acetoin			
Additional ATP consumption (AAC)	*Lc. lactis*	14 g/L titer	Liu et al. (2016a)
Biocompatible chemistry in addition to the inactivation of competing product pathway	*Lc. lactis*	5.8 g/L titer	Liu et al. (2016b)
2,3-butanediol			
Biocompatible chemistry	*Lc. lactis*	6.7 g/L titer	Liu et al. (2016d)
Diacetyl			
Inactivation of αldB gene and overexpression of ilvBN genes	*Lc. lactis*	Enhanced production	Benson et al. (2009)
Co-factor and promoter engineering	*Lc. lactis*	Enhanced production	Guo et al. (2012)
Biocompatible chemistry	*Lc. lactis*	87% production yield	Liu et al. (2016d)
Ethanol			
Gene knockout along with redirection of metabolic flux	*Lc. lactis*	a yield of 70%	Liu et al. (2016c)
3-hydroxypropionaldehyde			
Combined approach complemented by overexpression of glycerol dehydratase gene	*Lactobacillus diolivorans*	35.9 g/L production	Lindlbauer et al. (2017a) and Lindlbauer et al. (2017b)
3-hydroxypropionic acid			
Metabolic flux re-routing	*Limosilactobacillus reuteri*	10.6 g/L production	Dishisha et al. (2014)
1.3-Propanediol			
Pathway optimization	*Lb. diolivorans*	95% yield	Lindlbauer et al. (2017a)

(continued)

Table 8.1 (continued)

Metabolic engineering applications	The parent or engineered strain	Effect on the biosynthesis	References
Exopolysaccharides			
Overexpression of fbp gene	*Lc. lactis*	High level production	Looijesteijn et al. (1999)
Overexpression of galU using nisin inducible promoter	*Lc. lactis*	No significant increase	Boels et al. (2001)
Overexpression of pgm and galU genes	*S. thermophillus*	2-fold increase in EPS synthesis	Levander et al. (2002)
Evolutionary branching sucrase-like glucansucrases (GSs) in glycoside hydrolase family 70 (GH70) catalyze the synthesis of -glucans from sucrose enzyme	*Leuconostoc mesenteroides*	Promising for increased productivity	Yan et al. (2018)
Native type I-E CRISPR-Cas system	*Lb. crispatus*	Enhanced production	Hidalgo et al. (2019)
Isolation of new exopolysaccharide (EPS) produced by the *Lactococcus lactis* F-mou strain (LT898177.1) from Sahrawi camel milk	*Lc. lactis* F-mou	47-fold increase in EPS yield	Nehal et al. (2019)
Synthesizing glucan using sucrose by *Lactiplantibacillus plantarum* BR2 a newly isolated culture from Jack fruit	*Lb. plantarum* BR2	An yield of 2.8 ± 0.5 g/L	Sasikumar et al. (2017)

pure L-lactic acid by taking advantage of metabolic engineering pathways. In *Lb. helveticus*, L-lactic acid production was reached through the inactivation of the enzyme encoding D-lactate dehydrogenase, while this increase was attained in *Lc. lactis* by interfering with the genes encoding phosphofructokinase and pyruvate kinase enzymes (Davidson et al. 1995). L-lactic acid was produced from LAB by performing IdhD inactivation for the first time in *Lb. helveticus* (Bhowmik and Steele 1994). Inactivating the ldhD gene led to the suppression of *Lb. plantarum* (Ferain et al. 1996) while increasing the fabrication of L-lactic acid in *Lb. helveticus* and *Lb. johnsonii* (Kylä-Nikkilä et al. 2000). One of the additional observed results in *Lb. plantarum* is racemase activity. A racemic mixture of L- and D- forms of lactic acid shaped from lactic fermentation can occur together or separately (Goffin et al. 2005). It has been shown that L-lactate production, which is provided by the increase in phosphofructokinase activity, is increased up to two times (Papagianni and Avramidis 2011). This increase was also achieved by UV mutagenesis in *Lc. lactis* (Bai et al. 2004a). In addition to converting at least 90% of the sugars, it reacts to lactic acid, *Lc. lactis* is referred to as homofermentative. However, heterofermentative properties can also be seen in the aerobic environment or during the reactions of different carbohydrates such as galactose (Lohmeier-Vogel et al. 1986). Metabolic engineering approaches intending to multiply the lactic acid production levels of *Lc. lactis* by regulation in glycolysis and fermentation patterns have been developed (Chemical Economics Handbook 2009; Even et al. 2001). Phosphofructokinase has had a significant part to play in lactic acid metabolism in *Lc. lactis* and lactate fluxes decrease proportionally by phosphofructokinase activity (Papagianni et al. 2007). Furthermore, during the lactate formation of *Lc. lactis*, disruptions in the gene encoding the enzyme lactate dehydrogenase (Idh), the last task enzyme, cause pyruvate to be metabolized into mixed acids (Papagianni and Avramidis 2011). On the other hand, when mutant strains with altered lactate dehydrogenase activity were compared, it was revealed that there was no significant change in lactate production (Andersen et al. 2001). Research has been carried out on *Enterococcus faecalis* W11's ability to convert glycerol into lactate by using metabolic engineering strategies and biodiesel wastes as a carbon source. However, for W11 strain, components in biodiesel waste cause cell damage in bacteria and lactate dehydrogenase activity causes 4.1 times lower L-lactate production under anaerobic conditions compared to aerobic conditions (Doi 2015). In addition, L-lactic acid production of *Lacticaseibacillus paracasei* 7BL using lignocellulosic biomass with fed-batch strategy was achieved despite the risk that the lignocellulosic raw material could have a toxic effect on microorganisms (Kuo et al. 2015). Products expressed as lignocellulosic, such as agricultural wastes and food processing by-products, are promising due to their ability to be converted into different carbohydrates. Lignocellulosic materials are important resources due to their renewable nature and low cost. It is thought that lignocellulose will be one of the most frequently used substrates in the near future, with a better understanding of its importance (Cubas-Cano et al. 2018). Methods frequently evaluated when using lignocellulosic substrates include hydrolysis and fermentation (SHF) and simultaneous saccharification and fermentation (SSF). Unlike the first method, in the second method, there is a fermentation process carried out in the same vessel to reduce the cost. The advantages of the second method can be listed as less end-product inhibition, higher hydrolysis efficiency, shorter reaction times and lower enzyme loadings. Considering that cost is the most important factor in lactic acid production, lignocellulosic substrates come to the fore in the search for a more economical raw material. However, while the need for strains with high acid tolerance continues to be one of the factors that increase the cost, such as the ability to produce pure lactic acid, the option of genetic modification comes to the fore at the solution point (Cubas-Cano et al. 2018).

8.3.2 Acetaldehyde

Acetaldehyde is common within the food industry on account of its pleasing aroma and flavor. One of the most used foods is yogurt, which is a dairy product. The result was successful when the glyA gene was overexpressed in *Streptococcus thermophilus* to multiply the production rate of acetaldehyde. The glyA gene encodes the serine hydroxymethyltransferase (SHMT) enzyme with threonine aldolase (TA) activity, enabling it to convert threonine to acetaldehyde and glycine (Sharma et al. 2020). For *S. thermophilus*, acetaldehyde formation occurs only by SHMT activity. In cases where there is no alternative acetaldehyde synthesis pathway like this, the solution of increasing astealdehyde production by the overexpression of glyA was found to be successful (Chaves et al. 2002). Since it has been shown that acetaldehyde production increased by 82.0% as a result of replacing gene by a construct under the *Lc. lactis* PLAcA promoter, it is thought that the strategy of increasing acetaldehyde production can be carried out successfully. Another strategy to increase acetaldehyde production in *Lc. lactis* is to alter central carbon metabolism by replacing native pyruvate decarboxylase (pdc) with a PDC overexpressed from Z mobilis and native NADH oxidase (nox) (Bongers et al. 2005).

8.3.3 Acetoin

Production of acetoin was accomplished using dairy industry waste, through additional ATP consumption when *Lc lactis* was genetically modified by plasmid expression (Liu et al. 2016a). Another metabolic engineering strategy for acetoin production is called biocompatible chemistry, which is understood as the application of chemical reactions of a non-enzymatic nature with living organisms to produce metabolites such as acetoin. In *Lc. lactis* strain, changes in respiratory capacity of cells were chemically altered via pathways using ferric iron as a biocompatible chemistry strategy, resulting in high (3S)-acetoin production (Liu et al. 2016b).

8.3.4 Diacetyl

Diacetyl provides the butter flavor. Two different pyruvate molecules and the enzyme α-acetolactate synthase can together synthesize α-acetolactate. Diacetyl is obtained as a result of decarboxylation of α-acetolactate. Therefore, strategies are currently being tailored to enhance diacetyl production by increasing α-acetolactate. Diacetyl production may be increased by inactivating the gene encoding acetolactate decarboxylase enzyme of *Lc. lactis*. In *Lc. lactis*, this increase was achieved by the overexpression of α-acetolactate synthase genes. Pyruvate restricts production of diacetyl since α-acetolactate synthase genes show a minimal affinity towards it.

Therefore, by inactivating genes encoding acetolactate decarboxylase enzyme is performed. Because α-acetolactate synthase genes show higher affinity for pyruvate, diacetyl precursor α-acetolactate's conversion enzymatically to acetoin is inhibited after the genes are overexpressed (Benson et al. 2009). The so-called rational combination technology, which includes kinetic modeling of pyruvate metabolism, metabolic control analysis, and analysis to identify mutations, was intended to advance the steps of rate-limitation(Hoefnagel et al. 2002). On the other hand, a biovariant of *Lc. lactis* metabolizes citrate to produce diacetyl. It is emphasized that while this biovariant increases diacetyl production thanks to citrate lyase and oxaloacetate decarboxylase enzymes, it also enables the conversion of citrate to pyruvate without the need for a redox cofactor (Hugenholtz et al. 2000). However, due to the small amount of citrate in milk, metabolic engineering mostly focuses on the production of diacetyl from glucose and lactose. To achieve this, a strain that increases diacetyl production is designed from glucose with the NADH oxidase gene's (nox) overexpression in a α-acetolactate decarboxylase mutant (Hugenholtz et al. 2000).

8.3.5 Diacetyl and Butanediol

By affecting the redox in LAB, glycolysis and pyruvate pathways can be rerouting. In *Lc. lactis*, on the one hand, the nox gene was overexpressed, and on the other hand, the α-acetolactate decarboxylase enzyme was inactivated, resulting in an increase in diacetyl production (Hugenholtz et al. 2000). Strategies that enable the homolactic *Lc. lactis* strain to become a diacetyl-producing strain are listed as metabolic manipulation, respiration activation, and metal ion catalysis. Homo (S, S)-2,3-butanediol yield was obtained by advancing the reactions after binding the specific metabolic pathways using chemical catalysis (Sharma et al. 2020). *Lacticaseibacillus casei* is also utilized to obtain acetoin and diacetyl under fed-batch growth with whey permeate conditions (Nadal et al. 2009). Although the yield gained by this process is low, there is an approach that whey permeate can be used for this process in order to reduce the waste generated in cheese making (de Vos et al. 1998).

8.3.6 Ethanol

Modification of *Lc. lactis* through metabolic engineering has shown itself to be a major innovation in ethanol production from lactose as the lone product of fermentation. As a substrate for this process, it is aimed to use lactose especially in residual whey permeates. *Lc. lactis* CS4435 strain was used for this procedure (Liu et al. 2016c). It was accomplished through the presentation of the pyruvate decarboxylase and the alcohol dehydrogenase from *Zymomonas mobilis* with five inactivated genes to further improve ethanol production of *Lc. lactis* (Solem et al. 2013). Enhancement of ethanol production of *Lacticaseibacillus plantarum* was achieved by combining heterologous pyruvate decarboxylase with double crossover knockout of the LDH gene using *Sarcina ventriculi* (Liu et al. 2006).

8.3.7 Exopolysaccharides

Exopolysaccharides (EPSs) can be used especially in dairy products that have been fermented because of their effect upon taste and texture (Kodali and Sen 2008). EPS, whose production is low by LAB, is induced by the strategies of metabolic engineering. The production of EPs is biosynthetically completed in the order of 1-transport of sugar to the cytoplasm, 2-the synthesis of sugar-1-phosphates, 3-activation of sugars, 4- synthesis of EPS. In EPS biosynthesis, which includes a number of enzymes and genes and also where sugar nucleotide diphosphates act as EPS precursors, bacteria shape EPS diversity by being significantly affected by genetic structure and environment (Sharma et al. 2020). Glucose-1-phosphate (G1P) is the source of synthesis for sugar nucleotides. UDP-galactose and UDP-glucose are the main sugar nucleotides. G1P is formed from G6P by the phosphoglucomutase enzyme activity. UDP-glucose is synthesized from G1P with the UDP-glucose phosphorylase activity. The metabolism pathways, in which these two enzymes are involved, are the key points used in EPS production. For example, the overexpression of UDP-glucose phosphorylase in *Lc. lactis* caused an exponential increase in enzyme activity (Ruffing and Chen 2006). The overexpression of the gene that has the capacity to encode fructose bisphosphatase enzyme, the enzyme responsible for transformingfructose-1,6-diphosphate into fructose-6-phosphate, is one of the metabolic engineering procedures that leads to the synthesis of nucleotides from fructose and increased EPS. Another tactic of metabolic engineering used to heighten EPS production is to increase EPS yield by providing simultaneous overexpression of genes producing phosphoglucomutase and UDP-glucose phosphorylase enzymes of *S. thermophillus* (Levander et al. 2002).

8.3.8 1,3-propanediol, 3-hydroxypropionaldehyde and 3-hydroxypropionic Acid

1,3-PDO, obtained by the conversion of glycerol by many LAB including *Lactobacillus* species, namely, *Levilactobacillus brevis*, *Lentilactobacillus buchneri*, *Limosilactobacillus reuteri* and *Lb. diolivorans*, *Streptococcus sanguinis* and *Enterococcus malodoratus*, and whose production can be increased by the pathway optimization procedure of metabolic engineering, is a metabolite often used in cosmetics and adhesives (Lindlbauer et al. 2017a).

Glycerol dehydration occurs with the activation of glycerol dehydratase enzyme. It is stated that 3-HPA, which is an important platform chemistry, is formed as a result of this reaction. Metabolic engineering pathways are used to increase 3-HPA production. 3-HPA production is promoted in LAB strains by the overexpression of the gene that encodes the glycerol dehydratase enzyme (Lindlbauer et al. 2017b).

Enzymes that provide 3-HP synthesis are listed as phosphotransacylase (PduL), coenzyme-A acylating propionaldehyde dehydrogenase (PduP) and propionate kinase (PduW), and vitamin B_{12} is required as coenzyme. The advantage that *Limosilactobacillus reuteri* naturally produces B_{12} has highlighted this bacterium for use in 3-HP synthesis (Sabet-Azad et al. 2013). The ability of *Limosilactobacillus reuteri*, which has been shown to be able to metabolize glycerol, to use glycerol to produce HPA has been enhanced by metabolic engineering (Amin et al. 2013). In addition to the ability of *Limosilactobacillus* reuteri to produce vitamin B_{12}, genomic analyzes have shown that it has a significant level of pdu genes that enable the synthesis of the glycerol dehydratase enzyme (Morita et al. 2008). *Limosilactobacillus reuteri* synthesizes various chemicals by metabolizing glycerol due to the pdu genes it contains in its genetic material. These chemicals are 1-propanol, propionaldehyde, propionic acid and their hydroxyl derivatives: 3-HPA, 1,3 PD, 3-hydroxypropionic acid. Through metabolic engineering, the mutation of the catabolite repression element (CRE) in the *Limosilactobacillus reuteri* strain made it possible to induce of 3-HP and 1,3-PDO fabrication out of glycerol. When the dehydration pathway of glycerol is .examined metabolically, 3-HPA forms at a velocity 10 times faster than the development of 1,3-PDO and 3-HP from 3-HPA, which are the following steps. Therefore, by manipulating the metabolic flux rate of glycerol, 3-HPA accumulation is prevented and production of 1,3 PDO as well as 3-HP is ensured (Dishisha et al. 2014).

8.3.9 Low-Calorie Polyols

Reports have indicated that *S. avium* and *Lb. casei* strains can, in fact, metabolize xylitol, although they are unable to produce it through natural means (Sharma et al. 2020; London 1990). Production of xylitol from xylose was achieved in a recombinant *Lactococcus lactis* strain through the co-expression of the xylose reductase gene (XYL1) out of *Pichia stipitis* along with a xylose transporter from *Lb. brevis*. Glucose was used as an energy source. It was stated that with this strategy followed, the production of xylitol in the recombinant strain could not be improved as desired (Nyyssölä et al. 2005; Akinterinwa et al. 2008).

An important six-carbon sugar alcohol with numerous applications in various industrial processes, mannitol is used as an artificial sweetener, flavor regulator, anti-caking and texturizing agent. It can also be found in the pharmaceutical industry and in various clinical applications. In addition to its widespread use in masking the bitter flavor of certain antibiotics, mannitol may be utilized as an anti-carcinogen, an antioxidant, and even an osmotic diuretic agent (Hugenholtz and Smid 2002; Sharma et al. 2020; Akinterinwa et al. 2008; Ortiz et al. 2017). Under normal conditions, mannitol biosynthesis is noted to be essentially absent in *Lc. lactis*. It has been reported that fabrication of mannitol and mannitol-1-phosphate increase for a short time after the mannitol transport system inactivates in *Lactococcus lactis* strain with a defect in the lactate dehydrogenase (ldh) gene. Mannitol 1-phosphate accumulated in the cell up to a concentration of 76 mµ. Mannitol is formed out of fructose 6-phosphate through the collective working of mannitol-1-phosphate dehydrogenase and phosphatase. It is rapidly metabolized after glucose is depleted. In the reaction that reduces pyruvate to lactate, NAD+ ceases its regeneration, so the mannitol-1-phosphate dehydrogenase reaction is used (Sharma et al. 2020; Neves et al. 2000).

In one study, a mannitol-generating *lactis* strain was attained through interfering with two genes within the phosphoenolpyruvate (PEP)-mannitol phosphotransferase system (PTSMtl). The mtlA and mtlF genes in *Lc. lactis* FI9630 (a food-grade lactate dehydrogenase-deficient strain derived from MG1363) were independently deleted by double cross-recombination, resulting in two mutant strains (ldhmtlA and ldhmtlF). The new strains, FI10091 and FI10089, are considered practical for use within the food industry. Observations concluded that approximately 33% of the glucose was converted to hexitol and then to mannitol as the final product. Since mannitol transport in *L. lactis* is obstructed as a result of gene disruption, it was determined that the double mutants could not use mannitol even after complete glucose depletion (Gaspar et al. 2004). To better produce mannitol in *Lc. lactis*, it would be beneficial to conduct pathway engineering studies that will reduce the deviation of the metabolic flow to 2,3-butanediol, ethanol, and lactate biosynthesis (Sharma et al. 2020).

Synchronized cloning and stepwise inactivation of ldhD and ldhL through the system that replaces genes in the *Lb. fermentum* strain resulted in the fabrication of mannitol in tandem with a pure form of lactic acid or pyruvate. Mannitol and lactic acid yields from the solitary mutant strain GRL1030 were not affected through deleting the ldhD gene. Mutations found within the double mutant strain GRL1032 upset the glucose consumption rate, and the outcome was a reduction in cellular growth. In the double mutant strain, 2,3-butanediol and a bit of lactic acid was produced over and above the mannitol and pyruvate, as was expected (Aarnikunnas et al. 2003). In *Leuconostoc pseudomesenteroides*, it was determined that mannitol produced out of fructose increased from 74.0% to 85.0% (mol/mol) as a result of a synthesized approach which was comprised of both the inactivation of fructokinase activity and random mutagenesis. This was interpreted as a decrease in fructokinase activity, which may be related to a significant decrease in the outflow of fructose into the phosphoketolase pathway (Helanto et al. 2005).

As for *Lb. casei*, integrating the sorbitol-6-phosphate-dehydrogenase gene (gutF) into the chromosomal lactose (lac) operon led to the production of 0.024 mol sorbitol/mol glucose, a value which had been negligible in the parent strain. Since the cells are pre-grown in the medium containing lactose; sorbitol ends up being created out of the glucose. Sorbitol production improved to 0.043 moles per mole of glucose after further deletion of the LdhL gene. Some believe that this engineering technique could offer an alternate pathway for the regeneration of NAD+ (Nissen et al. 2005). With another metabolic engineering strategy developed, sorbitol

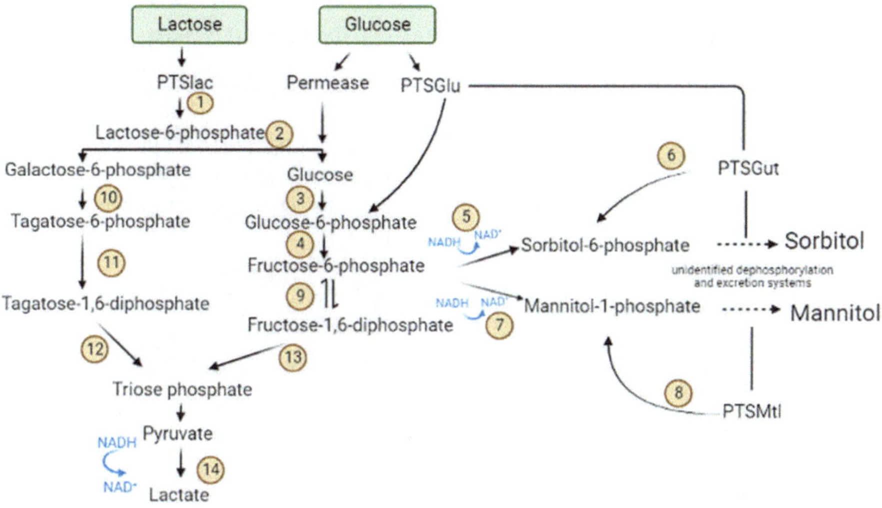

Fig. 8.2 The proposed pathways in the production of some polyols by metabolically engineered LAB. (Adapted from reference (Monedero et al. 2010). (1) phosphoenolpyruvate: lactose phosphotransferase system, (2) β-phosphogalactosidase, (3) glucose kinase, (4) phosphoglucose isomerase, (5) sorbitol-6 phosphate dehydrogenase, (6) phosphoenolpyruvate: sorbitol phosphotransferase system, (7) mannitol-1 phosphate dehydrogenase, (8) phosphoenolpyruvate: mannitol phosphotransferase system, (9) phosphofructokinase, (10) galactose-6 phosphate isomerase, (11) tagatose-6 phosphate kinase, (12) tagatose-1,6 diphosphate aldolase, (13) fructose-1,6bis phosphate aldolase, (14) lactate dehydrogenase

production was achieved in both *Lb. casei* and *Lb. plantarum* strains. mtlD (mannitol phosphate dehydrogenase) and ldh genes' inactivation along with the overexpression of stlDH (sorbitol dehydrogenase gene, SDH) in strain *Lb. plantarum*, increased bioconversion of fructose-6-phosphate to sorbitol-6-phosphate by SDH and 0.65 mol sorbitol/mol resulted in a theoretical yield of glucose. With the further improvement of sorbitol dephosphorylation and transport, more sorbitol production was realized (Ladero et al. 2007).

The proposed pathways in the production of some polyols by metabolically engineered LAB are shown in Fig. (8.2).

8.3.10 Vitamins

Even though LAB are classified as auxotrophic for many vitamins, a few strains can synthesize water-soluble B group vitamins, which may include folic acid, riboflavin, vitamin C, pyridoxal, and cobalamin (Wang et al. 2021). It has been reported that *Lb. plantarum* CRL 1973 and CRL 1970, *Lb. rhamnosus* CRL 1972 and *L. sakei* CRL 1978 strains from quinoa had high levels of folate; the *L. rhamnosus* CRL 1963 strain isolated from sourdough obtained from quinoa produces a significant

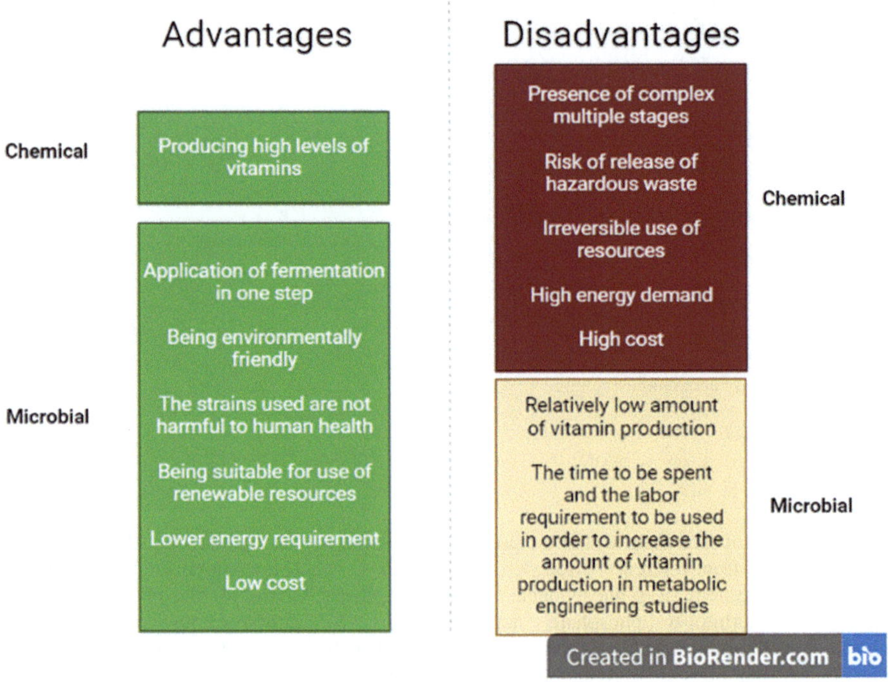

Fig. 8.3 Advantages and disadvantages of chemical and microbial methods in vitamin production. (Adapted from reference (Averianova et al. 2020)

amount of riboflavin (Carrizo et al. 2016). Obtaining new microbial strains which have been developed from metabolic engineering techniques to raise the rate of return of the produced vitamins is of great importance since the production of vitamins by the microbial method is limited compared to the chemical method (Averianova et al. 2020; LeBlanc et al. 2011) (Fig. 8.3). Metabolic engineering studies performed on some LAB strains to increase the production amount of the aforementioned vitamins are examined below.

The majority of the strains able to produce folic acid are in the genera *Streptococcus*, *Lactobacillus* and *Lactococcus* (Khalili et al. 2020). Folic acid synthesis pathways of LAB contain both Pterin and pABA branches. It has been reported that these two branches must function simultaneously in order for folic acid production to take place (Wegkamp et al. 2007). To accomplish this, adding p-Aminobenzoic acid (pABA) to the medium in which *Lactobacillus* strains are developed becomes necessary. LAB can use different substrates such as milk and its products, cereals in the production of vitamins. *Streptococcus thermophilus* ST-M6 and TH-4 strains have produced folate using a soy milk-based blended substrate (Albuquerque et al. 2017).

It has been reported that folate production increased three-fold in *Lc. lactis* NZ9000 strain, as the flux to the pathway and enzyme increased without feedback inhibition, with the overexpression of the gchI gene encoding GTP cyclohydrolase I, the primary enzyme in folate biosynthesis. The ratio of monoglutamyl-folate to polyglutamyl-folate was changed, and that led to more folate being released into the environment (Hugenholtz and Smid 2002). *Lc. lactis* subsp. *cremoris* MG1363 caused the extracellular level to grow to a rate nearly ten times the initial one in addition to the folate production level tripling as a result of overexpression of folKE, which encodes dimeric 2-amino-4-hydroxy-6-hydroxymethyldihydropteridine pyrophosphokinase and gchI genes. The majority of the folate generated through *Lc. lactis* accumulates intracellularly in the form of polyglutamyl, and such polyglutamylation may be deemed liable for retaining folate within the cell. The enhancement of the enzymatic capacity of folate synthetase/polyglutamyl folate synthetase was inadequate by the overexpression of folKE to convert excess folate to the polyglutamyl form, resulting in reduced folate uptake (Sybesma et al. 2003). According to the research, higher levels of intracellular folate may, in fact, be reached through overexpression of the gene cluster of folate biosynthesis simultaneously, which includes the folB, folKE, folP, folQ, folC genes, and pab genes for para-aminobenzoic acid (pABA) biosynthesis. Elevated pools of pABA have been shown to deter the enzyme folilpolyglutamate synthetase from being active, which is noteworthy considering its accountability for increased polyglutamate tails and a greater intracellular uptake (Wegkamp et al. 2007).

It is known that many microorganisms, including yeasts, can produce riboflavin. LAB are shown as interesting microorganisms that have the potential to increase their biosynthetic capacity with modern biotechnology methods in riboflavin production (Averianova et al. 2020; Thakur et al. 2016). Genes encoding riboflavin synthase in LAB are bunched upon a rib operon. RibC, RibB, RibA and RibH product enzymes assist GTP and 5-phosphate ribose in their conversion to riboflavin. Research involving the high production of riboflavin through the use of genetically engineered strains revealed that riboflavin synthesis of LAB may be affected as a result of some changes to be made in the DNA regions related to regulation. When a sliver of 1059-bp DNA was inserted into the upstream regulatory region of the rib operon of *Lb. plantarum*, the strain that had been engineered generated more riboflavin than the wild strain (Ge et al. 2020). In *Lc. lactis* JC017 strain, purine deficiency was found to provoke the overexpression of the riboflavin biosynthesis cluster ribABGH. Therefore, it has been stated that metabolic engineering studies to prevent purine biosynthesis will increase riboflavin production (Chen et al. 2017).

LAB is commonly used in dairy farming and engineering. By adding strains that produce riboflavin to fermented products like cheese, yoghurt and milk, riboflavin levels can be increased without any additional economic burden. In a study involving the biosynthesis of riboflavin in dairy products, it was determined that the riboflavin content increased significantly as a result of the fermentation process taking place in cow's milk alongside *Lc. lactis* and *Propionibacterium freudenreichii* ssp. *shermanii* as starter cultures (LeBlanc et al. 2011). In another study, riboflavin production was 2.36 mg/L in *Lb. fermentum* KTLF1 and 2.13 mg/L in *Lb. plantarum* (Thakur et al. 2016). Since a high level of riboflavin synthesis can be realized, *Lc. lactis* subsp. *cremoris* NZ9000 strain can serve as an effective model (Averianova et al. 2020; Burgess et al. 2004; Burgess et al. 2006). It was determined that *Lb. fermentum* MTCC 8711 strain produced 2.29 mg/L riboflavin at the end of a day in a special medium (Jayaraman and Kalaichelvan 2010). It has been reported that *Lb. acidophilus* produces more riboflavin than *Lc. lactis* in a medium containing whey protein (Guru and Viswanathan 2013).

In another study, riboflavin and folate production were simultaneously provided in *Lactococcus lactis* through the implementation of direct mutagenesis along with metabolic engineering (Sybesma et al. 2004). In a recent study, a mutualistic micro-organism community was examined together to increase riboflavin and folate production. A natural isolate of *Lb. plantarum* was cultured alongside an engineered *S. cerevisiae* strain. Vitamin-secreting bacteria were chosen based on whether or not they possessed the necessary biosynthetic genes in their genomes, while the auxotrophic *S. cerevisiae* strain was designed through the deletion of the RIB4 and RIB5 genes. Genomic, proteomic, and metabolomic analyses and flow balance analyses of the developed LAB showed that the metabolic regulation changed with the increasing amount of vitamin precursors. As a result of this study, an increase in the production of targeted vitamins was determined. It has also been shown how microbial metabolism acclimatizes to reciprocal lifestyle in the course of enhanced metabolite exchange (Konstantinidis et al. 2021).

Lb. reuteri is a gram-positive, heterofermentative LAB that has seen extensive and problem-free use within the food industry (Kandler and Weiss 1986). In a study that applied both genetic and physiological approaches in the process of metabolic engineering, it was noted that *Lb. reuteri* JCM1112 produced both B_{12} and folate. At the same time, it was determined that the folate/B_{12} mass ratio was less than the desired level for consumption by humans. In order to elevate this ratio, a natural fermentation process has been developed that significantly increases folate production (Santos et al. 2008).

Lactococcus lactis subsp. *cremoris MG1363* is able to produce vitamin K_2; *Latilactobacillus sakei* UONUMA strain can produce vitamins B_2, B_3 and B_6; the strains *Limosilactobacillus reuteri* CRL 1098 and *Lb. coryniformis* CRL 1001 are able to produce cobalamin (Liu et al. 2019; Oguro et al. 2017; Torres et al. 2018).

8.3.11 Succinic Acid

It has been reported that *Lb. plantarum* strain NCIMB 8826 produces a small measure of succinic acid naturally due to a deficiency in the tricarboxylic acid cycle. The strain of *Lb. plantarum* NCIMB 8826 (VL103) deficient in dehydrogenase was metabolically engineered to more efficiently produce succinic acid. The effect of overexpression or coexpression of prospective enzymes such as pyruvate carboxylase (PC), phosphoenolpyruvate carboxykinase (PEPCK) and malic enzyme on the creation of succinic acid was investigated. While supporting the knowledge that succinic acid biosynthesis mainly depends on the activity of the PC enzyme, it was determined that the PEPCK enzyme caused a large increase in biomass of *Lb.*

plantarum NCIMB 8826 (VL103). Coexpression of PC and PEPCK was determined to increase succinic acid production in *Lb. plantarum* VL103. When contrasted with the wild strain, the recombinant one was shown to generate 22 times more succinic acid (25.3% of glucose was converted to succinic acid) (Tsuji et al. 2013).

8.3.12 Alanine

Heterologous production of alanine in *Lc. lactis* NZ9000 was accomplished through the expression of the alanine dehydrogenase (ALDH) gene (alaD) from *Bacillus subtilis*. The overexpression of the heterologous alaD gene in *Lc. lactis* NZ9000 strain carrying pNZ273/ldhp/ald has reportedly increased the quantity of alanine to 52 mg/mL (26 times higher than parent strain) (Ye et al. 2010).

8.3.13 Pediocin PA-1

Pediocin PA-1, a bacteriocin, was produced in *Lc. lactis* MG1363 by the expression of pedA structural and pedB immune genes from native Pediococcus acidilactici K7 strain in one study (Srikanth and Halami 2008). And in another, it was reported that bacteriocin activity was increased in *Lactobacillus reuteri* KCTC 3679 by combining the α-amylase gene and pediocin PA-1 structural and immune genes (pedAB) from a bifidobacterial strain (Eom et al. 2010).

8.3.14 Vanillin

Metabolic rerouting can be achieved by providing metabolite-specific enzyme inducers in microbial cells. This strategy was used in a study to improve vanillin production in a LABl isolate, *Pediococcus acidilactici* BD16 strain. Vanillin manufacturing of *P. acidilactici* BD16 strain was found to be stimulated in a concoction which contained 150 g/L rice bran and 50 μg/L ferulic acid. By using rice bran as ferulate-containing substrate and ferulic acid as vanillin operon inducer, 1.26 g/L crude vanillin was produced in cells of bacteria (Kaur and Chakraborty 2013). It was detailed in a separate study how vanillin biosynthesis saw a slight increase when upon cloning a vanillin gene cassette which relayed the fcs and ech genes encoding feruloyl-CoA synthetase and enoyl-CoA hydratase in *P. acidilactici* BD16 strain. Following the metabolic engineering application, 4 g/L vanillin was retrieved from the recombinant cell culture. It has been noted that this amount is almost four times higher than the amount of vanillin obtained from the natural species (Kaur et al. 2014). In a study to improve aroma and flavor in wine, the production of various phenolic metabolites including vanillin was investigated in recombinant *P. acidilactici* BD16 (fcs+/ech+) strain (Kaur et al. 2015).

8.4 Metabolic Engineering of Yeasts in Industrial Applications for Use in the Industrial Applications

Yeasts are eukaryotic microorganisms that represent the Ascomycetes and Basidiomycetes phylum with biotechnological potential in the food industry, they contain a heterogeneous microbes group and belong to a group of organisms called fungi. Yeasts also include yeasts, molds and mushrooms (Rawat 2015; Rai and Kumaraswamy 2017). Yeasts have many potential effects in the food industry, medicine, agriculture, and bioengineering disciplines and have been utilized by civilizations for millennia (139, (Perricone et al. 2017b)). In food engineering, it is mainly used as a starter culture in the fermentation of products such as dairy products (cheese), grain-based products (sourdough), and alcoholic beverages (wine and beer). They have many significant roles in food fermentation, including the production of alcohol and organic acids, improving flavour, aroma and texture, enhancing the nutritional value, and reducing anti-nutritional factors and toxins. However, besides these positive effects, yeasts may also have some negative effects. The presence or growth of yeasts in some foods such as yoghurt, fruit juice, salad and mayonnaise may lead to health problems, spoilage in foods, and food losses (Rai and Kumaraswamy 2017; Johnson and Echavarri-Erasun 2011b; Lowes et al. 2000).

Metabolic engineering is defined as the intentional and focused modification of metabolic pathways that exist within an organism for greater understanding in order to exploit biological processes for chemical transformation, energy transfer, and supramolecular assembly (Rai et al. 2019). Metabolic engineering is the procedure of enhancing genetic and regulatory processes in cells to increase the cell's production of a particular substance. These processes are made up of chemical networks that execute a sequence of biochemical reactions and enzymes that allow cells to convert raw materials into molecules necessary for their survival (Rai et al. 2019). The purpose of metabolic engineering is to use molecular biology tools to analyze metabolic and other pathways to enhance cellular features for product improvement by developing and applying reasonable genetic alterations (Yang et al. 1998).

The framework of biotechnological applications is expanding day by day and is becoming more important in the food industry. Plant biotechnology (gene transfer in plants, plant genomic studies), aroma biotechnology (production of aroma substances by microbial or enzymatic means), and biotechnological applications of lipids (synthesis of structured lipids) are among these applications. These applications can be carried out by microorganisms (for example, the production of aroma substances by fermentation) as well as they can be used to improve the properties of microorganisms used as starter cultures (Kumar and Prasad 2011). Yeasts are one of the most frequently used microorganisms in biotechnological applications for many

Fig. 8.4 Engineering of the yeast genome

years (Fig. 8.4). The main fields where the yeasts are used are as follows: traditional yeast fermentation (beer, wine, bread and other fermentations), food and feed ingredients (amino acids, enzymes, flavors, pigments and organic acids), biocatalysis (pharmaceuticals, chiral chemical intermediates, biotransformations), heterologous protein production (protein pharmaceuticals, enzymes, hormones, vaccines, toxins), central biological research (genomics, functional genomics, molecular and cellular biology, pathway engineering, systems biology mechanisms), biomedical research (drug innovation, resistance and metabolism, elucidation of disease mechanisms), biocontrol (crop protection, food and feed safety, probiotics), and environmental biotechnology (bioremediation, pollutant degradation) (Johnson and Echavarri-Erasun 2011a). Metabolic engineering, which is used in many fields, has recently become popular in industrial applications. As far as the food service industry, the role of yeasts in industrial applications will be covered in the sections to come.

Metabolic engineering has been shown to be a useful approach for understanding enzyme processes and producing high-value bioactive compounds. Some researchers have applied metabolic engineering to develop strains of yeast in the production of bioactive compounds such as carotenoids, polyphenols, oligosaccharides, and folate (Fig. 8.5). Yamano et al. (1994) applied metabolic engineering to β-carotene and lycopene production in *Saccharomyces cerevisiae* about three decades ago (Aran and Biyoteknolojisi 2013). *S. cerevisiae*, which has an important place in industrial applications, cannot produce any carotenoid. However, it can synthesize diphosphate ergosterol from farnesyl via the sterol biosynthetic pathway. Based on this, the *crtE, crtB*, and *crtI* genes (for the biosynthesis of lycopene) and the *crtE, crtB, crtI*, and *crtY* genes (for the biosynthesis of β-carotene) of *Erwinia uredovora*, a gram-negative bacteria, were used and the accumulation of β-carotene and lycopene has been reported in the cells of *S. cerevisiae* (Aran and Biyoteknolojisi 2013). In another study, it was tried to increase the biosynthesis of resveratrol in *S. cerevisiae* yeast by using different Metabolic engineering methods. Without carrying resveratrol directly, a low-affinity, high-capacity bacterial araE transporter might increase resveratrol accumulation. The *araE* gene caused yeast cells to generate up to 2.44 times more resveratrol than control cells (Yamano et al. 1994). Researchers who developed a strain of *S. cerevisiae* that overproduce *p*-coumaric acid, which is a precursor for the biosynthesis of many secondary metabolites (polyphenols, flavonoids, and some polyketides), improved *p*-coumaric acid production 7.9-fold with a combination of several different ME methods. The developed *S. cerevisiae* strain was considered a reasonable option to produce *p*-coumaric acid-derived secondary metabolites such as polyphenols, flavonoids, and polyketides (Wang et al. 2011). Six flavonoids (quercetin, liquiritigenin, naringenin, resokaempferol, kaempferol and fisetin) were produced using metabolic engineering of *S. cerevisiae* utilizing

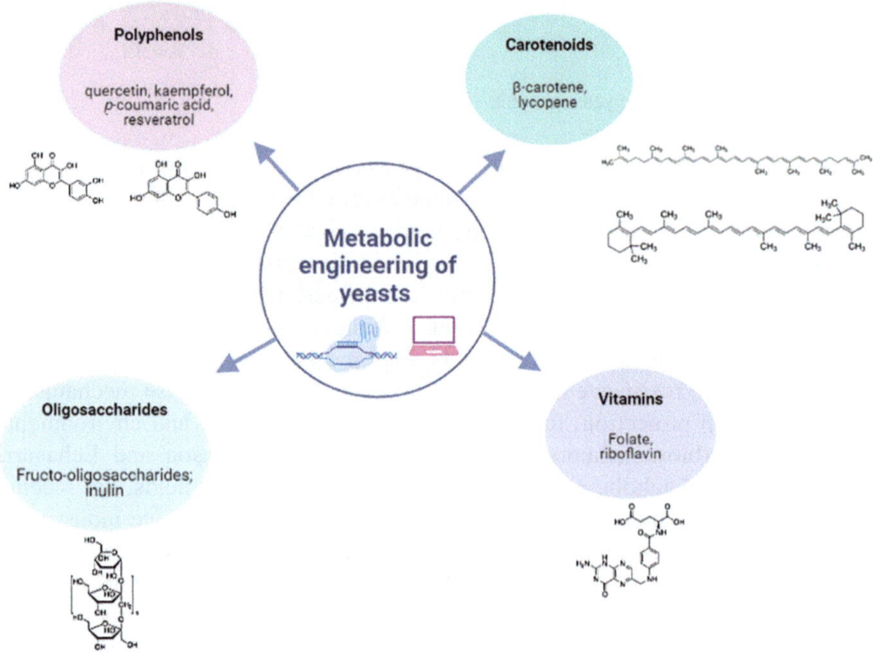

Fig. 8.5 Metabolic engineering of yeasts in the food industry

glucose as the sole carbon source. Moreover, quercetin and kaempferol were shown to have the highest extracellular quantities (Rodriguez et al. 2015). On the other hand, it has been reported that *S. cerevisiae* can be used for the production of top-value chemicals (organic acids, sugar alcohols, furans, bio hydrocarbons, glycerol and derivatives) from biorefinery carbohydrates via ME. *S. cerevisiae* is a good example of the development of sustainable bioprocesses due to its durability, simplicity of handling, genetic toolbox, and fitness in an industrial setting. The metabolic pathways for the production of organic acids are mainly ldh expression; integration of CDT-1, GH1–1, XYL1, XYL2, XYL3 and ldhA; expression of 12 glycolysis-related genes and LDH gene; expression of pyc2, mdh3, fumC, frdS1; fine-tuning of gene expression to enhance endogenous metabolic fluxes toward increasing levels of acetyl-CoA and malonyl-CoA. Moreover, expression of XYL1 (xylose reductase) from *Scheffersomyces stipitis*; CDT1 (cellodextrin transporter); GDH11(ßglucosidase) from the *Neurospora crassa*, BGL (β-glucosidase) from *Aspergillus aculeatus*, XylA (β-xylosidase) from *Aspergillus oryzae* and XYN (endoxylanase II) from *Trichoderma reesei* ME strategies for xylitol production in *S. cerevisiae* (Rodriguez et al. 2017; Guirimand et al. 2016; Lee et al. 2015).

Fructo-oligosaccharides (FOSs) are oligosaccharides found in a variety of plants, including onion, chicory, garlic, asparagus, banana, and artichoke (Baptista et al. 2021). FOSs have the benefits of lessened caloricity and sweetness and are considered safe for consumption since they are made up of short chains of fructose molecules with or without a terminal glucose residue (Sabater-Molina et al. 2009). Han et al. (2017) presented a bioprocess technique consisting of two stages for producing high-content FOSs from inulin via yeast fermentation. The bioprocess was carried out in a recombinant *Yarrowia lipolytica* strain in which the augmented endo-inulinase gene from *Aspergillus niger* was inserted into the chromosomal DNA using two linear plasmids. The showed *Yarrowia lipolytica* strain Enop56, in which endoilunase gene was overexpressed, resulted in a higher yield of fructo-oligosaccharides from inulin (Sabater-Molina et al. 2009). Recently, similar results have been obtained in studies with the same or different yeast species (*Y. lipolytica* and *S. cerevisiae*), regardless of one or two-stage bioprocesses (Han et al. 2017; Wang et al. 2016). It has been stated that using a recombinant *S. cerevisiae* strain, in which a heterologous endo-inulinase gene was expressed and the inherent invertase gene SUC2 was disrupted, increased the yield of FOS by 90% via one-step bioprocess (Han et al. 2017). Mao et al. (2019) showed that the endo-inulinase with a C-terminal fused inulin binding module had more sufficient inulinase activity, thermostability, and inulin binding capacity of the associated endo-inulinase than the endo-inulinase with an N-terminal fused inulin binding module and the native endo-inulinase (Wang et al. 2016).

For plants, β-Ionone is normally generated in small amounts. As a result, a viable alternative may be expressing plant biosynthetic pathway genes in microbial hosts and engineering the metabolic pathway/host to enhance metabolite synthesis (Mao et al. 2019). Using CRISPR/Cas9 technology, López et al. (2020) constructed numerous *S. cerevisiae* strains able to collect a greater number of β-ionone and its precursor, β-carotene. The findings showed that by employing rational metabolic engineering methods, high β-ionone production may be achieved. *Carotenoid cleavage dioxygenase 1* efficiency in high-accumulating β-carotene strains limits β-ionone synthesis in yeasts, according to experimental and modelling findings of that study (Paparella et al. 2021). By integrating lycopene-pathway genes into genomic DNA, some researchers metabolically engineered a synthetic lycopene pathway in *Pichia pastoris* X-33, which is a budding methylotrophic yeast and has been widely used in industrial applications. They reported that lycopene content was found to be 73.9 mg/L in a minimal medium (López et al. 2020). On the other hand, metabolic engineering is a method often used to improve the production of some vitamins. Dmytruk et al. (2011) showed that metabolic engineering of the yeast *Candida famata* (*Candida flareri*) enhanced riboflavin production (4.1-fold) (Bhataya et al. 2009). Also, they reported that the resistance to mycophenolic acid and fluorophenylalanine conferred by the *Debaryomyces hansenii* IMH3 and modified *ARO4* genes, respectively, were effectively employed as novel dominant selection markers for *C. famata* (Bhataya et al. 2009). Through using metabolic engineering of yeast (*S. cerevisiae*), it has been shown that the amount of folate in wine can increase by the overexpression of *FOL2*, but not *ABZ1* (Dmytruk et al. 2011).

Being able to enhance the levels of bioactive compounds is a significant result in terms of both human health and industrial applications. Although the studies show that biotechnological production of bioactive compounds is technically feasible, further metabolic engineering efforts are needed to explain specific pathways and yeast strains in industrial applications.

8.5 Conclusion

Metabolic engineering is characterized by using both systems and synthetic biology processes to use existing metabolic pathways or establish new ones efficiently. Lactic acid bacteria and yeast have already proven to be ideal microorganisms for this technology. In this context, the industry applies metabolic engineering in LAB and yeast to boost the production of various metabolites that are important for health and have technological value. By using different strains or genetic engineering and/ or multi-omics technology, the desired compound is produced at a higher rate, or a compound that cannot be produced at all has a chance to be produced.

Since recombinant gene technology is strongly limited for food applications by especially EU regulations and the negative consumer perspective towards genetically modified microorganisms (GMOs), microorganism modifications are better reserved for wider industrial use. So, it is thought that the use of LAB and yeast in the production of beneficial compounds in the food and healthcare industry will become more widespread. Therefore, it is essential to understand the strain-specific or yeast-specific metabolism to be revealed by further studies to avoid any adverse effects on human health by metabolic engineering.

Acknowledgment Figures 8.1, 8.2, 8.3 and 8.5 were created with BioRender.com.

Conflicts of Interest The authors declare no conflict of interest.

References

Aarnikunnas J, Von Weymarn N, Rönnholm K, Leisola M, Palva A (2003) Metabolic engineering of Lactobacillus fermentum for production of mannitol and pure L-lactic acid or pyruvate. Biotechnol Bioeng 82(6):653–663

Akinterinwa O, Khankal R, Cirino PC (2008) Metabolic engineering for bioproduction of sugar alcohols. Curr Opin Biotechnol 19(5):461–467

Albuquerque MAC, Bedani R, LeBlanc JG, Saad SMI (2017) Passion fruit by-product and fructooligosaccharides stimulate the growth and folate production by starter and probiotic cultures in fermented soymilk. Int J Food Microbiol 261:35–41

Alvarez-Martin P, Florez AB, Hernández-Barranco A, Mayo B (2008) Interaction between dairy yeasts and lactic acid bacteria strains during milk fermentation. Food Control 19(1):62–70

Amin HM, Hashem AM, Ashour MS, Hatti-Kaul R (2013) 1,2 Propanediol utilization by Lactobacillus reuteri DSM 20016, role in bioconversion of glycerol to 1,3 propanediol, 3-hydroxypropionaldehyde and 3-hydroxypropionic acid. J Genet Eng Biotechnol 11(1):53–59

Andersen HW, Pedersen MB, Hammer K, Jensen PR (2001) Lactate dehydrogenase has no control on lactate production but has a strong negative control on formate production in Lactococcus lactis. Eur J Biochem 268(24):6379–6389

Aran N, Biyoteknolojisi G (2013) NA Yayincilik (ed) Ankara: Nobel Akademik Yayıncılık;

Averianova LA, Balabanova LA, Son OM, Podvolotskaya AB, Tekutyeva LA (2020) Production of Vitamin B2 (Riboflavin) by microorganisms: an overview. Front Bioeng Biotechnol 8:8

Bai DM, Zhao XM, Li XG, Xu SM (2004a) Strain improvement and metabolic flux analysis in the wild-type and a mutant Lactobacillus lactis strain for L(+)-lactic acid production. Biotechnol Bioeng 88(6):681–689

Bai D, Yan Z, Wei Q, Zhao X, Li X, Xu S (2004b) Ammonium lactate production by Lactobacillus lactis BME5-18M in pH-controlled fed-batch fermentations. Biochem Eng J 19(1):47–51

Baptista SL, Costa CE, Cunha JT, Soares PO, Domingues L (2021) Metabolic engineering of Saccharomyces cerevisiae for the production of top value chemicals from biorefinery carbohydrates. Biotechnol Adv 47:107697

Benson KH, Godon J-J, Renault P, Griffin HG, Gasson MJ (2009) Effect of ilvBN-encoded#±–acetolactate synthase expression on diacetyl production in Lactococcus lactis. Appl Microbiol Biotechnol 45:107–111

Berlec A, Štrukelj B (2012) Generating a custom TA-cloning expression plasmid for Lactococcus lactis. BioTechniques 52(1):51–53

Bernardo MP, Coelho LF, Sass DC, Contiero J (2016) l-(+)-Lactic acid production by Lactobacillus rhamnosus B103 from dairy industry waste. Braz J Microbiol 47(3):640–646

Bhataya A, Schmidt-Dannert C, Lee PC (2009) Metabolic engineering of Pichia pastoris X-33 for lycopene production. Process Biochem 44(10):1095–1102

Bhowmik T, Steele J (1994) Cloning, characterization and insertional inactivation of the Lactobacillus helveticus D (−) lactate dehydrogenase gene. Appl Microbiol Biotechnol 41(4):432–439

Boels IC, Kranenburg RV, Hugenholtz J, Kleerebezem M, de Vos WM (2001) Sugar catabolism and its impact on the biosynthesis and engineering of exopolysaccharide production in lactic acid bacteria. Int Dairy J 11(9):723–732

Bongers RS, Hoefnagel MHN, Kleerebezem M (2005) High-level acetaldehyde production in Lactococcus lactis by metabolic engineering. Appl Environ Microbiol 71(2):1109–1113

Börner RA, Kandasamy V, Axelsen AM, Nielsen AT, Bosma EF (2018) Genome editing of lactic acid bacteria: opportunities for food, feed, pharma and biotech. FEMS Microbiol Lett 366(1):fny291

Burgess C, O'Connell-Motherway M, Sybesma W, Hugenholtz J, van Sinderen D (2004) Riboflavin production in Lactococcus lactis: potential for in situ production of vitamin-enriched foods. Appl Environ Microbiol 70(10):5769–5777

Burgess CM, Slotboom DJ, Geertsma ER, Duurkens RH, Poolman B, van Sinderen D (2006) The riboflavin transporter RibU in Lactococcus lactis: molecular characterization of gene expression and the transport mechanism. J Bacteriol 188(8):2752–2760

Carrizo SL, Montes de Oca CE, Laiño JE, Suarez NE, Vignolo G, LeBlanc JG et al (2016) Ancestral Andean grain quinoa as source of lactic acid bacteria capable to degrade phytate and produce B-group vitamins. Food Res Int 89(Pt 1):488–494

Chaves ACSD, Fernandez M, Lerayer ALS, Mierau I, Kleerebezem M, Hugenholtz J (2002) Metabolic engineering of Acetaldehyde production by *Streptococcus thermophilus*. Appl Environ Microbiol 68(11):5656–5662

Chemical Economics Handbook (2009) Lactic Acid, its salts and esters. Biomass Bioenerg 33:21–25

Chen J, Vestergaard M, Jensen TG, Shen J, Dufva M, Solem C et al (2017) Finding the needle in the Haystack-the use of microfluidic droplet technology to identify vitamin-secreting lactic acid bacteria. MBio 8(3):e00526-17

Cubas-Cano E, González-Fernández C, Ballesteros M, Tomás-Pejó E (2018) Biotechnological advances in lactic acid production by lactic acid bacteria: lignocellulose as novel substrate. Biofuels Bioprod Biorefin 12(2):290–303

Davidson B, Llanos R, Cancilla M, Redman N, Hillier A (1995) Current research on the genetics of lactic acid production in lactic acid bacteria. Int Dairy J 5(8):763–784

de Vos WM, Hols P, van Kranenburg R, Luesink E, Kuipers OP, van der Oost J et al (1998) Making more of milk sugar by engineering lactic acid bacteria. Int Dairy J 8(3):227–233

De Vuyst L, Leroy F (2020) Functional role of yeasts, lactic acid bacteria and acetic acid bacteria in cocoa fermentation processes. FEMS Microbiol Rev 44(4):432–453

Dishisha T, Pereyra LP, Pyo S-H, Britton RA, Hatti-Kaul R (2014) Flux analysis of the Lactobacillus reuteri propanediol-utilization pathway for production of 3-hydroxypropionaldehyde, 3-hydroxypropionic acid and 1,3-propanediol from glycerol. Microb Cell Factories 13(1):76

Dmytruk KV, Yatsyshyn VY, Sybirna NO, Fedorovych DV, Sibirny AA (2011) Metabolic engineering and classic selection of the yeast Candida famata (Candida flareri) for construction of strains with enhanced riboflavin production. Metab Eng 13(1):82–88

Doi Y (2015) L-lactate production from biodiesel-derived crude glycerol by metabolically engineered Enterococcus faecalis: cytotoxic evaluation of biodiesel waste and development of a glycerol-inducible gene expression system. Appl Environ Microbiol 81(6):2082–2089

Douglas GL, Klaenhammer TR (2011) Directed chromosomal integration and expression of the reporter gene gusA3 in Lactobacillus acidophilus NCFM. Appl Environ Microbiol 77(20):7365–7371

Eom JE, Moon SK, Moon GS (2010) Heterologous production of pediocin PA-1 in Lactobacillus reuteri. J Microbiol Biotechnol 20(8):1215–1218

Even S, Lindley ND, Cocaign-Bousquet M (2001) Molecular physiology of sugar catabolism in *Lactococcus lactis* IL1403. J Bacteriol 183(13):3817–3824

Ferain T, Hobbs JN Jr, Richardson J, Bernard N, Garmyn D, Hols P et al (1996) Knockout of the two ldh genes has a major impact on peptidoglycan precursor synthesis in Lactobacillus plantarum. J Bacteriol 178(18):5431–5437

Gaspar P, Neves AR, Ramos A, Gasson MJ, Shearman CA, Santos H (2004) Engineering Lactococcus lactis for production of mannitol: high yields from food-grade strains deficient in lactate dehydrogenase and the mannitol transport system. Appl Environ Microbiol 70(3):1466–1474

Gaspar P, Neves AR, Gasson MJ, Shearman CA, Santos H (2011) High yields of 2, 3-butanediol and mannitol in Lactococcus lactis through engineering of NAD+ cofactor recycling. Appl Environ Microbiol 77(19):6826–6835

Gaspar P, Carvalho AL, Vinga S, Santos H, Neves AR (2013) From physiology to systems metabolic engineering for the production of biochemicals by lactic acid bacteria. Biotechnol Adv 31(6):764–788

Ge Y-Y, Zhang J-R, Corke H, Gan R-Y (2020) Screening and spontaneous mutation of pickle-derived lactobacillus plantarum with overproduction of riboflavin, related mechanism, and food application. Foods 9(1):88

Goffin P, Deghorain M, Mainardi JL, Tytgat I, Champomier-Vergès MC, Kleerebezem M et al (2005) Lactate racemization as a rescue pathway for supplying D-lactate to the cell wall biosynthesis machinery in Lactobacillus plantarum. J Bacteriol 187(19):6750–6761

Goh YJ, Azcárate-Peril MA, O'Flaherty S, Durmaz E, Valence F, Jardin J et al (2009) Development and application of a upp-based counterselective gene replacement system for the study of the S-layer protein SlpX of Lactobacillus acidophilus NCFM. Appl Environ Microbiol 75(10):3093–3105

Guirimand G, Sasaki K, Inokuma K, Bamba T, Hasunuma T, Kondo A (2016) Cell surface engineering of Saccharomyces cerevisiae combined with membrane separation technology for xylitol production from rice straw hydrolysate. Appl Microbiol Biotechnol 100(8):3477–3487

Guo T, Kong J, Zhang L, Zhang C, Hu S (2012) Fine tuning of the lactate and diacetyl production through promoter engineering in Lactococcus lactis. PLoS One 7(4):e36296

Guru V, Viswanathan K (2013) Riboflavin production in milk whey using probiotic bacteria-Lactobacillus acidophilus and Lactococcus lactis. Indian J Fundam Appl Life Sci 3(4):169–176

Han YZ, Zhou CC, Xu YY, Yao JX, Chi Z, Chi ZM et al (2017) High-efficient production of fructo-oligosaccharides from inulin by a two-stage bioprocess using an engineered Yarrowia lipolytica strain. Carbohydr Polym 173:592–599

Helanto M, Aarnikunnas J, von Weymarn N, Airaksinen U, Palva A, Leisola M (2005) Improved mannitol production by a random mutant of Leuconostoc pseudomesenteroides. J Biotechnol 116(3):283–294

Hidalgo-Cantabrana C, Goh YJ, Pan M, Sanozky-Dawes R, Barrangou R (2019) Genome editing using the endogenous type I CRISPR-Cas system in Lactobacillus crispatus. Proc Natl Acad Sci 116(32):15774–15783

Hoefnagel MHN, Starrenburg MJC, Martens DE, Hugenholtz J, Kleerebezem M, Van S II et al (2002) Metabolic engineering of lactic acid bacteria, the combined approach: kinetic modelling, metabolic control and experimental analysis. Microbiology (Reading) 148(Pt 4):1003–1013
Hu C-Y, Chi D-J, Chen S-S, Chen Y-C (2011) The direct conversion of xylan to lactic acid by Lactobacillus brevis transformed with a xylanase gene. Green Chem 13(7):1729–1734
Hugenholtz J, Smid EJ (2002) Nutraceutical production with food-grade microorganisms. Curr Opin Biotechnol 13(5):497–507
Hugenholtz J, Kleerebezem M, Starrenburg M, Delcour J, de Vos W, Hols P (2000) Lactococcus lactis as a cell factory for high-level diacetyl production. Appl Environ Microbiol 66(9):4112–4114
Jayaraman K, Kalaichelvan G, Jayashree S (2010) Isolation, screening and characterization of riboflavin producing lactic acid bacteria from Katpadi, Vellore district. Recent Res Sci Technol 2(1)
Jensen PR, Hammer K (1998) Artificial promoters for metabolic optimization. Biotechnol Bioeng 58(2–3):191–195
Johnson EA, Echavarri-Erasun C (2011a) Chapter 3 – Yeast biotechnology. In: Kurtzman CP, Fell JW, Boekhout T (eds) The Yeasts, 5th edn. Elsevier, London, pp 21–44
Johnson EA, Echavarri-Erasun C, (eds). (2011b) Chapter 3 Yeast biotechnology
Kadam SR, Patil SS, Bastawde KB, Khire JM, Gokhale DV (2006) Strain improvement of Lactobacillus delbrueckii NCIM 2365 for lactic acid production. Process Biochem 41(1):120–126
Kandler O (1983) Carbohydrate metabolism in lactic acid bacteria. Antonie Van Leeuwenhoek 49(3):209–224
Kandler O, Weiss N (1986) Regular, non-sporing gram-positive rods. In: Sneath HA, Mair NS, Sharpe ME, Holt JG (eds) Bergey's manual of systematic bacteriology: Williams and Wilkins. Baltimore, pp 1208–1234
Kaur B, Chakraborty D (2013) Statistical media and process optimization for biotransformation of rice bran to vanillin using Pediococcus acidilactici. Indian J Exp Biol 51(11):935–943
Kaur B, Chakraborty D, Kumar B (2014) Metabolic engineering of Pediococcus acidilactici BD16 for production of vanillin through ferulic acid catabolic pathway and process optimization using response surface methodology. Appl Microbiol Biotechnol 98(20):8539–8551
Kaur B, Kumar B, Kaur G, Chakraborty D, Kaur K (2015) Application of recombinant Pediococcus acidilactici BD16 (fcs^+/ech^+) in malolactic fermentation. Appl Microbiol Biotechnol 99(7):3015–3028
Khalili M, Rad AH, Khosroushahi AY, Khosravi H, Jafarzadeh S (2020) Application of probiotics in folate bio-fortification of yoghurt. Probiotics Antimicrob Proteins 12(2):756–763
Kodali VP, Sen R (2008) Antioxidant and free radical scavenging activities of an exopolysaccharide from a probiotic bacterium. Biotechnol J 3(2):245–251
Konstantinidis D, Pereira F, Geissen EM, Grkovska K, Kafkia E, Jouhten P et al (2021) Adaptive laboratory evolution of microbial co-cultures for improved metabolite secretion. Mol Syst Biol 17(8):e10189
Kumar RR, Prasad S (2011) Metabolic engineering of bacteria. Indian J Microbiol 51(3):403–409
Kunji ER, Slotboom D-J, Poolman B (2003) Lactococcus lactis as host for overproduction of functional membrane proteins. Biochimica et Biophysica Acta (BBA)-Biomembranes 1610(1):97–108
Kuo Y-C, Yuan S-F, Wang C-A, Huang Y-J, Guo G-L, Hwang W-S (2015) Production of optically pure l-lactic acid from lignocellulosic hydrolysate by using a newly isolated and d-lactate dehydrogenase gene-deficient Lactobacillus paracasei strain. Bioresour Technol 198:651–657
Kylä-Nikkilä K, Hujanen M, Leisola M, Palva A (2000) Metabolic engineering of Lactobacillus helveticus CNRZ32 for production of pure L-(+)-lactic acid. Appl Environ Microbiol 66(9):3835–3841
Ladero V, Ramos A, Wiersma A, Goffin P, Schanck A, Kleerebezem M et al (2007) High-level production of the low-calorie sugar sorbitol by *Lactobacillus plantarum* through metabolic engineering. Appl Environ Microbiol 73(6):1864–1872

Lambert JM, Bongers RS, Kleerebezem M (2007) Cre-lox-based system for multiple gene deletions and selectable-marker removal in Lactobacillus plantarum. Appl Environ Microbiol 73(4):1126–1135

LeBlanc JG, Laiño JE, del Valle MJ, Vannini V, van Sinderen D, Taranto MP et al (2011) B-group vitamin production by lactic acid bacteria—current knowledge and potential applications. J Appl Microbiol 111(6):1297–1309

Lee JY, Kang CD, Lee SH, Park YK, Cho KM (2015) Engineering cellular redox balance in Saccharomyces cerevisiae for improved production of L-lactic acid. Biotechnol Bioeng 112(4):751–758

Leenhouts K, Buist G, Bolhuis A, Ten Berge A, Kiel J, Mierau I et al (1996) A general system for generating unlabelled gene replacements in bacterial chromosomes. Mol Gen Genet MGG 253(1):217–224

Levander F, Svensson M, Rådström P (2002) Enhanced exopolysaccharide production by metabolic engineering of Streptococcus thermophilus. Appl Environ Microbiol 68(2):784–790

Lindlbauer KA, Marx H, Sauer M (2017a) Effect of carbon pulsing on the redox household of Lactobacillus diolivorans in order to enhance 1,3-propanediol production. New Biotechnol 34:32–39

Lindlbauer KA, Marx H, Sauer M (2017b) 3-Hydroxypropionaldehyde production from crude glycerol by Lactobacillus diolivorans with enhanced glycerol uptake. Biotechnol Biofuels 10(1):295

Liu S, Nichols NN, Dien BS, Cotta MA (2006) Metabolic engineering of a Lactobacillus plantarum double ldh knockout strain for enhanced ethanol production. J Ind Microbiol Biotechnol 33(1):1–7

Liu J, Kandasamy V, Würtz A, Jensen PR, Solem C (2016a) Stimulation of acetoin production in metabolically engineered Lactococcus lactis by increasing ATP demand. Appl Microbiol Biotechnol 100(22):9509–9517

Liu J, Solem C, Jensen PR (2016b) Integrating biocompatible chemistry and manipulating cofactor partitioning in metabolically engineered Lactococcus lactis for fermentative production of (3S)-acetoin. Biotechnol Bioeng 113(12):2744–2748

Liu J, Dantoft SH, Würtz A, Jensen PR, Solem C (2016c) A novel cell factory for efficient production of ethanol from dairy waste. Biotechnol Biofuels 9:33

Liu J, Chan SHJ, Brock-Nannestad T, Chen J, Lee SY, Solem C et al (2016d) Combining metabolic engineering and biocompatible chemistry for high-yield production of homo-diacetyl and homo-(S,S)-2,3-butanediol. Metab Eng 36:57–67

Liu Y, van Bennekom EO, Zhang Y, Abee T, Smid EJ (2019) Long-chain vitamin K2 production in Lactococcus lactis is influenced by temperature, carbon source, aeration and mode of energy metabolism. Microb Cell Factories 18(1):129

Lohmeier-Vogel EM, Hahn-Hägerdahl B, Vogel HJ (1986) Phosphorus-31 NMR studies of maltose and glucose metabolism in Streptococcus lactis. Appl Microbiol Biotechnol 25(1):43–51

London J (1990) Uncommon pathways of metabolism among lactic acid bacteria. FEMS Microbiol Rev 7(1–2):103–111

Looijesteijn PJ, Boels IC, Kleerebezem M, Hugenholtz J (1999) Regulation of exopolysaccharide production by Lactococcus lactis subsp. cremoris by the sugar source. Appl Environ Microbiol 65(11):5003–5008

López J, Bustos D, Camilo C, Arenas N, Saa PA, Agosin E (2020) Engineering Saccharomyces cerevisiae for the overproduction of β-ionone and its precursor β-Carotene. Front Bioeng Biotechnol 8(1145):578793

Lowes KF, Shearman CA, Payne J, MacKenzie D, Archer DB, Merry RJ et al (2000) Prevention of yeast spoilage in feed and food by the yeast mycocin HMK. Appl Environ Microbiol 66(3):1066–1076

Lu W, Kong J, Kong W (2013) Construction and application of a food-grade expression system for Lactococcus lactis. Mol Biotechnol 54(2):170–176

Mao W, Han Y, Wang X, Zhao X, Chi Z, Chi Z et al (2019) A new engineered endo-inulinase with improved activity and thermostability: application in the production of prebiotic fructo-oligosaccharides from inulin. Food Chem 294:293–301

McKay LL, Baldwin KA (1990) Applications for biotechnology: present and future improvements in lactic acid bacteria. FEMS Microbiol Rev 7(1–2):3–14

Mierau I, Kleerebezem M (2005) 10 years of the nisin-controlled gene expression system (NICE) in Lactococcus lactis. Appl Microbiol Biotechnol 68(6):705–717

Mokoena MP (2017) Lactic acid bacteria and their bacteriocins: classification, biosynthesis and applications against uropathogens: a mini-review. Molecules (Basel, Switzerland) 22(8):1255

Monedero V, Pérez-Martínez G, Yebra MJ (2010) Perspectives of engineering lactic acid bacteria for biotechnological polyol production. Appl Microbiol Biotechnol 86(4):1003–1015

Morita H, Toh H, Fukuda S, Horikawa H, Oshima K, Suzuki T et al (2008) Comparative genome analysis of Lactobacillus reuteri and Lactobacillus fermentum reveal a genomic Island for reuterin and cobalamin production. DNA Res 15(3):151–161

Nadal I, Rico J, Pérez-Martínez G, Yebra MJ, Monedero V (2009) Diacetyl and acetoin production from whey permeate using engineered Lactobacillus casei. J Ind Microbiol Biotechnol 36(9):1233–1237

Nehal F, Sahnoun M, Smaoui S, Jaouadi B, Bejar S, Mohammed S (2019) Characterization, high production and antimicrobial activity of exopolysaccharides from Lactococcus lactis F-mou. Microb Pathog 132:10–19

Nes IF, Gabrielsen C, Brede DA, Diep DB (2015) Novel developments in bacteriocins from lactic acid bacteria. Biotechnology of lactic acid bacteria. pp 80–99

Neves AR, Ramos A, Shearman C, Gasson MJ, Almeida JS, Santos H (2000) Metabolic characterization of Lactococcus lactis deficient in lactate dehydrogenase using in vivo 13C-NMR. Eur J Biochem 267(12):3859–3868

Neves AR, Pool WA, Castro R, Mingote A, Santos F, Kok J et al (2006) The α-phosphoglucomutase of Lactococcus lactis is unrelated to the α-d-phosphohexomutase superfamily and is encoded by the essential gene pgmH. J Biol Chem 281(48):36864–36873

Nissen L, Pérez-Martínez G, Yebra MJ (2005) Sorbitol synthesis by an engineered Lactobacillus casei strain expressing a sorbitol-6-phosphate dehydrogenase gene within the lactose operon. FEMS Microbiol Lett 249(1):177–183

Nyyssölä A, Pihlajaniemi A, Palva A, von Weymarn N, Leisola M (2005) Production of xylitol from D-xylose by recombinant Lactococcus lactis. J Biotechnol 118(1):55–66

Oguro Y, Nishiwaki T, Shinada R, Kobayashi K, Kurahashi A (2017) Metabolite profile of koji amazake and its lactic acid fermentation product by Lactobacillus sakei UONUMA. J Biosci Bioeng 124(2):178–183

Okano K, Zhang Q, Shinkawa S, Yoshida S, Tanaka T, Fukuda H et al (2009) Efficient production of optically pure D-lactic acid from raw corn starch by using a genetically modified L-lactate dehydrogenase gene-deficient and alpha-amylase-secreting Lactobacillus plantarum strain. Appl Environ Microbiol 75(2):462–467

Okano K, Hama S, Kihara M, Noda H, Tanaka T, Kondo A (2017) Production of optically pure D-lactic acid from brown rice using metabolically engineered Lactobacillus plantarum. Appl Microbiol Biotechnol 101(5):1869–1875

Ortiz ME, Bleckwedel J, Fadda S, Picariello G, Hebert EM, Raya RR et al (2017) Global Analysis of Mannitol 2-Dehydrogenase in Lactobacillus reuteri CRL 1101 during Mannitol production through enzymatic, genetic and proteomic approaches. PLoS One 12(1):e0169441

Papagianni M, Avramidis N (2011) Lactococcus lactis as a cell factory: a twofold increase in phosphofructokinase activity results in a proportional increase in specific rates of glucose uptake and lactate formation. Enzym Microb Technol 49(2):197–202

Papagianni M, Avramidis N, Filiousis G (2007) Glycolysis and the regulation of glucose transport in Lactococcus lactis spp. lactis in batch and fed-batch culture. Microb Cell Factories 6:16

Paparella A, Shaltiel-Harpaza L, Ibdah M (2021) β-Ionone: its occurrence and biological function and metabolic engineering. Plan Theory 10(4):754

Perricone M, Gallo M, Corbo MR, Sinigaglia M, Bevilacqua A (2017a) Yeasts. The microbiological quality of food. Elsevier, pp 121–131
Perricone M, Gallo M, Corbo MR, Sinigaglia M, Bevilacqua A (2017b) Chapter 5 – Yeasts. In: Bevilacqua A, Corbo MR, Sinigaglia M (eds) The microbiological quality of food. Woodhead Publishing, pp 121–131
Pinto JP, Zeyniyev A, Karsens H, Trip H, Lolkema JS, Kuipers OP et al (2011) pSEUDO, a genetic integration standard for Lactococcus lactis. Appl Environ Microbiol 77(18):6687–6690
Rai AK, Kumaraswamy J (2017) Role of yeasts in food fermentation. In: Satyanarayana T, Kunze G (eds) Yeast diversity in human welfare Singapore. Springer
Rai AK, Pandey A, Sahoo D (2019) Biotechnological potential of yeasts in functional food industry. Trends Food Sci Technol 83:129–137
Rajan PJ, Madhavan KN (2008) Strain improvement of Lactobacillus delbrueckii using nitrous acid mutation for L-lactic acid production. World J Microbiol Biotechnol 24:3105–3109
Rawat S (2015) Food spoilage: microorganisms and their prevention. Asian J Plant Sci Res 5(4):47–56
Rodriguez A, Kildegaard KR, Li M, Borodina I, Nielsen J (2015) Establishment of a yeast platform strain for production of p-coumaric acid through metabolic engineering of aromatic amino acid biosynthesis. Metab Eng 31:181–188
Rodriguez A, Strucko T, Stahlhut SG, Kristensen M, Svenssen DK, Forster J et al (2017) Metabolic engineering of yeast for fermentative production of flavonoids. Bioresour Technol 245:1645–1654
Ruffing A, Chen RR (2006) Metabolic engineering of microbes for oligosaccharide and polysaccharide synthesis. Microb Cell Factories 5(1):25
Ruiz Rodríguez LG, Mohamed F, Bleckwedel J, Medina R, De Vuyst L, Hebert EM et al (2019) Diversity and functional properties of lactic acid bacteria isolated from wild fruits and flowers present in Northern Argentina. Front Microbiol 10:1091
Sabater-Molina M, Larqué E, Torrella F, Zamora S (2009) Dietary fructooligosaccharides and potential benefits on health. J Physiol Biochem 65(3):315–328
Sabet-Azad R, Linares-Pastén JA, Torkelson L, Sardari RR, Hatti-Kaul R (2013) Coenzyme A-acylating propionaldehyde dehydrogenase (PduP) from Lactobacillus reuteri: kinetic characterization and molecular modeling. Enzym Microb Technol 53(4):235–242
Santos F, Wegkamp A, de Vos WM, Smid EJ, Hugenholtz J (2008) High-level folate production in fermented foods by the B12 producer Lactobacillus reuteri JCM1112. Appl Environ Microbiol 74(10):3291–3294
Sasikumar K, Kozhummal Vaikkath D, Devendra L, Nampoothiri KM (2017) An exopolysaccharide (EPS) from a Lactobacillus plantarum BR2 with potential benefits for making functional foods. Bioresour Technol 241:1152–1156
Sharma A, Gupta G, Ahmad T, Kaur B, Hakeem KR (2020) Tailoring cellular metabolism in lactic acid bacteria through metabolic engineering. J Microbiol Methods 170:105862
Shinkawa S, Okano K, Yoshida S, Tanaka T, Ogino C, Fukuda H et al (2011) Improved homo L-lactic acid fermentation from xylose by abolishment of the phosphoketolase pathway and enhancement of the pentose phosphate pathway in genetically modified xylose-assimilating Lactococcus lactis. Appl Microbiol Biotechnol 91(6):1537–1544
Shirai K, Guerrero Legarreta I, Huerta Ochoa S, Saucedo-Castañeda G, Castillo A, Gonzalez R et al (2001) Effect of initial glucose concentration and inoculation level of lactic acid bacteria in shrimp waste ensilation. Enzym Microb Technol 28:446–452
Singhvi M, Joshi D, Adsul M, Varma A, Gokhale D (2010) d-(−)-Lactic acid production from cellobiose and cellulose by Lactobacillus lactis mutant RM2-24. Green Chem 12(6):1106–1109
Solem C, Jensen PR (2002) Modulation of gene expression made easy. Appl Environ Microbiol 68(5):2397–2403
Solem C, Dehli T, Jensen PR (2013) Rewiring Lactococcus lactis for ethanol production. Appl Environ Microbiol 79(8):2512–2518

Srikanth A, Halami PM (2008) Cloning of pediocin PA-1 and its immunity genes from Pediococcus acidilactici K7 using pAMJ shuttle vector into Lactococcus lactis MG1363. Indian J Biotechnol 7:550–553

Sybesma W, Starrenburg M, Kleerebezem M, Mierau I, de Vos WM, Hugenholtz J (2003) Increased production of folate by metabolic engineering of Lactococcus lactis. Appl Environ Microbiol 69(6):3069–3076

Sybesma W, Burgess C, Starrenburg M, van Sinderen D, Hugenholtz J (2004) Multivitamin production in Lactococcus lactis using metabolic engineering. Metab Eng 6(2):109–115

Thakur K, Tomar S (2015) Exploring indigenous lactobacillus species from diverse niches for riboflavin production. J Young Pharm 7:122–127

Thakur K, Tomar SK, De S (2016) Lactic acid bacteria as a cell factory for riboflavin production. Microb Biotechnol 9(4):441–451

Torres AC, Vannini V, Font G, Saavedra L, Taranto MP (2018) Novel pathway for corrinoid compounds production in Lactobacillus. Front Microbiol 9:2256

Tsuji A, Okada S, Hols P, Satoh E (2013) Metabolic engineering of Lactobacillus plantarum for succinic acid production through activation of the reductive branch of the tricarboxylic acid cycle. Enzym Microb Technol 53(2):97–103

Van Pijkeren J-P, Britton RA (2012) High efficiency recombineering in lactic acid bacteria. Nucleic Acids Res 40(10):e76-e

Van Pijkeren J-P, Neoh KM, Sirias D, Findley AS, Britton RA (2012) Exploring optimization parameters to increase ssDNA recombineering in Lactococcus lactis and Lactobacillus reuteri. Bioengineered 3(4):209–217

Wang HH, Isaacs FJ, Carr PA, Sun ZZ, Xu G, Forest CR et al (2009) Programming cells by multiplex genome engineering and accelerated evolution. Nature 460(7257):894–898

Wang Y, Halls C, Zhang J, Matsuno M, Zhang Y, Yu O (2011) Stepwise increase of resveratrol biosynthesis in yeast Saccharomyces cerevisiae by metabolic engineering. Metab Eng 13(5):455–463

Wang D, Li F-L, Wang S-A (2016) A one-step bioprocess for production of high-content fructooligosaccharides from inulin by yeast. Carbohydr Polym 151:1220–1226

Wang C, Sun J, Lassabliere B, Yu B, Zhao F, Zhao F et al (2019) Potential of lactic acid bacteria to modulate coffee volatiles and effect of glucose supplementation: fermentation of green coffee beans and impact of coffee roasting. J Sci Food Agric 99(1):409–420

Wang Y, Wu J, Lv M, Shao Z, Hungwe M, Wang J et al (2021) Metabolism characteristics of lactic acid bacteria and the expanding applications in food industry. Front Bioeng Biotechnol 9:378

Wee Y-J, Yun J-S, Kim D, Ryu H-W (2006) Batch and repeated batch production of L (+)-lactic acid by Enterococcus faecalis RKY1 using wood hydrolyzate and corn steep liquor. J Ind Microbiol Biotechnol 33(6):431–435

Wegkamp A, van Oorschot W, de Vos WM, Smid EJ (2007) Characterization of the role of para-aminobenzoic acid biosynthesis in folate production by Lactococcus lactis. Appl Environ Microbiol 73(8):2673–2681

Wisselink HW, Moers APHA, Mars AE, Hoefnagel MHN, Vos WM, Hugenholtz J (2005) Overproduction of Heterologous Mannitol 1-Phosphatase: a key factor for engineering mannitol production by *Lactococcus lactis*. Appl Environ Microbiol 71(3):1507–1514

Yamano S, Ishii T, Nakagawa M, Ikenaga H, Misawa N (1994) Metabolic engineering for production of β-Carotene and Lycopene in Saccharomyces cerevisiae. Biosci Biotechnol Biochem 58(6):1112–1114

Yan M, Wang B-H, Xu X, Chang P, Hang F, Wu Z et al (2018) Molecular and functional study of a branching sucrase-like glucansucrase reveals an evolutionary intermediate between two subfamilies of the GH70 Enzymes. Appl Environ Microbiol 84(9):e02810–e02817

Yang Y-T, Bennett GN, San K-Y (1998) Genetic and metabolic engineering. Electron J Biotechnol 1(3):134–141

Ye W, Huo G, Chen J, Liu F, Yin J, Yang L et al (2010) Heterologous expression of the Bacillus subtilis (natto) alanine dehydrogenase in Escherichia coli and Lactococcus lactis. Microbiol Res 165(4):268–275

Yi X, Zhang P, Sun J, Tu Y, Gao Q, Zhang J et al (2016) Engineering wild-type robust Pediococcus acidilactici strain for high titer L- and D-lactic acid production from corn stover feedstock. J Biotechnol 217:112–121

Yoshida S, Okano K, Tanaka T, Ogino C, Kondo A (2011) Homo-D-lactic acid production from mixed sugars using xylose-assimilating operon-integrated Lactobacillus plantarum. Appl Microbiol Biotechnol 92(1):67–76

Zhang J, Wu C, Du G, Chen J (2012) Enhanced acid tolerance in Lactobacillus casei by adaptive evolution and compared stress response during acid stress. Biotechnol Bioprocess Eng 17:283–289

Zheng J, Wittouck S, Salvetti E, Franz CM, Harris HM, Mattarelli P et al (2020) A taxonomic note on the genus Lactobacillus: description of 23 novel genera, emended description of the genus Lactobacillus Beijerinck 1901, and union of Lactobacillaceae and Leuconostocaceae. Int J Syst Evol Microbiol 70(4):2782–2858

Zhu Y, Zhang Y, Li Y (2009) Understanding the industrial application potential of lactic acid bacteria through genomics. Appl Microbiol Biotechnol 83(4):597–610

Chapter 9
Immobilization Techniques Applicable to Sourdough

Zeynep Agirbasli, Mensure Elvan, and Sebnem Harsa

9.1 Introduction

The term of immobilization is mostly used to define the physically fixed enzyme, cell or biocatalysts in a particular environment (Eş et al. 2015). Immobilization process can be applied for enzymes, cells or functional units in order to prevent them from moving freely and independently throughout the bioreactor system. Immobilized enzyme in a physically localized region is preferred due to its repeatability and protected catalytic activity (Mohamad et al. 2015). Main strategies used in immobilization are physical adsorption due to electrostatic forces, cross-linking, covalent binding and entrapment/encapsulation (DiCosimo et al. 2013; Mitropoulou et al. 2013; Nguyen and Kim 2017; Mohamad et al. 2015).

The most important immobilization techniques can be distinguished as follows: A film fermenter in which cellular material is encapsulated in a biological film. Other processes in which biological catalysts are entrapped or otherwise confined on or in a solid support. A process in which cells and enzymes are separated into one phase by a semi-porous membrane, and the process by which cells or enzymes are preferentially distributed among multiple fluid-phase environments (Guisan 2013).

There are advantages to developing immobilized enzymatic processes. Immobilization of enzymes on water-insoluble particles greatly improves enzyme stability, improves the ability to readily recover enzymes from products and substrates, enables reuse and continuous processing, and significantly improves economics. A disadvantage of immobilized enzymes is that the heterogeneous nature of such catalysts can lead to diffusion limitations, reducing or altering their activity. Using immobilized cells instead of batch-grown regular cells has similar advantages to immobilized enzymes. (1) Batch fermentation can be replaced by a continuous

Z. Agirbasli · M. Elvan · S. Harsa (✉)
Department of Food Engineering, Izmir Institute of Technology, Urla, Izmir, Turkey
e-mail: zeynepagirbasli@iyte.edu.tr; mensureelvan@iyte.edu.tr; sebnemharsa@iyte.edu.tr

E. B. Ceresino et al. (eds.), *Sourdough Microbiota and Starter Cultures for Industry*, https://doi.org/10.1007/978-3-031-48604-3_9

reaction. (2) Fixed cells allow the use of significantly higher cell densities, allowing process intensification. (3) Many metabolites and enzymes are only active inside cells in the stationary phase, but immobilized systems can maintain cells in this state (Bié et al. 2022).

Enzymes are usually expensive items in enzymatic technology processes. Worse, the catalytic activity is wasted at the end of the process. In fact, money often has to be spent to remove (i.e. discard) enzyme proteins from the product stream. Immobilization facilitates removal of the product from the stream, reusing the enzyme and reducing costs. The use of immobilized enzymes often facilitates continuous reactor development and increases productivity (Dwevedi 2016).

There are four principal methods available for immobilizing enzymes: Adsorption, Covalent coupling (covalent binding), Entrapment, Membrane incorporation (membrane confinement). Various immobilization methods have been applied so far including covalent binding, encapsulation, cross-linking, adsorption and entrapment (Maghraby et al. 2023).

Adsorption It is very simple and can reach high loadings of up to 1 gram of protein per gram of matrix. In this method, proteins and adsorbents are incubated under appropriate conditions such as pH, ionic strength, and temperature, then excess protein is washed away. Connections between the matrix and protein molecules are usually hydrophobic and/or ionic in the form of salt bridges. Therefore, it is important to ensure that no enzyme is released from the particles during use. Suitable sorbents include: Alumina and charcoal, ion exchange matrices, porous carbon clays, hydrous metal oxides, glasses, polymer resins (Jesionowski et al. 2014).

Covalent Attachment to Solid Supports Enzymes can be immobilized by covalent bonding to insoluble carriers. A variety of supporting materials were used including: porous glass ceramics, stainless steel, sand, charcoal, celulose, synthetic polymers (nylon), metallic oxides (TiO_2) (Chan et al. 2007).

Amino or carboxyl groups present in the enzyme protein are used during immobilization. In most cases, immobilization consists of at least two steps: carrier activation and specific coupling reactions. For many supports to be suitable for covalent attachment of enzyme molecules, functional groups capable of undergoing coupling reactions must be introduced or released. An example is liberation of primary amino groups by acid hydrolysis of nylon supports. Corning Glass Works used porous glass beads to immobilize enzymes such as glucose isomerase and lactase. Glass supports are silanized by treatment with 3-aminopropyltriethoxysilane $(C_6H_5O)3Si(CH_2)3NH_2$. This results in an activated support; ceramic-Si $(C_6H_5O)2(CH_2)3NH_2$ (Novick and Rozzell 2005).

Covalent Coupling The activation step is completed by the reaction of this support with gluteraldehyde. Here, one of the carbonyl groups reacts with the NH_2 groups of the support. Ceramic-Si $(C_2H_5O)2$ $(CH_2)3$ -N=CH- $(CH_2)3$ -CHO. Unreacted glutaraldehyde is washed away and the second carbonyl group reacts with the amino group of the enzyme. Other specific coupling reactions involve NH_2, OH or SH groups of enzyme proteins (Novick and Rozzell 2005).

Entrapment Enzymes can be immobilized within a three-dimensional polymer network by adding them to the monomers prior to gelation. The polymers used to form the gel can be either natural organic polymers or synthetic polymers. These are enzymes that can be immobilized in a three-dimensional polymer network by adding the enzyme to the monomers prior to gelation (Imam et al. 2021).

There are many natural polymers that can form gels under mild conditions. Materials such as the natural fibrous proteins collagen, gelatin, agar, and agarose (a polysaccharide found in red seaweed) can gel when aqueous solutions are cooled. Other polysaccharides such as alginates can be polymerized at lower temperatures (25 °C). Due to these very mild conditions, alginate is one of the most commonly used polymers. Synthetic and naturally occurring polymers can be used to entrap enzymes or cells within particles by: (1) Polymerizing a block and breaking this into particles (2) Moulding the polymer into a template (3) Bead forming by dropping cells and polymer into a Ca^{2+} containing medium in the case of alginate or into a hydrophobic phase for other polymers (Moo-Young 2019; Imam et al. 2021).

Microencapsulation Developed by Chang (1972), this method involves the production of microcapsules by interfacial polymerization, liquid-drying, or phase separation. For example, an aqueous enzyme solution is emulsified with a water-immiscible organic solvent to form aqueous microdroplets. A film is formed on the surface of the microdroplets by adding the polymer solution to the stirred emulsion. These microdroplets are released into the aqueous phase. This method has the advantage of high activity due to its high surface-to-volume ratio, but is susceptible to hypertonic solutions (Moo-Young 2019).

Cross-Linking with Bifunctional Reagents Enzymes can be immobilized by cross-linking enzyme molecules together to create particles. The most common cross-linking reagents are glutaraldehyde, dimethyladipimidate, dimethylsuberimidate, and aliphatic diamines. The first three directly connects enzymes through amino groups. The diamine cross-links the enzyme through its carboxyl groups after activation with carbodiimide. An example is the cross-linking enzyme glucose isomerase produced by NOVO upon glutaraldehyde treatment (Novick and Rozzell 2005; Moo-Young 2019).

Membrane Incorporation (Capture behind semi-permeable membranes): Enzymes can be encapsulated in semipermeable membranes, while low molecular weight substrates and products are allowed to pass freely, while high molecular weight enzymes are retained. The simplest method is to entrap the enzyme in hollow fibers of a semipermeable membrane and pass a substrate solution through it. This is the base of the hollow fiber reactor. Enzymes are retained within permeable membranes or hollow fiber units. Penicillin acylase, lactase, and aminoacylase were immobilized using hollow fiber units (Moo-Young 2019).

Immobilization of enzymes such as lipase, lactase, glucose isomerase and phytase have been used (DiCosimo et al. 2013; Hassan et al. 2019; Pragya et al. 2021); however, their applications in sourdough products is not explored in detail.

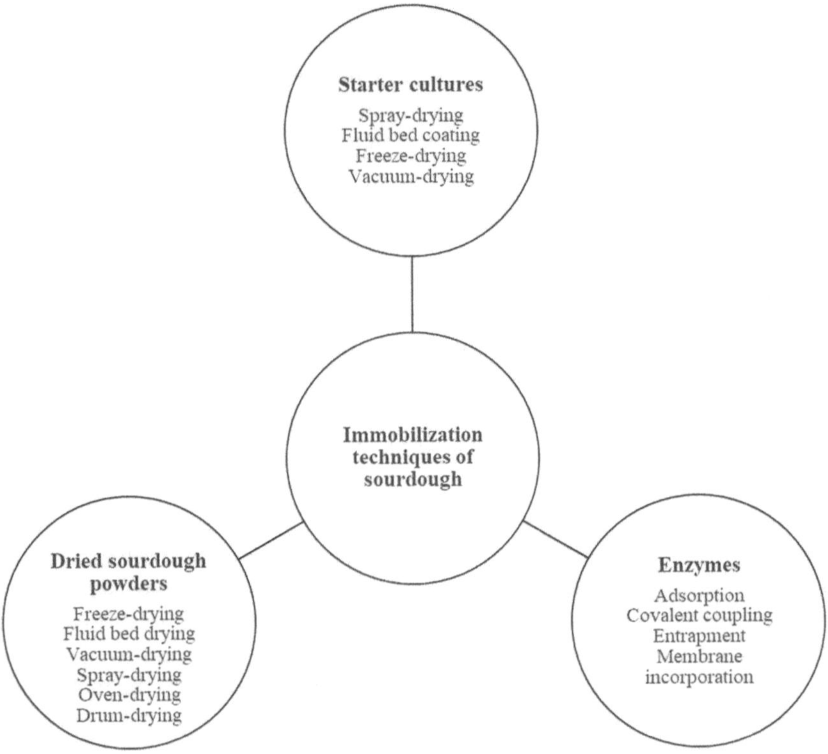

Fig. 9.1 Immobilization techniques for sourdough

Meanwhile, microencapsulation of microbial cells has been reviewed as an immobilization technology for sourdough products (Kosseva 2011; Mitropoulou et al. 2013; Rathore et al. 2013). Since it is one of the main methods of immobilization techniques, the terms immobilization and encapsulation have been used interchangeably (Kosseva 2011).

Therefore, this chapter mainly covers various immobilized systems, e.g. starter cultures, enzymes, and their usage and impact on physicochemical, microbiological and sensorial properties of sourdough (Fig. 9.1).

9.2 Immobilization Applications for Sourdoughs

9.2.1 Immobilized Starter Cultures Applicable to Sourdough

Fermentation during sourdough production affects aroma, flavor, taste and technological properties such as texture, dough volume and structure, and shelf-life of the bread. Production of sourdough with repeatable qualities from artisanal starter microorganism is dependent on several conditions such as stable environment temperature, ingredients and repeated use of starter cultures to produce mother culture. These conditions increase the need for sourdough starter microorganisms with enhanced properties (Mohd Roby et al. 2020). Various techniques such as spray-drying, fluid bed coating and freeze- or vacuum-drying have been evaluated for microencapsulation of starter cultures. Freeze-drying makes starter microorganisms more suitable for large scale production (Mantzourani et al. 2019). Therefore, the selection and incorporation of microorganisms and their immobilization techniques are crucial to ensure and preserve the quality of sourdoughs and sourdough products. In addition to this, adaptation of selected potential starter cultures to the environment of sourdough and fermentation is also crucial (Gul et al. 2020). A recent study used gum arabic and mineral water for encapsulation of liquid kombucha sourdough starter by spray-drying technique. In the study, sourdough breads made with encapsulated kombucha starter culture, liquid traditional sourdough starter and baker's yeast were compared for physical qualities. It was found that the loaf volume (976.7 ± 25.2 mL) and specific loaf volume (4.38 ± 0.12 mL/g) were increased compared to yeast bread, and the crumb firmness (116.07 ± 6.28 g) was decreased compared to traditional sourdough bread and yeast bread. (Mohd Roby et al. 2020).

Different food matrices have been tested for immobilization of starter culture of sourdough bread. Immobilization can affect the metabolic and fermentation activity, growth and stability of the cells (Plessas et al. 2012). A study investigated the immobilization of *Pediococcus pentosaceus* SP2 on wheat bran. In addition to wild-type microbiota, both free and immobilized cells were freeze-dried and used in the production of sourdough bread. The authors proposed this immobilized strain due to their stability on sourdough (Plessas et al. 2020). Brewery spent grains has been another immobilization matrix for microorganisms to be used in sourdough formation. A study used immobilized baker's yeast, kefir or *Lacticaseibacillus casei* cells on brewer's spent grains in breads made by either with sourdough or straight-dough technique. In comparison of commercial type bread, breads made by immobilized cells had lower volume. On the other hand, sourdough breads had 20–30% higher volume than straight dough breads. Compared to straight dough bread, sourdough bread had two times longer shelf-life. Sensory analysis did not show any significant difference between these two type breads made by immobilized cells (Plessas et al. 2007). Immobilization of *Lacticaseibacillus paracasei* K5 and *Lactobacillus bulgaricus* ATCC 11842 on wheat bran and their freeze-dried forms were studied for their effects on bread volume, organic acid content and antimicrobial effects (Mantzourani et al. 2019). Compared to bread prepared with wild type sourdough, total titratable acidity and organic acid content was higher in all sourdough breads. The shelf-life of immobilized sourdough breads was prolonged up to 14 day, whereas wild-type sourdough bread was spoiled at sixth day (Mantzourani et al. 2019). In another study, corn grains were evaluated for immobilization of kefir grains due to entrapment of (growing) cells inside the carrier's porous structure. These immobilized cells were used to prepare wheat sourdough bread (Plessas et al. 2012). Higher lactic acid production and improved fermentation rate were obtained by addition of immobilized kefir cells in contrast to free kefir cells. Immobilized kefir cells had produced sourdough bread similar to control sourdough bread in terms of aroma compounds, whereas the specific volume of sourdough breads produced by immobilized cells decreased by 20% w/w compared to control sourdough breads (Plessas et al. 2012). On the other hand, orange pulp as an immobilization matrix has been proposed for kefir stains; 10% addition of immobilized kefir to 500 g flour was studied for direct bread production method and sourdough bread production method. Aerobic fermentation of orange pulp by kefir had higher aroma compounds and esters compared to baker's yeast, which could be evaluated as advantageous for flavor of the end product. Likewise, breads made by sourdough technique had higher volatile compounds. The authors concluded that volatile compounds of breads could be enhanced by immobilized kefir on industrial waste (Plessas et al. 2008).

Another study focused on the microbiological, textural and physicochemical properties of sourdough bread by using freeze-dried *Lacticaseibacillus paracasei* (LP1 and LP2) and *Saccharomyces cerevisiae* (SC1 and SC2) as starter cultures (Aplevicz et al. 2014). It was found that bread volume was improved when LP and SC were used together compared to bread made with LP alone. Also, mixed culture of LP2 and SC1 had highest sensorial acceptability (Aplevicz et al. 2014). Recently, various cryoprotectans such as sucrose, trehalose and skim milk has been evaluated for their stabilization effect on microorganisms. A sourdough isolate *Latilactobacillus curvatus* N19 was freeze-dried with skim milk, lactose and sucrose, then fresh and freeze-dried forms stored at different temperatures (25 °C and 4 °C) were used in sourdough fermentation as starter cultures. The mixture of cryoprotectant skim milk, lactose and sucrose (20%/3.57%/10%) increased the survival of freeze-dried culture up to 98.6% (Gul et al. 2020). According to this study, low acidification occurred in fermentation with freeze-dried *Latilactobacillus curvatus* N19 stored at 25 °C, but the same pH value was observed in all dough samples at the end of the fermentation (Gul et al. 2020). In another research, a Brazilian grape sourdough isolate *Lacticaseibacillus paracasei* FNU was spray-dried with reconstituted skim milk and cheese whey as encapsulation material under 160 ± 2 °C inlet and 55–60 °C outlet air temperatures. As a result, 93.12 ± 2.15 g in 100 g cell viability were obtained by this encapsulation method (Ilha et al. 2015). In another research, sourdough microbial isolates *Limosilactobacillus fermentum* IAL 4541 and *Wickerhamomyces anomalus* IAL 4533 were encapsulated using different cryoprotectant agents, and encapsulated cells were freeze-dried at −50 °C under 0.15 mbar pressure for 24 h. Skim milk together with sodium glutamate resulted with 100% encapsulation efficiency and protected 87% of cells up to 1 year (Stefanello et al. 2019). Two mixtures from *Lacticaseibacillus paracasei* RN5, *Lactiplantibacillus plantarum* X2, *Lactobacillus brevis* LBRZ7, *Limosilactobacillus fermentum* LBRH10, *Lentilactobacillus buchneri* LBRZ6, *Fructilactobacillus sanfranciscensis* LSR and *Propionibacterium frendenreichii* ssp. *shermanii* NBIMCC 327 were prepared as sourdough starters. The starter culture mixture was immobilized in pectin-sodium alginate containing matrix and freeze-dried at −30 to −35 °C for 12–15 h. This immobilization step protected freeze-dried starter mixtures during 1 year storage up to 10^{12} CFU/g. When different concentrations of these freeze-dried starters were used to prepare wheat and rye sourdough breads, it was observed that bacterial spoilage was prevented until 96 h at room temperature and 37 °C by addition of 15% starter mixture whereas mold spoilage did not occur at 96 h and 120 h by addition of 15 and 20% starter mixture, respectively (Denkova et al. 2014).

One of the techniques used in pharmaceutical to produce capsules and tablets is granulation process and it is also used in agricultural, chemical and food industries (Pathare and Byrne 2011; Blaiotta et al. 2022). It is defined as "a size enlargement process in which small particles are formed into larger, physically strong agglomerates, in which primary particles are still distinguishable" (Pathare and Byrne 2011). It can be also applied for the delivery of probiotic microorganisms (Aponte et al. 2018). Although wet granulation has been preferred in pharmaceutical applications such as tablets and capsules, its incorporation to food industry is limited. In a novel study, lactic acid bacteria (LAB) strains were previously isolated from sourdough and used for wet granulated sourdough for the first time (Blaiotta et al. 2022). In the study, it was aimed to stabilize cell viability in sourdough matrix by using pharmaceutical grade corn starch (CS), microcrystalline cellulose and lactose monohydrate as a granulation matrix. For this purpose, different culture mixtures were granulated in different size. Resistance to the granulation process varied depending on the bacterial mixtures. Although the results were not significant, increased granule size had a positive effect on bacterial viability, and bacterial mixtures used in granulations affected the microbial load. After fresh and wet granulated cell mixtures were used for the sourdough fermentations, microorganism levels did not differ (Blaiotta et al. 2022).

Usage of waste products from various food applications as an immobilization matrix is one of the main strategies. A study used apple pomace for immobilization of sourdough isolate *Pediococcus acidilactici* LUHS29 (Bartkiene et al. 2017). Pomace was first suspended in 2 L water, sterilized and inoculated with 3% of *Pediococcus* culture. Immobilized cell culture was stabilized by spray-drying and used to prepare sourdough bread. After immobilization, *Pediococcus acidilactici* LUHS29 were detected as $7.95 \pm 0.03 \log_{10}$ CFU/g and maintained $7.0 \log_{10}$ CFU/g after 1 month storage at 20 °C (Bartkiene et al. 2017). The apple press cake obtained from the juice production as a waste has been evaluated for the improvement of antifungal activity on sourdough breads. After *Loigolactobacillus coryniformis* LUHS71, *Latilactobacillus curvatus* LUHS51, *Lentilactobacillus farraginis* LUHS206 and *Leuconostoc mesenteroides* LUHS225 strains were immobilized in apple press cake, it was used as a coating material of bread. Coating of sourdough bread with immobilized cells enhanced bread shelf-life and protected against mold formation (Bartkiene et al. 2019).

In a recent study, egg white protein nanoparticles were used with bioactive compounds, carvacrol, thymol and trans-cinnamaldehyde. After screening of their antifungal activities, these nanoencapsulated solution matrix were used as coating agents for the bread. While coating of bread did not alter the physicochemical properties of bread, it improved the shelf-life of bread due to microbiological analyses (Deseta et al. 2021).

9.2.2 Immobilized Enzymes Applicable to Sourdough

Several commercial enzymes have been produced by immobilization; however, their application on sourdoughs or producer strains previously isolated from sourdough is not explored until recently. Phytase is one of the significant enzymes in the baking industry. Recently, purified phytase enzyme which is significant to enhance nutritional value of sourdough breads has been immobilized on chitosan with iron nanoparticles. The application of immobilized phytase on green lentils, red lentils, peas, pinto beans, beans, brass, corn, dried corn, oat, rye, wheat, broad bean, chickpea and peanuts were evaluated. The results implied that the enzyme is more efficient on the degradation of phytic acid of dry beans (Onem and Nadaroglu 2018). Since these cereals and beans are the main ingredients of sourdoughs, this research shows potential applications of immobilized enzymes on bread-making (Onem and Nadaroglu 2018). Another study used beeswax as an encapsulation agent for α-amylase and used microencapsulated enzyme on gluten free bread formation (Haghighat-Kharazi et al. 2018). Due to the prevention of the degradation of enzyme during high acid and temperature conditions, gluten free breads had better textural qualities than free α-amylase added breads in terms of lower hardness and chewiness (Haghighat-Kharazi et al. 2018). In other research, addition of a novel α-amylase secreted in *Aspergillus oryzae* to dough enhanced bread volume and improved textural characteristic such as gumminess and cohesiveness. This enzyme was also immobilized using magnetic iron oxide nanoparticles and this application increased the tolerance of enzyme to temperature and low pH conditions (He et al. 2017). Glucose oxidase was encapsulated by using sodium alginate and chitosan. The microencapsulated enzyme was added to dough and examined for its effect on texture and sensory characteristic of Chinese steamed bread. It was found that microencapsulated enzyme decelerated oxidation of dough and this positively affected dough microstructure, textural and sensory properties of Chinese steamed bread. Also, addition of the enzyme leaded to longer dough development time, stability time and better gluten network formation (Zhang et al. 2018).

Singh and Satyanarayana (2008) studied the production of phytase by a thermophilic mold *Sporotrichum thermophile* in cane molasses medium and its application in bread manufacturing. Free and immobilized conidiospores in cane molasses medium were used for fermentations in shake flasks, stirred tank and air-lift fermenters. The presence of surfactant, e.g. Tween-80, found to enhance the phytase production in cane molasses medium. Ca-alginate support was suitable to immobilize conidiospores, and it was possible to use this support to produce sustained levels of phytase over five cycles. This phytase enzyme was applied to bread-making process; the bread contained reduced levels of phytic acid and a high-soluble phosphate. This thermostable enzyme was acid stable and protease insensitive, therefore it can find applications in food and feed industries.

Glucose oxidase (GO) enzyme was immobilized on chitosan (CS)-sodium tripolyphosphate (TPP) support with a high enzyme activity recovery (appr. 85%) and investigations have been carried out for the application of immobilized GO in order to improve bread-making quality based on wheat flour. Together with fungal alpha-amylase (62.5 U/kg flour), the obtained immobilized GO was also added (400 U/kg flour) and results were compared to the addition of $KBrO_3$ (60 mg/kg flour) as a control. Bread springiness slightly increased, specific volume of crumb increased by appr. 13.5% and hardness decreased by 19.00% compared to the control samples. Conclusions were withdrawn as that CSTPP-GO combined with fungal alpha-amylase showed potential application to improve bread-making quality of commercial wheat flour since slower release of H_2O_2 from CSTPP-GO was happened than free GO and GO was prevented from denaturing in dough. Gas retention properties were also increased with the application of immobilized GO, and fast oxidation of glucose oxidase did not happen during dough preparation. Therefore, dough rheological properties and bread quality were improved, and this makes immobilized GO enzyme a feasible alternative to $KBrO_3$ (Tang et al. 2014).

A unique antimicrobial approach was developed for the enhancement of the bread quality aspects and increasing the shelf-life of bread, e.g. against fungal growth. The hydrogen peroxide synthesis in bread through the glucose oxidase (GO) bioactivity, and oxidation of thiol protein bonds enhanced the dough rheology, volume and other characteristics by providing structural stability to the bread. GO enzyme purified from *Aspergillus niger* was immobilized on zinc oxide nanoparticles (ZnONPs) – GO/ZnONPs complex was created – and were employed during the mixing process of bread production. It was found that the use of GO/ZnONPs improved bread quality significantly, e.g. bread height, crumb color and volume, by more than 30% when compared with the controls. Microbiological quality of bread treated with the immobilized GO enzyme during the shelf-life was extended at a great extent. It was suggested that GO/ZnONPs bioconjugate had an industrial potential and might be also applied to other food products as well in order to extend the shelf-life (Khan et al. 2022).

9.2.3 Dried Sourdough Powders

In addition, sourdough powder production is necessary for domestic and industrial applications. Dried sourdoughs containing inactive yeasts and LAB are shelf-stable products (Brandt 2007; Santivarangkna et al. 2008). Drying processes strongly affect the flavor, shelf-life and microbial viability of the end product (Albagli et al. 2021). Limited number of studies explored the effect of drying process on sourdough. Although mostly freezing and freeze-dried sourdough applications are popular, high cost and energy requirements are the main disadvantages. Therefore, spray-drying, fluidized bed-drying and vacuum-drying techniques were explored for the production of dry sourdoughs (Reale et al. 2019). Freeze-drying method protects the cell number and sensory properties but requires cryoprotectants, high energy and more time. On the other hand, spray-drying is faster than freeze-drying, and does not require the use of cryoprotectants (Albagli et al. 2021).

During the sourdough immobilization, microbial viability can be decreased due to the process conditions. The drying process reduced the viable yeast and LAB levels in dried sourdough powders. One of the latest studies used both spray- and freeze-drying methods to improve microbiological and physicochemical properties of sourdough. Spray-drying was applied at 160 ± 5 °C and 90 ± 4 °C inlet and outlet temperatures, respectively, whereas freeze-drying was applied at −80 °C for 1 day. The LAB and yeast levels of fresh sourdough was approximately 8.7 and 8.6 log CFU/g, respectively. These numbers were detected as 8 log CFU/g and 4.9–5.0 log CFU/g in freeze-dried sourdough powders and spray-dried sourdough powders, respectively (Caglar et al. 2021). Another study found that spray-drying conditions (180 °C inlet and 90 °C outlet air temperatures) caused a reduction from 10^9 to 10^5 CFU/g in sourdough LAB viability (Tafti et al. 2013b). On the contrary, a comparative study on drying methods showed the difference of cell viability between spray-drying (different inlet (130–140 °C) and outlet air temperatures (54–75 °C)), drying (40 °C for 48 h under vacuum condition, −87 kPa), freezing (−18 °C for 24 h) and freeze-drying (−40 °C, under vacuum of 15 kPa for 48 h). The authors found that freezing reduced the cell viability up to 4 log, and spray-drying was detected as the best method for the protection of microbial culture in sourdough. They also concluded that inlet and outlet temperatures of spray-drying were also crucial to protect cells from thermal damage (Reale et al. 2019). In addition, the impacts of freeze-drying, spray-drying and oven-drying were evaluated in spontaneously and starter fermented sourdoughs. The oven-drying process at 37 ± 1 °C, freeze-drying process at −68 °C and spray-drying process at 135 ± 5 °C until moisture decreased to 4–5% in order to investigate the impact of drying on the number of LAB. Spray-drying protected LAB viability much better whereas oven-drying reduced LAB level because of the long-term drying (Ertop et al. 2018).

Since low temperatures can cause cell death, cryoprotective agents are needed to preserve viable microbial load of sourdough powders. Several sugar compounds such as trehalose, sucrose, lactose or skim milk as a diverse matrix have been used for the best results (Stefanello et al. 2018; Gul et al. 2020). The difference in microorganism viability with the addition of 0, 10 or 15% trehalose to sourdoughs prior to freeze-drying step (−80 °C) were studied by Stefanello et al. (2018). The results showed that while the number of microorganisms was preserved at a rate of 67% in the samples without trehalose, it was preserved at a rate of 81% in the samples with 15% trehalose, due to the inhibitory effect of trehalose on the development of ice crystals in the cell (Stefanello et al. 2018).

9.3 Immobilization Applications to Preserve Nutritional Value and Health Effects of Sourdough

Sourdough takes attention due to its nutritional features and beneficial health effects by increasing phenolics and sterols and decreasing anti-nutritive factors (Olojede et al. 2020; Fekri et al. 2020). As stated in Fig. 9.2, it improves the bioavailability of

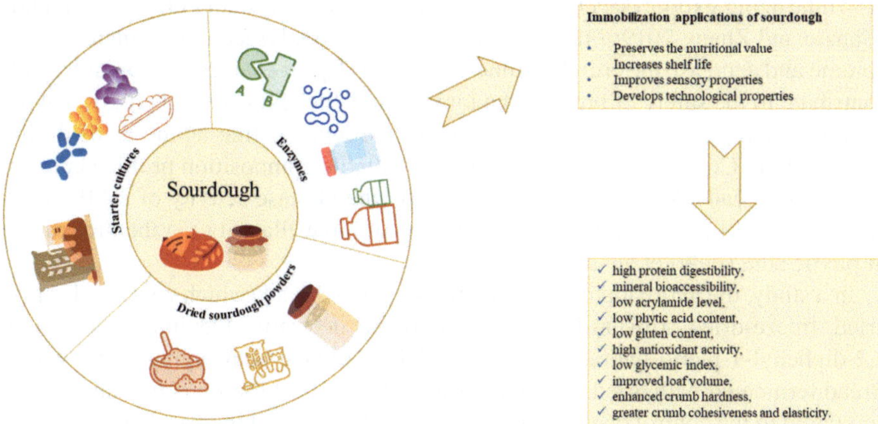

Fig. 9.2 Effect of immobilization applications on sourdough

some nutrients such as dietary fibers, vitamins and minerals (Leenhardt et al. 2005). Sourdough consumption can regulate gut microbiota and reduce irritable bowel syndrome symptoms. In addition, sourdough fermented bread intake resulted in less gas production and a low number of sulfate-reducing bacteria in irritable bowel syndrome donors (Costabile et al. 2014). Protein digestibility and release of bioactive peptides were higher in sourdough bread when compared to the baker's yeast fermented bread (Rizzello et al. 2019). Likewise, the phytase activity of sourdough LAB causes an increment of protein, amino acid and mineral bioavailability via the breakdown of phytic acid (Fekri et al. 2020). In addition, sourdough fermentation decreases the glycemic index of bakery products such as bread by lowering starch digestibility. During sourdough fermentation, the synthesis of free amino acids and their derivatives mask the lower salt content, which is important for patients with high blood pressure (Gobbetti et al. 2019). Sourdough-fermented bakery consumption becomes safe for celiac patients since gluten has been hydrolyzed during the fermentation (Gobbetti et al. 2014). In recent decades, sourdough has been used commonly for natural, tasty, healthy and nutritious foods (Palla et al. 2020).

Sugar-derived carbonyl complexes and asparagine form acrylamide during the Maillard reaction (Keramat et al. 2011). Sourdough fermented bread showed a lower acrylamide content when compared to the control bread (Forstova et al. 2014). When dried sourdough powders were used in lower amounts in bread-making, the acrylamide level of the bread was found to be lower (Brandt 2019). In a study, immobilization of *Pediococcus acidilactici* LUHS29 in apple pomace was conducted to use in barley sourdough fermented bread production to increase bread's functional value. The immobilized cell application resulted in high phenolic content and antioxidant activity, low acrylamide level, and retarded bread staling (Bartkiene et al. 2017).

Sourdough microbiota mainly consists of LAB (and yeasts). This bacterial group shows positive effects on flavor, texture, nutritional qualities and spoilage because of synthesizing exopolysaccharides, organic acids, enzymes and other compounds (Gänzle and Zheng 2019; Fraberger et al. 2020). Exopolysaccharides increase loaf volume and decrease the crumb hardness of bread (İspirli et al. 2020). Organic acids contribute to the safety of bread and a longer shelf-life (Plessas et al. 2020). LAB's peptidase and proteinase action can increase the level of free amino acids during fermentation. Consequently, enhancement of volatile composition profile occurs in dough and bread (Pétel et al. 2017). Likewise, the phytase activity of LAB causes an increment of protein, amino acid and mineral bioavailability via the breakdown of phytic acid (Fekri et al. 2020).

In a study, the antioxidative activity of sourdough bread (which contained spray-dried, freeze-dried or oven-dried sourdough) was determined as inhibition (%) by 2,2-diphenyl-1-picryl-hydrazyl-hydrate (DDPH) radical scavenging activity test. Bread fermented with dried sourdough showed higher inhibition percentages when compared to the control one. In addition, the inhibition level increased with increasing concentrations of supplemented dried sourdough (Ertop et al. 2018; Ertop and Tuğkan 2018). The antioxidative activity can alter phytochemicals during fermentation and LAB metabolism (Ertop et al. 2018; Vermeulen et al. 2007). Also, the crusts of bakery goods show antioxidant properties (Michalska et al. 2007). Proteolysis and acidification interaction occurs during sourdough fermentation and, thus, bioactive peptides are formed. These peptides also show antioxidant activity (Rizzello et al. 2008). Likewise, some sourdough LAB can synthesize antioxidant peptides during fermentation (Coda et al. 2012). According to a previous study, using different flours such as rice, spelt, wheat and wholemeal raised the free phenolic acid level and antioxidant activity (Di Nunzio et al. 2018).

There are limited studies on the preservation of the health effects and nutritional value of sourdough by immobilization. According to previous studies, sourdough exerts high nutritional value with unique sensory characteristics and health benefits. The application of an immobilization process can result in high protein digestibility, mineral bioaccessibility, low acrylamide level, low phytic acid content, high antioxidant activity and low glycemic index.

Sugar-derived carbonyl complexes and asparagine forms acrylamide during the Maillard reaction (Keramat et al. 2011). Sourdough fermented bread showed a lower acrylamide content when compared to the control bread (Forstova et al. 2014). In bread-making, the dried sourdough powders were applied in lower amounts, thus the resulting bread's acrylamide level would be lower (Brandt 2019).

9.4 Immobilization Applications to Develop Technological Properties of Sourdough

Sourdough fermentation, which is an ancient biotechnology, continues to gain interest because of its positive influences on the technological attributes of bakery goods. Sourdough fermented bakery products have improved loaf volume and crumb hardness. In gluten-free bread-making, sourdough fermentation plays a crucial role in improving the texture (Nionelli and Rizzello 2016).

Specific volume is one of the important parameters of the visual quality of bread. Based on previous studies, immobilized sourdough addition can increase the specific loaf volume (Clarke et al. 2002) or reduce the volume level (Gezginc and Kara 2019; Novotni et al. 2017). Lower pH conditions during sourdough fermentation cause inhibition of carbon dioxide production by baker's yeast, resulting in harder bread crumbs and lower volume (Rozyło et al. 2016). A study showed that dry sourdough addition increased the hardness of bread samples (Caglar et al. 2021). Due to the immobilization process, LAB and yeasts are stressed, injured and resulting in a lower cell counts (Reale et al. 2019). Therefore, rehydrating dry sourdough and refreshing LAB and yeasts may be beneficial to ensure that they adequately show their activity during sourdough fermentation (Caglar et al. 2021).

A novel study showed that 1–2% dry sourdough addition to wheat flour developed the physical characteristics of bread such as reducing gumminess, firmness, resilience, and cohesiveness; darkening the bread crust and crumb, as well as enhanced the sensory characteristics of bread (Codina et al. 2021). Spray-dried and frozen sourdough fermented bread slices demonstrated homogeneous textures (Reale et al. 2019). In previous studies, freeze-dried sourdoughs were used to produce gluten-free bread. The use of freeze-dried amaranth and buckwheat sourdough caused a reduction in crumb hardness (Rozyło et al. 2015). The freeze-dried rice sourdough fermented bread showed greater crumb cohesiveness and elasticity (Rozyło et al. 2016).

Leuconostoc mesenteroides LUHS225, *Loigolactobacillus coryniformis* LUHS71, *Lentilactobacillus farraginis* LUHS206, and *Latilactobacillus curvatus* LUHS51 strains were immobilized into apple by-products to use in sourdough bread-making. Based on the findings, the sourdough bread volume and porosity of the crumb were improved, and acrylamide content was reduced depending on the LAB strains (Bartkiene et al. 2019). In an earlier study, kefir grains were immobilized on corn and the immobilized cells were used for sourdough fermentation. The resulted bread showed high loaf volume and desirable sensory properties (Plessas et al. 2012).

The use of immobilized sourdough was used in bakery products both to shorten the fermentation time and to see the effect of dry sourdough on bread. According to the results of the studies carried out, bread texture and crust hardness varied according to the immobilized sourdough used.

In a novel study, immobilization of sourdough was carried out using four different methods: drying, freezing, freeze-drying and spray-drying. When all immobilized sourdough powders were used in sourdough production, dough leavening was delayed due to the long lag-phase duration (Reale et al. 2019). For that reason, immobilized sourdoughs are not applicable to inoculate directly in bread-making. During the stabilization process, the microbial levels decreased, microorganisms were injured and their metabolic activities were restricted (Lattanzi et al. 2014). Therefore, some pretreatments (e.g., rehydration of dried sourdough powders, refreshment of microorganisms) are needed to activate starter sourdough (Caglar et al. 2021). Moreover, spray-dried and frozen sourdough fermented bread slices demonstrated homogeneous textures. But dried sourdough fermented ones showed a nonuniform crumb structure. Based on the digital image analysis, bread slices illustrated compact and unleavened structures because of the low leavening power of frozen and dried sourdoughs (Reale et al. 2019).

The findings of Tafti et al. (2013a) showed that water absorption of dough increased with the addition of sourdough powders; however, another study demonstrated water absorption was not influenced by the freeze-dried or spray-dried sourdough addition (Caglar et al. 2021). That can be explained by the differentiation of sourdough drying process conditions and added sourdough powder concentration. In addition, the increasing concentration of freeze-dried and spray-dried sourdough powders addition caused a reduction in dough stability and an increment in the dough's degree of softening (Caglar et al. 2021). Low pH values of added sourdough powders resulted in a weak gluten structure (Nogueira et al. 2015) and therefore caused dough softening (Arendt et al. 2007).

9.5 Immobilization Applications to Improve Sensory Features of Sourdough

Sensory attributes of products are the most crucial parameters affecting their acceptance by consumers. The flavor and aroma of bread are influenced by raw materials and ingredients, enzymatic reactions during sourdough fermentation, and thermal reactions such as Maillard reactions and caramelization during baking (Cho and Peterson 2010).

Fermentation with spray-dried sourdough enhanced the sensorial characteristics of bread (Tafti et al. 2013a; Tafti et al. 2013b). The spray-dried sourdough powders provide stable sourdough starter cultures and help to create better-producing conditions (Kulp and Lorenz 2003). In a study, trehalose was used as a cryoprotectant to protect the predominant microbiota of sourdough (Stefanello et al. 2018). Because of the unreactive sugar profile, trehalose did not change the sensory properties of sourdough. Previously performed sensory analysis, trehalose showed no effect on the aroma and flavor intensity levels of the bread (Stefanello et al. 2018).

In a current study, the addition of different dried sourdough powders (freeze-dried and spray-dried) resulted in color differentiation in the bread samples (Caglar et al. 2021). These findings agreed with Ertop et al. (2018) that the drying methods for sourdough immobilization (spray-drying, freeze-drying and oven-drying) affected the color of powders. Melanoidin pigments occurred during the oven drying process due to the Maillard reaction. Thus, the color of dried sourdough was seen as significantly different from spray-dried and freeze-dried sourdoughs. On the other hand, in spray-drying, color-related chemical reactions were inhibited and, concomitantly, the white color remained (Ertop et al. 2018).

According to Reale et al. (2019), in bread-making, spray-dried and freeze sourdough application demonstrated good quality, but dried and freeze-dried sourdough usage resulted in a reduced volume. Spray-dried sourdough fermented bread was found more acceptable in appearance, flavor and texture than dried and freeze-dried sourdough fermented bread (Reale et al. 2019).

Specific volume is a parameter to show the visual quality of bread. Spray-dried sourdough containing fermented bread showed higher volume, and specific volume than the same level of freeze-dried sourdough powder containing bread. Increasing the level of spray-dried and freeze-dried sourdough powder addition caused a reduction in the specific volume of bread (Caglar et al. 2021). Because sourdough fermentation resulted in lower acidity that caused inhibition of the carbon dioxide production by yeasts, harder crumbs and lower volume occurred in the bread (Rozyło et al. 2016). The bread hardness increased with the decreasing volume (Clarke et al. 2002; Komlenić et al. 2010). The wheat bread containing the dried sourdough powder showed a higher level of hardness when compared to the control bread (produced without dried sourdough powder) (Caglar et al. 2021). In a previous study, the volumes of dried sourdough contained bread samples were found to be higher when compared to the control bread. The dried sourdough showed improving impacts due to the microbial interaction and fermentation. The highest volume of bread was obtained by adding oven-dried sourdough (Ertop et al. 2018). Freeze-dried amaranth and buckwheat sourdough addition improved bread volume and lowered crumb hardness; larger pores were seen in the bread crumb (Rozyło et al. 2015).

Acetic acid plays a crucial role in sourdough bread quality, such as shelf-life, rope inhibition, and flavor. Due to the sourdough drying techniques (spray-drying, lyophilization, drum-drying, spray granulation and fluidized bed-drying), dried sourdough may have a low level of acetic acid when compared to fresh sourdough (Brandt 2007). One of the drawbacks of freeze-drying is the loss of volatile compounds (Kirchhoff and Schieberle 2001). Dried sourdough gives a malty and buttery sensorial character to the bread compared to the fresh sourdough. On the other hand, fresh sourdough gives a more acidic flavor than dried sourdough (Brandt 2007). Moreover, dried sourdough powders can be produced from alternative flours to improve the taste and aroma of gluten-free bakery products (Brandt 2019). Based on the findings of a study, lyophilized rice sourdough improved the texture characteristics of gluten-free bread (Rozyło et al. 2016).

9.6 Immobilization Applications to Increase the Shelf-Life of Sourdough

Sourdough immobilization can be important to delay staling and prevent mold growth. During the freeze-drying process, addition of trehalose showed cryoprotective properties for LAB (Stefanello et al. 2018). LAB produce acids during fermentation, this helps to prevent the development of undesired microorganisms and increase the sensorial properties of the products (Leroy and De Vuyst 2004). While the pH value of fresh sourdough was found as 3.5, the pH value increased after freeze- and spray-dried processes. Titratable acidity of sourdoughs also increased after drying processes (Caglar et al. 2021). The pH value of sourdough that can be added to the bread dough should be about 4.0–4.5 (Torrieri et al. 2014). Before and after the drying process, the pH values of dried sourdoughs remained constant and available level to produce bread dough (Ertop et al. 2018).

According to the Caglar et al. (2021) study, dough samples were produced with different concentrations of freeze-dried and spray-dried sourdough powders. As the number of powder levels increased, the pH values of dough, degree of staling and specific volume decreased, and titratable acidity and hardness increased in dough samples. Additionally, the pH of resulting bread samples decreased with the increase in sourdough powder addition. Likewise, titratable acidity increments and pH reduction in bread samples were associated with the amount of dried sourdoughs used (Ertop et al. 2018).

In a study, the spray-dried sourdough powder was used to produce Sangak bread (which is an Iranian flat bread and is produced from wholemeal wheat flour). The dough and bread, including dried sourdough, had lower pH levels and higher titratable acidity values than the control dough and bread. Additionally, a higher level of spray-dried sourdough addition resulted in lower pH levels and higher titratable acidity values (Tafti et al. 2013b).

The moisture level of sourdough powders is vital to eliminate microbial growth and increase shelf-life; below 5% moisture content was stated as safe for the storage stability (Chavez and Ledeboer 2007). Previous studies have shown that the moisture levels of sourdough powders vary depending on the method used. A recent study found that the moisture level of freeze-dried sourdough powders (2.8%) was lower than that of spray-dried sourdough powders (5.2%) (Caglar et al. 2021). However, a previous study observed lower moisture content (4.8%) in spray-dried sourdough, while oven-drying and freeze-drying techniques resulted in moisture levels of more than 5% (Ertop et al. 2018). In another study, the moisture level of spray-dried sourdough was found to be 3.72% (Tafti et al. 2013b). In bakery products, moisture level and moisture retention are crucial during storage. Slow moisture loss is expected in bakery goods (Ertop et al. 2018). Based on the findings of the study, while control bread loses moisture rapidly, dried sourdough added fermented bread samples lost moisture slowly. The moisture content of all bread at the beginning and end of the storage period was higher compared to the control bread. In brief, dried sourdough showed the ability to slow the loss of moisture in bread (Ertop et al. 2018).

Bulk density affects the physical features of dried sourdough powder, such as wettability, instantizing and dispersibility. The bulk density of spray-dried sourdough was 0.6 g/cm^3 (Tafti et al. 2013b). Similarly, in a novel study, the bulk density values of spray-dried and freeze-dried sourdoughs varied from 0.525–0.775 g/cm^3 and 0.45–0.7 g/cm^3, respectively (Caglar et al. 2021). These values are in the range of skim milk powders.

It was previously reported that sourdough showed anti-stalling activity on white bread (Corsetti et al. 2008; Torrieri et al. 2014). For this reason, dried sourdough powder addition caused a lower staling degree in wheat bread samples produced by freeze-dried and spray-dried sourdough fermentation (Caglar et al. 2021). Sourdough LAB demonstrate proteolytic activity and cause acidic conditions, thus delay the bread staling. In addition, this microbial group affects the retrogradation of starch molecules by synthesizing enzymes and slows the bread staling rate (Arendt et al. 2007). α-Amylase has been found to disrupt the outer amylopectin branches, thereby producing too short chains of amylopectin to crystallize. This prevented the persistence of the amylopectin network, and staling was delayed (Goesaert et al. 2009).

Spray-dried sourdough added Sangak bread showed slower staling compared to the control bread fermented without sourdough. The control bread started to stale on the first day and showed a more rigid texture than bread containing dried sourdough. While 9% dried sourdough addition gave good sensorial attributes to bread during storage, 12% addition caused a negative influence (Tafti et al. 2013b).

Freeze-dried sourdough used in wheat germ fermentation resulted in the inhibition of lipase activity and increase of phytase activity in comparison with raw wheat germ. After sourdough fermentation, shelf-life of wheat germ increased and raffinose concentration and volatile compound level decreased (Rizzello et al. 2010). Due to the production of antifungal compounds by sourdough LAB, freeze-dried sourdough fermented wheat germ showed antifungal activity and an anti-stalling effect (Rizzello et al. 2011).

While wheat bread without dried sourdough showed mold growth after 3 days of storage, mold growth with the addition of dried sourdough was detected after 5 days. Considering that bread consumption can take 5 days after production, the use of dried sourdough is a practical way to extend the shelf-life of bread. Furthermore, a high level of dried sourdough addition (12%) was more effective against mold growth (Ertop et al. 2018). On the contrary, dried sourdough addition demonstrated no impact against molds, all bread samples were spoiled 3 days later (Tafti et al. 2013b). The high moisture level in the bread crumb and the proper temperature during storage can provide favorable conditions for mold growth.

9.7 Future Trends

Sourdough applications show many advantages: increased nutritional attributes decreased bread staling, lowered glycemic index, enhanced organoleptic characteristics, resistance against fungal spoilage and lowered chemical preservation (Plessas 2021). Considering these advantages, it will be essential to apply immobilization techniques to sourdough so that the positive effects of sourdough can continue, improve shelf-life and ensure its use by the commercial sector.

According to current market research, it is predicted that the sourdough market will reach 5.3 billion USD by 2030. Additionally, it is estimated to have a Compound

Annual Growth Rate (CAGR) of 8% during this time frame (Market Research Future 2022). Another market survey indicated that the sourdough market would grow at a CAGR of 9.37% to reach $2.36 billion between 2022 and 2026 (Research and Market 2022). In this regard, the sourdough market is a promising and growing industry. Dried sourdough preserving its nutritional, technological and sensory properties, can be quickly sold at markets and thus can be reachable to all consumers.

Sourdough is traditionally used in producing various bread, pizza, pasta, etc. Additionally, sourdough draws attention to functional food production with its nutritional values and unique organoleptic properties. It also plays an essential role in improving the texture properties of gluten-free bread. Since dried sourdough rather than fresh sourdough is more accessible to everyone, consumers are more likely to use it to produce both traditional and functional foods.

9.8 Conclusion

Immobilization of sourdough has several advantages over fresh sourdoughs, such as stable product quality, longer shelf-life and lower shipping cost. This chapter explained the immobilization techniques, mainly drying methods, and showed the possibility of dried sourdough production for bread-making. The effect of dried sourdough addition to dough and bread samples was investigated, and different literature analysis results came together within this scope. Based on the findings of studies, dried sourdough powders positively affect the nutritional value, technological properties, sensory characteristics, and shelf-life of the bread. On the other hand, due to the process condition of drying techniques, the viability of microorganisms is strongly influenced. The cryoprotectant addition provides high survivability in the freeze-drying method, but cryoprotectants increase the cost. Spray-drying method gives higher cell viability with more accessible application compared to freeze-drying. In general, sourdough microflora is diverse, and drying processes are different to protect microorganisms, so there is a need to investigate different drying conditions and starter cultures to obtain the best characteristics in the resulted foods. The limited data on the literature indicates that immobilization of enzymes in sourdough product is another area waiting to be enlightened.

References

Albagli G, do Monte Schwartz I, Amaral PF, Ferreira TF, Finotelli PV (2021) How dried sourdough starter can enable and spread the use of sourdough bread. LWT-Food Sci Technol 149:111888

Aplevicz KS, da Silva T, Fritzen-Freire CB, Amboni RD, Barreto PL, Sant'Anna, E. S. (2014) Effect of the incorporation of different freeze-dried cultures on the properties of sourdough bread. J Culin Sci Technol 12(4):354–367

Aponte M, Ungaro F, d'Angelo I, De Caro C, Russo R, Blaiotta G, Dal Piaz F, Calignano A, Miro A (2018) Improving in vivo conversion of oleuropein into hydroxytyrosol by oral granules containing probiotic *Lactobacillus plantarum* 299v and an *Olea europaea* standardized extract. Int J Pharm 543(1–2):73–82

Arendt EK, Ryan LAM, Bello FD (2007) Impact of sourdough on the texture of the bread. Food Microbiol 24:165–174

Bartkiene E, Vizbickiene D, Bartkevics V, Pugajeva I, Krungleviciute V, Zadeike D, Zavistanaviciute P, Juodeikiene G (2017) Application of *Pediococcus acidilactici* LUHS29 immobilized in apple pomace matrix for high value wheat-barley sourdough bread. LWT-Food Sci Technol 83:157–164

Bartkiene E, Bartkevics V, Lele V, Pugajeva I, Zavistanaviciute P, Zadeike D, Juodeikiene G (2019) Application of antifungal lactobacilli in combination with coatings based on apple processing by-products as a bio-preservative in wheat bread production. J Food Sci Technol 56(6):2989–3000

Bié J, Sepodes B, Fernandes PCB, Ribeiro MHL (2022) Enzyme immobilization and co-immobilization: main framework, advances and some applications. PRO 10:494

Blaiotta G, Romano R, Trifuoggi M, Aponte M, Miro A (2022) Development of a wet-granulated sourdough multiple starter for direct use. Foods 11(9):1278

Brandt MJ (2007) Sourdough products for convenient use in baking. Food Microbiol 24(2):161–164

Brandt MJ (2019) Industrial production of sourdoughs for the baking branch–an overview. Int J Food Microbiol 302:3–7

Caglar N, Ermis E, Durak MZ (2021) Spray-dried and freeze-dried sourdough powders: properties and evaluation of their use in bread making. J Food Eng 292:110355

Chan L, Cross HF, She JK, Cavalli G, Martins HFP, Neylon C (2007) *PLoS One* 2(11):e1164

Chavez BE, Ledeboer AM (2007) Drying of probiotics: optimization of formulation and process to enhance storage survival. Dry Technol 25:1193–1201

Cho H, Peterson DG (2010) Chemistry of bread aroma: a review. Food Sci Biotechnol 19(3):575–582

Clarke CI, Schober TJ, Arendt EK (2002) Effect of single strain and traditional mixed strain starter cultures on rheological properties of wheat dough and on bread quality. Cereal Chem 79:640–647

Coda R, Rizzello CG, Pinto D, Gobbetti M (2012) Selected lactic acid bacteria synthesize antioxidant peptides during sourdough fermentation of cereal flours. Appl Environ Microbiol 78:1087–1096

Codină GG, Sarion C, Dabija A (2021) Effects of dry sourdough on bread-making quality and acrylamidecontent. Agronomy 11(10):1977

Corsetti A, Gobbetti M, Balestrieri F, Paoletti F, Russi L, Rossi J (2008) Sourdough lactic acid bacteria effects on bread firmness and stalin. J Food Sci 63:347–351

Costabile A, Santarelli, S, Claus SP, Sanderson J, Hudspith BN, Brostoff J, Gibson, GR (2014) Effect ofbreadmaking process on in vitro gut microbiota parameters in irritable bowel syndrome. PLoS One 9(10):e111225

Denkova R, Georgieva L, Denkova Z (2014) Freeze-dried sourdough starters. J Food Packag Sci Technique Technol 4(1):74–80

Deseta ML, Sponton OE, Erben M, Osella CA, Frisón LN, Fenoglio C, Piagentini AM, Santiago L, G., & Perez, A. A. (2021) Nanocomplexes based on egg white protein nanoparticles and bioactive compounds as antifungal edible coatings to extend bread shelf-life. Food Res Int 148:110597

DiCosimo R, McAuliffe J, Poulose AJ, Bohlmann G (2013) Industrial use of immobilized enzymes. Chem Soc Rev 42(15):6437–6474

Dwevedi A (2016) Basics of enzyme immobilization, Chapter 2, pp 21–44 in Enzyme immobilization advances in industry, agriculture, medicine, and the environment, pg. 132

Ertop MH, Tuğkan Şİ (2018) Optimization of the amount of chickpea sourdough and dry yeast in wheat bread formulation: evaluation of physicochemical, sensory and antioxidant properties. Food Sci Technol Res 24(1):45–53

Ertop MH, İlter SM, Yilmaz F, Baltaci C, Gündoğdu A (2018) Quality properties of wheat breads incorporated with dried sourdoughs produced with different fermentation and drying methods. Food Sci Technol Res 24(6):971–980

Eş I, Vieira JDG, Amaral AC (2015) Principles, techniques, and applications of biocatalyst immobilization for industrial application. Appl Microbiol Biotechnol 99(5):2065–2082

Fekri A, Torbati M, Khosrowshahi AY, Shamloo HB, Azadmard-Damirchi S (2020) Functional effects of phytate-degrading, probiotic lactic acid bacteria and yeast strains isolated from Iranian traditional sourdough on the technological and nutritional properties of whole wheat bread. Food Chem 306:125620

Forstova V, Belkova B, Riddelova K, Vaclavik L, Prihoda J, Hajslova J (2014) Acrylamide formation in traditional Czech leavened wheat-rye breads and rolls. Food Control 38:221–226

Fraberger V, Unger C, Kummer C, Domig KJ (2020) Insights into microbial diversity of traditional Austrian sourdough. LWT-Food Sci Technol 127:109358

Gänzle MG, Zheng J (2019) Lifestyles of sourdough lactobacilli–do they matter for microbial ecology and bread quality? Int J Food Microbiol 302:15–23

Gezginc Y, Kara Ü (2019) The effect of exopolysaccharide producing Lactobacillus plantarum strain addition onsourdough and wheat bread quality. Foods is Qual. Assur. Saf 11(1):95–106

Gobbetti M, Rizzello CG, Di Cagno R, De Angelis M (2014) How the sourdough may affect the functionalfeatures of leavened baked goods. Food Micro 37:30–40

Gobbetti M, De Angelis M, Di Cagno R, Calasso M, Archetti G, Rizzello CG (2019) Novel insights onthe functional/nutritional features of the sourdough fermentation. Int J Food Micro 302:103–113

Goesaert H, Leman P, Bijttebier A, Delcour JA (2009) Antifirming effects of starch degrading enzymes inbread crumb. J Agri Food Chem 57(6):2346–2355

Guisan JM (2013) Immobilization of enzymes and cells, 3rd edn. Humana Press, Springer Protocols, p 377

Gul LB, Con AH, Gul O (2020) Storage stability and sourdough acidification kinetic of freeze-dried *lactobacillus curvatus* N19 under optimized cryoprotectant formulation. Cryobiology 96:122–129

Haghighat-Kharazi S, Milani JM, Kasaai MR, Khajeh K (2018) Microencapsulation of α-amylase in beeswax and its application in gluten-free bread as an anti-stalling agent. LWT-Food Sci Technol 92:73–79

Hassan ME, Yang Q, Xiao Z, Liu L, Wang N, Cui X, Yang L (2019) Impact of immobilization technology in industrial and pharmaceutical applications. *3*. Biotechnology 9(12):1–16

He L, Mao Y, Zhang L, Wang H, Alias SA, Gao B, Wei D (2017) Functional expression of a novel α-amylase from Antarctic psychrotolerant fungus for baking industry and its magnetic immobilization. BMC Biotechnol 17(1):1–13

Ilha EC, Da Silva T, Lorenz JG, de Oliveira Rocha G, Sant'Anna ES (2015) *Lactobacillus paracasei* isolated from grape sourdough: acid, bile, salt, and heat tolerance after spray drying with skim milk and cheese whey. Eur Food Res Technol 240(5):977–984

Imam HT, Marr PC, Marr AC (2021) Enzyme entrapment, biocatalyst immobilization without covalent attachment. Green Chem 23:4980–5005

İspirli H, Özmen D, Yılmaz MT, Sağdıç O, Dertli E (2020) Impact of glucan type exopolysaccharide (EPS) production on technological characteristics of sourdough bread. Food Control 107:106812

Jesionowski T, Zdarta J, Krajewska B (2014) Enzyme immobilization by adsorption: a review. Adsorption 20:801–821

Keramat J, LeBail A, Prost C, Jafari M (2011) Acrylamide in baking products: a review article. Food Bioprocess Technol 4:530–543

Khan J, Khurshid S, Sarwar A, Aziz T, Naveed M, Ali U, Makhdoom SI, Nadeem AA, Khan AA, Sameeh MY, Alharbi AA, Filimban FZ, Rusu AV, Göksen G, Trif M (2022) Enhancing bread quality and shelf-life via glucose oxidase immobilized on zinc oxide nanoparticles—A sustainable approach towards food safety. Sustain For 14:14255

Kirchhoff E, Schieberle P (2001) Determination of key aroma compounds in the crumb of a three-stage sourdough rye bread by stable isotope dilution assays and sensory studies. J Agric Food Chem 49:4304–4311

Komlenić DK, Ugarčić-Hardi Ž, Jukić M, Planinić M, Bucić-Kojić A, Strelec I (2010) Wheat dough rheology and bread quality effected by *lactobacillus brevis* preferment, dry sourdough and lactic acid addition. Int J Food Sci Technol 45(7):1417–1425

Kosseva MR (2011) Immobilization of microbial cells in food fermentation processes. Food Bioprocess Technol 4(6):1089–1118

Kulp K, Lorenz K (eds) (2003) Handbook of dough fermentations, vol 127. CRC Press

Lattanzi A, Minervini F, Gobbetti M (2014) Assessment of comparative methods for storing type-I wheat sourdough. LWT-Food Sci Technol 59(2):948–955

Leenhardt F, Levrat-Verny MA, Chanliaud E, Rémésy C (2005) Moderate decrease of pH by sourdoughfermentation is sufficient to reduce phytate content of whole wheat flour through endogenous phytase activity. J Agric Food Chem 53(1):98–102

Leroy F, De Vuyst L (2004) Lactic acid bacteria as functional starter cultures for the food fermentation industry. Trends Food Sci Technol 15:67–78

Maghraby YR, El-Shabasy RM, Ibrahim AH, Azzazy HME (2023) Enzyme immobilization technologies and industrial applications. ACS Omega 8:5184–5196. https://doi.org/10.1021/acsomega.2c07560

Mantzourani I, Terpou A, Alexopoulos A, Bezirtzoglou E, Plessas S (2019) Assessment of ready-to-use freeze-dried immobilized biocatalysts as innovative starter cultures in sourdough bread making. Foods 8(1):40

Market Research Future (2022) Sourdough Market Research Report – Global Forecast till 2030. Available at: https://www.marketresearchfuture.com/reports?utf8=%E2%9C%93&q=Sourdough+Market+Research+Report+-+Global+Forecast+till+2030 Access on: May, 2022

Michalska A, Ceglinska A, Amarowicz R, Piskula MK, SzawaraNowak D, Zielinski H (2007) Antioxidant contents and Antioxidative properties of traditional Rye breads. J Agric Food Chem 55:734–740

Mitropoulou G, Nedovic V, Goyal A, Kourkoutas Y (2013) Immobilization technologies in probiotic food production. J Nutr Metab 2013:1–15

Mohamad NR, Marzuki NHC, Buang NA, Huyop F, Wahab RA (2015) An overview of technologies for immobilization of enzymes and surface analysis techniques for immobilized enzymes. Biotechnol Biotechnol Equip 29(2):205–220

Mohd Roby BH, Muhialdin BJ, Abadl MMT, Mat Nor NA, Marzlan AA, Lim SAH, Mustapha NA, Meor Hussin AS (2020) Physical properties, storage stability, and consumer acceptability for sourdough bread produced using encapsulated kombucha sourdough starter culture. J Food Sci 85(8):2286–2295

Moo-Young M (2019) Comprehensive biotechnology, 3, Elsevier Publisher, pg. 4420

Nguyen HH, Kim M (2017) An overview of techniques in enzyme immobilization. Appl Sci Converg Technol 26(6):157–163

Nionelli L, Rizzello CG (2016) Sourdough-based biotechnologies for the production of gluten-free foods. Foods 5(3):65

Nogueira AC, Kussano JT, Steel CJ (2015) Sourdough reduces sodium in wheat flour doughs. Int J Food Sci Technol 50:2621–2629

Novick SN, Rozzell JD (2005) Immobilization of enzymes by covalent attachment. In: Barredo JL (ed) Microbial enzymes and biotransformations, vol 17, 1st edn. Humana Press, p 320

Novotni D, Špoljarić IV, Drakula S, Čukelj N, Voučko B, Ščetar M, Ćurić D (2017) Influence of barleysourdough and vacuum cooling on shelf life quality of partially baked bread. Food Tech Biotech 55(4):464

Olojede AO, Sanni AI, Banwo K (2020) Rheological, textural and nutritional properties of gluten-freesourdough made with functionally important lactic acid bacteria and yeast from Nigerian sorghum. LWT, 120:108875

Onem H, Nadaroglu H (2018) Immobilization of purified phytase enzyme from Tirmit (*Lactarius volemus*) on coated chitosan with iron nanoparticles and investigation of its usability in cereal industry. Iran J Sci Technol, Trans A: Sci 42(3):1063–1075

Palla M, Blandino M, Grassi A, Giordano D, Sgherri C, Quartacci MF, Giovannetti M (2020) Characterization and selection of functional yeast strains during sourdough fermentation of different cerealwholegrain flours. Sci Rep 10(1):12856

Pathare PB, Byrne EP (2011) Application of wet granulation processes for granola breakfast cereal production. Food Eng Rev 3(3):189–201

Pétel C, Onno B, Prost C (2017) Sourdough volatile compounds and their contribution to bread: A review. Trends Food Sci Technol 59:105–123

Plessas S (2021) Innovations in Sourdough Bread Making. *Fermentation* 7(1):29

Plessas S, Trantallidi M, Bekatorou A, Kanellaki M, Nigam P, Koutinas AA (2007) Immobilization of kefir and *lactobacillus casei* on brewery spent grains for use in sourdough wheat bread making. Food Chem 105(1):187–194

Plessas S, Koliopoulos D, Kourkoutas Y, Psarianos C, Alexopoulos A, Marchant R, Banat IM, Koutinas AA (2008) Upgrading of discarded oranges through fermentation using kefir in food industry. Food Chem 106(1):40–49

Plessas S, Alexopoulos A, Bekatorou A, Bezirtzoglou E (2012) Kefir immobilized on corn grains as biocatalyst for lactic acid fermentation and sourdough bread making. J Food Sci 77(12):C1256–C1262

Plessas S, Mantzourani I, Bekatorou A (2020) Evaluation of *Pediococcus pentosaceus* SP2 as starter culture on sourdough bread making. Foods 9(1):77

Pragya, Sharma KK, Kumar A, Singh D, Kumar V, Singh B (2021) Immobilized phytases: an overview of different strategies, support material, and their applications in improving food and feed nutrition. Crit Rev Food Sci Nutr:1–23

Rathore S, Desai PM, Liew CV, Chan LW, Heng PWS (2013) Microencapsulation of microbial cells. J Food Eng 116(2):369–381

Reale A, Di Renzo T, Preziuso M, Panfili G, Cipriano L, Messia MC (2019) Stabilization of sourdough starter by spray drying technique: new bread making perspective. LWT-Food Sci Technol 99:468–475

Research and Market (2022) Global Sourdough Market 2022–2026. Available at: https://www.researchandmarkets.com/reports/5575476/global-sourdough-market-2022-2026 Access on: May, 2022

Rizzello CG, Cassone A, Di Cagno R, Gobbetti M (2008) Synthesis of angiotensin I-converting enzyme (ACE)-inhibitory peptides and g-aminobutyric acid (GABA) during sourdough fermentation by selected lactic acid bacteria. J Agric Food Chem 56:6936–6943

Rizzello CG, Nionelli L, Coda R, De Angelis M, Gobbetti M (2010) Effect of sourdough fermentation on stabilization, and chemical and nutritional characteristics of wheat germ. *Food Chem* 119:1079–1089

Rizzello CG, Cassone A, Coda R, Gobbetti M (2011) Antifungal activity of sourdough fermented wheat germ used as an ingredient for bread making. Food Chem 127:952–959

Rizzello CG, Portincasa P, Montemurro M, Di Palo, DM, Lorusso MP, De Angelis, M, Gobbetti M (2019) Sourdoughfermented breads are more digestible than those started with baker's yeast alone: an in vivo challenge dissectingdistinct gastrointestinal responses. Nutr 11(12):2954

Rozyło R, Rudy S, Krzykowski A, Dziki D (2015) Novel application of freeze-dried Amaranth sourdough in gluten-free bread production. J Food Process Eng 38(2):135–143

Rozyło R, Rudy S, Krzykowski A, Dziki D, Siastała M, Polak R (2016) Gluten-free bread prepared with fresh and freeze-dried rice sourdough texture and sensory evaluation. J Texture Stud 47:443–453

Santivarangkna C, Kulozik U, Foerst P (2008) Inactivation mechanisms of lactic acid starter cultures preserved by drying processes. J Appl Microbiol 105(1):1–13

Singh B, Satyanarayana T (2008) Phytase production by *Sporotrichum thermophile* in a cost-effective cane molasses medium in submerged fermentation and its application in bread. J Appl Microbiol 105:1858–1865

Stefanello RF, Machado AAR, Cavalheiro CP, Santos MLB, Nabeshima EH, Copetti MV, Fries LLM (2018) Trehalose as a cryoprotectant in freeze-dried wheat sourdough production. LWT-Food Sci Technol 89:510–517

Stefanello RF, Nabeshima EH, Iamanaka BT, Ludwig A, Fries LLM, Bernardi AO, Copetti MV (2019) Survival and stability of *lactobacillus fermentum* and *Wickerhamomyces anomalus* strains upon lyophilisation with different cryoprotectant agents. Food Res Int 115:90–94

Tafti AG, Peighardoust SH, Behnam F, Bahrami A, Aghagholizadeh R, Ghamari M, Rafat SA (2013a) Effects of spray-dried sourdough on flour characteristics and rheological properties of dough. Czech J Food Sci 31(4):361–367

Tafti AG, Peighambardoust SH, Hesari J, Bahrami A, Bonab ES (2013b) Physico-chemical and functional properties of spray-dried sourdough in bread making. Food Sci Technol Int 19(3):271–278

Tang L, Yang R, Hua X, Yu C, Zhang W, Zhao W (2014) Preparation of immobilized glucose oxidase and its application in improving bread making quality of commercial wheat flour. Food Chem 161:1–7

Torrieri E, Pepe O, Ventorino V, Masi P, Cavella S (2014) Effect of sourdough at different concentrations on quality and shelf-life of bread. LWT-Food Sci Technol 56:508–516

Wang J, Chatzidimitriou E, Wood L, Hasanalieva G, Markellou E, Iversen PO, Rempelos L (2020) Effect of wheat species (Triticum aestivum vs T. spelta), farming system (organic vs conventional) and flour type(wholegrain vs white) on composition of wheat flour–Results of a retail survey in the UK and Germany–2.Antioxidant activity, and phenolic and mineral content. Food Chem: X, 6:100091

Vermeulen N, Czerny M, Gänzle MG, Schieberle P, Vogel RF (2007) Reduction of (E)-2-nonenal and (E, E)-2, 4-decadienal during sourdough fermentation. J Cereal Sci 45:78–87

Zhang T, Cui Q, Zhang F, Zhang L, Wang X (2018) Effects of microencapsulated glucose oxidase on wheat flour dough properties and Chinese steamed bread quality. Int J Food Sci Technol 53(7):1657–1665

Chapter 10
Sourdough-Based Starter Cultures for Fermentation in Agri-food Industry

Gabriel Albagli, Lillyan Nathalia Brito Silva, Nathália Moura Nunes, Daniel Perrone Moreira, Priscilla Filomena Fonseca Amaral, and Priscilla Vanessa Finotelli

10.1 Introduction

According to the Food and Agriculture Organization of the United Nations (FAO), since the '60s, there has been a wide expansion in the industrialization and globalization of food and agriculture products and, consequently, the consumption of processed, packed, and prepared foods has increased intensively (FAO 2017). Paradoxically to this trend, the search for sourdough products has grown over the last years, with a market projection of 3.5 billion dollars in the year 2025 (Research GV Global Sourdough Market Trends, Analysis Report, 2019-2025). Sourdough products tend to be more artisanal and natural and can be characterized by a better

G. Albagli
By&Bio – Byproducts to Bioproducts lab, Chemistry School, Federal University of Rio de Janeiro, Rio de Janeiro, RJ, Brazil

Laboratório de Nanotecnologia Biofuncional, College of Pharmacy, Federal University of Rio de Janeiro, Rio de Janeiro, RJ, Brazil

L. N. B. Silva · D. P. Moreira
Laboratório de Bioquímica Nutricional e de Alimentos, Chemistry Institute, Federal University of Rio de Janeiro, Rio de Janeiro, RJ, Brazil

N. M. Nunes
Laboratory of Food Science, Department of Basic and Experimental Nutrition, Nutrition Institute, Rio de Janeiro State University, Rio de Janeiro, RJ, Brazil

P. F. F. Amaral
By&Bio – Byproducts to Bioproducts lab, Chemistry School, Federal University of Rio de Janeiro, Rio de Janeiro, RJ, Brazil

P. V. Finotelli (✉)
Laboratório de Nanotecnologia Biofuncional, College of Pharmacy, Federal University of Rio de Janeiro, Rio de Janeiro, RJ, Brazil
e-mail: finotelli@pharma.ufrj.br

E. B. Ceresino et al. (eds.), *Sourdough Microbiota and Starter Cultures for Industry*, https://doi.org/10.1007/978-3-031-48604-3_10

flavor, texture, shelf-life, and health-promoting properties, which may be the reasons for its success (Montemurro et al. 2019).

The sourdough production process can be considered one of the oldest bioprocessing techniques and can be characterized by the simultaneous presence of lactic acid bacteria (LAB) and yeasts in a dough. This microbiota, known as the starter culture, can be obtained either by spontaneous fermentation or by selecting the species (De Vuyst et al. 2014), and can be classified into four different types. The oldest one, type I, is characterized by the fermentation of microorganisms originating from the flour, liquid, and tools used in its preparation and also from the environment where the starter is kept. Fermentations with type I starters tend to be slower. Type II sourdough starters are prepared with specific LABs and with or without selected yeasts, making it possible to set some characteristics in the final product. Type III sourdough starter is a dried type II sourdough, and type IV is a mixture of type I and II (Siepmann et al. 2019).

The downside of using type I sourdough starter in industries is the lack of control on the microbial population, since it may vary due to several factors, including temperature, fermentation time, and the ingredients used. A solution to this problem can be the usage of type II or III starter culture, which can be selected for specific activities and stability (Korcari et al. 2021).

The microbiota selection represents one of the biggest challenges for industrial sourdough breadmaking. The strains used as a starter culture, usually, are evaluated mainly based on the acidification, proteolytic activity, nitrogen metabolism, and synthesis of aromatic compounds (Carnevali et al. 2007). This microbial selection can impact the final characteristics of the product, such as flavor, texture, and nutritional quality, mainly due to the fermentation products. Several metabolites can be produced during fermentation by specific sourdough starter species, such as organic acids, exopolysaccharides, enzymes, volatile and bioactive compounds. In this sense, the starter cultures or isolated strains can be applied to obtain other products than the sourdough bread. In this chapter, we will address the microbial characteristics of sourdough starters, their metabolites produced during fermentation, and the possible uses for those cultures.

10.2 Microbiology of Sourdough Starter

The microbiology of sourdough starter can be remarkably diverse, including mostly lactic acid bacteria (LAB) and yeasts, usually in a proportion of 100:1 (LAB/yeasts) (Reale et al. 2019a, b). Most LAB species isolated from sourdough starter belong to one of the following genera: *Lactiplantibacillus, Levilactobacillus, Limosilactobacillus, Fructilactobacillus, Pediococcus, Leuconostoc,* and *Weisella*, with a few exceptions. Among yeasts, *Kazachstania, Torulopsis, Pichia, Candida, Yarrowia,* and *Saccharomyces* are the most frequent genera in this microbiota (Chavan and Chavan 2011; Liu et al. 2018).

Considering all manuscripts referring to sourdough in the literature, it is possible to identify at least 60 species of LAB being mentioned, including *Lactiplantibacillus plantarum* (formerly known as *Lactobacillus plantarum*), *Levilactobacillus brevis* (formerly known as *Lactobacillus brevis*), *Limosilactobacillus fermentum (*formerly known as *Lactobacillus fermentum), Fructilactobacillus sanfranciscensis (*formerly known as *Lactobacillus sanfranciscensis*), *Leuconostoc mesenteroides*, *Weissella cibaria*, and *Weissella confus*e (De Vuyst et al. 2014; De Vuyst et al. 2016; Harth et al. 2016; Gobbetti et al. 2016; Landis et al. 2021). However, usually a mature starter contains one to three dominant bacterial species (Landis et al. 2021).

A systematic search over the last 30 years of publications related to sourdough revealed that the yeast species that are often isolated from sourdough (present in at least 10 publications) are: *Saccharomyces cerevisiae, Kazachstania humilis, Yarrowia keelungensis, Torulaspora delbrueckii, Kazachstania unispora, Kazachstania exigua, Pichia kudriavzevii,* and *Kazachstania barnetti* (Arora et al. 2021). They also revealed that *Lactobacilli* species are the most abundant in sourdough, with more than 80 species identified. Nomadic and obligatory heterofermentative species are the ones most identified, including *L. plantarum* (identified in more than 45% of the studies), *L. brevis* (30% of the studies), *F. sanfranciscensis* (29% of the studies), and *L. fermentum* (18% of the studies). Obligately homofermentative are not often found, for example, *L. delbrueckii* was identified in only 3% of the studies. Homofermentative LAB mainly produces lactic acid as a fermentation product while heterofermentative LAB produces lactic acid, ethanol, acetic acid, and carbon dioxide. Damiani et al. (1996) characterized hetero- and homofermentative LAB from Italian wheat sourdough and observed that LABs are strongly related to the volatile profile and sensory attributes of the bread since heterofermentative LAB mainly produces ethyl acetate and some alcohols and aldehydes while homofermentative LAB synthesizes diacetyl and other carbonyls.

The genus *Lactobacillus* was proposed by Beijerinck in 1901, but approximately 20 years ago, genotypic, and chemotaxonomic criteria started to be used for the delineation of new bacterial species. Based on whole genomic sequences, Zheng et al. (2020) proposed a reclassification of the genus *Lactobacillus* into 25 genera including the emended genus *Lactobacillus*, which includes host-adapted organisms that have been referred to as the *Lactobacillus delbrueckii* group, *Paralactobacillus* and 23 novel genera. Many sourdough species have been reclassified by them, such as: *Levilactobacillus brevis, Lactiplantibacillus plantarum, Limosilactobacillus fermentum* (Zheng et al. 2020). These species are considered generally recognized as safe (GRAS) and, therefore can be applied as a starter in various food fermentations.

Lactiplantibacillus plantarum is classified as facultative heterofermentative bacilli, capable to utilize several carbohydrates, which allow them to adapt to different ecological environments. This species has been shown to affect human health through various beneficial activities, e.g., antimicrobial, antioxidative, antigenotoxic, anti-inflammatory, and immunomodulatory, in several *in vitro* and *in vivo* studies (Garcia-Gonzalez et al. 2021). Cai et al. (2021) also depicted a possible role of *L. plantarum* in protecting human reproductive ability and maintaining food safety. Various studies have demonstrated the ability of orally ingested *L. plantarum* to rapidly bind contaminants, leading to less intestinal absorption and increasing fecal excretion of dietary contaminants from the body (Cai et al. 2021). This species is a good option for industrial and commercial applications since in a freeze-dried format, *L. plantarum* is robust and stable at room temperature, enabling long shelf-lives (Nordström et al. 2021). They are often the first species of type I sourdough starters (Gänzle and Zheng 2019), because they are more likely to be present in a given environment (Calvert et al. 2021). Ventimiglia et al. (2015) identified twelve cultures from sourdoughs produced in western Sicily (southern Italy) that showed good acidifying abilities *in vitro* and *L. plantarum* PON100148 produced the highest concentrations of organic acids. The characteristics of the sourdoughs and the resulting breads indicated that the best productions were obtained in presence this strain.

Levilactobacillus brevis can be isolated from milk, cheese, sauerkraut, sourdough, faeces, mouth, and intestinal tract of humans and rats (Rönkä et al. 2003). It has been proved that *L. brevis* strains are potential probiotics and that *L. brevis* isolated from fermented food can be an effective biocontrol of mycotoxigenic fungus that are natural contaminants of raw materials used in the food and feed production (Somashekaraiah et al. 2021). Meignen et al. (2001) reported the effect of mixing *L. brevis* with baker's yeast *Saccharomyces cerevisiae* on bread quality. Yeast growth in dough and ethanol production were reduced but more glycerol and acetic acid were formed, maintaining lactic acid constant. When dough was fermented in parallel with single starters then, mixed and fermented again, the production of typical sour dough type aroma was enhanced.

Limosilactobacillus fermentum is known to enhance the immunologic response as well as prevent gastrointestinal and upper respiratory infections (Rodríguez-Sojo et al. 2021). It can be isolated from different food products, including fermented vegetables (Di Cagno et al. 2008), and dairy goods (Coton et al. 2008), and is also found in human microflora, being detected in breast milk (López-Huertas 2015), saliva (Dal Bello and Hertel 2006), and feces (Park et al. 2015). *L. fermentum* is able to produce exopolysaccharides (EPS) which can be secreted to the medium or remain attached to the cell wall. An investigative revision performed by Ale et al. in 2020 concluded that certain EPS structures play central roles in the bioactivity of the producing strain and, that the EPS structure seems to have an important impact. Moreover, there is increasing evidence that associates the health-promoting properties of these strains with the synthesis of these metabolites (Ale et al. 2020). Being GRAS strains of *L. fermentum* have been widely applied as a starter in various food fermentations, and as a food preservative agent. It is a key species in sourdough technology, contributing to flavor, texture, or health-promoting dough ingredients (Naghmouchi et al. 2020; De Vuyst et al. 2009).

Fructilactobacillus sanfranciscensis is a key bacterium in traditional (type 1) sourdough fermentations (Rogalski et al. 2021). *It* uses maltose as major carbohydrate source and its metabolism is not repressed by glucose (Corsetti and Settanni 2007). It is widely found in sourdoughs probably because it originates from insects

(Boiocchi et al. 2017). It has a great influence on the rheological properties of dough and bread characteristics, texture, aroma, sensory quality, shelf- life, EPS production, and gluten properties (Suo et al. 2021). An analysis of a gene deletion mutant indicated that *F. sanfranciscensis* accumulated reduced glutathione and γ-glutamylcysteine through glutathione reductase activity during proofing, which is beneficial to improve the taste and texture of type I sourdough bread (Tang et al. 2017).

In relation to yeast species, *Saccharomyces cerevisiae*, the most famous yeast species in human history, has been domesticated to produce a large diversity of fermented drinks (wine, beer, etc.) and foods (bread, cocoa, etc.) (*e.g.* Stewart 2014). *S. cerevisiae* is usually considered the most usual sourdough yeast because it is found in most starters and it is associated with the first leavened breads (Valmorri et al. 2010; Landis et al. 2021). However, because *S. cerevisiae* grows really fast when compared with other yeast species (Landis et al. 2021), its presence may be relatively recent and a reflection of rapid dispersal and invasion facilitated by human movement among bakeries and continents (Michel et al. 2022; Bigey et al. 2021). In addition to ethanol and CO_2, yeasts generate a large spectrum of metabolites during sourdough fermentation, responsible for the aroma profile, or so-called "fermented taste" in the production of bread. This profile consists of a complex mixture of flavor compounds, such as fusel alcohols (3-methyl-1-butanol, 2-methyl-1- butanol and 2-methyl-1-propanol) (Guerzoni et al. 2013). Damiani et al. (1996) showed that the presence of *Saccharomyces* and *Hansenula* during sourdough fermentation favored an increased synthesis of alcohols, esters and some carbonyl compounds compared to sourdough fermentation without the addition of yeasts, especially, ethanol, methylpropanol, 2- and 3-methylbutanol, and ethyl acetate.

Kazachstania humilis is an ascomycetous budding yeast known as a fermenting yeast in rye and wheat sourdoughs. This species is the second most commonly identified sourdough yeast and found in sourdoughs from every continent (Wittwer et al. 2022). Carbonetto et al. (2020) revealed neutral or negative asymmetric interaction outcomes between *K. humilis* and diverse LAB strains found in sourdoughs. When in co-culture, the yeasts´ population size decreased in the presence of LAB regardless of the strain, while the LAB's population size was rarely influenced by the presence of yeasts. However, a higher maltose depletion was shown in co-cultures compared to monocultures. In addition, LAB and *K. humilis* strains leavened dough as much as couples of LAB and *S. cerevisiae* strains, while *K. humilis* strains never leavened dough as much as *S. cerevisiae* when in monoculture. Rogalski et al. (2020) found that some *Fructilactobacillus sanfranciscensis* strains can be dominant in sourdough regardless of the presence/absence of *S. cerevisiae* and/or *K. humilis*. Other strains can be influenced depending on yeast species. Interestingly, in fermentations that were not inoculated with any yeast a spontaneous occurrence of *S. cerevisiae* or *K. humilis* was observed.

Torulaspora delbrueckii are fermentative yeasts found in wild and anthropic habitats (Ramírez and Velázquez 2018). *T. delbrueckii* has attracted interest in recent years, especially due to its biotechnological potential, arising from its flavor- and aroma-enhancing properties when used in wine, beer, or bread dough fermentation, as well as from its remarkable resistance to osmotic and freezing stresses (Fernandes et al. 2021). Co-cultures of *S. cerevisiae* and *T. delbrueckii* enhanced bread quality with superior aroma and improved sensorial attributes, with several volatile compounds being identified in the bread crumb leavened with both yeasts (Wahyono et al. 2016).

10.3 Microbial Metabolites Produced During Fermentation

Sourdough fermentation can generate unique products that will reflect on rheological properties (mainly related to texture), flavor, shelf-life, and nutritional quality. These characteristics depend on factors such as microbiota, fermentation processes (time and temperature, mainly), and flours. According to De Vuyst et al. (2014), the chemical composition and quality of the flour strongly influence sourdough microbiota and fermentation processes. Fermentation temperature is also an especially important factor that affects the microbiota of sourdough starters. At high temperatures, the concentration of lactic acid increases, favoring acidification and LAB reproduction, while at low temperatures yeast growth is promoted, as well as ethanol and flavor compounds production (Banu et al. 2011; Decock and Cappelle 2005; Katina et al. 2006).

The distinguishing aspect of the sourdough process is the association of the carbohydrate metabolism microorganisms (LAB and yeast) with the substantial contribution of endogenous enzymes to biochemical conversions at the dough stage (Gänzle 2014b). The substances resulting from these processes can contribute to improvements in the nutritional quality of bread (e.g., production of bioactive compounds), in addition to enhancements in sensorial aspects and increased shelf-life. Thus, the description of the microbial metabolites produced during fermentation and their contribution to sourdough attributes will be presented in this section to highlight the characteristics that may be interesting for the use of sourdough starter in agri-food industry.

10.3.1 Sensory Attributes

Sensory attributes, such as appearance, odor, flavor, taste, and texture, are often used to evaluate food quality (Chumngoen and Tan 2015). These characteristics may also influence the acceptability and preferences of consumers. The increase in sourdough consumption is mainly due to its differentiated organoleptic characteristics, such as improved crumb structure, volume, texture, and flavor properties compared to traditional doughs (Limbad et al. 2020).

During the sourdough fermentation and the backslopping process, the cultured microorganisms can produce a series of metabolites such as organic acids, exopolysaccharides, or enzymes, which will contribute to a positive effect on the texture of the bread. Some of these metabolites can potentially replace hydrocolloids, used as bread improvers, and cause a decline in pH, promoting increased proteases and amylases activities in flours (Arendt et al. 2007).

Exopolysaccharides (EPS) are homopolymers or heteropolymers with a wide diversity of structures, capable of modifying the sensory properties of foods (Ansorena and Astiasarán 2016). Sourdough LABs can produce EPS *in situ,* affecting the rheological properties of dough as well as loaf volume, crumb texture, and the handling characteristics of bread (Yazar and Tavman 2012). Some LAB species (*Lactobacillus acidophilus, Lactobacillus delbrueckii subsp. bulgaricus, Lactococcus lactis, L. plantarum, Lacticaseibacillus casei,* etc.) are known as strain-specific EPS producers. In fact, *Leuconostoc mesenteroides* strains, a LAB species, excretes dextran, authorized as a novel food ingredient for bakery products (Ansorena and Astiasarán 2016). These EPS improve the bread volume and reduce firmness. Dextran also acts as a replacement for gluten, binding the water of the dough (Wolter et al. 2014). Sourdough LAB affects the microstructure of dough not only by producing EPS but also by producing CO_2 and lowering pH because of lactic acid production (Yazar and Tavman 2012). Investigators are considering replacing synthetic food stabilizers by EPS, which could be provided by sourdough fermentation. EPS is being used, as a gelling agent and for suspending and thickening food, improving rheological properties (Suryawanshi et al. 2022).

Corsetti et al. (1998) observed that moderate hydrolysis of the starch and proteolysis of gluten occurs in sourdough due to the production of acetic and lactic acid by LAB during the fermentation process. Under acid conditions, the cereal native proteases can degrade prolamins, reducing gluten proteins, which will affect the rheology of wheat sourdoughs and, consequently, the texture of the bread (Thiele et al. 2004).

In 2017, Pétel et al. compiled a table on sourdough bread's volatile composition, listing 93 volatile compounds in wheat or rye sourdough bread for the first time. Hexanal, nonanal, 1-pentanol, 1,4-butanediol, 6-methyl-5-hepten-2-one and octyl acetate were the most cited. In sourdough starter and bread, 196 volatile compounds have already been detected. The profile of volatile organic compounds in the final bread is strongly affected by the starter culture, depending on the complex and dominant sourdough microbiota, and some technological properties, such as temperature and the flour type (Siepmann et al. 2019). Sourdough starter with *L. plantarum XL23* and *F. sanfranciscensis RL976* strains isolated during sourdough fermentations in Turkey resulted in faster acidification and more volatile organic compounds compared to control sourdough produced spontaneously. This research revealed the presence of 23 volatile organic compounds, including D-limonene, 2-pentyl furan, ethyl acetate, acetaldehyde, 3-methyl-butanal, and heptanal (Boyaci Gunduz et al. 2022). The development of starter culture combinations could be an attractive strategy for industrial applications since rapid acidification is essential to conduct sourdough fermentation in a short time.

Different cereals and their combinations can be explored in sourdough fermentation to evaluate the effects on flavor and the other sensorial attributes. Mukisa and colleagues used sourdough starter to ferment sorghum malt, a typical African cereal, to obtain a fermented beverage called Obushera and then identified the potential starter cultures for traditional fermented cereal products (Mukisa et al. 2017). Besides producing lactic acid, LAB also contributed to the production of other flavor compounds, notably acetate, ethanol, acetaldehyde, diacetyl, acetone, and acetoin. Yeasts mainly contributed to ethanol, acetaldehyde, methyl aldehydes, and methyl alcohols production. The starter culture combinations: *L. plantarum* + *L. lactis* + *S. cerevisiae*, or *L. plantarum* + *L. lactis* + *W. confusa* + *S. cerevisiae* were adopted for the fermentation of Obushera.

10.3.2 Nutritional Attributes

The main improvements in nutritional aspects are related to synthesizing bioactive compounds produced during the fermentation process, such as flavonoids, phenolic acid, lignans, folates, tocopherol, and sterols (Katina et al. 2006). This section will describe phenolic compounds in detail because they are considered a vital human dietary component and exhibit tremendous antioxidant activity and other health benefits (Kumar and Goel 2019).

Phenolic compounds are secondary plant metabolites characterized by one or more mono- or polyhydroxylated aromatic rings and a functional group present in their chemical structures (Pérez-Jiménez et al. 2010). Plant foods are the main sources of these compounds, which are of great interest to studies and the market because of their functionalities. In plant tissue, they act as defense agents against biotic and abiotic stresses and are responsible for imparting astringency, color, flavor, and aroma characteristic of grains, fruits, and vegetables (Tuladhar et al. 2021). In animal tissue, they can contribute to antioxidant activity because when consumed in the human diet, they reduce the risk of developing some chronic diseases (e.g., cardiovascular disorders, diabetes, and cancer) (Cory et al. 2018).

Phenolic acids are wheat's main phenolic compound class, which may contain derivatives of hydroxybenzoic acids, such as *p*-hydroxybenzoic, protocatechuic, vanillic, syringic, and gallic, and derivatives of hydroxycinnamic acids, such as ferulic, *p*-coumaric and caffeic. Among them, ferulic acid is the most abundant in wheat, mainly in the insoluble form (above 90%) (Wang et al. 2014). Comparing conventionally grown wheat to organic cultivars, Zuchowski et al. (2011) revealed higher concentrations of ferulic and *p*-coumaric acids in organic wheat flour. The limited nitrogen availability in organic flour contributes to a considerable improvement in plant food chemical composition, including the secondary metabolites that are intensively synthesized and act as an immune system for the plants (Straus et al. 2012).

In sourdough, the presence of phenolic acids implies a toxic environment to LAB and affects their competitiveness and composition (Sekwati-Monang et al. 2012). In turn, LAB promotes the phenolic acids biotransformation as a detoxification mechanism and to increase survivability (Sánchez-Maldonado et al. 2011; Filannino et al. 2018). An example of a resistant LAB commonly present in sourdough is *Lactobacillus* spp. (Lee et al. 2006).

Gänzle (2014a, b) has already stated that the metabolism of sourdough microbiota and the activity of cereal enzymes are interdependent. The activity of cereal enzymes is an important determinant of the sourdough microbial ecology. The acidification promoted by LAB during fermentation modulates the activity of cereal enzymes that are present mainly in the grain's germ and aleurone. In general, amylases and proteases provide substrates for bacterial growth, while cellulases and xylanases release phenolic compounds bound to the plant matrix, solubilizing them in the medium. In turn, soluble phenolic acids considered toxic to the survival of microorganisms are converted by decarboxylase and/or reductase enzymes, specific to each LAB species, into their respective metabolites (Sánchez-Maldonado et al. 2011; Gänzle 2014a, b). The metabolism of phenolic compounds by LAB is explained as an efficient mechanism to detoxify the environment, as a way of species survival (Fig. 10.1) (Gänzle 2014a, b; Ripari et al. 2019).

The LAB species of sourdough microbiota that can metabolize soluble phenolic acids present in cereals, commonly wheat, are *L. curvatus*, *L. plantarum*, *L. fermentum*, *F. sanfranciscensis*, *L. brevis*, *L. casei* and *L. reuteri* (Svensson et al. 2010; Sánchez-Maldonado et al. 2011; Filannino et al. 2014). The phenolics acids biotransformation is mediated by their reductases and decarboxylases (Gänzle 2014a, b; Santamaría et al. 2018). The hydroxybenzoic and hydroxycinnamic acids are decarboxylated to the corresponding phenol or vinyl derivatives. In turn, hydroxycinnamic acids and their vinyl derivatives are converted by reductases which hydrogenate the double bond (Van Beek and Priest 2000). The hydroxycinnamic acids that can be biotransformed are ferulic, *p*-coumaric, and caffeic acids. Ferulic acid can be reduced to dihydroferulic acid (by the action of reductases) or decarboxylated, resulting in 4-vinylguaiacol (by the action of phenolic acid decarboxylase),

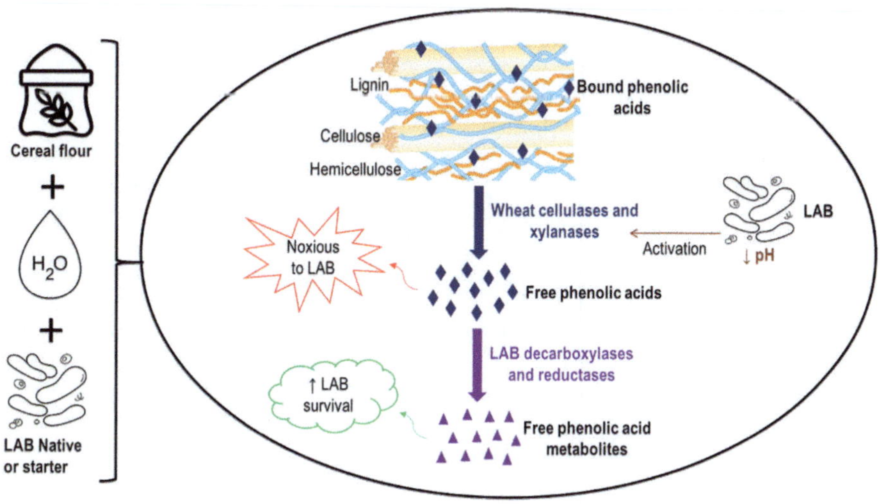

Fig. 10.1 Graphical summary of phenolic compounds bioconversion in the process of making sourdough fermentation

which can be further reduced to 4-ethylguaiacol (by the action of reductases). The *p*-coumaric acid can be reduced to dihydro-*p*-coumaric acid or decarboxylated to 4-vinylphenol, which can be further reduced to 4-ethylphenol. Caffeic acid, in turn, can be reduced to dihydrocaffeic acid or decarboxylated to 4-vinylcatechol, which can be further reduced to 4-ethyl catechol (Fig. 10.2) (Gänzle 2014a, b; Filannino et al. 2018).

The organic acids produced by LAB's metabolism contribute to the lower glycemic indexes of bread, reducing starch digestibility by turning it to resistant starch and lowering the postprandial glucose level (Katsi et al. 2021). Sourdough fermentation also reduces the phytate content of whole wheat bread, which also increases the bioavailability of minerals (Hansen and Schieberle 2005). Leenhardt et al. (2004) compared the degradation of phytate and magnesium in whole wheat bread produced with baker's yeast, with sourdough and with lactic acid and compared them to bread produced without any leavening or acidification agent (control). It was observed that after 4 h, 60% of the control phytic acid content disappeared, probably caused by endogenous phytase from the flour Baker's yeast dough presented a very similar phytate degradation profile, proving that phytate degradation could not be attributed to baker's yeast. In sourdough, 36% of the initial phytic acid content was hydrolyzed after 30 min and after 4 h, sourdough presented 2 times lower phytic acid concentration when compared to control samples. The pH-lowering effect caused by the production of acids during sourdough fermentation could be the reason for those results because the addition of lactic acid to dough resulted in almost the same degradation of phytic acid. Gänzle and De Vuyst (2004) also reported that fermentation increased the bioavailability of minerals due to phytate degradation through microbial and cereal enzymes.

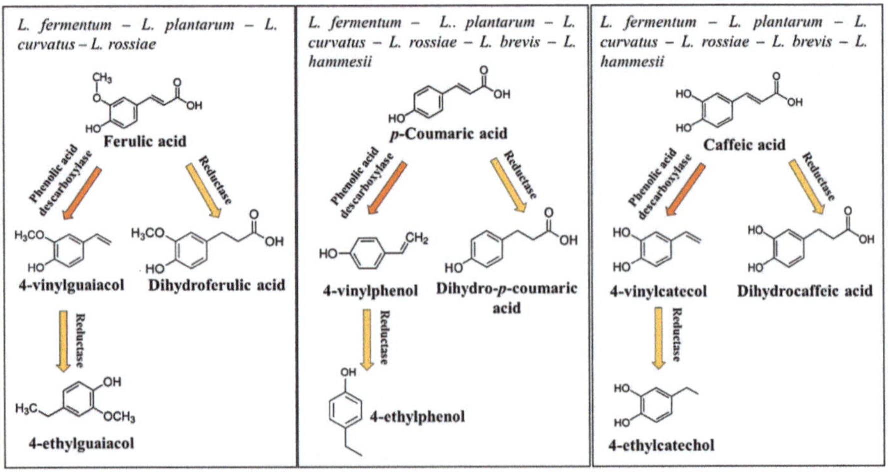

Fig. 10.2 Conversion of ferulic acid, p-coumatic acid and caffeic acid present in cereals by phenolic acid decarboxylase and reductase of LAB specific strains

Exploiting the potential of non-conventional flours (legumes and pseudo-cereals) and milling by-products (bran and germ) could add nutritional values to the sourdough. In the systematic review on sourdough fermentation, Arora et al. (2021) discuss that sourdough fermentation of legume flours increased the contents of free amino acids, γ-amino butyric acid (GABA), polyphenols, dietary fibers, and bioavailable minerals, promoted antioxidant activities, and the *in vitro* protein digestibility, and lowered the glycemic index (Arora et al. 2021). Chickpea, amaranth, quinoa, and buckwheat were the most suitable flours to be enriched with GABA (Coda et al. 2010). However, the addition of non-flour ingredients to sourdough production could require attention. The survival and growth of microorganisms is also determined by the presence of the appropriate substrates (De Vuyst et al. 2021).

Diverse ingredients could be incorporated to the sourdough in order to increase the nutritional value of the bread. Martău et al. (2021) associated sourdough fermentation with by-products/waste (apple pomace) and obtained a product with increased nutritional values. Results suggested that apple pomace may be an important source of organic acids and fermentable sugars for sourdough.

Bioactive peptides from fermented foods have recently attracted attention for their antioxidant properties (Singh et al. 2022). Bioactive peptides, produced by enzymatic action on native proteins of the grain, have a biological role in human health, including antimicrobial, anti-proliferative, antioxidant, blood pressure-lowering, antithrombotic, and mineral absorption activities (Graça et al. 2021). The bioactive peptides produced during sourdough fermentation were isolated and used in *ex vivo* assays to prove the ability to reduce intracellular reactive oxygen species production and decrease the expression of some inflammatory markers. LAB strains were combined with the flours and added with *S. cerevisiae* strains (Luti et al. 2021). The results indicate that combining flours, LAB and yeast led to the production of bakery products with beneficial effects on cell viability and oxidative and inflammatory status. Indeed, further analysis to identify the amino acid composition and sequence of peptides derived from these fermented breads is required to compare with peptides from other fermented foods.

10.3.3 Shelf-Life Extension

The fermentation process can stabilize and transform food materials by the metabolic activity of microorganisms into food products with an enhanced shelf-life and better nutritional or organoleptic properties (Srivastava 2018). Increased sourdough shelf-life is particularly due to LAB fermentation that produce antimicrobial (and antioxidant) metabolites, involved in important anti-bacterial and antifungal effects, namely organic acids (mostly acetic and lactic acids), hydrogen peroxide, carbon hydroxide, ethanol, diacetyl, γ-aminobutyric acid, propionic acid, benzoic acid, fatty acids, bacteriocins, and bacteriocin-like inhibitory substances, among others (Porto et al. 2017; Sidhu and Nehra 2021). Bacteriocins are peptides or proteins synthesized in the ribosomes of some LAB and released extracellularly to inhibit the growth of pathogenic microorganisms resistant to conventional antimicrobials (Strack et al. 2020). LAB have been related to inhibiting the growth of pathogenic microorganisms, to degrading or binding mycotoxins, and to probiotic function (Naidu et al. 1999; Todorov 2009).

Hajinia et al. (2021) investigated the potential of whole oat sourdough fermented with selected *Pediococcus pentosaceus* isolate as a quality-promoting ingredient in supplemented wheat bread. Fermented oat significantly reduced the mold expansion on the bread, besides, improved attributes such as crumb porosity and antioxidant capacity. Hence, they conclude that controlled fermented oat containing the selected isolate can be a great alternative for synthetic improvers and preservatives in bread production by industries.

Jin et al. (2021) isolated LAB and yeasts from Pyeongchang spontaneous sourdough (Gangwon-do, Korea) and created starters for sourdough bread to enhance bread quality; these breads were preservative-free since the developed starter has antifungal action against *Aspergillus flavus*. The combinations of *P. pentosaceus*, *S. cerevisiae*, *P. pentosaceus* and commercial yeast showed the highest inhibition rate after 4 days (56.4 ± 5.5%). The authors attribute the antifungal activity to production of acetic acid during fermentation due to its higher pK_a. With the new market perspective of preservative-free products, sourdough fermentation could be a potential bioprocess to increase said products on the market, in addition to its other benefits.

The metabolites produced during sourdough fermentation that improve bread shelf-life offer an interesting perspective for use as an additive in the food industry. The antimicrobial and antioxidant activities qualify these components as potential natural additives that, when used in processed foods, will provide protection against deterioration and contamination.

10.4 Sourdough Applications

10.4.1 Traditional Sourdough Uses

Sourdough starter is used to produce breads, panettone, pizza, buns, cakes, cookies, waffles, pancakes and others, being one of the oldest fermentation methods known. According to the website Business Wire (2020), pizza is expected to be the most lucrative growth in the next 5 years due to increasing consumer receptivity to new flavors and rising consumption on Europe.

Sandez Penidez et al. (2022) evaluated the capacity of sourdough composed with lactic acid bacteria from quinoa and amaranth to increase the antioxidant activity and total phenolic compounds in quinoa sourdough to select best performing strains to be used in elaboration of cookies. It was observed that the sourdough fermented by LAB showed higher total antioxidant activity than the control samples and that the concentration of total phenols significantly increased during fermentation by LAB in all sourdough samples prepared. Based on those results a principal component analysis were made and allowed the selection of three LAB strains (*L. plantarum* CRL 1964, CRL 1973 and *Lc. mesenteroides* subsp. *mesenteroides* CRL 2131) to be used as starter culture to manufacture quinoa cookies. Sourdough cookies had a higher amount of antioxidant activity and total phenol content than control cookies, suggesting that the sourdough cookie could maintain its characteristics for a longer time.

De Vero et al. (2021) collected two sourdough samples from bakeries in Italy to evaluate possible starter cultures to panettone manufacture, a total of 57 yeast and 20 LAB were isolated after basic microbiological screenings. One of the sourdough samples showed a high concentration of *K. humillis,* the other sourdough sample presented a higher concentration of *S. cerevisiae* followed by *K. humillis,* both yeasts widely described in sourdough starter. The LAB results showed the prevalence of *F. sanfranciscensis* in both sourdough samples, it was also observed the presence of *Furfurilactobacillus rossiae, L. plantarum, Lc. citreum* and *Lc. mesenteroides,* the two last LABs typical of the last stages of panettone fermentation process. The authors than conclude that the differences in microbiota and volatile profile of both samples were related to the backslopping process and frequency, thus showing it is possible to estimate the species contribution to flavor formation during the fermentation process.

10.4.2 Sourdough Use to Improve Gluten-Free Products Quality

According to Kamiński et al. (2020) gluten-free diet is the third most popular diet worldwide nowadays. Consumers are adopting this type of diet for different reasons such as gluten sensibility, skin allergy, celiac disease, or health benefits. The increasing popularity of gluten-free products has been boosting the sourdough industry. The majority of gluten related articles are about the usage of sourdough starter to produce and improving the quality of gluten-free products with different grains and legumes, such as sorghum, quinoa, corn, millet and acorn (Shiri et al. 2021; Falasconi et al. 2020; Carbó et al. 2020).

In gluten-free baking, the addition of hydrocolloids such as xanthan gum, guar gum and hydroxypropylmethylcellulose is essential to obtain acceptable product quality in terms of volume, texture and shelf-life (Lazaridou et al. 2007). As previously shown, many LABs can produce a wide variety of EPS, which can improve the technological as well as nutritional properties of gluten-free products, acting as prebiotics and hydrocolloids, respectively (Arendt et al. 2011). As an example of the importance of EPS production the paper by Schwab et al. (2008) produced sourdough breads with quinoa and sorghum using the strains *L. reuteri* LTH5448 and *Weissella cibaria* 10 M as starter culture, both microbes capable of producing EPS. The authors observed that breads produced with *W. cibaria* were softer than the ones containing no EPS, furthermore, the oligosaccharides produced during the fermentation were not digested by baker's yeast and were still present in the final bread, acting like a prebiotic.

10.4.3 Different Uses of Sourdough Microbiota

Since sourdough microbiota is vastly diverse and can be selected, it can be applied in different products. In this section we will present some papers and patents that used sourdough microbes to produce different products.

Once LAB exhibit antimicrobial and antifungal properties, hypothetically they can be isolated from spontaneous sourdough and be applied in a wide range of agri-food industries, such as baking (sourdough bread and other sourdough-based baking goods, e.g., biscuits, cookies, crackers, pastry, pizza and pasta), feed and pet food, dairy (yoghurts, cheeses, smoothies, etc), meat (dried sausages, etc), alcoholic beverages (e.g., beer, cider and wine) and non-alcoholic beverages (juices, refrigerants, non-fermented and fermented cereal drinks), and nutraceuticals (dietary supplements and food additives), and also in other industrial sectors such as cosmetics and pharmaceuticals (Bartkiene et al. 2020). Bartkiene et al. (2020) isolated and phenotypically characterized thirteen LAB strains species-level via PCR LAB from a spontaneous rye sourdough manufactured following traditional protocols. LAB isolates showed antifungal activities against *Aspergillus nidulans*, *Penicillium funiculosum*, and *Fusarium poae*, as well as other antimicrobial properties. Consequently, they emphasized the potential of sourdough biotechnology.

The patent WO1994019950A1, deposited by Williamson and Aylward (1993) describes the use of sourdough starter cultures as biocontrol agents for agriculture. The advantages of biocontrol agents over pesticides are: (i) the registration of this product is easier since these microorganisms are also used in the food industry; (ii) the probiotic effects of the bacteria used; and (iii) the method proposed is compatible with sustainable agriculture and relatively free from health risks, as opposed to some pesticides and chemicals traditionally used in agriculture. The patent describes the usage of a carrier (that can be lipids, such as coconut oil, vegetable oil and olive oil) to stabilize and maintain the sourdough microbiota on the target surface. A thickening agent (Gums, carrageenan, xanthan gum, etc.) can also be used in the composition. The biocontrol agent may include or be derived from a mixed culture of yeasts – *Candida krusei, S. cerevisiae, S. exiguss* and *Pichia saitoi* – and bacteria – *Lactiplantibacillus plantarum, L. casei, L. delbrueckii, Levilactobacillus brevis, Limosilactobacillus fermentum, L. pastorius, L. bucheri, L. acidophilus, Companilactobacillus alimentarius*, *Leuconostoc, Pediococcus* and *Streptococcus* – and a substrate for the mixed culture.

Fu et al. (2021) investigated the effect of yeasts in the metabolism of LAB to reduce wheat allergens. Three different microbial strains, previously isolated from sourdough starter, were used: *Pediococcus acidilactici XZ31*, *S. cerevisiae JM1*, and *Torulaspora delbrueckii JM4*. The LAB *P. acidilactici XZ31* was monocultured and co-cultured with the other two yeasts. The results showed that Tri a 28/19 allergens content decreased after 24 h fermentation. The co-culture fermentation accelerated the degradation of protein, and reduced the allergen content to a greater extent, revealing a synergistic effect between *Pediococcus acidilactici XZ31* and yeast degrading wheat allergens.

Mascia et al. (2015) produced wheat beer from a *S. cerevisiae* isolated from sourdough. The beer produced by sourdough yeast presented a higher value for alcohol content and a lower value for pH when compared to the control sample (beer produced by a commercial strain of *S. cerevisiae*). The authors justify these findings by the higher degree of fermentation of the sourdough beer (fermentations that occurs at 18–24 °C and takes 3–7 days), increasing the alcohol content, and lowering the pH. It was also observed that the sourdough beer presented a higher polyphenolic content, such as esters (responsible for fruity character of beer), than the control sample. The interaction of proteins with polyphenols, forming insoluble precipitates might be the reason for that.

A tendency to prefer the sourdough beer was indicated, probably due to the greater fruity character, although the consumer test showed no significant difference between sourdough beer and control beer.

Johansson et al. (2021) isolated ten different yeast species from wheat, rye, and barley sourdough to assess their potential as a fermentation source for low-alcohol beer brewing. From the ten yeasts isolated, two species (*Kazachtania servazzii* and *Pichia fermentans*) were selected for scale fermentation trial (production of wort). Besides, sensory analysis based on a series of trials regarding alcohol, maltose, flavour compounds production, stress tolerance and flocculence. *Pichia fermentans* was considered suitable for production of low-alcohol wheat beers due its production of spice aroma. While *Kazachtania servazzii* showed potential for lager-style beers due to its clean flavour profile and tolerance to low temperatures.

Kaya et al. (2022) investigated if the lactic acid bacteria present in sourdough starter had any probiotics characteristics (antifungal and antibacterial activities, low pH and bile salt tolerance ability, antibiotic resistance adherence to human gut cells, production of GABA and exopolysaccharides (EPS) production capability). Ten sourdough samples were used in their study that isolated *Lactococcus garvieae, Lactobacillus paracasei, Lactobacillus plantarum, Lactobacillus graminis* and *Pediococcus pentosaceus* from the sourdough starters. The EPS production was higher for *L. graminis, L. plantarum* and *L. paracasei.* EPS might balance the immune responses and affect the adhesion of probiotic strains. Microbial adhesion to the gut epithelial tissue is required for the microbial colonization and *L. paracasei* and *L. garvieae* presented the higher adhesive ability to HT-29 cells (a widely used colon cancer cell line). All the strains tested could grow at 0.3% bile and their survivability in low pH (pH 4.0) was high (>35%). The sourdough isolated strains inhibited the growth pathogenic microorganisms and the highest GABA producers were *L. plantarum* and *L. garvieae*. Therefore, all strains tested had some probiotic properties and could be used to produce a functional bread and improve consumers' health.

Arsoy et al. (2022) identified and investigated the antifungal activity of LAB isolated from Turkish spontaneous sourdough. It is worth mentioning that LABs are recognized as safe for human consumption and could be interesting their application as antifungal biocontrol agents in food production. Eight LAB strains, identified as *Weissella cibaria 908, Lactiplantibacillus plantarumsubsp. plantarum 2114, Leuconostoc pseudomesenteroides 2619, Lactiplantibacillus plantarum subsp. plantarum 2702, Fructilactobacillus sanfranciscensis 2709, Lactobacillus brevis 2216Y, Lactiplantibacillus pentosus Y118, and Lactiplantibacillus plantarumsubsp. plantarum Y201* were indicated as a potential biopreservative which may be used in food industry. The antifungal activity of them would be attributed to a synergic action of organic acid, especially lactic and acetic acid.

Kvass or Kvas, a traditional Eastern European fermented drink is produced from malt or sourdough bread. The beverage undergoes lactic fermentation, is sweetened with sucrose and is slightly alcoholic (0.5–1%). The microbiota, usually, present in this kind of beverage are *S. cerevisiae* and LAB, including the dextran-producing specie *Leuconostoc* spp. The association with *S. cerevisiae* improve aroma and organoleptic properties of kvass (Gänzle 2014a, b). The presence of high cell counts of viable lactic acid bacteria and the possibility to produce prebiotic carbohydrates during fermentation provides new opportunities for the development of functional, cereal-based fermented beverages (Dlusskaya et al. 2008).

Fois et al. (2018) analyzed starch digestibility and properties of fresh pasta made with semolina and liquid sourdough. The authors used *S. cerevisiae* and *Companilactobacillus alimentarius* as starter cultures, which increased the pasta firmness due to the interaction between starch and gluten promoted by the organic acids produced by those microorganisms. Both pasteurized samples (with and without sourdough) presented low bacteria and yeast counts, however the sourdough sample presented a lower level, probably due to the acidity that reinforces the lethality effect of pasteurization. The sourdough fermentation did not affect the total amount of starch digested in 180 min in the *in vitro* digestion, however, it significantly decreased the amount of slow digestible starch (between 20 and 120 min) from 49.8% in control, to 43.0% in sourdough pasta, and increased the amount of inaccessible digestible starch (between 120 and 180 min) from 34.5% in control to 40.2%. Starch from slow digestible starch is used on the calculation of glycemic index with rapid digestible starch. Since there was no significant difference between control and sourdough samples in relation to rapid digestible starch, the authors affirm that the predicted glycemic index of sourdough pasta is lower than the control sample.

Francesca et al. (2019) produced pizza dough using sourdough starter inoculated with facultative heterofermentative species (*Latilactobacillus curvatus* PON 100490, *Latilactobacillus graminis* PON 100244 and *Lactiplantibacillus plantarum* PON 100148) and obligate heterofermentative species (*Levilactobacillus brevis* PON 200571, *Furfurilactobacillus rossiae* PON 100500 and *Fructilactobacillus sanfransciscensis* PON 100336). It was observed eight classes of volatile compounds: acids, alcohols, aldehydes, esters, furans, ketones, phenols and terpenes, with the chemical complexity being similar to sourdough breads. In the sensorial aspect, all attributes evaluated for the pizzas were significantly different for judges. In general, the highest quality parameters were registered in the presence of obligate heterofermentative LAB, the highest scores at overall assessment were reached by pizzas produced with all selected obligate-facultative heterofermentative LAB in mixed culture.

Colomba, an Italian sweet leavened baked product was produced using type I sourdough starter. The pH registered during fermentation was low (3.55) when compared to other sourdoughs used for sweet, leavened products (3.7–4.1), however the final pH of the product was 5.37. Lactic and acetic acids were the major fermentation products, affecting the sensorial characteristics, however volatile organic compounds were also identified on the final product, the main volatile found were alcohols, aldehydes, esters and ketones. The colomba was characterized by a low LAB population, with only 10 species identified: *Latilactobacillus curvatus, Enterococcus* sp., *E. durans, E. faecalis, Leucon ostoc citreum, Leuconostoc pseudomesenteroides, Leuconostoc nesebteriudes* (2 biotypes), *Pediococcus pentosaceus* and *Streptococcus macedonicus*, all the microbes previously described on sourdough starter (Raimondi et al. 2017).

Considering all the potential of the sourdough starter microbiota in a fermentation process, it was possible to list diverse applications other than bakery products, yet the application is very sparse, and the publications seems to be repetitive, testing different ingredients to produce baked goods. Table 10.1 summarizes the different sourdough starter applications presented in this section.

10.4.4 Stabilization of Sourdough Microbiota for Different Applications

With the widely grown in the market in the last years, the stabilization of sourdough starter, through a drying process, could be an advantage for decreasing the maintenance expenses of sourdough starter while helping to keep high fermentation capacity and high cell viability in the bakery industry, being the main challenge of drying sourdough starter to keep the high cell viability after the process and the reconstitution to guarantee the final product characteristics. There are some papers in the literature concerning this challenge and for now there are no certainty of which is the better drying method as reviewed by Albagli et al. (2021).

The drying process can strongly affect microbial viability, flavour and shelf life of the final product (Brandt 2007). Hendek Ertop et al. (2018) reported that breads containing dried sourdough starter (freeze-dried, spray-dried, and oven-dried) showed higher volume, better bread shape and improved color when compared to control breads. Besides the drying process, the rehydration process and conditions are also very important to recover the dried cells of the microbes present in the sourdough starter, the rehydration media composition can strongly affect this process (Morgan et al. 2006). It has been shown that media containing cryoprotectant substances and a complex media can increase cell viability due to a high osmotic pressure controlling the process and due to providing nutrients and essential cell

Table 10.1 Different uses of sourdough starter

Objective	Agri-food products	Reference
Identify sourdough isolated LAB strains as potential antimicrobial agents on agri-food industry.	Antimicrobe	Bartkiene et al. (2020)
Use of sourdough starter cultures (*Candida krusei, S. cerevisiae, S. exiguss, Pichia saitoi, Lactiplantibacillus plantarum, L. casei, L. delbrueckii, Levilactobacillus brevis, Limosilactobacillus fermentum, L. pastorius, L. bucheri, L. acidophilus, Companilactobacillus alimentarius, Leuconostoc, Pediococcus* and *Streptococcus*) as biocontrol agents for agriculture	Biocontrol agente in agriculture	Williamson and Aylward (1993)
Investigate how three yeasts isolated from sourdough starter influences the metabolism of LAB to reduce wheat allergens on breads	Bread reduced of wheat allergens	Fu et al. (2021)
Produce wheat beer from a *S. cerevisiae* isolated from sourdough starter	Wheat beer	Mascia et al. (2015)
Investigate the usage of yeast species present in sourdough starter to produce low-alcohol beer	Low-alcohol beer	Johansson et al. (2021)
Investigate if LAB present in sourdough starter had probiotic characteristics (antifungal and antibacterial activities, low pH and bile salt tolerance ability, antibiotic resistance adherence to human gut cells, production of GABA and exopolysaccharides (EPS) production capability)	Probiotics	Kaya et al. (2022)
Identify and investigate the antifungal activity of LAB present in spontaneous sourdough starter	Antifungal	Arsoy et al. (2022)
Explore starch digestibility and properties of fresh pasta made with semolina and liquid sourdough	Pasta	Fois et al. (2018)
Evaluate pizza produced with different species of microbes present in sourdough	Pizza	Francesca et al. (2019)
Produce colomba with sourdough starter, identify present microbes and physic-chemical characteristics	Colomba	Raimondi et al. (2017)

components required to injured cells, respectively (Ray et al. 1971; Morgan et al. 2006). The rehydration temperature also affects cell recovery, as shown by Ray et al. (1971) a higher cell concentration was found using the range of 15–25 °C than 35 °C and 45 °C. As shown in a review by Albagli et al. (2021), freeze-drying has shown good results concerning cell viability, but cryoprotectant substances are needed, which increases cost of the process. The research lead by Stefanello and collaborators (2018) freeze-dried a starter using trehalose as a cryoprotectant substance and observed that increasing trehalose concentration (0–15% w/w) improved the microbial survival, however this cryoprotectant is expensive, making it not the best option for using, while others cryoprotectant being cheaper.

However, literature data points to spray-drying as the best drying process concerning cell viability, sensory and physical properties of the final product, being easier to scale up in industries than freeze-drying. Reale et al. (2019a, b) observed that a spray-dried sourdough starter presented higher LAB and yeast counts when compared to freeze-drying, oven-drying, and freezing. It was also observed that no statistical difference between breads made with spray-dried and control samples regarding sensorial characteristics.

The stabilization of sourdough starter could also improve its availability to the wide public, making it possible to be sold on markets or small bakeries with ease. It could also be used more easily as a food conserving method due to the production of bacteriocins after fermentation and to produce other foods, such as fermented beverages and pickle. This subject is relatively new, and it is expanding, with more publications every year, the drying methods, the rehydration of the dried powder can strongly influence the microbe viability, flavour and shelf life of the final product, however no concern of what is the best method to dry sourdough has been decided, needing more studies and testing.

10.5 Conclusions

Sourdough fermentation is considered the most traditional and effective tool to improve the nutritional and functional value of baking products. The microbiological diversity of the sourdough starter, with mainly Lactobacilli species and *Kazachstania*, *Torulopsis*, *Pichia*, *Candida*, *Yarrowia*, and *Saccharomyces*, enables the production of several metabolites during fermentation such as organic acids, exopolysaccharides, enzymes, volatile and bioactive compounds. So, there are possibilities of different applications, in addition to the bakery business, representing a new opportunity for the agri-food industry.

Sourdough starter cultures can be applied in a wide range of agri-food industries such as feed and pet food, dairy, meat, alcoholic and non-alcoholic beverages, nutraceuticals, and in other industrial sectors such as cosmetics and pharmaceuticals. Therefore, future efforts should be focused on exploring the microbiome of sourdough starter, including the drying process as an efficient strategy to reach stability, in order to expand its use.

References

Albagli G, Schwartz I d M, Amaral PFF, Ferreira TF, Finotelli PV (2021) How dried sourdough starter can enable and spread the use of sourdough bread. LWT 149:111888. https://doi.org/10.1016/J.LWT.2021.111888

Ale EC, Rojas MF, Reinheimer JA, Binetti AG (2020) Lactobacillus fermentum: could EPS production ability be responsible for functional properties? Food Microbiol 90:103465. https://doi.org/10.1016/J.FM.2020.103465

Ansorena D, Astiasarán I (2016) Fermented foods: composition and health effects. Encycl Food Heal:649–655. https://doi.org/10.1016/B978-0-12-384947-2.00285-3

Arendt EK, Moroni A, Zannini E (2011) Medical nutrition therapy: use of sourdough lactic acid bacteria as a cell factory for delivering functional biomolecules and food ingredients in gluten free bread. Microb Cell Factories 10:S15. https://doi.org/10.1186/1475-2859-10-S1-S15

Arendt EK, Ryan LAM, Dal Bello F (2007) Impact of sourdough on the texture of bread. Food Microbiol 24:165–174. https://doi.org/10.1016/j.fm.2006.07.011

Arora K, Ameur H, Polo A, Di Cagno R, Rizzello CG, Gobbetti M (2021) Thirty years of knowledge on sourdough fermentation: a systematic review. Trends Food Sci Technol 108:71–83. https://doi.org/10.1016/J.TIFS.2020.12.008

Arsoy ES, Gül LB, Çon AH (2022) Characterization and selection of potential antifungal lactic acid bacteria isolated from Turkish spontaneous sourdough. Curr Microbiol 79:148. https://doi.org/10.1007/S00284-022-02839-Z

Banu I, Vasilean I, Aprodu I (2011) Effect of select parameters of the sourdough rye fermentation on the activity of some mixed starter cultures. Food Biotechnol 25:275–291. https://doi.org/10.1080/08905436.2011.617251

Bartkiene E, Lele V, Ruzauskas M, Domig KJ, Starkute V, Zavistanaviciute P, Bartkevics V, Pugajeva I, Klupsaite D, Juodeikiene G, Mickiene R, Rocha JM (2020) Lactic acid bacteria isolation from spontaneous sourdough and their characterization including antimicrobial and antifungal properties evaluation. Microorganisms 8:64. https://doi.org/10.3390/MICROORGANISMS8010064

Bigey F, Segond D, Friedrich A, Guezenec S, Bourgais A, Huyghe L, Agier N, Nidelet T, Sicard D (2021) Evidence for two main domestication trajectories in Saccharomyces cerevisiae linked to distinct bread-making processes. Curr Biol 31:722–732.e5. https://doi.org/10.1016/J.CUB.2020.11.016

Boiocchi F, Porcellato D, Limonta L, Picozzi C, Vigentini I, Locatelli DP, Foschino R (2017) Insect frass in stored cereal products as a potential source of lactobacillus sanfranciscensis for sourdough ecosystem. J Appl Microbiol 123:944–955. https://doi.org/10.1111/jam.13546

Boyaci Gunduz CP, Agirman B, Gaglio R, Franciosi E, Francesca N, Settanni L, Erten H (2022) Evaluation of the variations in chemical and microbiological properties of the sourdoughs produced with selected lactic acid bacteria strains during fermentation. Food Chem X 14:100357. https://doi.org/10.1016/J.FOCHX.2022.100357

Brandt MJMJ (2007) Sourdough products for convenient use in baking. Food Microbiol 24:161–164. https://doi.org/10.1016/j.fm.2006.07.010

Business Wire (2020) $5.51 Bn Sourdough Market Insights, 2018–2027 – ResearchAndMarkets.com | Business Wire. In: $5.51 Bn Sourdough Mark. Insights, 2018–2027. https://www.businesswire.com/news/home/20200331005436/en/5.51-Bn-Sourdough-Market-Insights-2018-2027%2D%2D-ResearchAndMarkets.com. Accessed 5 Oct 2021

Cai SS, Zhou Y, Ye BC (2021) Reducing the reproductive toxicity activity of Lactiplantibacillus plantarum: a review of mechanisms and prospects. Environ Sci Pollut Res Int 28:36927–36941. https://doi.org/10.1007/S11356-021-14403-6

Calvert MD, Madden AA, Nichols LM, Haddad NM, Lahne J, Dunn RR, McKenney EA (2021) A review of sourdough starters: ecology, practices, and sensory quality with applications for baking and recommendations for future research. PeerJ 9:e11389. https://doi.org/10.7717/PEERJ.11389

Carbó R, Gordún E, Fernández A, Ginovart M (2020) Elaboration of a spontaneous gluten-free sourdough with a mixture of amaranth, buckwheat, and quinoa flours analyzing microbial load, acidity, and pH. Food Sci Technol Int 26:344–352. https://doi.org/10.1177/1082013219895357

Carbonetto B, Nidelet T, Guezenec S, Perez M, Segond D, Sicard D (2020) Interactions between Kazachstania humilis yeast species and lactic acid bacteria in sourdough. Microorganisms 8. https://doi.org/10.3390/MICROORGANISMS8020240

Carnevali P, Ciati R, Leporati A, Paese M (2007) Liquid sourdough fermentation: industrial application perspectives. Food Microbiol 24:150–154. https://doi.org/10.1016/j.fm.2006.07.009

Chavan RS, Chavan SR (2011) Sourdough technology-a traditional way for wholesome foods: a review. Compr Rev Food Sci Food Saf 10:169–182. https://doi.org/10.1111/j.1541-4337.2011.00148.x

Chumngoen W, Tan FJ (2015) Relationships between descriptive sensory attributes and physicochemical analysis of broiler and Taiwan native chicken breast meat. Asian-Australas J Anim Sci 28:1028–1037. https://doi.org/10.5713/AJAS.14.0275

Coda R, Rizzello CGG, Gobbetti M (2010) Use of sourdough fermentation and pseudo-cereals and leguminous flours for the making of a functional bread enriched of γ-aminobutyric acid (GABA). Int J Food Microbiol 137:236–245. https://doi.org/10.1016/j.ijfoodmicro.2009.12.010

Corsetti A, Gobbetti M, Balestrieri F, Paoletti F, Russi L, Rossi J (1998) Sourdough lactic acid bacteria effects on bread firmness and staling. J Food Sci 63:347–351. https://doi.org/10.1111/j.1365-2621.1998.tb15739.x

Corsetti A, Settanni L (2007) Lactobacilli in sourdough fermentation. Food Res Int 40:539–558. https://doi.org/10.1016/j.foodres.2006.11.001

Cory H, Passarelli S, Szeto J, Tamez M, Mattei J (2018) The role of polyphenols in human health and food systems: a mini-review. Front Nutr 5:87. https://doi.org/10.3389/FNUT.2018.00087

Coton M, Berthier F, Coton E (2008) Rapid identification of the three major species of dairy obligate heterofermenters Lactobacillus brevis, Lactobacillus fermentum and Lactobacillus parabuchneri by species-specific duplex PCR. FEMS Microbiol Lett 284:150–157. https://doi.org/10.1111/J.1574-6968.2008.01206.X

Dal Bello F, Hertel C (2006) Oral cavity as natural reservoir for intestinal lactobacilli. Syst Appl Microbiol 29:69–76. https://doi.org/10.1016/J.SYAPM.2005.07.002

Damiani P, Gobbetti M, Cossignani L, Corsetti A, Simonetti MS, Rossi J (1996) The sourdough microflora. Characterization of hetero- and homofermentative lactic acid bacteria, yeasts and their interactions on the basis of the volatile compounds produced. LWT – Food Sci Technol 29:63–70. https://doi.org/10.1006/fstl.1996.0009

De Vero L, Iosca G, La China S, Licciardello F, Gullo M, Pulvirenti A (2021) Yeasts and lactic acid bacteria for panettone production: an assessment of candidate strains. Microorganisms 9:1093. https://doi.org/10.3390/MICROORGANISMS9051093

De Vuyst L, Comasio A, Van Kerrebroeck S (2021) Sourdough production: fermentation strategies, microbial ecology, and use of non-flour ingredients. Crit Rev Food Sci Nutr 63:2447–2479. https://doi.org/10.1080/10408398.2021.1976100

De Vuyst L, Harth H, Van Kerrebroeck S, Leroy F (2016) Yeast diversity of sourdoughs and associated metabolic properties and functionalities. Int J Food Microbiol 239:26–34. https://doi.org/10.1016/j.ijfoodmicro.2016.07.018

De Vuyst L, Van Kerrebroeck S, Harth H, Huys G, Daniel H-MM, Weckx S (2014) Microbial ecology of sourdough fermentations: diverse or uniform? Food Microbiol 37:11–29. https://doi.org/10.1016/j.fm.2013.06.002

De Vuyst L, Vrancken G, Ravyts F, Rimaux T, Weckx S (2009) Biodiversity, ecological determinants, and metabolic exploitation of sourdough microbiota. Food Microbiol 26:666–675. https://doi.org/10.1016/j.fm.2009.07.012

Decock P, Cappelle S (2005) Bread technology and sourdough technology. Trends Food Sci Technol 16:113–120

Di Cagno R, Surico RF, Siragusa S, De Angelis M, Paradiso A, Minervini F, De Gara L, Gobbetti M (2008) Selection and use of autochthonous mixed starter for lactic acid fermentation of carrots, French beans or marrows. Int J Food Microbiol 127:220–228. https://doi.org/10.1016/J.IJFOODMICRO.2008.07.010

Dlusskaya E, Jänsch A, Schwab C et al (2008) Microbial and chemical analysis of a kvass fermentation. Eur Food Res Technol 227:261–266. https://doi.org/10.1007/s00217-007-0719-4

Falasconi I, Fontana A, Patrone V, Rebecchi A, Garrido GD, Principato L, Callegari ML, Spigno G, Morelli L (2020) Genome-assisted characterization of lactobacillus fermentum, Weissella cibaria, and Weissella confusa strains isolated from sorghum as starters for sourdough fermentation. Microorganisms 8:1–16. https://doi.org/10.3390/MICROORGANISMS8091388

FAO (2017) The future of food and agriculture – trends and challenges, Rome

Fernandes T, Silva-Sousa F, Pereira F, Rito T, Soares P, Franco-Duarte R, Sousa MJ (2021) Biotechnological importance of Torulaspora delbrueckii: from the obscurity to the spotlight. J Fungi 7:712. https://doi.org/10.3390/JOF7090712

Filannino P, Di Cagno R, Gobbetti M (2018) Metabolic and functional paths of lactic acid bacteria in plant foods: get out of the labyrinth. Curr Opin Biotechnol 49:64–72. https://doi.org/10.1016/J.COPBIO.2017.07.016

Filannino P, Gobbetti M, De Angelis M, Di Cagno R (2014) Hydroxycinnamic acids used as external acceptors of electrons: an energetic advantage for strictly heterofermentative lactic acid bacteria. Appl Environ Microbiol 80:7574–7582. https://doi.org/10.1128/AEM.02413-14

Fois S, Piu PP, Sanna M, Roggio T, Catzeddu P (2018) Starch digestibility and properties of fresh pasta made with semolina-based liquid sourdough. LWT 89:496–502. https://doi.org/10.1016/J.LWT.2017.11.030

Francesca N, Gaglio R, Alfonzo A, Corona O, Moschetti G, Settanni L (2019) Characteristics of sourdoughs and baked pizzas as affected by starter culture inoculums. Int J Food Microbiol 293:114–123. https://doi.org/10.1016/J.IJFOODMICRO.2019.01.009

Fu W, Liu C, Meng X, Tao S, Xue W (2021) Co-culture fermentation of Pediococcus acidilactici XZ31 and yeast for enhanced degradation of wheat allergens. Int J Food Microbiol 347:109190. https://doi.org/10.1016/J.IJFOODMICRO.2021.109190

Gänzle MG (2014a) BREAD | sourdough bread. In: Encyclopedia food microbiology, 2nd edn, pp 309–315. https://doi.org/10.1016/B978-0-12-384730-0.00045-8

Gänzle MG (2014b) Enzymatic and bacterial conversions during sourdough fermentation. Food Microbiol 37:2–10. https://doi.org/10.1016/j.fm.2013.04.007

Gänzle MG, De Vuyst L (2004) Sourdough fermentation: from fundamentals to applications. Cereal Foods World 49:42–43

Gänzle MG, Zheng J (2019) Lifestyles of sourdough lactobacilli – do they matter for microbial ecology and bread quality? Int J Food Microbiol 302:15–23. https://doi.org/10.1016/j.ijfoodmicro.2018.08.019

Garcia-Gonzalez N, Battista N, Prete R, Corsetti A (2021) Health-promoting role of Lactiplantibacillus plantarum isolated from fermented foods. Microorganisms 9:1–30. https://doi.org/10.3390/MICROORGANISMS9020349

Gobbetti M, Minervini F, Pontonio E, Di Cagno R, De Angelis M (2016) Drivers for the establishment and composition of the sourdough lactic acid bacteria biota. Int J Food Microbiol 239:3–18. https://doi.org/10.1016/j.ijfoodmicro.2016.05.022

Graça C, Lima A, Raymundo A, Sousa I (2021) Sourdough Fermentation as a Tool to Improve the Nutritional and Health-Promoting Properties of Its Derived-Products. Ferment 7:246. https://doi.org/10.3390/FERMENTATION7040246

Guerzoni ME, Serrazanetti DI, Vernocchi P, Gianotti A (2013) Physiology and biochemistry of sourdough yeasts. In: Handbook on sourdough biotechnology. Springer, Boston, MA, pp 155–181

Hajinia F, Sadeghi A, Sadeghi Mahoonak A (2021) The use of antifungal oat-sourdough lactic acid bacteria to improve safety and technological functionalities of the supplemented wheat bread. J Food Saf 41:e12873. https://doi.org/10.1111/JFS.12873

Hansen A, Schieberle P (2005) Generation of aroma compounds during sourdough fermentation: applied and fundamental aspects. Trends in Food Science and Technology, In

Harth H, Van Kerrebroeck S, De Vuyst L (2016) Community dynamics and metabolite target analysis of spontaneous, backslopped barley sourdough fermentations under laboratory and bakery conditions. Int J Food Microbiol 228:22–32. https://doi.org/10.1016/j.ijfoodmicro.2016.04.011

Hendek Ertop M, Ilter SMM, YilMaz F, BaltaCi C, Gündogdu A (2018) Quality properties of wheat breads incorporated with dried sourdoughs produced with different fermentation and drying methods. Food Sci Technol Res 24:971–980. https://doi.org/10.3136/fstr.24.971

Jin J, Nguyen TTH, Humayun S, Park SH, Oh H, Lim S, Mok IK, Li Y, Pal K, Kim D (2021) Characteristics of sourdough bread fermented with Pediococcus pentosaceus and Saccharomyces cerevisiae and its bio-preservative effect against Aspergillus flavus. Food Chem 345:128787. https://doi.org/10.1016/J.FOODCHEM.2020.128787

Johansson L, Nikulin J, Juvonen R, Krogerus K, Magalhães F, Mikkelson A, Nuppunen-Puputti M, Sohlberg E, de Francesco G, Perretti G, Gibson B (2021) Sourdough cultures as reservoirs of maltose-negative yeasts for low-alcohol beer brewing. Food Microbiol 94:103629. https://doi.org/10.1016/J.FM.2020.103629

Kamiński M, Skonieczna-Żydecka K, Nowak JK, Stachowska E (2020) Global and local diet popularity rankings, their secular trends, and seasonal variation in Google trends data. Nutrition 79-80:110759. https://doi.org/10.1016/j.nut.2020.110759

Katina K, Heiniö RL, Autio K, Poutanen K (2006) Optimization of sourdough process for improved sensory profile and texture of wheat bread. LWT – Food Sci Technol 39:1189–1202. https://doi.org/10.1016/j.lwt.2005.08.001

Katsi P, Kosma IS, Michailidou S, Argiriou A, Badeka AV, Kontominas MG (2021) Characterization of artisanal spontaneous sourdough wheat bread from Central Greece: evaluation of physico-chemical, microbiological, and sensory properties in relation to conventional yeast leavened wheat bread. Foods (Basel, Switzerland) 10. https://doi.org/10.3390/FOODS10030635

Kaya Y, Erten T, Vurmaz M, İspirli H, Şimşek Ö, Dertli E (2022) Comparison of the probiotic characteristics of lactic acid bacteria (LAB) isolated from sourdough and infant feces. Food Biosci 47:101722. https://doi.org/10.1016/J.FBIO.2022.101722

Korcari D, Secchiero R, Laureati M, Marti A, Cardone G, Rabitti NS, Ricci G, Fortina MG (2021) Technological properties, shelf life and consumer preference of spelt-based sourdough bread using novel, selected starter cultures. LWT 151:112097. https://doi.org/10.1016/J.LWT.2021.112097

Kumar N, Goel N (2019) Phenolic acids: natural versatile molecules with promising therapeutic applications. Biotechnol Rep (Amsterdam, Netherlands) 24:e00370. https://doi.org/10.1016/J.BTRE.2019.E00370

Landis EA, Oliverio AM, McKenney EA, Nichols LM, Kfoury N, Biango-Daniels M, Shell LK, Madden AA, Shapiro L, Sakunala S, Drake K, Robbat A, Booker M, Dunn RR, Fierer N, Wolfe BE (2021) The diversity and function of sourdough starter microbiomes. Elife 10:1–24. https://doi.org/10.7554/ELIFE.61644

Lazaridou A, Duta D, Papageorgiou M, Belc N, Biliaderis CG (2007) Effects of hydrocolloids on dough rheology and bread quality parameters in gluten-free formulations. J Food Eng 79:1033–1047. https://doi.org/10.1016/J.JFOODENG.2006.03.032

Lee HC, Jenner AM, Low CS, Lee YK (2006) Effect of tea phenolics and their aromatic fecal bacterial metabolites on intestinal microbiota. Res Microbiol 157:876–884. https://doi.org/10.1016/J.RESMIC.2006.07.004

Leenhardt F, Levrat-Verny MA, Chanliaud E, Rémésy C (2004) Moderate decrease of pH by sourdough fermentation is sufficient to reduce phytate content of whole wheat flour through endogenous phytase activity. J Agric Food Chem 53:98–102. https://doi.org/10.1021/JF049193Q

Limbad M, Maddox NG, Hamid N, Kantono K (2020) Sensory and physicochemical characterization of sourdough bread prepared with a coconut water kefir starter. Foods 9:1165. https://doi.org/10.3390/foods9091165

Liu T, Li Y, Sadiq FA, Yang H, Gu J, Yuan L, Lee YK, He G (2018) Predominant yeasts in Chinese traditional sourdough and their influence on aroma formation in Chinese steamed bread. Food Chem 242:404–411. https://doi.org/10.1016/j.foodchem.2017.09.081

López-Huertas E (2015) Safety and efficacy of human breast milk lactobacillus fermentum CECT 5716. A mini-review of studies with infant formulae. Benef Microbes 6:219–224. https://doi.org/10.3920/BM2014.0091

Luti S, Galli V, Venturi M, Granchi L, Paoli P, Pazzagli L (2021) Bioactive properties of breads made with sourdough of hull-less barley or conventional and pigmented wheat flours. Appl Sci 11:3291. https://doi.org/10.3390/APP11073291

Martău GA, Teleky BE, Ranga F, Pop ID, Vodnar DC (2021) Apple pomace as a sustainable substrate in sourdough fermentation. Front Microbiol 12:3850. https://doi.org/10.3389/FMICB.2021.742020/BIBTEX

Mascia I, Fadda C, Dostálek P, Karabín M, Zara G, Budroni M, Del Caro A (2015) Is it possible to create an innovative craft durum wheat beer with sourdough yeasts? A case study. J Inst Brew 121:283–286. https://doi.org/10.1002/jib.215

Meignen B, Onno B, Gélinas P, Infantes M, Guilois S, Cahagnier B (2001) Optimization of sourdough fermentation with lactobacillus brevis and baker's yeast. Food Microbiol 18:239–245. https://doi.org/10.1006/FMIC.2000.0395

Michel E, Masson E, Bubbendorf S, Lapicque L, Nidelet T, Segond D, Guézenec S, Marlin T, deVillers H, Rué O, Onno B, Legrand J, Sicard D, bakers the participating (2022) Artisanal and farmers bread making practices differently shape fungal species diversity in French sourdoughs bioRxiv 679472. https://doi.org/10.1101/679472

Montemurro M, Coda R, Rizzello CG (2019) Recent advances in the use of sourdough biotechnology in pasta making. Foods (Basel, Switzerland) 8. https://doi.org/10.3390/FOODS8040129

Morgan CA, Herman N, White PA, Vesey G (2006) Preservation of microorganisms by drying; a review. J Microbiol Methods 66:183–193. https://doi.org/10.1016/j.mimet.2006.02.017

Mukisa IM, Byaruhanga YB, Muyanja CMBK, Langsrud T, Narvhus JA (2017) Production of organic flavor compounds by dominant lactic acid bacteria and yeasts from Obushera, a traditional sorghum malt fermented beverage. Food Sci Nutr 5:702–712. https://doi.org/10.1002/FSN3.450

Naghmouchi K, Belguesmia Y, Bendali F, Spano G, Seal BS, Drider D (2020) Lactobacillus fermentum: a bacterial species with potential for food preservation and biomedical applications. Crit Rev Food Sci Nutr 60:3387–3399. https://doi.org/10.1080/10408398.2019.1688250

Naidu AS, Bidlack WR, Clemens RA (1999) Probiotic spectra of lactic acid bacteria (LAB). Crit Rev Food Sci Nutr 39:13–126. https://doi.org/10.1080/10408699991279187

Nordström EA, Teixeira C, Montelius C, Jeppsson B, Larsson N (2021) Lactiplantibacillus plantarum 299v (LP299V ®): three decades of research. Benef Microbes 12:441–465. https://doi.org/10.3920/BM2020.0191

Park JS, Shin E, Hong H, Shin HJ, Cho YH, Ahn KH, Paek K, Lee Y (2015) Characterization of lactobacillus fermentum PL9988 isolated from healthy elderly Korean in a longevity village. J Microbiol Biotechnol 25:1514–1522. https://doi.org/10.4014/JMB.1505.05015

Pérez-Jiménez J, Neveu V, Vos F, Scalbert A (2010) Systematic analysis of the content of 502 polyphenols in 452 foods and beverages: an application of the phenol-explorer database. J Agric Food Chem 58:4959–4969. https://doi.org/10.1021/JF100128B

Pétel C, Onno B, Prost C (2017) Sourdough volatile compounds and their contribution to bread: a review. Trends Food Sci Technol 59:105–123. https://doi.org/10.1016/j.tifs.2016.10.015

Porto MCW, Kuniyoshi TM, Azevedo POS, Vitolo M, Oliveira RPS (2017) Pediococcus spp.: an important genus of lactic acid bacteria and pediocin producers. Biotechnol Adv 35:361–374. https://doi.org/10.1016/J.BIOTECHADV.2017.03.004

Raimondi S, Amaretti A, Rossi M, Fall PA, Tabanelli G, Gardini F, Montanari C (2017) Evolution of microbial community and chemical properties of a sourdough during the production of Colomba, an Italian sweet leavened baked product. LWT 86:31–39. https://doi.org/10.1016/J.LWT.2017.07.042

Ramírez M, Velázquez R (2018) The yeast Torulaspora delbrueckii: an interesting but difficult-to-use tool for winemaking. Ferment 4:94. https://doi.org/10.3390/FERMENTATION4040094

Ray B, Jezeski JJ, Busta FF (1971) Effect of rehydration on recovery, repair, and growth of injured freeze-dried Salmonella anatum. Appl Microbiol 22:184–189. https://doi.org/10.1128/aem.22.2.184-189.1971

Reale A, Di Renzo T, Boscaino F, Nazzaro F, Fratianni F, Aponte M (2019a) Lactic acid bacteria biota and aroma profile of italian traditional sourdoughs from the irpinian area in Italy. Front Microbiol 10:1621. https://doi.org/10.3389/fmicb.2019.01621

Reale A, Di Renzo T, Preziuso M, Panfili G, Cipriano L, Messia MC (2019b) Stabilization of sourdough starter by spray drying technique: new breadmaking perspective. LWT 99:468–475. https://doi.org/10.1016/j.lwt.2018.10.016

Research GV Global Sourdough Market Trends, Analysis Report, 2019-2025. https://www.grandviewresearch.com/industry-analysis/sourdough-market. Accessed 3 Feb 2022

Ripari V, Bai Y, Gänzle M (2019) Metabolism of phenolic acids in whole wheat and rye malt sourdoughs. Food Microbiol 77:43–51. https://doi.org/10.1016/j.fm.2018.08.009

Rodríguez-Sojo MJ, Ruiz-Malagón AJ, Rodríguez-Cabezas ME, Gálvez J, Rodríguez-Nogales A (2021) Limosilactobacillus fermentum CECT5716: mechanisms and therapeutic insights. Nutrients 13:1–22. https://doi.org/10.3390/NU13031016

Rogalski E, Ehrmann MA, Vogel RF (2020) Role of Kazachstania humilis and Saccharomyces cerevisiae in the strain-specific assertiveness of Fructilactobacillus sanfranciscensis strains in rye sourdough. Eur Food Res Technol 2469(246):1817–1827. https://doi.org/10.1007/S00217-020-03535-7

Rogalski E, Ehrmann MA, Vogel RF (2021) Strain-specific interaction of Fructilactobacillus sanfranciscensis with yeasts in the sourdough fermentation. Eur Food Res Technol 247:1437–1447. https://doi.org/10.1007/S00217-021-03722-0/TABLES/2

Rönkä E, Malinen E, Saarela M, Rinta-Koski M, Aarnikunnas J, Palva A (2003) Probiotic and milk technological properties of lactobacillus brevis. Int J Food Microbiol 83:63–74. https://doi.org/10.1016/S0168-1605(02)00315-X

Sánchez-Maldonado AF, Schieber A, Gänzle MG (2011) Structure-function relationships of the antibacterial activity of phenolic acids and their metabolism by lactic acid bacteria. J Appl Microbiol 111:1176–1184. https://doi.org/10.1111/J.1365-2672.2011.05141.X

Sandez Penidez SH, Velasco Manini MA, LeBlanc JG, Gerez CL, Rollán GC (2022) Quinoa sourdough-based biscuits with high antioxidant activity fermented with autochthonous lactic acid bacteria. J Appl Microbiol 132:2093–2105. https://doi.org/10.1111/JAM.15315

Santamaría L, Reverón I, de Felipe FL, de Las Rivas B, Muñoz R (2018) Unravelling the reduction pathway as an alternative metabolic route to Hydroxycinnamate decarboxylation in lactobacillus plantarum. Appl Environ Microbiol 84. https://doi.org/10.1128/AEM.01123-18

Schwab C, Mastrangelo M, Corsetti A, Gänzle M (2008) Formation of oligosaccharides and polysaccharides by lactobacillus reuteri LTH5448 and Weissella cibaria 10M in sorghum sourdoughs. Cereal Chem 85:679–684. https://doi.org/10.1094/CCHEM-85-5-0679

Sekwati-Monang B, Valcheva R, Gänzle MG (2012) Microbial ecology of sorghum sourdoughs: effect of substrate supply and phenolic compounds on composition of fermentation microbiota. Int J Food Microbiol 159:240–246. https://doi.org/10.1016/j.ijfoodmicro.2012.09.013

Svensson L, Sekwati-Monang B, Lutz DL, Schieber R, Gänzle MG (2010) Phenolic acids and flavonoids in nonfermented and fermented red sorghum (Sorghum bicolor (L.) Moench). J Agric Food Chem 58:9214–9220. https://doi.org/10.1021/JF101504V/SUPPL_FILE/JF101504V_SI_001.PDF

Shiri A, Ehrampoush MH, Yasini Ardakani SA, Shamsi F, Mollakhalili-Meybodi N (2021) Technological characteristics of inulin enriched gluten-free bread: effect of acorn flour replacement and fermentation type. Food Sci Nutr 9:6139–6151. https://doi.org/10.1002/FSN3.2567

Sidhu PK, Nehra K (2021) Purification and characterization of bacteriocin Bac23 extracted from lactobacillus plantarum PKLP5 and its interaction with silver nanoparticles for enhanced antimicrobial spectrum against food-borne pathogens. LWT 139:110546. https://doi.org/10.1016/J.LWT.2020.110546

Siepmann FB, Sousa de Almeida B, Waszczynskyj N, Spier MR (2019) Influence of temperature and of starter culture on biochemical characteristics and the aromatic compounds evolution on type II sourdough and wheat bread. LWT 108:199–206. https://doi.org/10.1016/j.lwt.2019.03.065

Singh BP, Aluko RE, Hati S, Solanki D (2022) Bioactive peptides in the management of lifestyle-related diseases: current trends and future perspectives. Crit Rev Food Sci Nutr 62:4593–4606. https://doi.org/10.1080/10408398.2021.1877109

Somashekaraiah R, Mottawea W, Gunduraj A, Joshi U, Hammami R, Sreenivasa MY (2021) Probiotic and antifungal attributes of Levilactobacillus brevis MYSN105, isolated from an Indian traditional fermented food Pozha. Front Microbiol 12:1728. https://doi.org/10.3389/FMICB.2021.696267/BIBTEX

Srivastava RK (2018) Enhanced shelf life with improved food quality from fermentation processes. J Food Technol Preserv 2:1–7

Stefanello RF, Machado AAR, Pasqualin Cavalheiro C, Bartholomei Santos ML, Nabeshima EH, Copetti MV, Fries LLM (2018) Trehalose as a cryoprotectant in freeze-dried wheat sourdough production. LWT – Food Sci Technol 89:510–517. https://doi.org/10.1016/j.lwt.2017.11.011

Stewart GG (2014) Saccharomyces | Saccharomyces cerevisiae. In: Encyclopedia food microbiology, 2nd edn, pp 309–315. https://doi.org/10.1016/B978-0-12-384730-0.00292-5

Strack L, Carli RC, da Silva RV, Sartor KB, Colla LM, Reinehr CO (2020) Food biopreservation using antimicrobials produced by lactic acid bacteria. Res Soc Dev 9:e998986666. https://doi.org/10.33448/RSD-V9I8.6666

Straus S, Bavec F, Turinek M, Slatnar A, Rozman C, Bavec M (2012) Nutritional value and economic feasibility of red beetroot (Beta vulgaris L. ssp. vulgaris Rote Kugel) from different production systems. African J Agric Res 7:5653–5660. https://doi.org/10.5897/AJAR12.1519

Suo B, Chen X, Wang Y (2021) Recent research advances of lactic acid bacteria in sourdough: origin, diversity, and function. Curr Opin Food Sci 37:66–75. https://doi.org/10.1016/J.COFS.2020.09.007

Suryawanshi N, Naik S, Jujjawarapu SE (2022) Exopolysaccharides and their applications in food processing industries. Food Sci Appl Biotechnol, [Sl] 5(1):22–44. https://doi.org/10.30721/fsab2022.v5.i1.165

Tang KX, Zhao CJ, Gänzle MG (2017) Effect of glutathione on the taste and texture of type I sourdough bread. J Agric Food Chem 65:4321–4328. https://doi.org/10.1021/acs.jafc.7b00897

Thiele C, Grassl S, Gänzle M (2004) Gluten hydrolysis and depolymerization during sourdough fermentation. J Agric Food Chem 52:1307–1314. https://doi.org/10.1021/JF034470Z

Todorov SD (2009) Bacteriocins from Lactobacillus plantarum – production, genetic organization and mode of action: produção, organização genética e modo de ação. Brazilian J Microbiol 40:209–221. https://doi.org/10.1590/S1517-83822009000200001

Tuladhar P, Sasidharan S, Saudagar P (2021) Role of phenols and polyphenols in plant defense response to biotic and abiotic stresses. In: Biocontrol agents and secondary metabolites. Woodhead Publishing, pp 419–441. https://doi.org/10.1016/B978-0-12-822919-4.00017-X

Valmorri S, Tofalo R, Settanni L, Corsetti A, Suzzi G (2010) Yeast microbiota associated with spontaneous sourdough fermentations in the production of traditional wheat sourdough breads of the Abruzzo region (Italy). Antonie Van Leeuwenhoek 97:119–129. https://doi.org/10.1007/S10482-009-9392-X

Van Beek S, Priest FG (2000) Decarboxylation of substituted cinnamic acids by lactic acid bacteria isolated during malt whisky fermentation. Appl Environ Microbiol 66:5322–5328. https://doi.org/10.1128/AEM.66.12.5322-5328.2000

Ventimiglia G, Alfonzo A, Galluzzo P, Corona O, Francesca N, Caracappa S, Moschetti G, Settanni L (2015) Codominance of Lactobacillus plantarum and obligate heterofermentative lactic acid bacteria during sourdough fermentation. Food Microbiol 51:57–68. https://doi.org/10.1016/j.fm.2015.04.011

Wahyono A, Lee S-B, Kang W-W, Park H-D (2016) Improving bread quality using co-cultures of Saccharomyces cerevisiae, Torulaspora delbrueckii JK08, and Pichia anomala JK04. Ital J Food Sci 28:298–313. https://doi.org/10.14674/1120-1770/IJFS.V147

Wang T, He F, Chen G (2014) Improving bioaccessibility and bioavailability of phenolic compounds in cereal grains through processing technologies: a concise review. J Funct Foods 7:101–111. https://doi.org/10.1016/J.JFF.2014.01.033

Williamson MA, Aylward, James Harrison (1993) Biocontrol agents for use in agriculture – Google Patents

Wittwer AE, Sicard D, Howell KS (2022) Kazachstania humilis. Trends Microbiol 30:1012–1013. https://doi.org/10.1016/J.TIM.2022.05.007

Wolter A, Hager A-S, Zannini E, Czerny M, Arendt EK (2014) Influence of dextran-producing Weissella cibaria on baking properties and sensory profile of gluten-free and wheat breads. Int J Food Microbiol 172:83–91. https://doi.org/10.1016/j.ijfoodmicro.2013.11.015

Yazar G, Tavman Ş (2012) Functional and technological aspects of sourdough fermentation with Lactobacillus sanfranciscensis. Food Eng Rev 4:171–190. https://doi.org/10.1007/s12393-012-9052-1

Zheng J, Wittouck S, Salvetti E, Franz CMAP, Harris HMB, Mattarelli P, O'Toole PW, Pot B, Vandamme P, Walter J, Watanabe K, Wuyts S, Felis GE, Gänzle MG, Lebeer S (2020) A taxonomic note on the genus Lactobacillus: description of 23 novel genera, emended description of the genus Lactobacillus Beijerinck 1901, and union of Lactobacillaceae and Leuconostocaceae. Int J Syst Evol Microbiol 70:2782–2858. https://doi.org/10.1099/IJSEM.0.004107

Zuchowski J, Jonczyk K, Pecio L, Oleszek W (2011) Phenolic acid concentrations in organically and conventionally cultivated spring and winter wheat. J Sci Food Agric 91:1089–1095. https://doi.org/10.1002/JSFA.4288

Chapter 11
Sourdough Lactic Acid Bacteria as Functional Microorganisms for Technological Purposes

Zühal Alkay and Enes Dertli

11.1 Introduction

The typical properties of sourdough which relies on the activities of lactic acid bacteria (LAB) and yeasts, are primarily originated from its natural microflora. These microbial species keep the sourdough metabolically active and it can be also reactivate the dough for different applications (Corsetti and Settanni 2007). The microorganisms involved in sourdough were discovered at around early 1900s (Catzeddu 2019). Sourdough starters are low in terms of microbial variety (particularly phylogenetic variety) when compared to physical systems such as soil, intestines and many environmental systems. Sourdough thus ensure a comparatively basic system for studying and testing theories of microbial ecology (Calvert et al. 2021). Therefore, since about 1990, a regular research efficiency aiming to rediscover the role of sourdough as leavening agent and to ensure the indispensable technology transmission has begun. The first central point of the research was how sourdough fermentation affects taste, rheology, and shelf life. Because the technological effects of microbial interactions in this kind of a mixed ecosystem are important. At the same time, the microbial variety and particularly the succession of dominant lactic acid bacterial populations has led these studies to tests structure of sourdough (Arora et al. 2021). The diversity of the sourdough habitat is dependent on factors such as ecological and technological, metabolic adaptation to sourdough stress situations, nutritional interactions between microorganisms such as lactobacilli and

Z. Alkay
Food Engineering Department, Engineering Faculty, Necmettin Erbakan University, Konya, Turkey

E. Dertli (✉)
Department of Food Engineering, Faculty of Chemical and Metallurgical Engineering, Yıldız Technical University, İstanbul, Turkey

E. B. Ceresino et al. (eds.), *Sourdough Microbiota and Starter Cultures for Industry*, https://doi.org/10.1007/978-3-031-48604-3_11

yeasts, and the intrinsic robustness or weakness of microorganisms also have an impact on the microbial variety of sourdough (Gobbetti et al. 2016).

This chapter ensure an overview of the technological purposes of sourdough LABs with a focus on production of exopolysaccharides and enzymes, synthesis of antifungal and antimicrobial compounds, and formation of proteolytic and bioactive compounds.

11.2 LAB in Sourdough

Sourdough is formed by the fermentation of dough which is prepared with flour and water. Because flour is not sterilized, the creation and number of specific microorganism species rely on some parameters. These are combinations of present substrates and particular technological factors (Corsetti and Settanni 2007). LAB and yeasts dominate the habitat of sourdough environment. However, different microorganisms as acetic acid bacteria may be present at the starting of fermentation (Li et al. 2016; Ripari et al. 2016). Because the sourdough ecosystem has a hard-pressed habitat, the microbiota must adapt to factors such as changing carbohydrates, nutrient substance, and low pH (Martín-Garcia et al. 2021). Sourdough has the feature of being used as a starter culture in products such as dough and bread. In fact, it is a mixture of flour, effective yeast and LAB. After a while, a change occurs in the structure of the dough which is left to itself without adding yeast, and gas blisters create in it. Thus, the self-opening dough becomes soft and causes odor variances. These changes in dough can be caused by microorganisms of the environment as well as the flour and the water. Sourdough is a self-fermenting dough that creates a sour taste. It is known that sourdough includes homofermentative and heterofermentative LAB and yeast in changing rates and compounds (Yıldız et al. 2019). The formation of sourdough lactobacilli from obligate and facultatively heterofermentative and obligately homofermentative species can be attributed to type 0, type I, type II and type III sourdough types (Corsetti and Settanni 2007). LAB species in sourdough types are shown in Fig. 11.1 (Yıldız et al. 2019).

Type I sourdough are often used and preserved in artisanal bakeries that produce traditional bread. In artisan bakery applications, fermentation conditions such as preservation, ambient temperature, long of fermentation time, refresh rate include efficient control of the fermentation microbiota. For this reason, most of the sourdoughs in bakeries contain *Fructilactobacillus sanfranciscensis* strain (formerly known as *Lactobacillus sanfranciscensis*). *Frul. sanfranciscensis* strain as obligate heterofermentative LAB is common in both San Francisco sourdough and Panettone production. In addition, it is the dominant strain in numerous conventional Italian bakery products and snacks such as Pandoro, Colomba, and Cornettobrioche (De Vuyst and Vancanneyt 2007). Optimum pH values of LAB in sourdough range from 5 to 6 (Gobbetti and Gänzle 2012). Household sourdoughs require less stringent situation for fermentation and

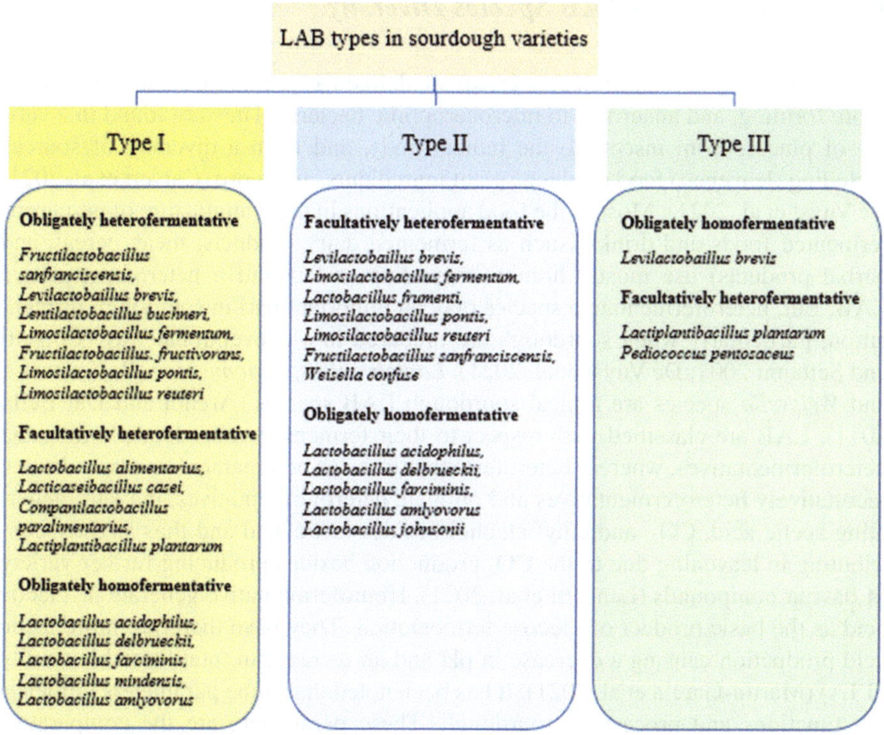

Fig. 11.1 LAB types in Type I, Type II and Type III sourdoughs. (Adapted from Yıldız et al. (2019))

care. Thus, the fermentation microbiota becomes more dependent on regional environmental situations, production applications, and materials. Another type of sourdough is Type II sourdough. Temperature-controlled fermentation, flour, water, and certain strains of acid-permissive LAB species is required to form this dough. Acid-permissive LAB species are *Lactobacillus amylovorus*, *Levilactobacillus brevis*, *Limosilactobacillus fermentum*, *Lactiplantibacillus plantarum,* and *Limosilactobacillus reuteri*. *Furfurilactobacillus rossiae* provides rapid acidification and produces desirable organoleptic properties (De Vuyst et al. 2021). LAB strains such as *Lactiplantibacillus plantarum* may be related to more than one environment. Hence, they are the first colonizers of spontaneous sourdough starters (Calvert et al. 2021). On the other hand, *Limosilactobacillus reuteri* has been reported to live in the gastrointestinal system of vertebrates or some sourdoughs (Gobbetti et al. 2016). Type III sourdough type is formed by combining Type I and Type II sourdough production methods. It is applied not only by artisan, but also by some bakerys with industrial characteristics. Type III sourdough is a type of sourdough that is dried with the help of spray or tumble sentence corrected dryer after fermentation (De Vuyst et al. 2021).

11.2.1 Sourdough LAB Species Diversity

LAB, belonging to the Firmicutes group are defined as gram (+), catalase (−), non-spore forming, and anaerobic to microaerophilic bacteria. They are found in a variety of places, from insects to the human body, and from a diversity of sources including (leavened) food products, plant ingredients, and more (Calvert et al. 2021; De Vuyst et al. 2021). Most of the food applications in the manufacture of numerous fermented foods and drinks (such as fermented dairy products, meat, cereal, and herbal products) use mostly homofermentative or facultative heterofermentative LAB. But, heterofermentative species play an important part in sourdough fermentation, particularly when sourdough are prepared in a conventional way (Corsetti and Settanni 2007; De Vuyst et al. 2021). *Lactobacillus, Leuconostoc, Pediococcus,* and *Weissella* species are typical sourdough LAB species (Arendt and Dal Bello 2011). LAB are classified with respect to their fermentation products, homo- and heterofermentatives, whereas heterofermentatives can be separated into two classes, facultatively heterofermentatives and obligate heterofermentatives, the later generating acetic acid, CO_2, and ethyl alcohol besides lactic acid and thus slightly contributing in leavening due to the CO_2 production besides producing further variety of flavour compounds (Lancetti et al. 2021). Homofermentative generate just lactic acid as the basic product of glucose fermentation. They have therefore more lactic acid production causing a decrease in pH and an increase in total titratable acidity (TTA) (Martín-Garcia et al. 2021). It has been noted that some parameters influence the functions and process of sourdough. These parameters are the comparative abundance, activity, and variety of heterofermentative and homofermentative LAB (Calvert et al. 2021). LAB species have fermentative properties. Especially in the case of homofermentative LAB species, pyruvate is produced from flour saccharides via EMP (Embden-Meyerhof-Parnas pathway) to generate ATP and reducing power. Then, redox balance is achieved by producing lactic acid from the produced pyruvate. In the event of obligateheterofermentative LAB species, acetic acid which are produced from pyruvate and acetyl-phosphate, lactic acid, ethyl alcohol, and CO_2 are generated from flour saccharides. For this, the phosphogluconate pathway is used to reduce ATP and power (De Vuyst et al. 2021). When the available literature was reviewed, it was noted that LABs in sourdough were affected by some factors. These factors are flour, distinct dough substances, and the habitat in which it is manufactured (Suo et al. 2021).

Sourdough is an original food habitat. It not only selects LAB strains that adapt to their environment, but also contains the LAB groups unique to each sourdough (Arendt and Dal Bello 2011). So far, it has been stated that the genus *Lactobacillus* includes 261 species. These species has recently been divided into 25 genera. Between them, more than 60 species have been identified from distinct sourdoughs (Suo et al. 2021). Although many new *Lactobacillus* species were originally isolated from sourdough, a few are thought to be native to sourdough environment (Catzeddu 2019). Bacteria can be found in flour, water or bowl, and are added continuously to maintain the starter. When the starter matures, the variety of bacterial species in the sourdough is reduced (Calvert et al. 2021). Some protocols are used for sourdough fermentation. For this, a microbial consortium exists primarily of obligate and facultative heterofermentative LAB (Arendt and Dal Bello 2011). In the order of relative abundance, 527 backslopping sourdoughs examined in one study are examples of obligate heterofermentative sourdough LAB types. These are *Fructilactobacillus sanfranciscensis* (47%), *Levilactobacillus brevis* (17%), and *Limosilactobacillus fermentum* (12%). Also, facultative heterofermenative and homofermentative LAB types such as *Lactiplantibacillus plantarum* (43%), *Pediococcus pentosaceus* (14%), and *Companilactobacillus paralimentarius* (13%) may occur (De Vuyst et al. 2021). It was noted that *Fructilactobacillus sanfranciscensis* dominates in the habitat of bakery and wheat sourdough produced in Italy and in traditional Greek wheat sourdough. Although *Lactiplantibacillus pentosus* and *Lactiplantibacillus plantarum* species were dominant in sourdough breads produced in the south, *Levilactobacillus brevis* was the dominant type in Turkish and Portuguese sourdough breads (Catzeddu 2019). The predominant LAB strains in the last decade isolated from sourdoughs are given in Table 11.1.

11.2.2 Areas Where LAB Contribute to the Technological Superiority of Sourdough

Microbes play a significant part in improving the properties of sourdough used to produce wheat flour foods and it was recorded that various works were realized on the microbiota of sourdough (Suo et al. 2021). Metabolites produced by microorganisms in sourdough can be effective on many factors such as technological performance of dough, nutrient properties, taste profile, shelf life, and all bread quality (Catzeddu 2019). Thanks to the metabolites produced by LAB, the quality of sourdough bakery foods products is improved. These features are as follows (Oshiro et al. 2021):

1. They produce metabolites to sourdough.
2. The role of metabolites produced by LAB in sourdough is quite different from the properties of sourdough yeasts. While the yeasts produce predominanty carbon dioxide throughout the fermentation of the dough, the LAB provide predominantly the production of acids that create sour flavor.
3. Also, the acid produced by LAB effects the texture of baked products because it changes the structure of gluten, which is an important protein complex and effective on the physical characteristics of wheat dough.
4. Acetic acid (shortening and hardening) and lactic acid (stabbing to be more elastic) cause some changes in the gluten structure.
5. Finally, the sourdough LAB contributes to the taste and shelf life of the emerging products.

The adaptations of some of the LAB to the sourdough environment are as follows:

Table 11.1 Dominant LAB strains in the last decade isolated from sourdoughs

Sourdough kind	Predominant lactic acid bacteria	References
Wheat bran spontaneous sourdough	*Levilactobacillus brevis, Lactiplantibacillus plantarum, Latilactobacillus curvatus, Latilactobacillus sakei, Leu. mesenteroides, Leuconostoc citreum, Pediococcus pentosaceus*	Manini et al. (2016)
Spontaneous rye sourdough	*Leu. mesenteroides, Lactiplantibacillus plantarum, Ent. pseudoavium, Lacticaseibacillus casei, Latilactobacillus curvatus, Lentilactobacillus farraginis, Pediococcus pentosaceus, Pediococcus acidilactici, Lacticaseibacillus paracasei, Loigolactobacillus coryniformis, Levilactobacillus brevis, Liquorilactobacillus uvarum*	Bartkiene et al. (2019)
Chickpea sourdough	*W. paramesenteroides, P. pentosaceus, Enterococcus mundtii, Enterococcus durans, Lactococcus garvieae, Weissella cibaria,*	Sáez et al. (2018)
Chia sourdough	*L. rhamnosus, Ent. casseliflavus, Ent. faecium, Enterococcus hirae, Enterococcus mundtii, Enterococcus devriesei, Lactococcus lactis, Enterococcus hermanniensis, Enterococcus italicus, W. cibaria*	Maidana et al. (2020)
Spontaneously fermented amaranth sourdough	*Lact. rhamnosus, Enterococcus casseliflavus, Ent. Mundti, Pediococcus pentosaceus, Lact. plantarum*	Ruiz Rodríguez et al. (2016a)
Whole grain oat flour sourdough	*Leu. argentinum, P. pentosaceus, W. cibaria, Loigolactobacillus coryniformis*	Hüttner et al. (2010)
Turkish Spontaneous Sourdough	*Weissella cibaria, Leuconostoc pseudomesenteroides, Lactiplantibacillus plantarum, Levilactobacillus brevis, Fructilactobacillus franciscensis*	Arsoy et al. (2022)
Argentinian Wheat Sourdough	*L. pentosus, Companilactobacillus paralimentarius, Lactiplantibacillus plantarum, Levilactobacillus brevis, Limosilactobacillus fermentum, P. acidilactici, Ent. faecium*	Lancetti et al. (2021)
Wheat and sorghum sourdough	*Lacticaseibacillus casei, Lentilactobacillus buchneri, Weissella cibaria*	Galle et al. (2011)
Turkish sourdough	*Lactiplantibacillus plantarum, Lactiplantibacillus paraplantarum, Latilactobacillus curvatus, Lactobacillus rossiae, Fructilactobacillus sanfranciscensis, Levilactobacillus brevis, Companilactobacillus paralimentarius, Weissella paramesenteroides, Leu. mesenteroides, Leu. pseudomesenteroides W. cibaria.*	Dertli et al. (2016)
Chinese traditional sourdough	*Fructilactobacillus sanfranciscensis, Companilactobacillus paralimentarius, Companilactobacillus crustorum, L. mindensis, L. rossiae, L. helveticus, Lactiplantibacillus plantarum, Levilactobacillus brevis, Leu. mesenteroides, P. pentosaceus Ent. faecalis*	Liu et al. (2016)
Spontaneously fermented quinoa sourdough	*Ent. hermanniensis, Enterococcus casseliflavus, Enterococcus mundtii, Lactococcus lactis, Leuconostoc citreum, Lb. plantarum, Leuconostoc mesenteroides, Levilactobacillus brevis*	Ruiz Rodríguez et al. (2016b)
Brazilian grape sourdough	*Lacticaseibacillus paracasei, Levilactobacillus brevis, Lb. delbrueckii*	Aplevicz et al. (2014)

(continued)

Table 11.1 (continued)

Sourdough kind	Predominant lactic acid bacteria	References
Spontaneous hull-less barley sour dough	*Pediococcus (dominant species), Latl. curvatus, Levl. brevis, Lacp. plantarum, Lim. fermentum, Lb. musae, Coml. paralimentarius, Leu. mesenteroides, Lim. equigenerosi*	Çakır et al. (2020a)
Traditional Sicilian Sourdough	*Lactiplantibacillus plantarum, Levilactobacillus brevis, Lb. rhamnosus*	Pino et al. (2022)
Spontaneous Rye Sourdough	*Leu. citreum, Lactiplantibacillus plantarum, Latilactobacillus graminis, Lactococcus lactis subsp. cremoris, W. confusa, W. cibaria*	Ispirli et al. (2018)
Einkorn sourdough	*Companilactobacillus crustorum, Pediococcus, Levl. brevis, Lacp. paraplantarum, Lacp. plantarum, Limosilactobacillus fermentum, Latilactobacillus curvatus*	Çakır et al. (2020b)
Spontaneously Fermented Sorghum Sourdough	*L. pentosus, Lacp. plantarum, L. ailmentaruis, Micrococcus varians.*	Ewuoso et al. (2020)
Typical Bulgarian Sourdough	*Lacp. plantarum, Lacp. plantarum subsp. plantarum, Ent. faecium, Ent. durans, Pediococcus parvulus*	Petkova et al. (2021)
Iranian Traditional Sourdoughs	*Lactobacillus and Lactiplantibacillus plantarum, Lacticaseibacillus casei, L. alimentarius*	Chandal and Mahjub (2020)
Spontaneously Fermented Greek Wheat Sourdoughs	*Lactiplantibacillus plantarum, Levilactobacillus brevis, Companilactobacillus paralimentarius, Lvb. zymae, Latilactobacillus curvatus, Ltb. sakei, Leuconostoc citreum, Ln. mesenteroides, Lactococcus lactis and Fructilactobacillus sanfranciscensis*	Syrokou et al. (2021a)
Ya'an sourdoughs	*Lactiplantibacillus plantarum, Lacticaseibacillus pantheris, Leu. citreum, W. viridescens, Leu. pseudomesenteroides, Lactococcus lactis, L. raffinolactis, Leu. mesenteroides*	Liu et al. (2018)
Korean sourdough	*Fructilactobacillus sanfranciscensis, Latilactobacillus curvatus, Levilactobacillus brevis*	Baek et al. (2021)
Traditional Austrian sourdough	*Levilactobacillus brevis, Loigolactobacillus coryniformis, Latilactobacillus curvatus, Lentilactobacillus diolivorans, L. gallinarum, Levilactobacillus hammesii, L.actobacillus harbinensis, Lactobacillus kimchii, Lentilactobacillus kisonensis, L. otakiensis, Levl. parabrevis, L. parabuchneri, Lac. paracasei, Coml. paralimentarius, L. perolens, L. plantarum, L. pontis, L. rossiae, L. sakei, L. sanfranciscensis,L. senmaizukei, L. spicheri L. vaccinostercus, L. xiangfangensis, Leu. citreum, Leu. mesenteroides, Leu. pseudomesenteroides, P. parvulus, P. pentosaceus, Streptococcus dentisani, Streptococcus salivarius, W. cibaria, W. viridescens*	Fraberger et al. (2020)
Traditional Italy sourdough	*Lacp. plantarum, Frul. sanfranciscensis, Coml. paralimentarius, L. rossiae, P. pentosaceus, Leuconostoc spp., Weissella cibaria*	Reale et al. (2019)

(i) It is the carry of sourdough particular carbohydrates like maltose and fructose to a unique central metabolism. The most abundantly fermented carbohydrate is maltose.
(ii) Fructose is an significant alternative electron acceptor.
(iii) Proteolytic activity and/or arginine deiminase pathway.
(iv) It has specific stress responses.
(v) Production of antimicrobial or antifungal compounds (De Vuyst and Vancanneyt 2007).

11.2.2.1 EPS and Biosynthesis of Enzymes

EPS (exopolysaccharides) are natural biopolymers produced by a few LAB species, various acetic acids, and microalgae. EPSs are either in the form of capsule or secretion in their surroundings (Dertli et al. 2016; De Vero et al. 2021). These polymers can vary both in their monosaccharide composition and in the degree of branching spacing (Abarquero et al. 2022; Harper et al. 2022). EPS have crucial roles emphasizing resistance at the one cell grade (Dertli et al. 2016). Generally, the polymers generated by EPS are not used as an energy resource by the maker microorganism, they only control the physicochemical properties of the cell surface in LAB ecology and protect the bacterial cells from abiotic or biotic stress (Abarquero et al. 2022; Korcz and Varga 2021). EPS are separated into two class in terms of form, the homopolysaccharides containing one kind of sugar monomer and the heteropolysaccharides consisting of two or more sugar monomers. For homopolymeric EPS production is required only one gene set (called *gtf* or *ftf*). However, for the production of heteropolymeric EPS is required a mixed EPS gene consisting of several genes (Dertli et al. 2016). It has been noted that *Fructilactobacillus, Lacticaseibacillus, Lactiplantibacillus, Lactobacillus, Lactococcus, Latilactobacillus, Lentilactobacillus, Leuconostoc, Limosilactobacillus, Pediococcus, Streptococcus*, and *Weissella* species can synthesize various EPS (Korcz and Varga 2021; Păcularu-Burada et al. 2020). The concentrations of synthesized EPS depend on various factors. These are the nutritional composition of the substrates, optimum fermentation conditions, strains, and their biochemical properties (Păcularu-Burada et al. 2020). It has been stated that EPS produced from lactobacilli related to grain contributes to some functions such as sucrose metabolism, defense against environmental attacks, and biofilm formation in gut habits (Gänzle 2014; Galle and Arendt 2014).

Microbial EPS from sucrose may be generated in sourdough by the efficiency of glycosyltransferases as well as polysaccharides which are take place naturally in cereal grains, flour, and dough (De Vero et al. 2021). EPS might increase the structural and nutritional characteristic of the product by replacing hydrocolloids used as texturizing, prevent staling, or as prebiotic additives in bread making by improving dough rheology (Bounaix et al. 2009; Zotta et al. 2006) and furthermore, the EPS increases the production of lactate, acetate and ethanol by promoting the metabolic activity of LAB strains (Martín-Garcia et al. 2021), and dietary fiber content (Gänzle

2014). In addition, it has been suggested that the preservative influence of EPS opposed to low pH and EPS production can result in osmotic and active benefits (Galle and Arendt 2014). EPS produced by LAB can induce various activities. These are useful antibacterial (Nehal et al. 2019) and antioxidant efficiencies (Bomfim et al. 2020) that forward increase their utilization possible in the food industry. Also, EPS can form networks with other dough ingredients and connect water. This has positive effects such as structure, volume, and rheology on bread. In addition to these features, the use of additives can be reduced or prevented by the application of EPS derived from LAB in foodstuffs, meeting both production optimization and customer needs (Korcz and Varga 2021). Therefore, the determination of EPS production properties of sourdough LAB is very important to reflect their all part in the fermentation process (Dertli et al. 2016).

Classification of EPS synthesized by LAB

LABs are capable of synthesizing cell wall structural polysaccharides (like peptidoglycan and lipoteichoic acids) and extracellular polymers. It consists of repeating units of carbohydrate derivatives or carbohydrate derivatives of capsular polysaccharides (CPS) covalently attached to the cell surface, as well as EPS which can compose a loosely bound layer that may be excreted into the habitat (Abarquero et al. 2022; Torino et al. 2015). It is very significant to figure out the microbial biosynthesis mechanism of EPS in order to figure out the variety of this polymer category and to utilize it successfully (Korcz and Varga 2021). EPS from LAB are quite various and are categorized according to some criteria (Torino et al. 2015). Classification of EPS synthesized by LAB are shown in Fig. 11.2 (Korcz and Varga 2021; Galle and Arendt 2014).

a. **Homopolysaccharides (HoPS)**

HoPS consists of glucose, fructose, or galactose, which presents as single monosaccharides in the repeating unit structure of this EPS. These EPS are called as glucans, fructans, or polygalactans (Galle and Arendt 2014; Torino et al. 2015). The biosynthesis of HoPS is either cell wall dependent or LAB utilize the extracellular synthesis pathway to synthesize HoPS using extracellular glucansucrases (GTF) and fructansucrases (FTF) (Abarquero et al. 2022; Galle and Arendt 2014; Bounaix et al. 2009), requiring sucrose as a special substrate. LAB use the extracellular synthesis pathway to synthesize HoPS using extracellular GTF or FTF. The energy which is required for the reactions of GTF/FTF is acquired from the breakdown of the osidic bond in sucrose. Therefore, HoPS may be produced in larger quantities than the energy-consuming biosynthesis of HePS (Abarquero et al. 2022; Galle and Arendt 2014; Bounaix et al. 2009). Several strains of LAB from sourdough bacteria were shown to be able to produce EPS from sucrose via the action of GTF/FTF. *Frul. sanfranciscensis* strain was demonstrated to produce EPS from fructose during fermentation of wheat and rye which was reported to positively affect the rheological characteristic of the dough (Arena et al. 2020).

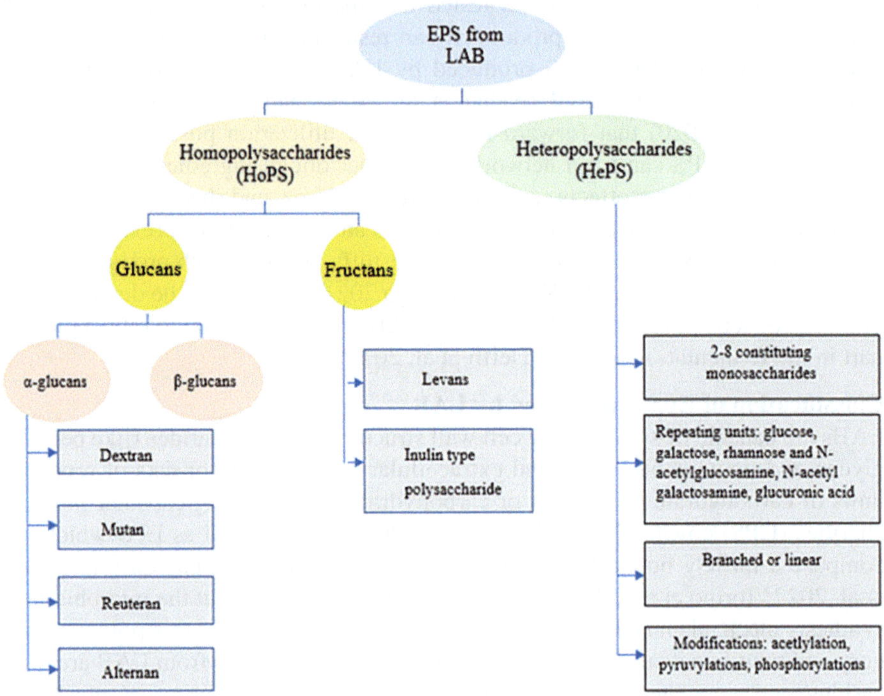

Fig. 11.2 Classification of EPSs synthesized by LABs. (Adapted from Korcz and Varga (2021))

Types of HoPS

Although recently galactan production was reported for LABs, in general LAB produces two types of HoPS, i.e. glucans and fructans. The glycosidic linkages between the glucose units results in the formation of four types of glucans produced by LAB strains which are dextran, mutan, alternan, and reuteran (Bounaix et al. 2009). For fructans two different structures were reported which are levan (β-D-Fru (2,6)) and inulin (β-D-Fru (2,1)). Figure 11.3 shows the structure of glucans produced by LAB (Abarquero et al. 2022).

Polygalactans include a pentameric repeat of galactose. These polymers are uncommon. Merely, has been defined in two strains such as *Lactococcus lactis* subsp. *lactis* H414 and *Lb. delbrueckii* subsp. *bulgaricus* CRL 406 and 142 (Torino et al. 2015).

Structure of glucansucrases determines the glycosidic connection type, level of branching, size, mass, and structure of polymers. Thus, it can synthesize various α-glucans with distinct physicochemical properties like solubility, viscosity, and other properties (Torino et al. 2015). Distinct acceptor molecules are utilized by glucansucrases, including maltose, isomaltose, and glucan which results in polymerization. It is known that the kind of glucan produced is defined by the enzyme (Galle and Arendt 2014). The acceptor carbohydrates for glucansucrase can be glucose, maltose, isomaltose, and lactose (Galle and Arendt 2014). Looking at the

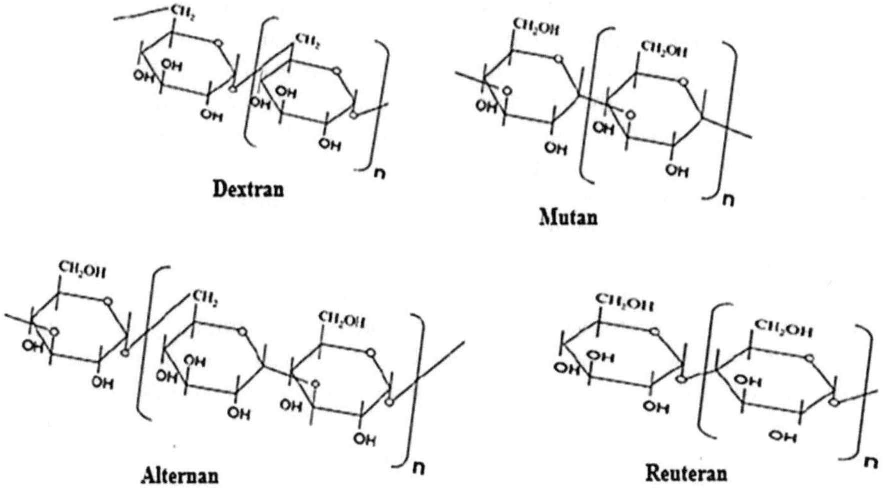

Fig. 11.3 Structure of glucans produced by LAB. (Adapted from Abarquero et al. (2022))

studies, α-glucan emergence by glucansucrases has been observed in bacterial genera like *Leuconostoc, Lactobacillus, Streptococcus, Pediococcus*, and *Weissella* (Torino et al. 2015). On the other hand, levan and inulin are fructans produced by fructansucrases. They are linked by β-(2,6) and β-(2,1) glycosidic bonds, respectively (Bounaix et al. 2009) and it is secreted for synthesis of high molecular fructans or low molecular FOS (Galle and Arendt 2014). It has been noted that a number of heterooligosaccharides are formed in wheat sourdough fermentation because the *Frul. sanfranciscensis* strain uses the raffinose, maltotriose, maltose, xylose, or raffinose fructosyl receptors of levansucrase. Lactosucrose or lactulose is composed in the presence of receptors such as lactose and fructose. While glucansucrase is found only in LAB species, it has been noted that bacterial fructansucrases are widely found in different bacterial species (Galle and Arendt 2014).

b. **Heteropolysaccharides (HePS)**

HePS are made up of repeating units of different monosaccharides. It is a more complex group of polysaccharides structurally and can be seen as branched or unbranched at the C2, C3, C4 or C6 positions (Abarquero et al. 2022). The synthesis of heteropolysaccharides either depends on the monosaccharide subunits of phosphate/sulfate groups or is complex and energy-intensive, including charged groups that are part of the subunit of glucuronic acid (Harper et al. 2022). HePS synthesis occur in 5 steps, beginning inside the cell and finishing outside the cell. First, the sugar is carried to the cytoplasm until glucose-1-P is acquired and sugar nucleotides are synthesized. The repeating units are then synthesized through the sequential addition of activated sugar residues with particular GTF that catalyze the glycosidic bond with binding to the glycosyl carrier lipid. The repeating units are then replaced by the Wzx-protein, which has flippase activity, for the extracellular surface. In the last step, extracellular polymerization of HePSs takes place under the influence of Wzy protein and the support in the media or the bacterial wall is released and the bond connecting the lipid carrier to the lipid carrier is broken via HePSs (Abarquero et al. 2022). They are synthesized from intracellular nucleotide pioneers, generally in the range of 104–106 Da and manufactured in mg.L^{-1} amounts (Harper et al. 2022). In general, it was emphasized that HePS production levels by LAB were lower than HoPS production levels. The reason for this is that it can be said that EPS production endures the strain, the compound of the culture medium, microbial culture conditions like pH, temperature, and incubation time (Torino et al. 2015).

It has been noted that a number of heterooligosaccharides are formed in wheat sourdough fermentation because the *Frul. sanfranciscensis* strain uses the raffinose, maltotriose, maltose, xylose, or raffinose fructosyl receptors of levansucrase. Lactosucrose or lactulose is composed in the presence of receptors such as lactose and fructose. While glucansucrase is found only in LAB species, it has been noted that bacterial fructansucrases are widely found in different bacterial species (Galle and Arendt 2014).

Galle et al. (Galle et al. 2011) fermented wheat and sorghum sourdoughs with *Lactobacillus casei* and *Lactobacillus buchneri* strains in their study. Synthesized heteropolysaccharide levels in these sourdough varieties were investigated. According to their study, they concluded that heteropolysaccharides reduce the resistance of the dough against deformation, and as a result, it causes an increase in the specific volume of gluten-free bread. It has been stated that this effect on dough rheology and stability can improve the technological properties of sugar-reduced products (Sahin et al. 2019).

High molecular weight HePS such as EPS in sourdough fermented by *Weissella confusa* QS813 and *Frul. sanfranciscensis* have a positive effect on bread property owing to their controlling properties of the thermal stability of starch granules. This was attributed to HePS-induced delay of amylose crystal development, resulting in a softer bread crumb (Sun et al. 2022).

Types of HePS

Kefiran is the most widely studied type of HePS found in microorganisms. It is found in kefir grains. It contains about the same amounts of D-glucose and D-galactose residues. Branched hexa or heptasaccharide is a water-soluble branched glucogalactan consisting of repeating units. Parameters such as sugar compound, chain length, sugar connections, branching frequency, and molecular mass are factors that affect the technological characteristic of EPS. Because EPS are complex and their structural analysis is difficult, the relationship between their structure and mission is not still completely understood. Only its contribution to food products can be attributed specifically to its interaction with proteins (Abarquero et al. 2022).

In situ EPS Production

It has been reported that LAB strains produce EPS during sourdough leavening in the presence of sucrose. Various factors affect the amount of EPS produced *in situ*. These are various parameters like fermentation time, pH, sucrose level, and fermentation substrate (Torino et al. 2015). In recent studies, it was noted that glucan production of *Limosilactobacillus reuteri* from sucrose was higher in soft doughs, likely due to better propagation of substrates and extracellular glucansucrases (Kaditzky and Vogel 2008). In addition, EPS efficiency increased when sucrose was added to the fermented dough. Oligo- and polysaccharides are formed from sucrose with the release of glucose or fructose. Thus, the formation of organic acid is affected (Torino et al. 2015). *In situ* EPS formation has been of interest to sourdough LAB strains. The main source of this interest is the effect of EPS production properties on the technological and physicochemical characteristic of sourdough and bread due to their functional roles (Dertli et al. 2016).

It has been noted at times that *in situ* EPS production is more effective than direct EPS addition in improving bread formulation (Galle et al. 2012). However, it has been stated in some studies that the opposite is the case (Kaditzky et al. 2008). In a study, it was stated that the use of in situ EPS production by LAB in making steamed bread has positive effects on dough and product quality (Tang et al. 2018; Sha et al. 2023). Recently, Huang et al., (Huang et al. 2023) reported that *in situ* EPS formed in red bean sourdough increased the viscoelastic property of the dough, its gas holding capacity, and the dough fermentation height. As a result, after 7 days of storage, they observed a decrease in baking loss and firmness, an increase in specific volume and in shelf life. Studies have shown that *in situ* EPS production during sourdough fermentation has the potential to be used as a bread improver for less use of additives in food products. It has been emphasized that this can replace hydrocolloids (Galle and Arendt 2014). Zhang et al., (2023) evaluated the potential of *in situ* EPS formation during sourdough fermentation to replace hydrocolloids in gluten-free sourdough. For this purpose, the effect of EPS producer *Weissella cibaria* NC516.11 on the chemical and rheological properties of sourdough and quality of buckwheat bread was evaluated. As a result, it was noted that the strain used improved the rheological and viscoelastic properties of the sourdough and improved the texture of the bread. Studies have shown that LAB strains or fermentation conditions maximize *in situ* EPS production. However, it was stated that acid production should be optimized for its effect on the acceptable volume, crumb texture, and flavor of the breads. For this purpose, it was emphasized that when EPS producing strains are used in dough applications, the pH of the metabolites at the end of fermentation and the molar ratio of lactate/acetate should be taken into account (Torrieri et al. 2014).

11.2.2.2 Proteolytic Activity and Specific Enzymes

Various studies have been suggested as whether the source of proteolytic activity during sourdough fermentation are LAB, yeasts, or native grain enzymes (Zotta et al. 2006). LAB are characterized by multiple amino acid auxotrophies and requires substrate-derived proteases/activity of proteolytic systems to accommodate nitrogen metabolism. The difference among sourdoughs and chemically acidified doughs is based on the activity of cereal endogenous proteinases, which is why their natural protein is degraded. The occurrence of actuation of aspartate-proteinases from cereals with varying optimum pH efficiency (between 3.0 and 4.5) has been attributed to acid formation produced by sourdough LAB. Moreover, selected sourdough lactobacilli may have to hydrolyze albumins, globulins, and gliadins throughout fermentation (Gobbetti and Gänzle 2012). Figure 11.4 gives an overview of amino acid metabolism and proteolysis that occurs in wheat and rye sourdoughs.

One of the principal features affecting the overall character of sourdough bread is the degradation of protein during sourdough fermentation. Proteolysis releases free amino acids and their metabolism and involves essential biochemical situations that will lend to the overall character of sourdough. These events take place the following steps (Reale et al. 2021):

(i) Primary hydrolysis: Efficiency of endogenous proteases (which breaks grain proteins into middle/minor peptides) is carried out by acidifying the dough by LAB.
(ii) Secondary proteolysis: Peptides hydrolyzed by LAB proteinases cause the accumulation of amino acids in the dough.
(iii) Amino acids are leaved throughout dough fermentation, sustaining bacterial growth and serving as pioneers of flavor compounds.

The development of amino acids during sourdough fermentation is complex. It has been noted that these amino acids depend on the metabolic necessity and proteolytic activities of LAB (Rollan et al. 2005). In sourdough, gluten proteolysis and medium hydrolysis of starch depend on the production of acetic acid and lactic acid by LAB. As a result of their production, a decrease in pH occurs. With the acidification of the environment, gluten proteins swell and higher solubility occurs. In addition, the intramolecular forces increase, the protein diffuses, and the hydrophobic portion is more exposed. As a result, the large protein aggregate undergoes proteolysis.

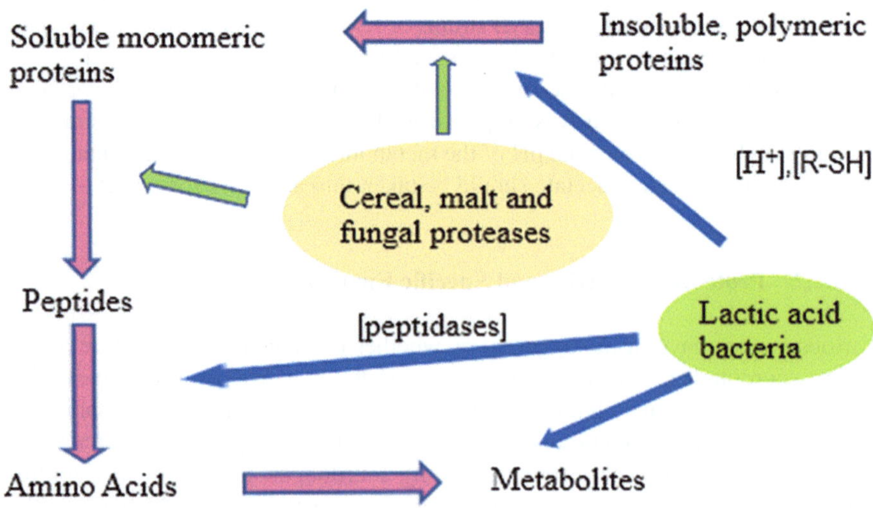

Fig. 11.4 Amino acid metabolism and proteolysis that occurs in wheat and rye sourdoughs (Gänzle 2014)

Thus, it creates a more extensible, consistent, and less elastic emulsion (Siepmann et al. 2018). Rohr lich and Hertel (Rohrlich and Hertel 1966), linked the rise in protein solubility and the presence of dialyzable peptides to the proteolytic activity during sourdough leavening. Moreover, the proteolytic activity of LAB in sourdough can also be demonstrated by varying amounts of certain amino acids (Spicher and Nierle 1988). In addition, disruption and depolymerization of proteins during sourdough fermentation may also rely on bacterial metabolic action and cereal enzymes (Gänzle 2014). The determination of the collection of free amino acids and peptides after proteolysis and fermentation has been indirectly investigated in studies. The basic proteolytic action was initially attributed to some endogenous enzymes. These enzymes are aminopeptidases, carboxypeptidases, and endopeptidases (Di Cagno et al. 2002). Cereal flour, yeasts, and LAB are contribute to proteolytic events in different ways by containing proteases and peptidases (Zotta et al. 2006). Peptidase activities reduce smaller peptides to amino acids. Thus, it can conduce proteolysis during sourdough fermentation and cause desired texture or flavor changes in bread making (Rollan et al. 2005). Free amino acids and oligopeptides accumulate in mature sourdough through the effect of combining endogenous carboxypeptidase activity in flour with endo- and exo-peptidases in LAB and these are liable for the hydrolysis of great/middle-sized peptides in it (Reale et al. 2021). Further hydrolysis of peptides to amino acids is accomplished by the intracellular/extracellular peptidases of LAB. Studies have shown that most sourdough LABs and especially *Frul. sanfranciscensis* strain have been reported to prefer peptide uptake rather than amino acids carry (Rosell et al. 2015).

Strain-specific enzymatic activities of LAB and yeasts in sourdough are another important point. In addition, the activation and release of endogenous grain enzymes during fermentation are important factors (Hajinia et al. 2021). Menezes et al., (2019) compared the fructan content of sourdough and commercial leavened breads in their study. As a result, they stated that sourdough bread decreased by 72% compared to commercial leavened bread. It has been reported that this reduction is due to specific enzymes responsible for the degradation of fructans in some types of LAB (Menezes et al. 2021). Also, LAB in sourdough improves the invertase activity of yeasts by providing acidic conditions, and thus can increase the breakdown of fructans (Menezes et al. 2019). Because most LAB strains contain some specific enzymes that are very important in FODMAP metabolism. For example, α-galactosidase, 1,6-α-glucosidase, levansucrase, and sucrose phosphorylase are a few of these enzymes (Menezes et al. 2019). Thanks to the specific enzymes produced by the sourdough microbiota, the texture and staling of the bread can be delayed (Alkay et al. 2022).

11.2.2.3 Antifungal Activity

Recent consumer interests to minimally processed foods increased the attention anti-mold properties of LAB (Gobbetti et al. 2005). Mold growth is one of the most common reasons of spoilage in various foods and feeds and producing potentially harmful secondary metabolites called mycotoxins (Arsoy et al. 2022; De Vero et al. 2021). The toxicity and pathogenicity of some species with mold growth may cause serious health risks for consumers as well as infections or allergies in sensitive individuals. Therefore, food safety should be ensured at all phases from food production to consumption and appropriate measures must be taken to limit quality loss. One of the best candidates for these measures are LAB due to their GRAS status (Arsoy et al. 2022). LAB inherently dominate the microflora of numerous foods during storage or fermentation and can inhibit mold growth by producing some metabolites under natural growth conditions (Hassan et al. 2016). Facultative and obligate heterofermentative lactobacilli produce most of the antifungal activity from LAB (Corsetti and Settanni 2007). Different types of LAB show antifungal properties. These species were *Lacticaseibacillus casei, Loigolactobacillus coryniformis, Lacticaseibacillus paracasei, Lacticaseibacillus rhamnosus, Lactiplantibacillus plantarum, Pediococcus pentosaceus, Pediococcus acidilactici*, and *Lactococcus lactis* (Debonne et al. 2020). LAB are usually considered as protective microorganisms, both responsible for the production of lactate and acetate, which play role as protective agents as well as for releasing other active compounds during fermentation. It has been stated that the synergistic influence between pH and antifungal metabolites is liable for the protective activity of LAB (Bartkiene et al. 2019). It shows its protection effect mainly by forming organic acids (lactic and acetic acid). In addition, it can be caused by many other antimicrobial compounds such as different types of bioactive molecules like fatty acids, H_2O_2, reuterin, bacteriocins, phenylactic, p-OH-phenyllactic acids, CO_2, ethyl alcohol, diacetyl, gamma aminobutyric acid, propionic acid) (Bartkiene et al. 2019; Arsoy et al. 2022).

a. **Organic acid**

In general, LAB have the ability to quickly acidify the raw material by producing organic acids. This acidification results in an improved shelf life and microbial safety (Martín-Garcia et al. 2021). Undissociated forms are exhibited in the production of organic acids during sourdough leavening. Thus, they provide great protection against pathogens and spoilage-causing microorganisms. The production of organic acids in sourdough fermentation delays the molding of sourdough breads and extends their shelf life. It also has the ability to restrict yeasts and bacteria. Because propionic acid preferably restricts molds and bacteria, co-fermentation of propionic acid bacteria, LAB and yeasts appears to be an encouraging native sourdough-based biotechnological process to delay mold growth. Fungistatic impacts are primarily owing to acetic acid rather than lactic acid production. Acetic acid has higher separation purpose (pKa) than lactic acid. Thus, lactic acid has less restrictive activity on mold growth compared to other organic acids (acetic acid and propionic acid) (Bartkiene et al. 2019; Sadiq et al. 2019). Heterofermentative LAB specifically release organic acids with mold inhibiting properties (Melini and Melini 2018). In this group, *Fructilactobacillus sanfranciscensis* CB1 shows the broadest spectrum of antifungal activity owing to the production of a blend of acetic, caproic, formic, propionic, butyric and n-valeric acids. Between these acids, it can be emphasized that caproic acid have the potential to be the organic acid with the highest anti-mold efficiency (Corsetti and Settanni 2007). Alkay et al. (2020) noted that heterofermentative LAB strains isolated from sourdough have high antifungal properties against various mold species. In another study, İspirli and Dertli (İspirli and Dertli 2022) stated that some LAB strains isolated from sourdough had antifungal properties. The antifungal properties of LAB strains depend on the metabolites they produce and the synergistic effect between the metabolites. It is seen that the antifungal activity of these microorganisms varies according to factors such as temperature, pH, and incubation time (Bartkiene et al. 2019).

Organic acid production of LAB strain is of great importance in determining mycotoxigenic mold limitation properties. However, the sort and amount of these acids may be different from strain to strain. The main part of these acids is produced by LAB as a by-product of the acidification procedure, in contrast to the active synthesis of metabolic compounds aimed at limiting the growth of molds. Lactic acid is the main organic acid produced by LAB. It is an organic acid that is commonly found in nature in two types (L and D). Of these, L lactic acid is known as a safe protective by the FDA. In addition, lactic acid and acetic acid have been identified as the main antifungal compound produced by types of *Enterococcus faecium, Lacticaseibacillus rhamnosus,* and *Lactiplantibacillus plantarum.* It has been stated that lactic acid and acetic acid are the most important antimicrobials generated and characterized by LAB. These are acids that are bioactive in protonated shape at low pH. Organic acids produced by lowering the pH activate other mold inhibiting compounds. Thus, it imparts inhibition properties to LAB (Nasrollahzadeh et al. 2022).

b. **Phenyllactic Acid (PLA)**

Another organic acid is phenyllactic acid (2-hydroxy-3-phenyl propionic acid) (Nasrollahzadeh et al. 2022). It has attracted attention because it is a metabolite produced by specific LAB (Mani-López et al. 2022). The increase in MIC values of PLA against mold strains was attributed to the complex cellular structure. In the studies, it is seen that the mechanism of action of PLA against mold species has been little studied. However, several studies have shown that PLA interferes with the proton gradient and inhibits cellular enzymes by interacting synergistically with other metabolites (Rajanikar et al. 2021). It is derived from the catabolism of phenylalanine and has natural antibacterial properties. In addition, it is metabolized by the glycolytic enzyme and lactate dehydrogenase throughout fermentation (Nasrollahzadeh et al. 2022). PLA has been noted to be non-cytotoxic to animals and human cell lines (Rajanikar et al. 2021). PLA has a molecular weight of 166 g/moL and a melting temperature of approximately 98 °C. It exists in two isomeric types (L and D) and is an aromatic compound with an asymmetric carbon atom. Although these two isoforms have antimicrobial effects, it has been reported that the L-isoform is a stronger inhibitor than the D-form (Rajanikar et al. 2021). The PLA biosynthesis pathway in LAB is as follows:

(i) wherein the precursor phenylalanine is the transamination reaction of the aminotransferase enzyme and

(ii) metabolized first to phenylpyruvic acid (PPA) by reduction to PLA by a dehydrogenase (Kadyan and Pradhan 2020).

Fermentation of some foods by LAB strains with antifungal properties reduced the use of chemical preservatives in foods (Nasrollahzadeh et al. 2022). Studies have shown that there is a powerful relationship between the PLA concentration in the cell-free supernatant and its mold limitation property. Therefore, it has been seen recently that research mostly includes the improvement of PLA produced by microorganisms (bacteria and yeast) and their application to biotechnology (Valerio et al. 2016). Sourdough can be also among these technological applications of PLA. Although the contribution of PLA to the restriction of mold growth on bread remains uncertain, the number of reports describing increasing its production in sourdough is increasing day by day (Debonne et al. 2020). Phenyllactic acid (PLA) has a broad spectrum of bacterial limitation for many organisms. It is an organic acid that can be effective on gram (−), gram (+) bacteria, and molds also high MICs (Rajanikar et al. 2021) have been determined at least for the inhibition of fungi (Wu et al. 2021). LABs with the ability to produce PLA are *Leuconostoc mesenteroides, Lactiplantibacillus plantarum, Lacticaseibacillus rhamnosus, Fructilactobacillus sanfranciscensis, Lactobacillus acidophilus, Companilactobacillus alimentarius, Lactobacillucoccusicius, Lactobacillus alimentococcus, Lactobacillus acidilactici* strains (Debonne et al. 2020). In addition, many fermented foods such as bakery products, flours, and cereals have been reported to contain PLA. It has been noted in studies that foods containing PLA have a longer shelf life than foods without PLA. It has been stated that this feature is owing to the capability of PLA to restrict the growth of detrimental microorganisms, for instance *Penicillium verrucosum, Penicillium citrinum, Penicillium roqueforti, Listeria monocytogenes, Aspergillus* spp., *Fusarium sporotrichioides*, and *Staphylococcus epidermidis*) (Wu et al. 2021). PLA producing strains were used in bread and sourdough bread to prevent "bread rope" and mold deterioration caused by spore-forming bacteria (*Bacillus* spp.) (Valerio et al. 2016).

c. **Reuterin**

Reuterin (β-hydroxy propionaldehyde) is mainly excreted by *Liml. reuteri* and has a low molecular weight composed of acrolein and is a multicompound system. However, it can be secreted by some other bacterial species and genera (Nasrollahzadeh et al. 2022). Other reuterin-producing LAB are *Levilactobacillus brevis, Loigolactobacillus coryniformis, Secundilactobacillus collinoides*, and *Lentilactobacillus buchneri* species (Kadyan and Pradhan 2020). Reuterine exhibits broad antimicrobial including antifungal activity against various gram (+) and gram (−) bacteria, bacterial spores, molds, yeasts, and protozoa (Nasrollahzadeh et al. 2022; Houssni et al. 2022; Riesute et al. 2021; Chen et al. 2021). The substance of reuterin generated by LAB can be affected by some factors such as growth conditions and culture medium (Nasrollahzadeh et al. 2022). To produce this compound, cells must be cultured under anaerobic situations with glycerol as the substrate. The produced active reuterin appears as a mixture of 3 different forms, *i.e* monomeric, hydrated monomeric, and cyclic dimeric forms of 3-HPA. It can work in a wide pH range. It can also stay active in the existence of diverse enzymes (Mishra et al. 2021). Reuterin metabolism from glycerol in *Liml. reuteri* has been attributed to the presence of Glycerol/diol dehydrase, which regulates the biotransformation of glycerol to 3-Hydroxy propionaldehyde (HPA) and 1,3-propanediol. Although until recently, reuterin was thought to contain 3-HPA, its hydrate and dimer, at present, another compound called 'acrolein' has been added to the reuterin compound mix and is formed as spontaneous dehydration of 3-HPA in aqueous form (Kadyan and Pradhan 2020). Acrolein makes it advantageous to use reuterin as a food preservative due to its potential toxicity. However, usage of *Liml. reuteri* DSM 20016 resulted in the formation of acrolein below the limits of the tolerable side according to WHO limits in food products (Drakula et al. 2019). The mechanism of the formation of reuterin is shown in Fig. 11.5.

When the literature studies were examined, it was stated that the *L.reuteri* strain isolated from sourdough generally did not synthesize reuterin (Messens and De Vuyst 2002). But, in a study, it was emphasized that 1,3-propanediol production in sorghum sourdough offered the possibility of reuterin production (Sekwati-Monang and Gänzle 2011). It has been stated that the presence of reuterin in sourdough may cause the formation of acrolein in bread. That is why it is suspicious (Drakula et al. 2022). But in some studies, it has been reported that sourdough breads prepared with *Liml. reuteri* have a good crumb texture and a long shelf life. *Liml. reuteri* holds the status of QPS (assumption of quality safety) granted by the European Food Safety Authority (EFSA) (Drakula et al. 2019). Also, various inhibitory activities of reuterin against microorganisms were investigated in studies. In addition, studies on the antifungal activity of reuterin have been limited due to the lack of investigation of inhibitory lead molecules found in the reuterine (Mishra et al. 2021). In another study, it was reported that a new strain of *Liml. reuteri* isolated from whole wheat sourdough could inhibit *Aspergillus niger* growth and biosynthesis of B1, B2, G1 and G2 aflatoxin (Păcularu-Burada et al. 2020).

d. **Fatty acids**

Organic acids with a carboxyl group and a methyl group at both ends are called fatty acids. A number of studies have reported that fatty acids have potential antifungal activity and are the basic antifungal metabolites that protects foods fermented by LAB. Antifungal fatty acids cause disruption of membrane integrity by breaking down the lipid bilayers of membranes. Thus, it leads to the breakdown of cells and the release of intracellular proteins and electrolytes (Nasrollahzadeh et al. 2022). Fatty acids have an antifungal effect due to their detergent-like activity (Riesute et al. 2021). Although the antifungal mechanisms of fatty acids are currently unknown, it has been stated that they lend to the taste characteristic of fermented foods (Chen et al. 2021). For instance, corolic acid was reported to be one of the effective fatty acids due to its effects on inhibition of the development of bread contaminants. This acid is a hydroxy fatty acid provided from peroxides (Houssni et al. 2022).

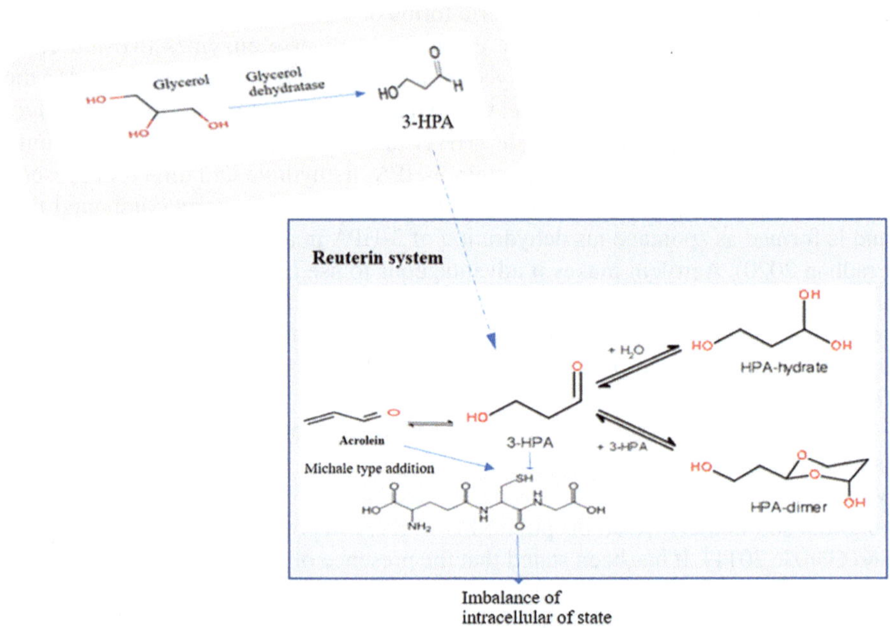

Fig. 11.5 The formation of reuterin system

Fatty acids are deposited by both plants and a wide spectrum of microorganisms such as bacteria, actinomycetes, and molds. The use of fatty acids as an active antimicrobial agent against various food spoilage bacteria, moulds, and yeasts may be encountered in the literature. In a study, the relations between the forms of hydroxy unsaturated fatty acids (HUFA) and their linked antifungal efficiency were emphasized. In addition, eighteen carbon unsaturated fatty acid chains with a hydroxyl group near the center have been shown to exhibit intensive antifungal activity. Antifungal activity can be violently influenced by carbon chain length. As the carbon chain length increases, the antifungal efficiency of fatty acids decreases. It has been noted that medium chain fatty acids generally elicited maximum antifungal activity (Mishra et al. 2021). The presence of at least one hydroxyl group and some degree of unsaturation in the carbon parts of fatty acids exhibits antifungal activity against many yeasts and molds (Kadyan and Pradhan 2020).

Some LAB strains can produce unsaturated fatty acids with important antifungal properties during fermentation. In the study of Vasilica et al., (Vasilica et al. 2022), *Lactiplantibacillus plantarum* ATCC 8014 and insect flour were fermented for 48 h and in the fermentation of sourdough SFA (saturated fatty acids) and MUFA (monosaturated fatty acids) groups were positively affected, but PUFA (polysaturated fatty acids) group decreased. The increase in fatty acid content during fermentation has been attributed to the different fatty acids conversion of LAB in various ways. These pathways are isomerization, hydration, dehydration, and saturation (Vasilica et al. 2022). In the study of Pastor et al., (Pastor et al. 2022) fatty acid profiles of normal commercial leavened bread and sourdough breads were compared. Bread crumbs, bread crust, and bread slices were analyzed separately. It was stated that the fatty acids found in the highest amount in all bread types were C18:2n6c, C18:1n9c, C16:0, C18:0, and C18:3n3. In addition, low amounts of short-chain and single-chain fatty acids were noted to be detected. As a result, it was emphasized that fatty acids are a good chemical descriptor for the classification of bread crumbs (Pastor et al. 2022).

e. **Cyclic Dipeptides**

Cyclic dipeptides (CDPs) often referred to as 2,5-dioxopiperazines are another class of low molecular weight (MW) peptide-derived compounds associated with various activities such as antimicrobial, antiviral, and antitumoral (Kadyan and Pradhan 2020). Protease enzyme treatment can be used to determine the peptide structure of active compounds. It has been noted that smaller peptides generally have stronger antifungal activity. Their mechanism of action is revealed when the peptides bind to the carpet-like lipid bilayers and pierce the channels in them, disrupting the function. Besides dissolving the membrane, they are also able to act via peptide-lipid interaction resulting in phase separation. Peptides obtained from LAB have high heat stability and low molecular weight. They have been observed to have high potential to replace chemical protectives usually used in bakery (Nasrollahzadeh et al. 2022).

Studies have reported that the ability of sourdough LAB species to synthesize cyclic peptides is mediated by cyclopeptidase entase enzymes and non-ribosomal peptide synthases. This enzymatic synthesis is supported by the occurrence of acidification during fermentation (Houssni et al. 2023). Niku-Paavola et al., (Niku-Paavola et al. 1999) stated in that cyclo (Gly-L-Leu) produced by *Lactiplantibacillus plantarum* strain limited the growth of *Fusarium avenaceum*growth. It was emphasized that this antifungal mechanism of action is seen in the form of binding to lipid bilayers and puncturing channels by disrupting the functions and structural integrity of the plasma membrane of the targeted molds (Houssni et al. 2023).

f. **Volatile compounds and others**

Volatile organic compounds, ethanol, diacetyl, and carbon dioxide formation are produced by LAB during heterofermentative fermentation process. Volatile compounds show antifungal activity in synergism with other metabolites. The antimicrobial effect of CO_2 is as follows. In the first step, it inhibits biocatalytic decarboxylation and then its accumulation occurs in the membrane lipophilic bilayer. Faulty permeability is caused by the development of anaerobiosis. Another compound with antifungal effect is ethanol. It shows its effect against molds by denaturing proteins and dissolving lipids against molds (Kadyan and Pradhan 2020). The formation of volatile substances in sourdoughs is due to the activities of LAB and yeasts. Parameters affecting their efficiency can determine the amount of metabolite formed. Generally, LAB are responsible for acidification. Therefore, their concentration can increase significantly during sourdough leavening (De Luca et al. 2021). In fact, microbiological fermentation in sourdough has been the most suitable route for the production of volatile compounds. Heterofermentative LAB produce more esters such as ethyl acetate, hexyl acetate, ethyl hexanoate, and isopentyl acetate, while homofermentative fermentations produce 2,3-butanedione (ketone), acetaldehyde (aldehyde), and 2-methylbutanol (alcohol) (Martín-Garcia et al. 2021).

Another acid that affects the aroma and the structure of the final products is acetic acid (Corsetti and Settanni 2007). In addition to these parameters, it has been stated that heterofermentative LAB partially contributes to the dough fermentation process. The formation of organic acids and volatile compounds by LAB contributes to a pleasant aroma and flavor in bread. In one study, most of the compounds identified in sourdough and not all volatile compounds detected by instrumental analysis had a detectable aroma. However, it has been stated that it strongly affects the bread flavor. These compounds were organic acids, alcohols, esters, and carbonyls (Settanni et al. 2013). The lactic and acetic acid formed throughout the fermentation of sourdough can lend to the taste of the rye and rye mixture bread. It is pointed out that it is especially important that sourdough fermentation breaks down flour proteins and forms intermediate products. Therefore, free amino acids and free sugars as pioneers of bread crust flavor have attracted a lot of attention today. Throughout cooking, some aldehydes are created from amino acids and reducing sugars. These are alanine (acetaldehyde), valine (isobutyric aldehyde), leucine (isovaleric aldehyde), isoleucine (2-methylbutanol), methionine (methionel), and phenylalanine (phenylacetaldehyde) compounds (Spicher and Nierle 1988). Further, it has been reported that the formation of aromatic volatile compounds during sourdough leavening is primarily due to either bacterial and yeast metabolism or enzymatic or autoxidation of flour lipids and Maillard reaction (Pizarro and FrAnCO Jr 2017).

The taste from the volatile components in sourdough was related to both the conversion to glutamate of glutamine released from cereal proteins by peptidases and the conversion of arginine to ornithine by LAB during baking (Akamine et al. 2023). Many factors affect the aroma profile of sourdough bread. These are parameters such as mixing, fermentation, baking, storage, flour type, dough yield, fermentation temperature, or fermentation time. These parameters affect the microbial diversity of the sourdough and, as a result, create complex interactions, affecting the taste of the bread (Warburton et al. 2022).

11.2.2.4 Antibacterial Activity

LABs show a wide spectrum of antimicrobial efficiency. Among these activities, the production of lactic acid and acetic acid are undoubtedly the most important (De Vuyst and Leroy 2007). The antimicrobial impact of LAB shows itself in the form of lactic acid production and a decrease in pH. It has been noted that various antimicrobial compounds, are produced by the LAB that may be classified as low molecular mass (LMM) compounds such as lactic acid, H_2O_2, CO_2, or diacetyl (2,3-butanoid) (Saranraj et al. 2013). The growth of microorganisms may be inhibited by the production of these metabolites (Gobbetti et al. 2005). Antibacterial efficiency is a sought-after character in selected LAB strains (Ibrahim et al. 2021). Antibacterial substances produced by some strains of LAB are indicated to help control spoilage and growth of pathogenic microorganisms if they are resistant to both physical conditions and baking conditions of bread (Şimşek et al. 2006).

(a) **Bacteriocins**

Bacteriocins have received major interest in the food industry owing to their potential in food protection. Many bacteriocins produced by LAB are safe and active native inhibitors of pathogenic and food spoilage bacteria (Saranraj et al. 2013). Bacteriocins are mainly composed of thermostable protein substances with antibacterial properties. It shows its effect in the form of binding of phosphate residues in the cell membranes of target cells, forming pores and activating autolysis, which disrupts bacterial cell walls. Bacteriocins consist of 20–60 amino acids and belong to various cationic and hydrophobic peptide groups (Zapaśnik et al. 2022). The bacteriocins produced by LAB have been considered ideal for food use due to several factors. These factors are;

1. side products of LABs have been recognized as GRAS (generally considered safe) by the FDA;
2. they are scentless and uncolored;
3. they do not affect the organoleptic and sensory properties in food.
4. bacteriocins differ from conventional antibiotics in that they are cleared by the digestive tract by proteolytic enzymes. Also, bacteriocins have bioengineering potential for improved performance (Ibrahim et al. 2021).

Studies have shown that various LAB bacteriocins have inhibitory effects against different food pathogens. *Lactobacilli* bacteriocin XH1 exhibit antimicrobial activity by inhibiting the proliferation of *Staphylococcus aureus* and *Escherichia coli*. Also, plantarisin P1053 produced by *Lpb. plantarum* PBS067 had wide-spectrum antibacterial efficiency against a gram (+) bacteria *Staphylococcus aureus* and gram (−) bacteria *Escherichia coli*. Because of these properties, bacteriocins have been noted as potent and promising antimicrobials. Therefore, the use of bacteriocin produced by LAB in food safety practices has become important and these strains were evaluated as promising (Ibrahim et al. 2021). Bacteriocins have various properties and the most important of these properties are their activity against natural structures such as bacteria, molds, viruses, parasites and biofilms. Alvarez-Sieiro et al. (2016) stated in their study that bacteriocins produced by LAB should be based on three basic classes. These classes can be divided into 3 categories according to their general structure and features (Zapaśnik et al. 2022; Mora-Villalobos et al. 2020).

According to their common structure and characteristic, bacteriocins may be divided into three various group:

- Group I (lantibiotics): peptides with unusual amino acids and low molecular weight, such as modified, heat-stable lanthionine, which is a non-canonical amino acid. They are extensible with a net positive charge (subclass A, nisin eg). They can be either negatively or uncharged spherical (subclass B, mersacidin eg).

- Group II: unmodified thermostable, low molecular weight, lanthionine-free heat stable peptides. Subclasses (subclass A, pediocin e.g.; subclass B, lactococcine plantaricin e.g., subclass C, acidosine e.g.) are highly dependent on activity.
- Group III: Its activities take different aspects like bactericidal or bacteriostatic effects and are effective on species related to the producer strain. They are large and thermolabile peptides that are not very well characterized. These are lytic proteins, usually grouped as murein-hydrolases (eg, helveticin), and their activities take different directions like bactericidal or bacteriostatic effects and are effective on species related to the producer strain (Zapaśnik et al. 2022; Mora-Villalobos et al. 2020).

One of the most studied lantibiotics is nisin and its antibacterial activity has been proven against pathogenic bacteria such as *Listeria* sp., *Clostridium* sp., and *Bacillus* sp. (Licandro et al. 2020). Nisin is used as protectives to prevent spoilage and growth of pathogenic microorganisms in food products. Nisin used as the common antibacterial mechanism of bacteriocins relies on deterioration of the cell membrane through pore creation or a "detergent impact" (Mora-Villalobos et al. 2020). Nisin is the one LAB bacteriocin accepted for use in food by the FDA (Ibrahim et al. 2021). The most important benefit of bacteriocins produced by LABs is their effects in opposition to opportunistic and pathogenic bacteria, including antibiotic-resistant strains. The synergy of several bacteriocins with antibiotics contributes to the reduction of their concentration and negative side effects. For all that, it has been determined that some bacteria species (for example; *Listeria monocytogenes* and *Staphylococcus aureus*) may improve resistance to bacteriocins. However, it is important to realize that this is minimal compared to the resistance of conventional antibiotics (Zapaśnik et al. 2022).

Bacteriocin can be important for the competition between bacteria in the sourdough flora (Messens and De Vuyst 2002). Bacteriocin production does not extend the fungi-free shelf life of sourdough-based products, but it positively affects the stability of sourdough by preventing the contamination of pathogenic bacteria (Syrokou et al. 2021b). To date, only a few bacteriocins produced by sourdough LAB include bavaricin A of *Lactobacillus bavaricus* MI401, plantarisin ST31 of *Lactiplantibacillus plantarum* ST31, or a bacteriocin-like inhibitory substance (BLIS C57) of *Fructilactobacillus sanfranciscensis* C57 contains or has been characterized (Gobbetti et al. 2005).

11.2.2.5 Antirope Characteristics

Ropiness initially appears as an unpleasant fruity odor and extracellular slimy structural features. Therefore, bacterial deterioration observed in bread occurs as a result of the enzymatic breakdown of crumbs that become soft and sticky. The main specie that play a role in bread spoilage is generally *Bacillus subtilis*. *Bacillus licheniformis, Bacillus pumilus*, and *Bacillus cereus* species have also been reported (Pepe et al. 2003) as agents causing ropiness, in addition to *Bacillus subtilis* species. It has been reported that these microorganisms occur naturally in the raw material, bakery atmosphere, and on the equipment surfaces. During the baking of bread, not only bacterial spores remain, but mold spores can also survive. Only vegetative forms of microorganisms are destroyed during baking. Considering these factors, it still constitutes a major source of economic concern for the bakery industry (Cizeikiene et al. 2013).

Bacillus endospores can also be found in flour. Contamination of wheat flour with *Bacillus* endospores is related to the stable formation of these organisms and the presence of *Bacillus*, a member of the seed-derived endophytic microbial community. It has been reported that rop formation in bread can develop rapidly in hot and humid conditions. It was emphasized that either chemical preservatives or sourdough should be used to prevent this. Sourdough prevents deterioration by forming the inhibitory activity of acid and undissociated organic acids (Li et al. 2020). *B. subtilis* spores, which are heat resistant, can present in places where the temperature is maximum between 97 and 101 °C. In addition, it has the ability to survive for a few minutes in the bread crumb kernel during baking. Water activity, pH and temperature during storage are significant in spore germination and growth of vegetative cells of *Bacillus* spp. (Pepe et al. 2003). Although sourdourg leavening in bread making is more mixed and time consuming compared to the plain dough process, it has many advantages.

In one study, LABs with bacteriocin-like antirope activity were screened. It was reported that *Lactiplantibacillus plantarum* E5 and *Leuconostoc mesenteroides* A27 that were selected from this screening inhibited the formation of rope in bread for more than 15 days (Gálvez et al. 2014). Bacteriocin production of LAB from sourdough isolates has been reported to have activity against rop-forming bacilli, although their activity in bread is limited. The limited activity in bread is due to the fact that bacteriocins show change reactions in sourdough and are inactivated by proteases and thiol (Li et al. 2020).

11.2.2.6 Bioactive Compounds

The effect of species-specific transformation of amino acids on the sensory property of bread have been the topic of a few review articles. The effect of peptides on bread quality is in the form of amino acids and products of microbial amino acid metabolism, flavor active compounds, flavor precursors or bioactive substances. It seems that it is important to collect peptides and amino acid metabolites with antioxidant, antihypertensive or anticancer activities in sourdough fermentation. In a study, it was reported that the species-specific peptide status of sourdough *Lactobacillus* significantly affected the accumulation of bioactive peptides in sourdough rye yeast (Gänzle 2014). Still other studies have focused on the fact that bioactive peptides may have antibacterial (Luz et al. 2017), antiproliferative (Rizzello et al. 2015), and antioxidant activities (Luti et al. 2020). Although degenerative diseases of aging may have different causes, interest in antioxidant peptides moving as inhibitors of lipid peroxidation has increased due to the characterization of unusualness in inflammatory response and oxidative stress. Antioxidant peptides generally occur from two to twenty amino acidic residues. Their molecular mass is <6.0 kDa. The antioxidant efficiency of peptides is based on amino acid composition, comfort and hydrophobic. Besides antioxidant properties, bioactive peptides may also have the ability to modulate inflammatory processes (Galli et al. 2018).

Bioactive compounds are formed during fermentation. It reduces gluten sensitivity, other compounds involved in grain-related digestive problems and pathologies like gastrointestinal syndromes. During the preparation of sourdough and dough, there is a decrease in components such as vitamin E, tocopherol, and tocotrienol. This is because it is air sensitive. Another important point for bioactive peptides, LABs are the most suitable microorganism species for bioactive peptide enrichment (De Vero et al. 2021). Rizzello et al. (2007, 2008) examined the effect on the release of Angiotensin I-Converting Enzyme (ACE)-inhibitory peptides by selecting sourdough lactobacilli with proteinase and peptidase activities specific for cereal proteins. These were used successfully during sourdough fermentation. The use of these peptides to avoid hypertension and other therapeutic aims has been mentioned as important for functional foods. In addition, another study has supported that LABs in fermentation of various flours such as wheat, barley, amaranth, soybean or rye increase the concentration of the anti-cancer peptide lunasin (De Vero et al. 2021). Recently, the potential of sourdough LAB to release lunasin, a potent antitumoral peptide, has been exploited during the fermentation of cereal flours. It has been observed in studies that the use of extracts from fermented legume pulps exhibited a significant inhibitory effect (up to 70%) on the proliferation of human adenocarcinoma Caco-2 cells (Rizzello et al. 2015). Moreover, the capability of LAB to leave antioxidant peptides during the fermentation of different cereal flours has been demonstrated. In another study, the effect of purified peptides on mouse fibroblasts artificially exposed to oxidative stress was noted in the form of *ex vivo* antioxidant activity (Coda et al. 2012). Additionally, LAB selected for proteolytic efficiency were used in the fermentation of wheat and rye to produce antihypertensive peptides. These LAB have been found to exhibit potent ACE inhibitory activity either by fermenting flour under semifluid conditions or using whole wheat flour (Nionelli and Rizzello 2016).

Luti et al. (2021) screened a large number of LAB isolates obtained from commercial wheat flour were screened. Of these isolates, *Companilactobacillus farciminis*, *Furfurilactobacillus rossiae*, and *Fructilactobacillus sanfranciscensis* have been identified. These 3 strains were chosen because of their ability to produce bioactive peptides endowed with antioxidant and anti-inflammatory activity. It has been observed that the same bioactive peptides are preserved in breads prepared with sourdough (Luti et al. 2021). The increase of bioactive peptides by LAB during food fermentation depends on amino acid sequences. Bioactive peptides that may have a positive effect on human health, are natural protein fragments with cryptic amino acid sequences. It is known to affect the cardiovascular, digestive-immune and nervous systems (Galli et al. 2018).

11.3 Conclusion

Sourdough technology is a biological process that has been used for years. It has many advantages due to its complex and unique composition. It has the potential to be an attractive product among consumers because it has positive effects on texture, rheology, extending shelf life, and health. This microbial dynamic in sourdough is mainly due to LAB and yeasts. LAB, which are among them and found in the microflora of sourdough, improve the technological, sensory and nutritional properties of bread and sourdough-based products, thanks to the metabolites they produce. Therefore, its importance in terms of sourdough technology has been emphasized, considering that the effects of the metabolites produced by LABs in the sourdough ecosystem on sourdough should be known in more depth allowing to select microbial cultures with specific functional features.

References

Abarquero D, Renes E, Fresno JM, Tornadijo ME (2022) Study of exopolysaccharides from lactic acid bacteria and their industrial applications: a review. Int J Food Sci Technol 57(1):16–26

Akamine IT, Mansoldo FR, Vermelho AB (2023) Probiotics in the sourdough bread fermentation: current status. Fermentation 9(2):90

Alkay Z, Kılmanoğlu H, Durak MZ (2020) Prevention of sourdough bread mould spoliage by antifungal lactic acid bacteria fermentation. Avrupa Bilim ve Teknoloji Dergisi 18:379–388

Alkay Z, Yılmaz MT, Can AM, İspirli H, Dertli E (2022) The effect of flours of different immature cereal grains on sourdough and sourdough bread: microbiological, rheological, textural and sugar profiles. J Food Process Preserv 46(11):e17097

Alvarez-Sieiro P, Montalbán-López M, Mu D, Kuipers OP (2016) Bacteriocins of lactic acid bacteria: extending the family. Appl Microbiol Biotechnol 100:2939–2951

Aplevicz KS, Mazo JZ, Ilha EC, Dinon AZ (2014) Isolation and characterization of lactic acid bacteria and yeasts from the Brazilian grape sourdough. Braz J Pharm Sci 50:321–327

Arena MP, Russo P, Spano G, Capozzi V (2020) From microbial ecology to innovative applications in food quality improvements: the case of sourdough as a model matrix. J Multidiscip Sci 3(1):9–19

Arendt E, Dal Bello F (eds) (2011) Gluten-free cereal products and beverages. Elsevier

Arora K, Ameur H, Polo A, Di Cagno R, Rizzello CG, Gobbetti M (2021) Thirty years of knowledge on sourdough fermentation: a systematic review. Trends Food Sci Technol 108:71–83

Arsoy ES, Gül LB, Çon AH (2022) Characterization and selection of potential antifungal lactic acid bacteria isolated from Turkish spontaneous sourdough. Curr Microbiol 79(5):1–11

Baek HW, Bae JH, Lee YG, Kim SA, Min W, Shim S et al (2021) Dynamic interactions of lactic acid bacteria in Korean sourdough during back-slopping process. J Appl Microbiol 131(5):2325–2335

Bartkiene E, Lele V, Ruzauskas M, Domig KJ, Starkute V, Zavistanaviciute P et al (2019) Lactic acid bacteria isolation from spontaneous sourdough and their characterization including antimicrobial and antifungal properties evaluation. Microorganisms 8(1):64

Bomfim VB, Neto JHPL, Leite KS, de Andrade Vieira É, Iacomini M, Silva CM et al (2020) Partial characterization and antioxidant activity of exopolysaccharides produced by Lactobacillus plantarum CNPC003. LWT 127:109349

Bounaix MS, Gabriel V, Morel S, Robert H, Rabier P, Remaud-Simeon M et al (2009) Biodiversity of exopolysaccharides produced from sucrose by sourdough lactic acid bacteria. J Agric Food Chem 57(22):10889–10897

Çakır E, Arıcı M, Durak MZ (2020a) Biodiversity and techno-functional properties of lactic acid bacteria in fermented hull-less barley sourdough. J Biosci Bioeng 130(5):450–456

Çakır E, Arıcı M, Durak MZ, Karasu S (2020b) The molecular and technological characterization of lactic acid bacteria in einkorn sourdough: effect on bread quality. J Food Meas Charact 14(3):1646–1655

Calvert MD, Madden AA, Nichols LM, Haddad NM, Lahne J, Dunn RR, McKenney EA (2021) A review of sourdough starters: ecology, practices, and sensory quality with applications for baking and recommendations for future research. PeerJ 9:e11389

Catzeddu P (2019) Sourdough breads. In: Flour and breads and their fortification in health and disease prevention. Academic, pp 177–188

Chandal M, Mahjub R (2020) Identification, characterization, and investigation of the antimicrobial activities of autochthonous lactic acid bacteria isolated from Iranian traditional sourdoughs. Avicenna J Pharm Res 1(1):1–4

Chen H, Yan X, Du G, Guo Q, Shi Y, Chang J et al (2021) Recent developments in antifungal lactic acid bacteria: application, screening methods, separation, purification of antifungal compounds and antifungal mechanisms. Crit Rev Food Sci Nutr 63:1–15

Cizeikiene D, Juodeikiene G, Paskevicius A, Bartkiene E (2013) Antimicrobial activity of lactic acid bacteria against pathogenic and spoilage microorganism isolated from food and their control in wheat bread. Food Control 31(2):539–545

Coda R, Rizzello CG, Pinto D, Gobbetti M (2012) Selected lactic acid bacteria synthesize antioxidant peptides during sourdough fermentation of cereal flours. Appl Environ Microbiol 78(4):1087–1096

Corsetti A, Settanni L (2007) Lactobacilli in sourdough fermentation. Food Res Int 40(5):539–558

De Luca L, Aiello A, Pizzolongo F, Blaiotta G, Aponte M, Romano R (2021) Volatile organic compounds in breads prepared with different sourdoughs. Appl Sci 11(3):1330

De Vero L, Iosca G, Gullo M, Pulvirenti A (2021) Functional and healthy features of conventional and non-conventional sourdoughs. Appl Sci 11(8):3694

De Vuyst L, Leroy F (2007) Bacteriocins from lactic acid bacteria: production, purification, and food applications. Microb Physiol 13(4):194–199

De Vuyst L, Vancanneyt M (2007) Biodiversity and identification of sourdough lactic acid bacteria. Food Microbiol 24(2):120–127

De Vuyst L, Comasio A, Kerrebroeck SV (2021) Sourdough production: fermentation strategies, microbial ecology, and use of non-flour ingredients. Crit Rev Food Sci Nutr 63:1–33

Debonne E, Vermeulen A, Bouboutiefski N, Ruyssen T, Van Bockstaele F, Eeckhout M, Devlieghere F (2020) Modelling and validation of the antifungal activity of DL-3-phenyllactic acid and acetic acid on bread spoilage moulds. Food Microbiol 88:103407

Dertli E, Mercan E, Arıcı M, Yılmaz MT, Sağdıç O (2016) Characterisation of lactic acid bacteria from Turkish sourdough and determination of their exopolysaccharide (EPS) production characteristics. LWT-Food Sci Technol 71:116–124

Di Cagno R, De Angelis M, Lavermicocca P, De Vincenzi M, Giovannini C, Faccia M, Gobbetti M (2002) Proteolysis by sourdough lactic acid bacteria: effects on wheat flour protein fractions and gliadin peptides involved in human cereal intolerance. Appl Environ Microbiol 68(2):623–633

Drakula S, Novotni D, Mustač NČ, Voučko B, Krpan M, Hruškar M, Ćurić D (2019) A simple HS-SPME/GC-MS method for determination of acrolein from sourdough to bread. Food Anal Methods 12(11):2582–2590

Drakula S, Mustač NČ, Novotni D, Voučko B, Krpan M, Hruškar M, Ćurić D (2022) Optimization and validation of a HS-SPME/GC–MS method for the analysis of gluten-free bread volatile flavor compounds. Food Anal Methods 15:1–16

Ewuoso MO, Animashaun OH, Adejumo AA (2020) Lactic acid bacteria and yeasts in spontaneously fermented sorghum sourdough. Am J Microbiol Res 8(2):63–72

Fraberger V, Unger C, Kummer C, Domig KJ (2020) Insights into microbial diversity of traditional Austrian sourdough. LWT 127:109358

Galle S, Arendt EK (2014) Exopolysaccharides from sourdough lactic acid bacteria. Crit Rev Food Sci Nutr 54(7):891–901

Galle S, Schwab C, Arendt EK, Gänzle MG (2011) Structural and rheological characterisation of heteropolysaccharides produced by lactic acid bacteria in wheat and sorghum sourdough. Food Microbiol 28(3):547–553

Galle S, Schwab C, Dal Bello F, Coffey A, Gänzle MG, Arendt EK (2012) Influence of in-situ synthesized exopolysaccharides on the quality of gluten-free sorghum sourdough bread. Int J Food Microbiol 155(3):105–112

Galli V, Mazzoli L, Luti S, Venturi M, Guerrini S, Paoli P et al (2018) Effect of selected strains of lactobacilli on the antioxidant and anti-inflammatory properties of sourdough. Int J Food Microbiol 286:55–65

Gálvez A, López RL, Pulido RP, Burgos MJG (2014) Natural antimicrobials for food biopreservation. In: Food biopreservation. Springer, New York, NY, pp 3–14

Gänzle MG (2014) Enzymatic and bacterial conversions during sourdough fermentation. Food Microbiol 37:2–10

Gobbetti M, Gänzle M (eds) (2012) Handbook on sourdough biotechnology. Springer Science & Business Media

Gobbetti M, De Angelis M, Corsetti A, Di Cagno R (2005) Biochemistry and physiology of sourdough lactic acid bacteria. Trends Food Sci Technol 16(1–3):57–69

Gobbetti M, Minervini F, Pontonio E, Di Cagno R, De Angelis M (2016) Drivers for the establishment and composition of the sourdough lactic acid bacteria biota. Int J Food Microbiol 239:3–18

Hajinia F, Sadeghi A, Sadeghi Mahoonak A (2021) The use of antifungal oat-sourdough lactic acid bacteria to improve safety and technological functionalities of the supplemented wheat bread. J Food Saf 41(1):e12873

Harper AR, Dobson RC, Morris VK, Moggré GJ (2022) Fermentation of plant-based dairy alternatives by lactic acid bacteria. Microb Biotechnol 15(5):1404–1421

Hassan YI, Zhou T, Bullerman LB (2016) Sourdough lactic acid bacteria as antifungal and mycotoxin-controlling agents. Food Sci Technol Int 22(1):79–90

Houssni IEL, Zahidi A, Khedid K, Hassikou R (2022) A review of spontaneous sourdough as a functional ingredient for improving the sensory and nutritional quality of wheat bread. J Mater Environ Sci 13(1):9–28

Houssni IE, Khedid K, Zahidi A, Hassikou R (2023) The inhibitory effects of lactic acid bacteria isolated from sourdough on the mycotoxigenic fungi growth and mycotoxins from wheat bread. Biocatal Agric Biotechnol 50:102702

Huang C, Huang J, Zhang B, Omedi JO, Chen C, Zhou L et al (2023) Rheo-fermentation dough properties, bread-making quality and aroma characteristics of red bean (Vigna angularis) sourdough induced by LAB Weissella confusa QS813 strain fermentation. Foods 12(3):605

Hüttner EK, Dal Bello F, Arendt EK (2010) Identification of lactic acid bacteria isolated from oat sourdoughs and investigation into their potential for the improvement of oat bread quality. Eur Food Res Technol 230(6):849–857

Ibrahim SA, Ayivi RD, Zimmerman T, Siddiqui SA, Altemimi AB, Fidan H et al (2021) Lactic acid bacteria as antimicrobial agents: food safety and microbial food spoilage prevention. Foods 10(12):3131

İspirli H, Dertli E (2022) Isolation of distinct Lactobacillaceae spp. with functional characteristics from traditional sourdough samples. Akademik Gıda 20(3):211–219

Ispirli H, Demirbaş F, Yüzer MO, Dertli E (2018) Identification of lactic acid bacteria from spontaneous rye sourdough and determination of their functional characteristics. Food Biotechnol 32(3):222–235

Kaditzky S, Vogel RF (2008) Optimization of exopolysaccharide yields in sourdoughs fermented by lactobacilli. Eur Food Res Technol 228(2):291–299

Kaditzky S, Seitter M, Hertel C, Vogel RF (2008) Performance of Lactobacillus sanfranciscensis TMW 1.392 and its levansucrase deletion mutant in wheat dough and comparison of their impact on bread quality. Eur Food Res Technol 227:433–442

Kadyan S, Pradhan D (2020) Antifungal lactic acid bacteria (LAB): potential use in food systems. In: Novel strategies to improve shelf-life and quality of foods. Apple Academic Press, pp 73–94

Korcz E, Varga L (2021) Exopolysaccharides from lactic acid bacteria: techno-functional application in the food industry. Trends Food Sci Technol 110:375–384

Lancetti R, Sciarini L, Pérez GT, Salvucci E (2021) Technological performance and selection of lactic acid bacteria isolated from argentinian grains as starters for wheat sourdough. Curr Microbiol 78(1):255–264

Li Z, Li H, Bian K (2016) Microbiological characterization of traditional dough fermentation starter (Jiaozi) for steamed bread making by culture-dependent and culture-independent methods. Int J Food Microbiol 234:9–14. https://doi.org/10.1016/J.IJFOODMICRO.2016.06.024

Li Z, Siepmann FB, Tovar LER, Chen X, Gänzle MG (2020) Effect of copy number of the spoVA2mob operon, sourdough and reutericyclin on ropy bread spoilage caused by Bacillus spp. Food Microbiol 91:103507

Licandro H, Ho PH, Nguyen TKC, Petchkongkaew A, Van Nguyen H, Chu-Ky S et al (2020) How fermentation by lactic acid bacteria can address safety issues in legumes food products? Food Control 110:106957

Liu T, Li Y, Chen J, Sadiq FA, Zhang G, Li Y, He G (2016) Prevalence and diversity of lactic acid bacteria in Chinese traditional sourdough revealed by culture dependent and pyrosequencing approaches. LWT-Food Sci Technol 68:91–97

Liu A, Jia Y, Zhao L, Gao Y, Liu G, Chen Y et al (2018) Diversity of isolated lactic acid bacteria in Ya'an sourdoughs and evaluation of their exopolysaccharide production characteristics. LWT 95:17–22

Luti S, Mazzoli L, Ramazzotti M, Galli V, Venturi M, Marino G et al (2020) Antioxidant and anti-inflammatory properties of sourdoughs containing selected Lactobacilli strains are retained in breads. Food Chem 322:126710

Luti S, Galli V, Venturi M, Granchi L, Paoli P, Pazzagli L (2021) Bioactive properties of breads made with sourdough of Hull-less barley or conventional and pigmented wheat flours. Appl Sci 11(7):3291

Luz C, Saladino F, Luciano FB, Mañes J, Meca G (2017) In vitro antifungal activity of bioactive peptides produced by lactobacillus plantarum against aspergillus parasiticus and Penicillium expansum. LWT-Food Sci Technol 81:128–135

Maidana SD, Ficoseco CA, Bassi D, Cocconcelli PS, Puglisi E, Savoy G et al (2020) Biodiversity and technological-functional potential of lactic acid bacteria isolated from spontaneously fermented chia sourdough. Int J Food Microbiol 316:108425

Mani-López E, Arrioja-Bretón D, López-Malo A (2022) The impacts of antimicrobial and antifungal activity of cell-free supernatants from lactic acid bacteria in vitro and foods. Compr Rev Food Sci Food Saf 21(1):604–641

Manini F, Casiraghi MC, Poutanen K, Brasca M, Erba D, Plumed-Ferrer C (2016) Characterization of lactic acid bacteria isolated from wheat bran sourdough. LWT-Food Sci Technol 66:275–283

Martín-Garcia A, Riu-Aumatell M, López-Tamames E (2021) Influence of process parameters on sourdough microbiota, physical properties and sensory profile. Food Rev Intl 39:1–15

Melini V, Melini F (2018) Strategies to extend bread and GF bread shelf-life: from sourdough to antimicrobial active packaging and nanotechnology. Fermentation 4(1):9

Menezes LAA, Molognoni L, de Sá Ploêncio LA, Costa FBM, Daguer H, Dea Lindner JD (2019) Use of sourdough fermentation to reducing FODMAPs in breads. Eur Food Res Technol 245:1183–1195

Menezes LAA, De Marco I, Oliveira N, dos Santos N, Costa Nunes C, Leite Cartabiano CE, Molognoni L et al (2021) Reducing FODMAPs and improving bread quality using type II sourdough with selected starter cultures. Int J Food Sci Nutr 72(7):912–922

Messens W, De Vuyst L (2002) Inhibitory substances produced by lactobacilli isolated from sourdoughs—a review. Int J Food Microbiol 72(1–2):31–43
Mishra B, Mishra AK, Kumar S, Mandal SK, Nsv L, Kumar V et al (2021) Antifungal metabolites as food bio-preservative: innovation, outlook, and challenges. Meta 12(1):12
Mora-Villalobos JA, Montero-Zamora J, Barboza N, Rojas-Garbanzo C, Usaga J, Redondo-Solano M et al (2020) Multi-product lactic acid bacteria fermentations: a review. Fermentation 6(1):23
Nasrollahzadeh A, Mokhtari S, Khomeiri M, Saris PE (2022) Antifungal preservation of food by lactic acid bacteria. Foods 11(3):395
Nehal F, Sahnoun M, Smaoui S, Jaouadi B, Bejar S, Mohammed S (2019) Characterization, high production and antimicrobial activity of exopolysaccharides from Lactococcus lactis F-mou. Microb Pathog 132:10–19
Niku-Paavola ML, Laitila A, Mattila-Sandholm T, Haikara A (1999) New types of antimicrobial compounds produced by lactobacillus plantarum. J Appl Microbiol 861:29–35
Nionelli L, Rizzello CG (2016) Sourdough-based biotechnologies for the production of gluten-free foods. Foods 5(3):65
Oshiro M, Zendo T, Nakayama J (2021) Diversity and dynamics of sourdough lactic acid bacteriota created by a slow food fermentation system. J Biosci Bioeng 131(4):333–340
Păcularu-Burada B, Georgescu LA, Bahrim GE (2020) Current approaches in sourdough production with valuable characteristics for technological and functional applications. Annals of the University Dunarea de Jos of Galati Fascicle VI-Food Technology 44(1):132–148
Pastor K, Zorlu G, Osman S, Sevdık YÖ, Kojıć J, Nastıć N, Senyuva H (2022) Chemometric fatty acidomics to distinguish between yeast and sourdough breads from Serbia and Turkey. J Food Nutr Res 61(4)
Pepe O, Blaiotta G, Moschetti G, Greco T, Villani F (2003) Rope-producing strains of Bacillus spp. from wheat bread and strategy for their control by lactic acid bacteria. Appl Environ Microbiol 69(4):2321–2329
Petkova M, Stefanova P, Gotcheva V, Angelov A (2021) Isolation and characterization of lactic acid bacteria and yeasts from typical Bulgarian sourdoughs. Microorganisms 9(7):1346
Pino A, Russo N, Solieri L, Sola L, Caggia C, Randazzo CL (2022) Microbial consortia involved in traditional Sicilian sourdough: characterization of lactic acid bacteria and yeast populations. Microorganisms 10(2):283
Pizarro F, FrAnCO F Jr (2017) Volatile organic compounds at early stages of sourdough preparation via static headspace and GC/MS analysis. Curr Res Nutr Food Sci J 5(2):89–99
Rajanikar RV, Nataraj BH, Naithani H, Ali SA, Panjagari NR, Behare PV (2021) Phenyllactic acid: a green compound for food biopreservation. Food Control 128:108184
Reale A, Di Renzo T, Boscaino F, Nazzaro F, Fratianni F, Aponte M (2019) Lactic acid bacteria biota and aroma profile of italian traditional sourdoughs from the irpinian area in Italy. Front Microbiol 10:1621
Reale A, Di Stasio L, Di Renzo T, De Caro S, Ferranti P, Picariello G et al (2021) Bacteria do it better! Proteomics suggests the molecular basis for improved digestibility of sourdough products. Food Chem 359:129955
Riesute R, Salomskiene J, Moreno DS, Gustiene S (2021) Effect of yeasts on food quality and safety and possibilities of their inhibition. Trends Food Sci Technol 108:1–10
Ripari V, Gänzle MG, Berardi E (2016) Evolution of sourdough microbiota in spontaneous sourdoughs started with different plant materials. Int J Food Microbiol 232:35–42. https://doi.org/10.1016/J.IJFOODMICRO.2016.05.025
Rizzello CG, De Angelis M, Di Cagno R, Camarca A, Silano M, Losito I et al (2007) Highly efficient gluten degradation by lactobacilli and fungal proteases during food processing: new perspectives for celiac disease. Appl Environ Microbiol 73(14):4499–4507
Rizzello CG, Cassone A, Di Cagno R, Gobbetti M (2008) Synthesis of angiotensin I-converting enzyme (ACE)-inhibitory peptides and γ-aminobutyric acid (GABA) during sourdough fermentation by selected lactic acid bacteria. J Agric Food Chem 56(16):6936–6943

Rizzello CG, Hernández-Ledesma B, Fernández-Tomé S, Curiel JA, Pinto D, Marzani B et al (2015) Italian legumes: effect of sourdough fermentation on lunasin-like polypeptides. Microb Cell Factories 14(1):1–20

Rohrlich M, Hertel W (1966) Untersuchungen über den Eiweißabbau im Sauerteig. Brot und Gebäck 20(6):109–113

Rollan G, De Angelis M, Gobbetti M, De Valdez GF (2005) Proteolytic activity and reduction of gliadin-like fractions by sourdough lactobacilli. J Appl Microbiol 99(6):1495–1502

Rosell CM, Bajerska J, El Sheikha AF (eds) (2015) Bread and its fortification: nutrition and health benefits. CRC Press

Ruiz Rodríguez L, Vera Pingitore E, Rollan G, Martos G, Saavedra L, Fontana C et al (2016a) Biodiversity and technological potential of lactic acid bacteria isolated from spontaneously fermented amaranth sourdough. Lett Appl Microbiol 63(2):147–154

Ruiz Rodríguez L, Vera Pingitore E, Rollan G, Cocconcelli PS, Fontana C, Saavedra L et al (2016b) Biodiversity and technological-functional potential of lactic acid bacteria isolated from spontaneously fermented quinoa sourdoughs. J Appl Microbiol 120(5):1289–1301

Sadiq FA, Yan B, Tian F, Zhao J, Zhang H, Chen W (2019) Lactic acid bacteria as antifungal and anti-mycotoxigenic agents: a comprehensive review. Compr Rev Food Sci Food Saf 18(5):1403–1436

Sáez GD, Saavedra L, Hebert EM, Zárate G (2018) Identification and biotechnological characterization of lactic acid bacteria isolated from chickpea sourdough in northwestern Argentina. LWT 93:249–256

Sahin AW, Zannini E, Coffey A, Arendt EK (2019) Sugar reduction in bakery products: current strategies and sourdough technology as a potential novel approach. Food Res Int 126:108583

Saranraj P, Naidu MA, Sivasakthivelan P (2013) Lactic acid bacteria and its antimicrobial properties: a review. Int J Pharm Biol Arch 4(6):1124–1133

Sekwati-Monang B, Gänzle MG (2011) Microbiological and chemical characterisation of ting, a sorghum-based sourdough product from Botswana. Int J Food Microbiol 150:115–121

Settanni L, Ventimiglia G, Alfonzo A, Corona O, Miceli A, Moschetti G (2013) An integrated technological approach to the selection of lactic acid bacteria of flour origin for sourdough production. Food Res Int 54(2):1569–1578

Sha HY, Wang QQ, Li ZJ (2023) Comparison of the effect of exopolysaccharide-producing lactic acid bacteria from sourdough on dough characteristics and steamed bread quality. Int J Food Sci Technol 58(1):378–386

Siepmann FB, Ripari V, Waszczynskyj N, Spier MR (2018) Overview of sourdough technology: from production to marketing. Food Bioprocess Technol 11(2):242–270

Şimşek Ö, Çon AH, Tulumogˇlu Ş (2006) Isolating lactic starter cultures with antimicrobial activity for sourdough processes. Food Control 17(4):263–270

Spicher G, Nierle W (1988) Proteolytic activity of sourdough bacteria. Appl Microbiol Biotechnol 28(4):487–492

Sun X, Wu S, Li W, Koksel F, Du Y, Sun L, Fang Y, Hu Q, Pei F (2022) The effects of cooperative fermentation by yeast and lactic acid bacteria on the dough rheology, retention and stabilization of gas cells in a whole wheat flour dough system–a review. Food Hydrocoll 137:108212

Suo B, Chen X, Wang Y (2021) Recent research advances of lactic acid bacteria in sourdough: origin, diversity, and function. Curr Opin Food Sci 37:66–75

Syrokou MK, Tziompra S, Psychogiou EE, Mpisti SD, Paramithiotis S, Bosnea L et al (2021a) Technological and safety attributes of lactic acid bacteria and yeasts isolated from spontaneously fermented Greek wheat sourdoughs. Microorganisms 9(4):671

Syrokou MK, Stasinopoulou P, Paramithiotis S, Bosnea L, Mataragas M, Papadopoulos GK et al (2021b) The effect of incubation temperature, substrate and initial pH value on plantaricin activity and the relative transcription of PLN genes of six sourdough derived Lactiplantibacillus plantarum strains. Fermentation 7(4):320

Tang X, Liu R, Huang W, Zhang B, Wu Y, Zhuang J et al (2018) Impact of in situ formed exopolysaccharides on dough performance and quality of Chinese steamed bread. LWT 96:519–525

Torino MI, Font de Valdez G, Mozzi F (2015) Biopolymers from lactic acid bacteria. Novel applications in foods and beverages. Front Microbiol 6:834

Torrieri E, Pepe O, Ventorino V, Masi P, Cavella S (2014) Effect of sourdough at different concentrations on quality and shelf life of bread. LWT-Food Sci Technol 56(2):508–516

Valerio F, Di Biase M, Lattanzio VM, Lavermicocca P (2016) Improvement of the antifungal activity of lactic acid bacteria by addition to the growth medium of phenylpyruvic acid, a precursor of phenyllactic acid. Int J Food Microbiol 222:1–7

Vasilica BTB, Chiş MS, Alexa E, Pop C, Păucean A, Man S et al (2022) The impact of insect flour on sourdough fermentation-fatty acids, amino-acids, minerals and volatile profile. *Insects* 13(7):576

Warburton A, Silcock P, Eyres GT (2022) Impact of sourdough culture on the volatile compounds in wholemeal sourdough bread. Food Res Int 161:111885

Wu H, Guang C, Zhang W, Mu W (2021) Recent development of phenyllactic acid: physicochemical properties, biotechnological production strategies and applications. Crit Rev Biotechnol 43:1–16

Yıldız B, Ceyhan T, Çakıcı A (2019) Sourdough yeasts in bread making: a review. Int J Food Eng Res 5(2):33–46

Zapaśnik A, Sokołowska B, Bryła M (2022) Role of lactic acid bacteria in food preservation and safety. Foods 11(9):1283

Zhang J, Yao Y, Li J, Ju X, Wang L (2023) Impact of exopolysaccharides-producing lactic acid bacteria on the chemical, rheological properties of buckwheat sourdough and the quality of buckwheat bread. Food Chem 425:136369

Zotta T, Piraino P, Ricciardi A, McSweeney PL, Parente E (2006) Proteolysis in model sourdough fermentations. J Agric Food Chem 54(7):2567–2574

Chapter 12
Yeast Strains from Sourdough as Potential Clean-Label Starters for Fermentation Processes

Oana Emilia Constantin, Bogdan Păcularu-Burada, and Gabriela-Elena Bahrim

12.1 Introduction

Bread and dough-like products are among the most popular grain foods. Compared to unleavened bread and yeast bread, sourdough bread plays a significant role in the diet due to its improved sensory and nutritional properties (Gobbetti et al. 2019). Cereals are important dietary components that offer, at the same time, some health benefits through the intake and metabolism of carbohydrates, proteins, lipids, dietary fibers, vitamins, and minerals. These are the main reasons why cereals are frequently involved in baked goods' production based on wheat, corn, rice, or barley. A continuously increasing demand of bakery products or similar foods, correlated with the consumers' requirements regarding different types of products in terms of sensory, nutritional, and functional properties, has led to the identification of alternative and underutilized raw materials comprising pseudocereals, legumes or tubers, such as sorghum, millet, cassava (Font de Valdez et al. 2010). The preparation of bread involves the use of an "acid dough" or sourdough, obtained from flour and water, a mixture that is then subjected to natural fermentation by lactic acid bacteria (LAB) and yeasts, which are already found in the specific microbiota of the raw materials (Gobbetti et al. 2019; Boudaoud et al. 2021). Around the world, sourdoughs have been characterized by a diverse range of microbial species, including over 90 LAB species and 40 yeast species (De Vuyst et al. 2017; Comasio et al. 2020). Each microbial group typically comprises just one or two dominating microbial species. The sourdoughs' microbiota seemed to be often dominated by species belonging to the new taxonomic genera of *Fructilactobacillus* spp., *Lactiplantibacillus* spp., *Levilactobacillus* spp., and *Latilactobacillus* spp.

O. E. Constantin · B. Păcularu-Burada · G.-E. Bahrim (✉)
Faculty of Food Science and Engineering, "Dunărea de Jos" University of Galați, Galați, Romania
e-mail: econstantin@ugal.ro; bogdan.pacularu@ugal.ro; gabriela.bahrim@ugal.ro

E. B. Ceresino et al. (eds.), *Sourdough Microbiota and Starter Cultures for Industry*, https://doi.org/10.1007/978-3-031-48604-3_12

(*former Lactobacillus spp.*) according to Zheng et al. (2020). Regarding the yeast strains, the most prevalent genera and species identified were the following: *Saccharomyces (S. cerevisiae), Kazachstania (Ka. humilis, Ka. exigua), Wickerhamomyces (W. anomalus,* formerly *Hansenula anomala, Pichia anomala), Torulaspora (Toso. delbrueckii),* and *Pichia (P. kudriavzevii)* (Lhomme et al. 2016; Van Kerrebroeck et al. 2017; Carbonetto et al. 2018; Carbonetto et al. 2020). However, the qualitative and quantitative microbial diversity is strongly linked with the microorganisms found in the raw materials' microbiota, the strains' abundance could be modified thereafter by the fermentation processes and their parameters, respectively by the processing conditions, in order to stimulate the competition between species and the biochemical activity of the microorganisms and enzymes with different origins (endogenous, exogenous or from microbial sources).

In baking, as numerous researchers reported (De Vuyst et al. 2017; Papadimitriou et al. 2019; Siepmann et al. 2019; Ma et al. 2021), four different sourdough fermentation processes are defined, depending on each sourdough's particularities, namely:

- **Sourdough type 0** – obtained by a spontaneous fermentation process assured by the yeast and LAB strains from the flours' wild microbiota.
- **Sourdough type I** – obtained by traditional methods, involving a mature sourdough obtained in a previous fermentation process, in which the valuable microorganisms dominate the ecosystem, after a natural selection, and ensure a stable inoculum for subsequent batches. The main yeast species identified in sourdough type I were *S. cerevisiae, Ka. exigua, Ka. (C.) humilis, P. kudriavzevii, Toso. delbrueckii,* and *W. anomalus.*
- **Sourdough type II** – uses selected exogenous starter cultures that need an extended fermentation time (2–5 days) at temperatures higher than 30 °C. The fermented product is then used as an additive in bread manufacturing process, as a source of competitive microorganisms, respectively as an enhancer of bioactive compounds with functional effects. This sourdough is dominated by *S. cerevisiae*, and it is used in a liquid state which increases its applications in industrial bakeries.
- **Sourdough type III** – produced after various drying processes or liquid pasteurization applied on the fermented products to ensure microbiological stability. The most frequent drying procedures utilized in type III sourdough manufacture are freeze-drying, spray-drying, and drum drying.
- **Sourdough type IV** – obtained by an extended fermentation with selected starter cultures or after a spontaneous fermentation ensured by sourdough ingredients' microbiota. This sourdough is a combination between sourdough types I and II.

Generally, the sourdough yeasts ferment carbohydrate-containing flour to produce ethanol and CO_2, enabling dough's rising. Additionally, based on their metabolic diversity, the yeast strains synthesize compounds (organic acids, peptides, aroma compounds, etc.) that impart unique taste to the bread, improving also the functionality, shelf-life and safety (Pétel et al. 2017; Cakir et al. 2021). As it has been already stated, homofermentative or heterofermentative LAB are also a predominant part of the sourdoughs' microbiota. The homofermentative LAB strains convert hexoses exclusively into lactic acid, whereas the heterofermentative ones convert the available sugars into lactic acid, acetic acid, ethanol, and CO_2 (Gänzle 2015; Arora et al. 2021). Acid production is mainly responsible for the sour flavor of bread, other important benefits being related to the products' shelf-life extension and safety assurance. LAB may alter the yeast strains' metabolism and vice versa. The interactions between yeasts and LAB strains may potentially affect the biochemical activities of the biocatalysts with impact on postbiotics' composition and effects of the fermented products. Thus, the breads' volatile profile was more complex when fermentation involved both yeast and LAB (Xu et al. 2019b) compared to the fermentation carried out only with yeast starters. Consequently, co-cultures of *F. (L.) sanfranciscensis* or *Lpb. plantarum* and *S. cerevisiae* produced a higher concentration of 2,3-methyl-1-butanol, 2-methyl-propanoic acid, 3-methyl-butanoic acid, and 2-phenylethanol (Gänzle 2014). In another instance, the maltose metabolism by *Ka. humilis* and heterofermentative LAB co-cultures was improved, compared with the same compound's behavior when LAB single cultures were studied. Such outcomes could be justified by the fact that *Ka. humilis* is a maltose-negative strain, and heterofermentative LAB strains that are usually found in sourdough can use maltose, thus developing a mutual relationship (Carbonetto et al. 2020). The importance of the microbial variety, as well as the interactions between the existing microorganisms in the sourdough, are important on improving the bread's quality (Liu et al. 2018; Yan et al. 2019a), but the relationships between starter cultures' microbial composition and the volatile compounds is still intensively studied. Therefore, volatile compounds are a significant part of ready-to-eat products' overall quality, also improving the consumers' acceptability. Microbial fermentation is an approachable biotechnological tool that can be used to enrich the sourdoughs' volatiles, even though the chemical acidification is applied in food industry to improve the overall aroma profile. The main classes of volatile compounds released as an effect of the symbiotic interactions between the yeast and LAB strains, including also the implication of the indigenous cereal enzymes, are acids, alcohols, ketones, esters, lactones, and aldehydes (Coda et al. 2014). Oro et al. (2018) demonstrated that the yeast volatiles possess antimicrobial potential against food spoilage microorganisms.

Selected or wild strains of yeasts and LAB can be used as valuable sourdough starters, requiring as fermentation substrates only water and flour from different raw materials. A complex characterization of the sourdoughs' microbial composition is difficult to be entirely elucidated because each sourdough ecosystem is unique regarding its microbial populations, metabolic potential, and functionality, respectively the interactions that occur between strains due to the variations in sourdoughs' ingredients used in the recipes, the environmental or biotechnological factors and bioprocesses design (Rafieenia et al. 2022; Bazalová et al. 2022). Considering the latest information reported in the scientific literature in the sourdough field, this chapter provides an overview of the yeasts populations identified in various sourdoughs, including a systematic analysis that highlights their potential to be used as unconventional starters aiming to increase the metabiotic potential with impact on improving the functional, bio-preservative and sensorial attributes of clean-label products, ensuring also the fermented products' diversification.

12.2 Microbial Diversity of Yeasts from Sourdoughs' Microbiota

Sourdoughs' microbial populations can be maintained by backslopping using water, different flour mixtures with a mature sourdough (Carbonetto et al. 2020). The microbial populations' stability in sourdough, especially cohabitation and symbiotic relationships between LAB and yeasts, are not fully clarified, and therefore this behavior may be beneficial for sourdoughs' stability and functionality (Minervini et al. 2014; De Vuyst et al. 2017; Carbonetto et al. 2018). The synergistic functionality between LAB and yeast strains in sourdough can be classified based on one's species impact on the others, respectively negative-negative, negative-neutral, negative-positive, positive-neutral, or positive-positive biological interactions, correlated with the fermentation environment's physicochemical intrinsic and extrinsic factors (Carbonetto et al. 2020). The spontaneously fermented sourdoughs create a stable microbiological community that contributes to the yeasts' species diversification and selection. Moreover, in the sourdoughs analyzed worldwide, approximately 80 yeast species were identified, the most frequent being depicted in Fig. 12.1 (De Vuyst et al. 2017; Comasio et al. 2020; Arora et al. 2021; De Vuyst et al. 2021).

The in-depth characterization of the microbiota's composition from the sourdoughs is mainly influenced by the geographical origin or regional specificity in which they were made (De Vuyst et al. 2017). Breads' manufacturing practices also influence the compositional quality of the sourdough microbiota, correlated with the physicochemical and biotechnological conditions, i.e., the fermentation

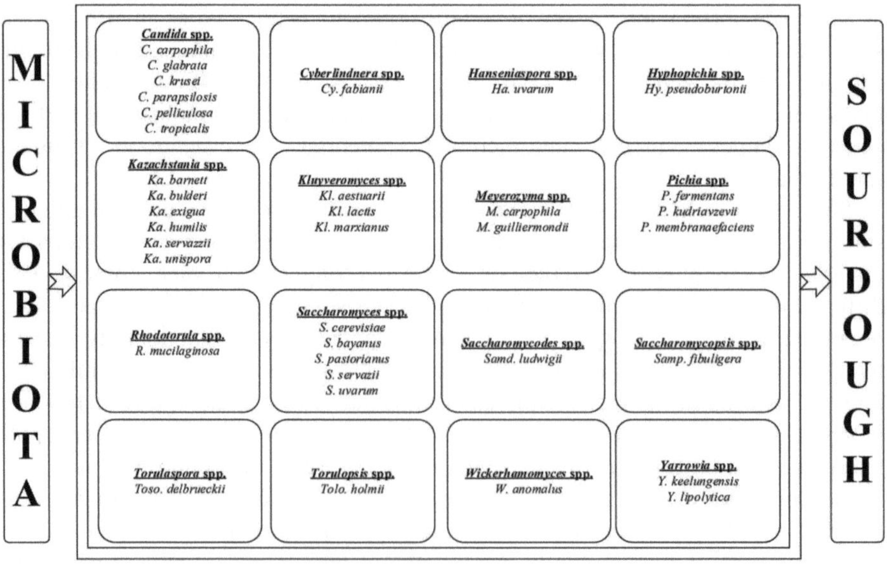

Fig. 12.1 The yeast strains' diversity in sourdoughs

substrates' richness and availability regarding carbohydrates, amino acids, and vitamins; temperature, pH, water activity and redox potential; processing parameters, yeast starters inoculum correlated with the endogenous microorganisms from the fermentation substrate. A particular importance in establishing a correct composition of the microbial populations rely on the sampling and identification techniques (Huys et al. 2013). The yeast species frequently isolated from sourdoughs or used in similar fermentation processes are summarized in Table 12.1, considering different particularities of the microbial strains or fermented products.

As it was shown in Fig. 12.1 and Table 12.1, the yeast species belonging to *Saccharomyces* spp., *Candida* spp., *Kazachstania* spp., *Torulopsis* spp., *Yarrowia* spp., *Wickerhamomyces* spp., *Kluyveromyces* spp., and *Pichia* spp. represent the most common genera. It seems that the yeast strains belonging to *S. cerevisiae* were frequently found in the sourdough because these yeast are able to rapidly metabolize simple carbohydrates (maltose and glucose) for multiplication and alcoholic fermentation (Lahue et al. 2020). The *S. cerevisiae* strains' abundance is supported, on one hand by the existence of these wild strains in the flour or, on the other hand, by baker's yeast utilization for sourdough making (Arora et al. 2021). Although *Ka. exigua* or *Ka. humilis* have been found in stable partnerships with LAB, able to produce sweet baked products, *S. cerevisiae* and *W. (P.) anomalus* were also often isolated from the same microbial ecosystems. The previously-mentioned species are characterized by an improved adaptability on extrinsic factors such as temperature, pH, and osmotolerance (De Vuyst et al. 2016; Pino et al. 2022). Furthermore, the strains belonging to *Ka. unispora* and *Toso. delbrueckii* might be considered minor sourdough microbiota's components in the fermented products dominated by *S. cerevisiae* (Vrancken et al. 2010). Many yeasts are important for their adaptability and diversity of metabolic potential with technological and functional properties. Among them, the non-conventional yeast *Y. lipolytica*, called the workhorse in biotechnology, is able to grow on a large variety of substrates. As such, the agri-food by-products' valorization can be assessed by *Y. lipolytica* fermentation (Gottardi et al. 2021; Madzak 2021).

12.3 Biochemical Activity of Sourdough's Yeasts as Promoters of Clean-Label Starters

12.3.1 Sourdough Yeasts in Fermentation Processes

Sourdough fermentation is one of the most approachable biotechnological process able to improve the rheology, sensory, shelf-life, and nutritional properties of different formulations consisting of wheat, rice, corn, and various pseudocereals. Similar sourdough fermentation processes were successfully designed by various agri-food by-products valorization to obtain value-added functional fermented products (Arora et al. 2021).

Table 12.1 Yeast species isolated or used for sourdough fermentations

Yeast species	Sourdough type/ Manufacturing conditions	Flour type	Geographic region	Reference(s)
S. cerevisiae	Local commercial manufacturer	Rice	Auckland (New Zealand)	Yang et al. (2021)
	Korean traditional sourdough/ Laboratory conditions	Rice	Seongnam (Korea)	Park et al. (2017)
	Type I/ Local manufacturer	Wheat	Shandong Province (China)	Yan et al. (2019b)
	Laboratory conditions	Sorghum	Kwara State (Nigeria)	Olojede et al. (2020)
	Laboratory conditions	Hull-less barley, wheat	Istanbul (Turkey)	Cakir et al. (2021)
S. servazii	Laboratory conditions	Quinoa	Santiago (Chile)	Franco et al. (2020)
Samp. fibuligera, *S. cerevisiae*, Commercial yeast	Laboratory conditions	Wheat	Pyeongchang (South Korea)	Jin et al. (2021)
S. cerevisiae, *W. anomalus*, *Toso. delbrueckii*, *P. kluyveri*, *C. boidinii*, *C. diddensiae*	Artisanal bakeries	Maiorca wheat	Sicily region (Italy)	Pino et al. (2022)
S. cerevisiae, *Ka. (C.) humilis*	Local commercial manufacturer	Rye, wheat	Kansai region (Japan)	Fujimoto et al. (2019)
S. cerevisiae, *Ka. (C.) humilis*, *W. anomalus*	Type I traditional sourdough	Wheat	Shandong province (China)	Yan et al. (2019a)
Ka. humilis, *Ka. exigua*, *S. cerevisiae*, *S. cerevisiae*, *S. cerevisiae Zeus*, *S. cerevisiae Lievitalia (Lesaffre Italia)*	Local commercial manufacturer	Wheat	Pistoia and Livorno (Italy)	Palla et al. (2019)
S. cerevisiae, *Ka. (C.) humilis*, *P. kudriavzevii*, *W. anomalus*, *P. membranifaciens*, *Samp. fibuligera*	Chinese traditional sourdough	Wheat	Gansu, Shandong, Shanxi, Heilongjiang, Henan, Hebei, Shannxi, and Anhui provinces (China)	Liu et al. (2016); Liu et al. (2018)

(continued)

Table 12.1 (continued)

Yeast species	Sourdough type/ Manufacturing conditions	Flour type	Geographic region	Reference(s)
Ka. humilis, Ka. bulderi, S. cerevisiae	Synthetic sourdough medium (SSM)	–	Montpellier (France)	Lhomme et al. (2015); Urien et al. (2019); Boudaoud et al. (2021)
C. carpophila, C. glabrata, C. parapsilosis, Hyphopicchia pseudoburtonii, Hy. burtonii, Ka. humilis, Ka. bulderi, Ka. exigua, Ka. humilis, Ka. saulgeensis, Ka. unispora, M. guilliermondii, M. farinosa, P. kudriavzevii, P. occidentalis, R. mucilaginosa, S. cerevisiae, Scheffersomyces stipitis, Toso. delbrueckii, W. anomalus	Local commercial manufacturers	Wheat	Alsace, Aquitaine, Bourgogne, Bretagne, Champagne Ardenne, Franche-Comté, Ile de France, Pays de Loire, Poitou Charentes, Provence Alpes Côte d'Azur, Rhone-Alpes (France)	Urien et al. (2019)
Ka. servazzii, S. cerevisiae	Local traditional manufacturers	Wheat, rye	Alsace, Aquitaine, Auvegne, Ile-de-France, Midi-Pyrénées, Pays de la Loir, Rhône-Alpes (France)	Lhomme et al. (2015)
C. carpophila, Hy. pseudoburtonii, Ka. unispora, Ka. exigua, Ka. bulderi, Ka. (C.) humilis, S. cerevisiae, Toso. delbrueckii	Local commercial manufacturers	Wheat, rye	Poitou-Charentes, Lot-et-Garonne, Rhône-Alpes, Pays de la Loire, Picardie (France)	Lhomme et al. (2016)
Cy. fabianii, Ha. uvarum, Ka. servazzii, Kl. marxianus, P. fermentans, P. kudriavzevii, S. pastorianus, S. cerevisiae, Samd. ludwigii, Toso. delbrueckii	Laboratory conditions	Wheat (whole-grain Graham), wheat (organic), wheat, rye (organic), rye, rye malt, barley, oat	Espoo (Finland)	Johansson et al. (2021)

(continued)

Table 12.1 (continued)

Yeast species	Sourdough type/ Manufacturing conditions	Flour type	Geographic region	Reference(s)
Kl. marxianus, Kl. lactis, Kl. aestuarii	Traditionally fermented sourdough	Whole wheat	East Azarbaijan (Iran)	Fekri et al. (2020)
Cy. fabianii (syn. *P. fabianii*), *Ka. bulderi/Ka. (C.) humilis, W. anomalus* (syn. *P. anomala*)	Laboratory conditions	White wheat	Brussels (Belgium)	Van Kerrebroeck et al. (2016)
S. cerevisiae, Samp. fibuligera, Toso. delbrueckii, W. anomalus	Traditional	Qu, rice	Shangqiu and Henan (China)	Li et al. (2016)
Ka. humilis, Ka. unispora, P. fermentans, S. cerevisiae, S. uvarum, Toso. delbrueckii	Traditional	Rye, wheat	Upper Austria, Lower Austria, Carinthia, Salzburg, Styria, Vienna (Austria)	Fraberger et al. (2020)
Ka. humilis, S. cerevisiae, W. anomalus	Type I sourdough, Laboratory conditions	Wheat	Wuxi (China)	Xu et al. (2019b)
P. kudriavzevii, Samp. fibuligera, M. guilliermondii, Toso. delbrueckii	Laboratory conditions	High-gluten wheat flour	Yihai Karry (China)	Xu et al. (2019a)
S. cerevisiae, P. membranifaciens	Local commercial manufacturers	Whole wheat	Ankara (Turkey)	Karaman et al. (2018)
Kl. marxianus, S. cerevisiae, Toso. delbrueckii	Local commercial manufacturers	Wheat	Black Sea and Aegean regions (Turkey)	Arici et al. (2018)
S. cerevisiae	Laboratory conditions	White sorghum	Ibadan (Nigeria)	Ogunsakin et al. (2017)
Ka. humilis, P. fermentans, P. membranifaciens, S. cerevisiae, W. anomalus	Laboratory conditions	Wheat	Aetolia-Acarnania, Arkadia, Thessaloniki, Viotia, Salamis Island (Greece)	Syrokou et al. (2020); Syrokou et al. (2021)
S. cariocanus	Type III sourdough / Laboratory conditions	Wheat	Brussels (Belgium)	Comasio et al. (2021)
P. cecembensis, S. cerevisiae, W. anomalus	Type II / Laboratory conditions			

Yeasts metabolize the simple carbohydrates that can be originated from the flours used or by sweeteners addition, especially sucrose. The aerobic conditions facilitate the yeast cells multiplication and biomass accumulation, that thereafter are delivered as postbiotics in the fermentation ecosystem after yeasts' autolysis. The resulted postbiotics impact the microbial nutrition or the functional characteristics (Nguyen et al. 2022). The enzymatic system of yeasts, which is mainly responsible for the alcoholic fermentation, acts intracellularly without external changes with the environment. The yeast strains' catalysts consist of complexes made between enzymes and coenzymes, that ensure the esterification of carbohydrates, the transfer of phosphorus groups, redox reactions, isomerizations, and decarboxylations. Specifically, yeasts use the Embden-Meyerhof-Parnas (EMP) pathway to convert flour saccharides (maltose, sucrose, glucose, and fructose) into pyruvate, generating both ATP and reducing power ($NADH + H^+$), and then pyruvate is converted into ethanol and carbon dioxide (alcoholic fermentation), regenerating the cofactor NAD^+ used in the upper part of the EMP pathway (De Vuyst et al. 2021). Carbohydrates' fermentation needs intracellular transport into the yeast cell, that occurs through passive diffusion due to the gradient concentration. This process is ensured when the concentration of carbohydrates outside the cell is higher than inside the cell, the receptor proteins being also involved. The ability of *W. anomalus* and *S. cerevisiae* to use maltose and sucrose, the two principal carbohydrates in flour, may be considered an advantage of these species. It is already known that the *W. anomalus* strains are osmotolerant, respectively the *S. cerevisiae* strains' adaptability at low pH values, that may be an advantage for *W. anomalus*, as the content of glucose made from sucrose can increase due to yeast's invertase activity (Vrancken et al. 2010). The competition for nutrients between *S. cerevisiae* and acid-tolerant *W. anomalus* may justify the *S. cerevisiae* strains' relatively low abundance in spontaneously fermented foods. Moreover, a symbiosis between the *S. cerevisiae* and *K. marxianus* co-cultures ensure an improved CO_2 production (Struyf et al. 2017a; b). Yeasts also contribute to products' flavor by producing a variety of flavor-active metabolites, especially alcohols from amino acids. The 3-methyl-1-butanol results from the catabolic or anabolic processes involving L-leucine in the metabolic pathways (Pétel et al. 2017).

In sourdough ecosystems the interactions between yeast and LAB strains determine stable microbial associations (Banik et al. 2019). These consortia reflect the metabolic features of the microorganisms involved, revealing the processes' complexity during sourdough fermentations. Sometimes, different types of relationships between the microbial strains can develop because there is a continuous competition for nutrients. As such, metabiosis and symbiosis can be established between yeast and LAB cells. The yeast's capacity to metabolize lactic and acetic acid, formed after LAB strains' activity, can be considered an important factor that defines metabiosis and limitates the strains' coexistence. The symbiotic relationships ensure that yeasts promote the LAB growth by providing them specific growth factors. Yeasts' respiration that uses oxygen create the favorable conditions for LAB, which are facultatively anaerobic. In turn, LAB produces acids, which maintain an acidic pH, favoring the normal development of alcoholic fermentation. The control of these interrelationships during the fermentation processes is important to achieve the targeted effects.

The mutualistic interactions between maltose-negative and maltose-positive strains, as some authors described (Li et al. 2016; De Vuyst et al. 2016; Carbonetto et al. 2018; Harth et al. 2018; Comasio et al. 2020), are the following:

- *Ka. (C.) humilis* (maltose-negative) and *F. (L.) sanfranciscensis* (maltose-positive)*;*
- *Ka. exigua* (maltose-negative) and *F. (L.) sanfranciscensis* or *L. brevis* (maltose-positive);
- *Ka. barnettii (*maltose-negative) and F. (*L.) sanfranciscensis* (maltose-positive);
- *S. cerevisiae* (maltose-positive) and *L. plantarum* (prefers glucose and fructose).

Thus, some yeasts that cannot use maltose, such as *Ka. (C.) humilis* and *Ka. exigua,* are able to create a nutritional mutualism with the heterofermentative *F. (L.) sanfranciscensis*, absorbing glucose and supplying amino acids via hydrolyzing sucrose and glucofructans. Maltose is transported in *F.* (*L.*) *sanfranciscensis* cells, where maltose phosphorylase act for maltose conversion into glucose and glucose-1-phosphate. (De Vuyst et al. 2016). The mutualistic interactions between *Ka. (C.) humilis* and *F. (L.) sanfranciscensis* are active at temperatures between 25–30 °C and pH 4.2–6.7 even though *F. (L.) sanfranciscensis* optimal growth is reached at 32 °C. A stable coexistence of *S. cerevisiae*, *C. milleri (Ka. (C.) humilis)*, and *F. (L.) sanfranciscensis,* despite the substrate competition, was demonstrated in Lagaccio and Panettone sourdough (Venturi et al. 2012). Other maltose-based interactions between LAB and yeasts able to use maltose were determined for some sourdough yeasts such as S. *cerevisiae* and *W. anomalus*, as it was previously established (De Vuyst et al. 2016). Because *S. cerevisiae* preferentially utilizes maltose, low rates of maltose consumption by homofermentative LAB are likely to preserve the cohabiting *S. cerevisiae* population (Oshiro et al. 2021). Moreover, *L. plantarum* has a stimulating effect on *S. cerevisiae* because the yeast cannot use lactose as a carbon source, using galactose and lactic acid instead (Sieuwerts et al. 2018). In sourdoughs, because co-cultures depleted more maltose than monocultures, maltose-positive bacteria may release more glucose, which yeasts can then use. LAB could also convert maltose into organic acids, CO_2, and ethanol (Carbonetto et al. 2018; Carbonetto et al. 2020). Some *Kl. marxianus* strains were able to produce ethanol or 2-phenylethanol from lactose, therefore contributing to the dairy by-products' biovalorization. The concentration of lactose or amino acids from the fermentation medium affect the fermentation yield (Kurniawati et al. 2022; Valdez Castillo et al. 2022).

The activity of LAB (homo- and heterofermentative) and yeast strains, and their interactions, can exert influence on the volatiles' formation. As a result of the microbiological processes, a series of organic substances accumulated impacts the volatile organic compounds' profile (De Luca et al. 2021). Yeasts and LAB form volatile compounds during the substrates' carbohydrates fermentation and the degradation of amino acids via the Ehrlich pathway, creating specific compounds due to their catabolism (Van Wyk et al. 2018). The flours' fermentation by yeasts and/or LAB,

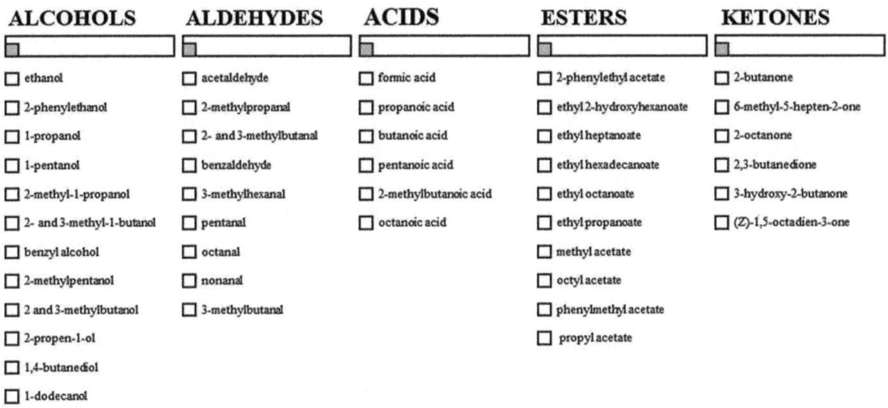

Fig. 12.2 Main volatile and non-volatile compounds produced by yeast during fermentation (Pétel et al. 2017; Liu et al. 2018; Van Wyk et al. 2018; De Luca et al. 2021)

their enzymatic activity, and the thermal reactions can influence the aromatic profile of the fermented products (Pico et al. 2015). Also, the ingredients used for fermentation substrate's formulation are crucial in developing the obtained product's aromatic profile.

More than 300 volatile and non-volatile organic compounds with different chemical structures, such as organic acids, aldehydes, ketones, alcohols, esters, and sulphur compounds, are formed during the fermentation of flours or after the fermented products' processing. The main volatiles produced during fermentations, that can be identified in the fermented products are depicted in Fig. 12.2.

A large number of aldehydes were identified in sourdough, and sourdough bread; among them were acetaldehyde (green apple odor), 2-methylpropanal (floral odor), 2- and 3-methylbutanal (malty odor), benzaldehyde (bitter almond odor), 3-methylhexanal (sweet odor), pentanal, octanal, nonanal, and 3-methylbutanal (fruity odor) (Xu et al. 2019b; Yan et al. 2019b; De Luca et al. 2021). Acetaldehyde was exclusively found in sourdough bread, with the highest concentrations in sourdoughs fermented with *S. cerevisiae.* The volatiles produced by yeast metabolism were reduced in regular yeast-leavened bread. Strecker aldehydes 2-methylpropanal, 2-methylbutanal, and 3-methylbutanal are produced via amino acid conversion by yeasts (Jensen et al. 2011; Pétel et al. 2017).

Between the alcohols reported in the literature and summarized in Fig. 12.2, the concentration of 3-methyl-1-butanol (fruity whiskey aroma) is usually high in sourdoughs enriched with yeasts. The Ehrlich pathway in yeast cells was used to produce 3-methyl-1-butanol and 2-methyl-1-butanol during fermentation from flour's amino acids (Birch et al. 2014; De Luca et al. 2021). Thus ethanol, 3-methyl-1-butanol, 2-phenylethanol (coffee, chocolate aroma), 2-methyl-1-propanol (sweet odor), 2,3-pentanedione (fruity; sweet; coffee odor), ethyl octanoate (fruity apricot-like odor), 2-octanone (sweet odor), dihydro-5-pentyl-2(3H)-furanone (coconut-like odor), octanal (fruit-like odor), benzaldehyde, and 2-pentylfuran were found in high concentrations in steamed bread made with *S. cerevisiae* (Liu et al. 2018). In addition, 2-phenylethanol has been linked to the flavor of wheat crumbs, and it is thought to be produced by phenylalanine fermentation in yeast cells via the Ehrlich pathway (Dai et al. 2021). *S. cerevisiae* is well-known for its ability to produce 2-phenylethanol, and the concentration of 2-phenylethanol rises with prolonged dough fermentation and the presence of *S. cerevisiae*. It was demonstrated that some non-conventional yeasts can produce 2-phenylethanol and that some of them have a better capability for producing aroma metabolites than *S. cerevisiae* (de Jesús Rodríguez-Romero et al. 2020). Other sourdough-originated yeasts were able to produce 2-phenylethanol, such as *Kl. marxianus, Kl. lactis, P. fermentans, P. anomala, P. membranaefaciens, C. utilis, C. tropicalis, Y. lipolytica,* and *P. kudriavzevii* (Martínez-Avila et al. 2018; Mierzejewska et al. 2019; Mitri et al. 2022). The steamed bread fermented by *C. (Ka.) humilis* was characterized by an improved amount of ethanol, 3-methyl-3-buten-1-ol, (E,E)-2,4-decadienal. The presence of 2-methyl-1-propanol, 3-(methylthio)-1-propanol, and 3-methyl-1-butanol was identified in *W. anomalus* steamed bread, whereas 1-butanol, 1-pentanol, and 1-hexanol was found in similar products fermented with *S. fibuligera* (Liu et al. 2018). Moreover, various ethyl acetate derivatives with specific aromas were determined in yeasts or yeasts and LAB fermented products. The type and the concentration of esters in the fermented products composition are influenced by the substrate composition, the inoculum type and dimension, or by the extrinsic and intrinsic physicochemical and biological parameters of the bioprocess. Large quantities of ethyl acetate derivatives were found in *S. cerevisiae* and *P. kudriavzevii* fermented sourdough breads, whereas *Ka. humilis* produced high levels of isopropyl myristate. Hexyl acetate and ethyl decanoate were synthesized by *W. anomalus* (De Vero et al. 2021).

The short and medium-chain fatty acids (SCFAs, MCFAs) and classified as non-volatile compounds, identified in sourdoughs, respectively lactic, acetic, formic, propanoic, butanoic, octanoic, and pentanoic acid or their corresponding derivatives, are produced by LAB strains, but some yeasts were reported also as organic acids producers. During alcoholic fermentation by baker's yeast, the lactic acid formed can accumulate from pyruvic acid. The acetic acid plays a role in esters' formation, its high concentrations may change the products' taste. In fermented steamed sourdough bread by *Ka. (C.) humilis*, acetic acid, butanoic acid, octanoic acid, and decanoic acid were identified, as Fig. 12.2 shows (Liu et al. 2018). Also, 2-methylpropanoic acid and 3-methylbutanoic acid were detected in the steamed bread fermented with *P. kudriavzevii,* after carbon's oxidation to corresponding acids instead of alcohols in Ehrlich's metabolic pathway (Hazelwood et al. 2008).

Other important bioactive compounds produced by yeasts during fermentations are the exopolysaccharides and gamma-aminobutyric acid (GABA). Firstly, exopolysaccharides are extracellular carbohydrate-based polymers with varying structural complexity. As such, homopolysaccharides are made up of a single monomer, whereas heteropolysaccharides comprise differently structured monomers (Gientka et al. 2015; Rahbar Saadat et al. 2021). Yeast strains belonging to *Candida* spp., *Pichia* spp., *Kluyveromyces* spp., *Rhodotorula* spp., can produce extracellular polysaccharides. The production of exopolysaccharides is linked to yeast strains' metabolism, the biopolymer's structure and physical properties are influenced by various biotechnological parameters, including the culture medium's composition and fermentation conditions such as pH, temperature, and oxygen concentration (Rusinova-Videva et al. 2010). Secondly, GABA is a L-glutamic acid-derived amino acid catalyzed by glutamate decarboxylase. GABA is recognized and studied for its health-related properties (Cui et al. 2020). Lactic acid bacteria strains are often studied in the literature for GABA synthesis, the most representative producers being *Levilactobacillus (Lev.) brevis*, *Lpb. plantarum*, and *Streptococcus thermophilus* (Mancini et al. 2019; Han et al. 2020; Park et al. 2021), but there have been a few information about the use of yeast strains for GABA production (Zhang et al. 2020; Li et al. 2022).

12.3.2 Yeasts in Further Food-Related Bioconversion Processes

The yeast strains can be performant starter cultures in bioconversion processes due to the enzymatic modifications induced on the substrates, that enhance the functional properties, thus making these microbial strains suitable candidates to be exploited in industry (Phukon et al. 2022). In bioconversion processes, also known as biotransformation, the molecules available in the substrate are converted to other compounds (bioderivatives) by activity of the microbial enzymes, mainly released extracellular. On the opposite, in fermentation processes the nutrients from the substrate are used by microorganisms to growth, multiplication and generate novel metabolites through specific metabolic pathways (Hosoglu et al. 2018; Maurya et al. 2021). Yeast strains belonging to different genera were able to convert complex substrates into compounds with improved bioavailability and bioaccessibility. The enzymatic and functional properties of yeasts could be maximized by co-cultivation with other microorganisms, or genetic engineering can be applied to obtain recombinant yeasts with superior functionality (Rai et al. 2019; Terpou and Rai 2022). Therefore, in the last years, the scientific studies were focused on a few unconventional and versatile strains unexploited enough towards such objectives.

Kl. marxianus yeast strains gained researchers' interest in the last decades due to the diverse applications of this yeast species in foods, feed, nutraceuticals, pharmaceuticals, and cosmetics. These beneficial properties are assured by the *Kl. marxianus* strains' extracellular proteins, therefore, the results could be positively or negatively affected by the biotechnological conditions used (strain type, inoculum, medium composition, fermentation parameters). As such, inulinases able to hydrolyze inulin into fructose and fructo-oligosaccharides, lactases (β-galactosidase) responsible for lactose bioconversion into glucose and galactose, β-glucosidases, pectinases, galacturonases or lipases, were reported in the literature and summarized by Karim et al. (2020), emphasizing the utilization of the final products as single cell proteins, single cell oils, emulsifiers, bakery ingredients or prebiotics. Statistical optimized experimental designs were constructed for inulinase's production with a commercial strain which is associated with the sourdoughs' microbiota, respectively *Kl. marxianus* ATCC 36907, in a solid-state fermentation system with dried beer residue, minerals, sugarcane molasses, and yeast extract. The optimal fermentation conditions lead to an improved inulinase production yield of 238.38 U/gram dry solids, considering 72 h of fermentation at 37.50 °C, pH 4.5, 47.10% (w/w) sugarcane molasses, and 0.03% (w/w) yeast extract (Yupanqui-Mendoza et al. 2022). The β-galactosidase enzymatic catalysis was improved when a recombination using *Kluyveromyces* spp. and *Escherichia coli* (*E. coli)* strains was assessed, at 20 °C for 48 h in Terrific Broth supplemented with inducers, respectively with 0.50 mM isopropyl β-D-1-thiogalactopyranoside which led to 108.77 U/mL, or 5 g/L lactose which determined an enzymatic activity of 93.10 U/mL. Furthermore, it was concluded that the β-galactosidase activity was improved when an increased concentration of proteins, respectively metal ions were added into the fermentation medium, the enzymatic synthesis being determined also at low temperatures (4 °C). This β-galactosidase produced by *Kluyveromyces* spp. could be easily purified by its affinity with cellulose, result that extends the applications in dairy due to the enzyme's optimum activity in milk at normal pH and temperature values (Gennari et al. 2021).

Plackett-Burman and Box-Behnken experimental designs were used as well to screen and optimize the fermentation processes conducted by selected yeast strains, respectively *S. cerevisiae*, isolated from grapes, and *C. tropicalis*, able to synthesize endoglucanase, β-glucosidase, and xylanase, as Qadir et al. (2018) reported. The experimental results of this study suggested that an alkaline treatment with 1% NaOH of the sugarcane bagasse used as the fermentation substrate facilitated the synthesis of the studied enzymes. Hence, a mixed inoculum with both yeast strains (0.50 v/w) used to ferment the alkali treated substrate at 35 °C for 94 h determined an improvement only for endoglucanase (9.81 IU/mL), the selected biotechnological factors included in the Box-Behnken design were not significant for the β-glucosidase and xylanase. Laurent et al. (2020) highlighted that *S. cerevisiae* strains used as starters in industry can differently modify the concentrations of fructans and fermentable dietary fibers from wheat due to some strains' capacity to produce invertases which reduce the side-effects of fermentable oligosaccharides, disaccharides, monosaccharides and polyols (FODMAPs) on irritable bowel syndrome patients. *S. cerevisiae* strains used in industrial bakeries determined reduced levels of residual fructose after wheat dough fermentation, whereas the strains belonging to the same species, originated from other environments, were characterized by a low invertase activity, result that justified the reduced specific volume of the obtained breads, therefore containing increased amounts of fermentable dietary fibers. Moreover, the high-pressure treatment (200–600 MPa) applied to the industrial baker's yeast, *S. cerevisiae,* cells was beneficial upon proteolytic activity improvement correlated with a faster cells' permeabilization and autolysis. Precisely, the study conducted by Dimopoulos et al. (2021) demonstrated that 200 MPa applied for 40 min, respectively 400 MPa for 10 min, accelerated cells' autolysis, permeabilization, and endogenous proteins' hydrolysis. A pressure of 600 MPa improved the yeast's permeabilization from the cellular suspensions but damaged the proteases by inactivating them after 15 min of treatment, consequently an increased high-pressure treatment may cause negative effects.

Feruloyl esterases are responsible for cinnamic acids' release from vegetal raw materials, *Aspergillus* spp. being often used for such objectives. Interestingly, the genes that ensure the bioactives' catalysis were transferred into *P. pastoris* genome to increase this strain's industrial applications (Dilokpimol et al. 2018).

Y. lipolytica is a versatile biotechnological tool which can be successfully used to obtain value-added products due to this yeast strain's capacity to use the nutrients from various industrial by-products to fulfill its nutritional requirements (Madzak 2021). The technological benefits of *Y. lipolytica* utilization are correlated with specific enzymes' production, especially lipases and proteases. Various saturated and unsaturated fatty acids, their esters or methyl esters were bioconverted by extra- or intracellular lipases, respectively by cell-bounded enzymes active at temperatures between 4–55 °C. Additionally, acidic or alkaline extracellular proteases exert their positive effects on proteinaceous fermentation substrates made from foods by-products (Gottardi et al. 2021). A solid-state fermentation medium based on soybean flour was fermented with *Y. lipolytica* strain aiming to identify the optimal fermentation conditions to maximize the enzymes synthesis, respectively lipase and protease. The results of this study suggest that the enzymatic properties of this yeast strain were strongly affected by the nutrients included in the medium's composition (glucose or soybean oil, yeast extract, bactopeptone or diammonium phosphate), the fermentation time, temperature, and the bioreactor type had an impact on the results (do Nascimento et al. 2021). Therefore, the bioconversion efficiency of *Y. lipolytica* can differ among the microbial strains, as Janek et al. (2020) and Hapeta et al. (2020) observed. These authors demonstrated that the enzymatic activities of some yeast strains are related to the nitrogen availability, thus the optimal lipase production conditions (37 °C, pH 8.0) were identified for a *Y. lipolytica* strain genetically modified by overexpression of lipase production genes hence, increasing the strains' applications in foods, feeds, nutraceuticals, and cosmetics. The proteases produced by *Y. lipolytica* during bovine colostrum's fermentation enriched the fermented product with bioactive peptides with antioxidant properties that also intensified the cells' proliferation (Cotârleţ et al. 2020a; Gaspar-Pintiliescu et al. 2020). Coconut and pork fat was successfully bioconversed into medium-chain fatty acids with antimicrobial and antioxidant activity by the same yeast strain after solid-state fermentations, as Parfene et al. (2013) and Cotârleţ et al. (2020b) reported.

12.3.3 Yeast Cells' Autolysis in Food Applications

Yeast autolysis is an endogenous, and irreversible lytic process that occurs after cell death and is caused by intracellular enzymes (Martínez et al. 2018). The compounds released from the breakdown of internal constituents enrich the fermentation medium (Tao et al. 2014), thus becoming a cost-effective process to obtain high-value metabiotics, respectively prebiotics and postbiotics. Yeast autolysis can be designed as an industrial process to benefit from yeasts' autolysates, but because it is a slow process, numerous physical techniques for cell inactivation have been investigated over time to improve its efficiency (Comuzzo et al. 2017). Therefore, the process can be accelerated by drying or continuous cycles of freezing – pre-thawing of the cells. Yeasts' autolysis is influenced by various intrinsic or extrinsic factors, such as the culture medium's pH, its composition, yeast cells' concentration, respectively the growth phase of the microorganisms.

A yeast extract or a yeast autolysate for food industry applications was obtained by three different processes: autolysis, plasmolysis, and hydrolysis. The autolysis process involves salt addition to the cells' suspension so that the process of osmosis occurs, and water is removed from the cell. The cell wall is destroyed, and the endogenous enzymes inside the yeast cells are released and begin to hydrolyze the cell contents by autolysis (Fig. 12.3). During autolysis, yeast proteins are solubilized and hydrolyzed, and the cells are heated to decompose the cells further. The autolytic process is completed by collecting and concentrating the soluble fraction.

The yeast extracts achieved are flavor products and nutrients stabilized after the concentration of the yeast cells' constituents such as: proteins, peptides, amino acids, nucleotides, sugars, vitamins, and flavor compounds (Tiefenbacher 2017). Compared to brewer's yeast, baker's yeast tends to have a higher level of carbohydrates and a lower level of protein, even though both belong to *Saccharomyces* spp. Autolysis of the yeast cells can be accomplished by two methods, respectively induced or natural autolysis (Wang et al. 2018). Induced autolysis is used in food industry to extract nucleotides, amino acids, lipids, and other bioactive compounds from yeast cells' as it was previously described (Lombardi et al. 2015; Rakowska et al. 2017).

Autolysis occurs in *S. cerevisiae* when fermentable carbohydrates are depleted during the yeast's stationary phase (Lo et al. 2021). Because amino acids, peptides, and vitamins are released during cells' lysis, the sourdough flavor and functionality are enhanced (De Vuyst et al. 2021). Furthermore, ornithine and glutathione can result from lysed yeast cells, affecting the taste and aroma of the bread (Yan et al.

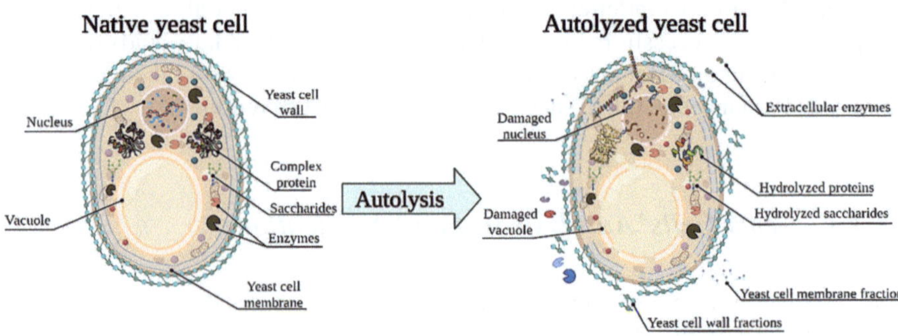

Fig. 12.3 Yeast cell's modifications during autolysis

2019b). A *Kl. marxianus* strain isolated from milk kefir grains was used for cells' autolysis, this process being induced by LAB as an effect of the lactate produced, which stimulates *Kl. marxianus* to release proteins and peptides (Lo et al. 2021).

12.3.4 Metabiotic Potential of Yeast Strains from Sourdough with Functional Impacts

Yeast strains' metabolites have multiple functional and technological properties with impact on foods' overall quality and consumers' health estate. Nowadays, a novel concept of metabiotics (probiotics, prebiotics, postbiotics, and paraprobiotics) is developing using the complex potential of microorganisms and their metabolism upon technological, safety and functionality of fermented foods (Singh et al. 2018).

The main categories of biotic compounds, resulting from yeasts' biochemistry, and their obtaining methods were highlighted in Fig. 12.4. The beneficial effects assured by their functionality or technological implications will be discussed.

Prebiotics include dietary fibers which generally cannot be disrupted by the enzymatic reactions after gastric digestion. Thus, these carbohydrate-based macromolecules exert health-promoting properties for the host supporting the intestinal microbiome's selective growth and development, therefore preventing dysbiosis which is the main factor for multiple diseases (Rai et al. 2019). Oligosaccharides are frequently found in various food matrices or by-products from peanuts, wheat, coconut, potatoes or oranges. To increase the prebiotic effect of the oligosaccharides, their extraction and purification was assessed by hydrolysis (acid, alkali, enzymatic) or by controlled fermentations with selected yeasts strains belonging to *Saccharomyces* spp. or *Pichia* spp. The production yield, as well as the technological and functional properties of the oligosaccharides made by microbial fermentations are strongly linked with both intrinsic and extrinsic factors comprising the nutritional composition of the substrate and its particle size, temperature, pH, yeast strains used their biochemical properties (Awasthi et al. 2022).

One of the major and intensively studied paraprobiotic produced by *S. cerevisiae* and/or *S. cerevisiae* var. *boulardii* yeasts are the β-glucans. They gained an increasing interest due to the prebiotic potential with immunomodulatory and anticarcinogenic effects (Chan and Liu 2022). Yeast strains can be used as β-glucans sources because they store this macromolecular prebiotic, made of D-glucose monomers linked by 1,3- or 1,6-glycosidic bonds, into the cell walls. Hence, yeast β-glucans seemed to regulate the intestinal microbiome being also involved in type 2 diabetes treatment (Sivieri et al. 2022). The β-glucans from *S. cerevisiae* were included in a food supplement formulation containing selenium, zinc, magnesium, and silymarin, this composition had beneficial effects upon obesity or weight loss, respectively on the diabetes symptoms (Nehmi et al. 2021). *S. cerevisiae* var. *boulardii* strain was subjected to metabolic engineering techniques to enhance its prebiotic properties.

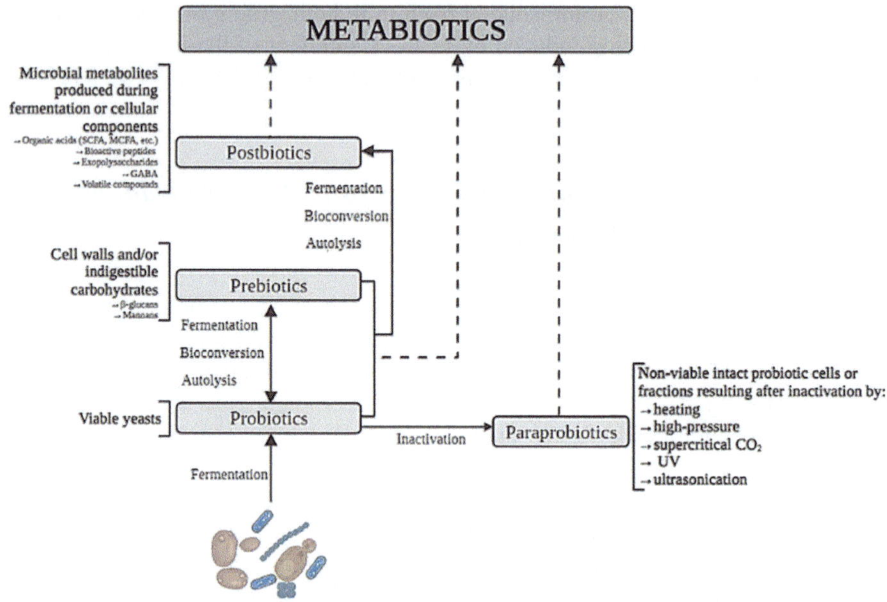

Fig. 12.4 The main classes of yeast-derived metabiotics with industrial applications (Rai et al. 2019; Cuevas-González et al. 2020; Żółkiewicz et al. 2020; Shruthi et al. 2022)

The results of this study, reported by Kwak et al. (2022), showed that a 52% enhancement of the mannan's production was obtained after the overexpression of the synthetases, phosphatases, and mannoproteins involved in the process. Furthermore, the mannan stored in the cell walls improved the antibacterial effect against *Bacteroides thetaiotaomicron* and *Clostridioides difficile*, the co-aggregation with *Salmonella enterica* being also improved.

The β-glucan from *S. cerevisiae* originated from spent brewer's slurry was extracted after induced autolysis of the dried yeast cells (50 °C for 48 h), an extended thermal treatment being then applied at 121 °C for 6 h to attain the β-glucan's disruption from the cell walls. The conclusions of this study confirmed the impact of purification methods (deproteinization and defatting) or particle size resulting after different milling techniques, on the technological characteristics respectively on the glucose and cholesterol levels reduction (Fu et al. 2022). The optimal combination of parameters for the high-pressure treatment (0.039 MPa, 108 °C, 5 min) that was applied firstly improved the cell walls interaction with the alkali solution (0.85% v/w) used in the next step of the β-glucan extraction process. These combined treatments allowed a purification and extraction yield over 78% for the obtained β-glucan (Tian et al. 2019). *S. cerevisiae* strains isolated from grapes or soil, respectively their cells (4% w/v) were involved in isomalto-oligosaccharides' purification. After 1 h of fermentation at 30 °C with a *S. cerevisiae* strain, the purification level of the isomalto-oligosaccharides increased from 67% to 91% due to glucose metabolism (Rengarajan and Palanivel 2020). Yeast cell walls were used for the extraction of mannan at 100 °C during 2 h using 1% (w/v) NaOH and a 10:1 ratio between the yeast cells and the alkali solution. The acid purification methods applied thereafter determined a mannan's purification of 96.15%. Furthermore, the prebiotic mannan's functionality was increased by chemical derivatization. Nevertheless, the carboxymethylation, sulfation or phosphorylation of the yeast cell walls' mannan improved the inhibition potential against free radicals, a lipid peroxidation inhibition being also determined (Liu and Huang 2018). Rengel dos Passos et al. (2021) demonstrated that the valorization of cheese whey can be maximized using a biotechnological optimized approach which involve a probiotic strain of *S. cerevisiae* var. *boulardii* and lactase. Following a step-by-step optimization process the authors determined the optimal conditions, respectively a fermentation temperature of 29.50 °C and 0.14 (w/w) lactase addition in the substrate, to assess galacto-oligosaccharides amounts over 56 g/L correlated with 10^7 CFU/mL viable probiotic cells. Probiotic yeast strains able to metabolize the residual monosaccharides from the fermentation medium are sustainable ways to purify the undigestible prebiotics obtained. Moreover, *C. utilis* ATCC 9950 commercial strain was cultivated in a potato juice water-based medium supplemented with glycerol to evaluate the binding capacity of the yeast β-glucan individually or in combination with the cell walls. Results suggested that the mixture of *C. utilis* cell walls and β-glucan significantly reduced some mycotoxins' (zearalenone, ochratoxin A, and aflatoxin B1) levels. The mycotoxins were differently entrapped by the products synthesized by the yeast strain. Precisely, the pH value of the medium or the proteins from the cell walls, as well as the mycotoxin's structure directly influenced the experimental results obtained (Bzducha-Wróbel et al. 2019). Humic acid is a common decontamination agent with various industrial applications. Fortunately, its decontamination power upon aflatoxin B1 was enhanced by encapsulation in commercial *S. cerevisiae* cell walls. The results of this study conducted by Hamza et al. (2019) showed that the mycotoxin's binding capacity depends also on the strain used in the process, the composition of the cell walls being different among yeast strains. Consequently, maximum absorption rates were observed during gastrointestinal *in vitro* digestion, respectively after 10 min of simulated gastric digestion or after 1 h of intestinal digestion. These decontamination effects in the *in vitro* models were associated to the glucan mannan lipids and their particle size. In another study, Young et al. (2020) evaluated comparatively the behavior of active dry yeast (*S. cerevisiae*) native cells, respectively of some yeast cells walls' fractions obtained by different preliminary treatments to be used as encapsulation matrices for curcumin delivery. Significant differences were observed at different concentrations of bile salts in the simulated intestinal fluid, rather between the differently treated encapsulation yeast-based systems. Though, high concentrations of bile salts (25 mg/mL) and lipase (10 mg/mL) determined structural changes upon the microencapsulation system by breaking it down, when the native yeast or cell walls' fractions were used.

Probiotic microorganisms consist of living cells able to exert health-promoting effects on the host considering a long-term administration in an adequate dose. Hence, before including a yeast strain into probiotics, specific features must be evaluated to ensure its biotherapeutic effects. One of the most important features of the probiotic yeasts is related to the cells' viability throughout the gastrointestinal tract (Shruthi et al. 2022). Performant yeast strains can be originated from the specific microbiota of diverse traditional fermented products based on vegetables, tubers, cereals or millets. These raw materials fulfill the nutritional requirements of yeasts and permit the development of a stable microbial population, especially after multiple backsloppings applied in sourdough technology (Ilango and Antony 2021). *S. cerevisiae* var. *boulardii* strain's viability at different pH values was enhanced by encapsulation (Sen and Mansell 2020). Whey protein concentrate combined with water-soluble chitosan (29:1 w/v) was a suitable combination to microencapsulate the probiotic *Kl. marxianus* strain previously isolated from whey. This combination maintained 95% probiotic yeast's viability after *in vitro* gastrointestinal digestion (Vanden Braber et al. 2020). Therefore, various yeast species belonging to *Kluyveromyces* spp. (*Kl. lactis*, *Kl. marxianus*), *Pichia* spp. (*P. fermentans*, *P. kudriavzevii, P. membranaefaciens*) or *Y. lipolytica* were tested for the probiotic traits including the auto-aggregation capacity, antibiotic resistance, growth at different temperatures, hydrophobicity, antioxidant properties or antimicrobial spectrum against *E. coli*, *Staphylococcus aureus*, *Candida* spp., *Aspergillus flavus* (*A. flavus*) or *Penicillium* spp. These competitive probiotic yeasts offer innovative perspectives for the enrichment and diversification of foods, nutraceuticals and cosmetics (Fernandez-Pacheco Rodríguez et al. 2018; Merchán et al. 2020). Dairy applications were identified also for some potential probiotic *Kl. lactis* or *Toso. delbrueckii* yeast strains isolated from cheese, able to overcome *Salmonella* spp. growth on *in vivo* experimental models (Andrade et al. 2021). The mechanisms behind the antipathogenic effect against *Salmonella* spp. were elucidated by Gut et al. (2022) which determined that some bioactive peptides produced during the metabolic activity of some *Kl. lactis* or *S. unisporus* strains were responsible for this positive effect.

Probiotic *S. cerevisiae* strains were studied and selected to produce a probiotic beer. The outcomes of this study bring forth the importance of inoculum strains used for the fermentation process. Precisely, for the beers fermented by the co-culture of *S. cerevisiae* purchased as freeze-dried starter and *L. paracasei*, it seemed that some antagonistic relationships occurred during the mixed fermentation, which directly affected both yeast and *L. paracasei* viability and abundance. Hence, a probiotic beer was produced by a primary fermentation with *L. paracasei* (36 °C, 24 h) followed by *S. cerevisiae*'s fermentation at 20 °C for 5 days (Silva et al. 2020). The iso-α-acids from hop negatively affect the probiotic lactobacilli viable counts, as Alcine Chan et al. (2019) suggested, when the co-inoculation with probiotic freeze-dried commercial starter of *S. cerevisiae* was intended. Fortunately, the isomerization of the hop extract seemed to be beneficial for both microbial strains involved in beer's fermentation. The study conducted by Pereira de Paula et al. (2021) demonstrates that probiotic wheat beer can be produced only with *S. cerevisiae* var. *boulardii* from the commercial product Floratil® 200 when the manufacturing process is monitored and adapted to ensure increased sensorial acceptability correlated with a probiotic cells' viability of 10^8 CFU/mL. Probiotic coffee was obtained by a mixed fermentation with *Lacticaseibacillus* (*Lcb.*) *rhamnosus* and *S. cerevisiae* var. *boulardii* co-culture, the probiotic yeasts being able to support the viable counts of lactobacilli higher than 10^7 CFU/mL (Chan et al. 2021). Adzuki beans and lentils sprouts were made by soaking the seeds in *S. cerevisiae* var. *boulardii* inoculum and incubation at 30 °C for 4 respectively 8 h, followed by a germination for 4 days. This protocol assured a viability of 10^7 CFU/g sprouts for the probiotic yeast strain that was isolated from a probiotic pharmaceutical product, other important functional characteristics of these probiotic carriers being related to the improvement of the antioxidant and bioactive properties, a reduced glycemic index, as well as an increased microbiological safety associated with the reduced populations of molds and coliforms (Swieca et al. 2019).

Moreover, Martínez et al. (2019) reported that probiotic yeast strains used for wine fermentation were able to metabolize and consequently minimize the concentration of aflatoxin M1 from milk, with degradation ratios ranging between 33–86%. The authors concluded that the aflatoxin's metabolism was different among the yeast strains studied, thus, the most performant ones being a *Kl. marxianus* strain, respectively a *S. cerevisiae* var. *boulardii* strain. Additionally, these previously mentioned strains were capable to reduce the aflatoxin's toxicity from the fermented medium after 24 or 48 h of fermentation. The inhibition of *A. flavus* and *A. parasiticus* growth, as well as a reduction of aflatoxin B1 concentration was determined when two strains of *S. cerevisiae* or *P. kudriavzevii* classified as potential probiotic candidates were tested by Sampaolesi et al. (2019). This study reinforces the sustainable utilization of beer fermentation residue as an important source of yeast strains with techno-functional characteristics. The probiotic effect of the yeast strain *S. cerevisiae* var. *boulardii* was improved by inulin addition (1–2% w/v) in an ice cream recipe, this adaptation of the manufacturing process enhanced the viability of the analyzed yeast strain that remained constant (~10^6 CFU/g) after 4 months of storage at −18 °C. The prebiotic agent used improved the technological properties of the designed ice cream and protected the probiotic cells upon mechanical stress or cells' modifications during freezing.

Another major food safety concern is attributed to patulin as the main contaminant in apples and apple-derived products. This mycotoxin is synthesized by various mold strains but a probiotic commercial yeast strain belonging to *P. guilliermondii* seemed to reduce more than 90% of patulin's content when the optimal conditions proposed by Luo et al. (2022) are followed, respectively 24 h of incubation at 30 °C using 10^8 CFU/mL. The above-mentioned authors highlighted that the fermentation temperature is a decisive biotechnological factor in order to assess the patulin's decontamination. As such, incubation temperatures over 40 °C can negatively alter the probiotic yeast strain's activity. Furthermore, it was demonstrated that *P. guilliermondii* cells were able to entrap patulin into the cells' walls, by this mechanism being ensured an increased stress resistance to patulin concentrations up to 100 mg/L. *P. kluyveri* or multiple *S. cerevisiae* probiotic strains isolated from cocoa beans or alcoholic beverages were studied as adjunct starter cultures for the fermentation of a maize-based beverage, the results suggesting that this dairy-free fermented product had more than 10^6 CFU/mL, after 28 days of refrigerated storage, its properties being positively affected by a symbiotic culture of *S. cerevisiae* and *L. paracasei* probiotic strains (Menezes et al. 2018).

Conclusively, the identification of novel sourdough-associated yeast strains isolated from various microbial ecosystems, able to increase the technological and functional properties of foods comprising dairy products, beverages based on whey, tea, or fruit juices respectively chocolate, or cornflakes was an intensively studied topic in the last decades (Chan and Liu 2022). Thus, there are various innovative perspectives for the yeasts applications in food industry that need to be investigated for new clean-label functional foods development.

Postbiotics exert direct or indirect benefits on the consumers' health, these compounds include cells' metabolites (organic acids, bioactive peptides, exopolysaccharides, enzymes, phenolics, etc.) or cells' fractions liberated extracellularly during the fermentation processes respectively after different treatments during processing (Żółkiewicz et al. 2020). Postbiotics were efficient in specific disorders' treatment such as metabolic diseases, diabetes or obesity. The selective proliferation of the beneficial microorganisms in the gut is ensured by some postbiotic compounds responsible also for an improved immune response or antipathogenic effects. Postbiotics have beneficial properties *in vitro*, used as biopreservatives, antibiofilms, sensorial and texture improvers (Peluzio et al. 2021; Li et al. 2021a). Currently, the LAB strains are the most studied microorganisms due to their various postbiotics with therapeutic effects but, fortunately, the yeast strains gained the researchers' interest because they possess the necessary mechanisms to synthesize postbiotics with similar or superior functional effects. Firstly, the numerous volatile compounds (alcohols, aldehydes, ketones, acids, and esters) produced by yeasts fermentations can be included into postbiotics because they have important antimicrobial features against food spoilage microorganisms, apart from the well-known utilizations as aroma enhancers (Oro et al. 2018). Genetic engineering protocols can be assessed to improve the volatiles' abundance in the fermented products (Morita et al. 2022). Yeast strains belonging to *Pichia* spp., *Kl. marxianus*, *Y. lipolytica*, or *Issatchenkia orientalis* isolated from kimchi were screened for their antimicrobial effects upon *Penicillium digitatum* (*Pen. digitatum*) contamination on citrus, to conclude that this mold inhibition potential is determined by different mechanisms specific to each of the yeast strains. Specifically, the inhibition ratios determined by the yeast strains on *Pen. digitatum* varied between 18–57% due to the biofilm formation of *K. marxianus* and *P. kudriavzevii*, this strain being able to synthesize volatiles that inhibited the studied mold strain. The extracellular hydrolytic enzymes (chitinase, glucanases, lipases and proteases) produced by *Y. lipolytica* were responsible for the antimicrobial effect against molds or pathogen bacteria (Parfene et al. 2013; Cotârleț et al. 2020b; Delali et al. 2021).

Various volatiles respectively alcohols, aldehydes, esters, or organic acids were determined in a yeast-fermented soy whey supplemented with 2% or 10% (w/v) salt when selected yeast strains of *Zygosaccharomyces rouxii* (*Z. rouxii*), *C. versatilis*, and *Toso. delbrueckii* were used to produce a fermented soy sauce (Zhou et al. 2022). *S. cerevisiae* strain's capacity to be used as a starter culture in pineapple juice fermentation was assessed by Chanprasartsuk and Prakitchaiwattana (2022), to highlight that, considering the concentration of ethanol and volatiles produced during fermentation, the *S. cerevisiae* strains previously isolated from pineapple peels' can be successfully used for such objectives, thus, a mixed inoculum with other yeast strains may improve the technological and sensorial properties of the fermented fruit juice. The *S. cerevisiae* strain isolated from a brewing by-product synthesized 2,3-butanediol and ethanol able to bind the mycotoxins produced by *A. flavus* and *A. parasiticus* strains and inhibited the molds' growth. Furthermore, the previously mentioned postbiotics determined a retarded proliferation of *Salmonella enteritidis* or *E. coli* (Sampaolesi et al. 2022). The antipathogenic effect against *Salmonella* spp. determined by *K. lactis* and *S. unisporus* individual strains, included in the milk kefir grains' microbial consortium, were attributed to some postbiotic enzymes belonging to peroxidases and dehydrogenases (Gut et al. 2022).

The yeast species capable of producing GABA belong to *Kluyveromyces* spp., *Saccharomyces* spp., and *Pichia* spp. (Li et al. 2021b; Zhang et al. 2022a). Other GABA-producing yeast of *Kl. marxianus*, *S. cerevisiae* and *P. kudriavzevii* strains were reported in the literature (Li et al. 2022). Zhang et al. (2022a) reported that a yeast strain, respectively *Kl. marxianus*, produced a high concentration of GABA after okara's fermentation, thus developing innovative GABA-enriched products. The apple juice fermented with *S. cerevisiae* strain was enriched with GABA to obtain a functional fermented beverage (Sun et al. 2022). Similarly, a co-culture made of *S. cerevisiae* and *L. plantarum* was used to formulate new functional mulberry fermented beverages. The results reported in this study demonstrated that the mixed inoculum was beneficial for an improved GABA content, whereas the L-glutamate was metabolized by the involved starters to obtain functionalized beverages (Zhang et al. 2020).

Exopolysaccharides from microbial sources have previously been utilized to improve the texture of baked goods, also increasing the products' shelf life. Traditional sourdough yeasts such as *Kl. aestuarii*, *Kl. lactis*, *S. cerevisiae*, and *S. servazii* strains were used as alternative starters in bread making, producing high concentration of exopolysaccharides when they were tested as single cultures. The symbiotic relationship between *Enterococcus faecium* and *Kl. aestuarii* enhanced the bread's composition with phenolic compounds that contribute to the antioxidant capacity (Fekri et al. 2020). Furthermore, some microbial exopolysaccharides, such as dextran, levan, and ropy capsular polysaccharides, can overcome the negative effects of chemical additives due to the antifungal properties of these postbiotic macromolecular compounds (Tinzl-Malang et al. 2015). Some unconventional yeast strains were studied in the literature for the beneficial health-promoting effects of the postbiotic exopolysaccharides. It was reported that a cold adapted *R. mucilaginosa* strain, isolated from leaves, was active at 4 °C and produced a β-D-glucan composed of glucose and mannose units. This heteropolysaccharide was characterized by anticarcinogenic and antioxidant effects *in vitro* (Hamidi et al. 2020). Another heteropolysaccharide composed of mannose and glucose was characterized as well by DPPH radical scavenging activity, thermal stability, and prebiotic effect, nevertheless this exopolysaccharide produced by a *S. cerevisiae* strain isolated from rice can be considered an important postbiotic due to the previously mentioned features (Liu et al. 2022). Probiotic yeast strains belonging to *Kl. marxianus* and *P. kudriavzevii* isolated from dairy products demonstrated postbiotic and paraprobiotic effects due to some exopolysaccharides able to act on *in vitro* colon cancer cells (Rahbar Saadat et al. 2020). The exopolysaccharides' yield, structure, and features are related to the cultivation conditions, being as well strain-dependent. Gientka et al. (2016) demonstrated that *Candida* spp. strains' nutritional requirements differ among the studied yeasts. The available carbon, respectively sorbitol, glycerol, maltose, sucrose, or lactose, from the fermentation medium determined various exopolysaccharides' structures and production yields, improving these biopolymers' applications in foods, pharmaceuticals, and cosmetics.

The volatile compounds produced during yeasts' fermentations enhance the sensorial, technological, and functional properties of industrial products from various industries. Hence, the biotechnological processes involving yeast strains must be carefully monitored to reach the desired objectives. It was demonstrated that a *Kl. marxianus* strain, previously known as an important 2-phenylethanol industrial producer, modified its metabolic and functional properties at high levels of alcohol. A concentration of 3 g/L 2-phenylethanol, added in the fermentation medium in the exponential growth phase, modified the yeast cells' membrane permeability, growth, metabolism, and exopolysaccharides' biosynthesis. It seemed that the action of 2-phenylethanol was time-dependent and *Kl. marxianus* strain originated from dairy had adaptation mechanisms to remain viable, despite the changes that occurred (Balbino et al. 2021). Furthermore, a *Z. rouxii* strain isolated from beans' chili paste was studied by Zhang et al. (2022b) regarding the exopolysaccharides' biosynthesis, to conclude that a complex biopolymer made of mannose, galactose, and glucose was produced by this yeast strain. This exopolysaccharide was successfully used to improve the *Lactococcus lactis* strain's storage stability in food industry, extending the symbiotic products' availability on the market.

Considering all the health-related properties determined by various sourdough yeasts strains' prebiotics, probiotics, or postbiotics, these microorganisms should be considered also valuable candidates for paraprobiotics production ("ghost probiotics", non-viable microbial cells – intact or broken after thermal, high-pressure, supercritical CO_2 extraction, ultrasonic and ohmic treatments, respectively crude cell extracts with complex chemical composition) with multiple applications in various fields to increase the life's quality (Cuevas-González et al. 2020; Păcularu-Burada et al. 2021).

Nevertheless, the novel isolated yeast strains need to be approved for utilization in foods and feed, considering specific legislation requirements. As such, European Food Safety Authority (EFSA) attributed the *Qualified Presumption of Safety* (QPS) status to the yeast species considering specific criteria as following: taxonomy, pathogenic properties, and final intended utilization (Herman et al. 2019). The QPS list updated on July 2022 stated that viable cells of *Kl. lactis*, *Kl. marxianus*, *S. cerevisiae* var. *boulardii* or *Z. rouxii* can be administrated as is, respectively added in foods or feed without restrictions. Therefore, only non-viable cells or biomass of *W. anomalus* or *Y. lipolytica* are allowed for the same purposes (EFSA BIOHAZ et al. 2022). The Food and Drug Administration (FDA) is responsible for the *Generally Recognized as Safe* (GRAS) status for the microorganisms and their products used in food industry in the United States. The partial list of GRAS yeasts and derivatives approved by FDA include dried *S. cerevisiae* and *C. utilis*, respectively extracts, proteins, invertase or glycans produced by *S. cerevisiae*. *Y. lipolytica* can be used as citric acid producer, whereas *Kl. lactis* is allowed for lactase production and further utilization in dairy (FDA, 2022). Thus, various yeasts species were reported by Yabaci Karaoglan et al. (2022) as safe for food applications, respectively *Ka. servazzii*, *P. fermentans*, *P. kluyveri*, *P. kudriavzevii*, *Toso. delbrueckii*, *W. animalus*, and *Z. rouxii*. The lists of QPS and GRAS microorganisms are made based on a scientific basis, thus, these species will be probably listed soon as safe in order to allow the obtainment of clean-label foods and feed, taking in accord their beneficial implications in bioprocessing and positive effects *in vitro* and also *in vivo*.

12.4 Concluding Remarks and Future Perspectives

Sourdoughs reunite a diverse and complex microbial ecosystem in which yeast strains contribute with their own properties to increase the fermented products' overall techno-functional properties. This chapter reviewed the yeast strains' origins from different sourdoughs produced worldwide, highlighting their metabolic characteristics upon the flavors, enzymes, antimicrobials, autolysates, or exopolysaccharides obtainment. The sourdough yeast-derived metabiotics (pre-, pro-, post-, and paraprobiotics) were discussed in this chapter, to highlight and extend the yeast strains' industrial applications by clean-label fermentation processes. Furthermore, considering the latest scientific trends, studying yeasts as paraprobiotics seems to offer new perspectives. Generally regarded as safe microorganisms, the yeasts could be exploited as performant cell factories to manage the bioprocesses by metabolic engineering due to the strains' biochemical diversity and adaptability. Such approaches are challenging insights that could extend their benefits and applications.

References

Alcine Chan MZ, Chua JY, Toh M, Liu SQ (2019) Survival of probiotic strain *Lactobacillus paracasei* L26 during co-fermentation with *S. cerevisiae* for the development of a novel beer beverage. Food Microbiol 82:541–550. https://doi.org/10.1016/j.fm.2019.04.001

Andrade GC, Andrade RP, Oliveira DR et al (2021) *Kluyveromyces lactis* and *Torulaspora delbrueckii*: probiotic characterization, anti-*Salmonella* effect, and impact on cheese quality. LWT Food Sci Technol 151:112240. https://doi.org/10.1016/j.lwt.2021.112240

Arici M, Ozulku G, Yildirim RM, Sagdic O, Durak MZ (2018) Biodiversity and technological properties of yeasts from Turkish sourdough. Food Sci Biotechnol 27(2):499–508. https://doi.org/10.1007/s10068-017-0282-0

Arora K, Ameur H, Polo A, Di Cagno R, Rizzello CG, Gobbetti M (2021) Thirty years of knowledge on sourdough fermentation: a systematic review. Trends Food Sci Technol 108:71–83. https://doi.org/10.1016/j.tifs.2020.12.008

Awasthi MK, Tarafdar A, Gaur VK et al (2022) Emerging trends of microbial technology for the production of oligosaccharides from biowaste and their potential application as prebiotic. Int J Food Microbiol 368:109610. https://doi.org/10.1016/J.IJFOODMICRO.2022.109610

Balbino TR, da Silveira FA, Ventorim RZ et al (2021) Adaptive responses of *Kluyveromyces marxianus* CCT 7735 to 2-phenylethanol stress: alterations in membrane fatty-acid composition, ergosterol content, exopolysaccharide production and reduction in reactive oxygen species. Fungal Genet Biol 151:103561. https://doi.org/10.1016/J.FGB.2021.103561

Banik A, Mondal J, Rakshit S et al (2019) Amelioration of cold-induced gastric injury by a yeast probiotic isolated from traditional fermented foods. J Funct Foods 59:164–173. https://doi.org/10.1016/j.jff.2019.05.039

Bazalová O, Cihlář JZ, Dlouhá Z et al (2022) Rapid sourdough yeast identification using panfungal PCR combined with high resolution melting analysis. J Microbiol Methods 199:106522. https://doi.org/10.1016/J.MIMET.2022.106522

Birch AN, Petersen MA, Hansen ÅS (2014) Review: aroma of wheat bread crumb. Cereal Chem 91(2):105–114. https://doi.org/10.1094/CCHEM-06-13-0121-RW

Boudaoud S, Aouf C, Devillers H, Sicard D, Segond D (2021) Sourdough yeast-bacteria interactions can change ferulic acid metabolism during fermentation. Food Microbiol 98:103790. https://doi.org/10.1016/j.fm.2021.103790

Bzducha-Wróbel A, Bryła M, Gientka I et al (2019) *Candida utilis* ATCC 9950 cell walls and β(1,3)/(1,6)-glucan preparations produced using agro-waste as a mycotoxins trap. Toxins (Basel) 11(4):192. https://doi.org/10.3390/toxins11040192

Cakir E, Arici M, Durak MZ (2021) Effect of starter culture sourdough prepared with lactobacilli and *Saccharomyces cerevisiae* on the quality of hull-less barley-wheat bread. LWT Food Sci Technol 152:112230. https://doi.org/10.1016/j.lwt.2021.112230

Carbonetto B, Ramsayer J, Nidelet T, Legrand J, Sicard D (2018) Bakery yeasts, a new model for studies in ecology and evolution. Yeast 35(11):591–603. https://doi.org/10.1002/yea.3350

Carbonetto B, Nidelet T, Guezenec S, Perez M, Segond D, Sicard D (2020) Interactions between *Kazachstania humilis* yeast species and lactic acid bacteria in sourdough. Microorganisms 8(2):240. https://doi.org/10.3390/microorganisms8020240

Chan MZA, Liu SQ (2022) Fortifying foods with synbiotic and postbiotic preparations of the probiotic yeast, *Saccharomyces boulardii*. Curr Opin Food Sci 43:216–224

Chan MZA, Toh M, Liu SQ (2021) Growth, survival, and metabolic activities of probiotics *Lactobacillus rhamnosus* GG and *Saccharomyces cerevisiae* var. *boulardii* CNCM-I745 in fermented coffee brews. Int J Food Microbiol 350:109229. https://doi.org/10.1016/j.ijfoodmicro.2021.109229

Chanprasartsuk O, Prakitchaiwattana C (2022) Growth kinetics and fermentation properties of autochthonous yeasts in pineapple juice fermentation for starter culture development. Int J Food Microbiol 371:109636. https://doi.org/10.1016/j.ijfoodmicro.2022.109636

Coda R, Cagno RD, Gobbetti M, Rizzello CG (2014) Sourdough lactic acid bacteria: exploration of non-wheat cereal-based fermentation. Food Microbiol 37:51–58. https://doi.org/10.1016/j.fm.2013.06.018

Comasio A, Verce M, Van Kerrebroeck S, De Vuyst L (2020) Diverse microbial composition of sourdoughs from different origins. Front Microbiol 11:1212. https://doi.org/10.3389/fmicb.2020.01212

Comasio A, Van Kerrebroeck S, De Vuyst L (2021) Lemon juice and apple juice used as source of citrate and malate, respectively, enhance the formation of buttery aroma compounds and/or organic acids during Type 2 and Type 3 sourdough productions performed with *Companilactobacillus crustorum* LMG 23699. Int J Food Microbiol 339:109020. https://doi.org/10.1016/j.ijfoodmicro.2020.109020

Comuzzo P, Calligaris S, Iacumin L, Ginaldi F, Voce S, Zironi R (2017) Application of multi-pass high pressure homogenization under variable temperature regimes to induce autolysis of wine yeasts. Food Chem 224:105–113. https://doi.org/10.1016/j.foodchem.2016.12.038

Cotârleţ M, Vasile AM, Gaspar-Pintiliescu A et al (2020a) Tribiotication strategy for the functionalization of bovine colostrum through the biochemical activities of artisanal and selected starter cultures. CyTA J Food 18(1):274–280. https://doi.org/10.1080/19476337.2020.1745287

Cotârleţ M, Stănciuc N, Bahrim GE (2020b) *Yarrowia lipolytica* and *Lactobacillus paracasei* solid state fermentation as a valuable biotechnological tool for the pork lard and okara's biotransformation. Microorganisms 8:1098. https://doi.org/10.3390/microorganisms8081098

Cuevas-González PF, Liceaga AM, Aguilar-Toalá JE (2020) Postbiotics and paraprobiotics: from concepts to applications. Food Res Int 136:109502. https://doi.org/10.1016/j.foodres.2020.109502

Cui Y, Miao K, Niyaphorn S, Qu X (2020) Production of gamma-aminobutyric acid from lactic acid bacteria: a systematic review. Int J Mol Sci 21(3):995. https://doi.org/10.3390/ijms21030995

Dai J, Xia H, Yang C, Chen X (2021) Sensing, uptake and catabolism of L-phenylalanine during 2-phenylethanol biosynthesis via the Ehrlich pathway in *Saccharomyces cerevisiae*. Front Microbiol 12:601963. https://doi.org/10.3389/fmicb.2021.601963

de Jesús Rodríguez-Romero J, Aceves-Lara CA, Silva CF, Gschaedler A, Amaya-Delgado L, Arrizon J (2020) 2-Phenylethanol and 2-phenylethylacetate production by nonconventional yeasts using tequila vinasses as a substrate. Biotechnol Rep 25:e00420. https://doi.org/10.1016/j.btre.2020.e00420

De Luca L, Aiello A, Pizzolongo F, Blaiotta G, Aponte M, Romano R (2021) Volatile organic compounds in breads prepared with different sourdoughs. Appl Sci 11(3):1330. https://doi.org/10.3390/app11031330

De Vero L, Iosca G, La China S, Licciardello F, Gullo M, Pulvirenti A (2021) Yeasts and lactic acid bacteria for Panettone production: an assessment of candidate strains. Microorganisms 9(5):1093. https://doi.org/10.3390/microorganisms9051093

De Vuyst L, Harth H, Van Kerrebroeck S, Leroy F (2016) Yeast diversity of sourdoughs and associated metabolic properties and functionalities. Int J Food Microbiol 239:26–34. https://doi.org/10.1016/j.ijfoodmicro.2016.07.018

De Vuyst L, Van Kerrebroeck S, Leroy F (2017) Chapter two – microbial ecology and process technology of sourdough fermentation. In: Sariaslani S, Gadd GM (eds) Advances in applied microbiology. Academic, pp 49–160

De Vuyst L, Comasio A, Kerrebroeck SV (2021) Sourdough production: fermentation strategies, microbial ecology, and use of non-flour ingredients. Crit Rev Food Sci Nutr:1–33. https://doi.org/10.1080/10408398.2021.1976100

Delali KI, Chen O, Wang W et al (2021) Evaluation of yeast isolates from kimchi with antagonistic activity against green mold in citrus and elucidating the action mechanisms of three yeast: *P. kudriavzevii*, *K. marxianus*, and *Y. lipolytica*. Postharvest Biol Technol 176:111495. https://doi.org/10.1016/j.postharvbio.2021.111495

Dilokpimol A, Mäkelä MR, Varriale S et al (2018) Fungal feruloyl esterases: Functional validation of genome mining based enzyme discovery including uncharacterized subfamilies. New Biotechnol 41:9–14. https://doi.org/10.1016/J.NBT.2017.11.004

Dimopoulos G, Limnaios A, Aerakis E et al (2021) Effect of high pressure on the proteolytic activity and autolysis of yeast *Saccharomyces cerevisiae*. Innov Food Sci Emerg Technol 74:102865. https://doi.org/10.1016/j.ifset.2021.102865

do Nascimento FV, de Castro AM, Secchi AR, MAZ C (2021) Insights into media supplementation in solid-state fermentation of soybean hulls by *Yarrowia lipolytica*: impact on lipase production in tray and insulated packed-bed bioreactors. Biochem Eng J 166:107866. https://doi.org/10.1016/j.bej.2020.107866

Efsa Biohaz P, Koutsoumanis K, Allende A, Alvarez-Ordonez A, Bolton D, Bover-Cid S, Chemaly M, Davies R, De Cesare A, Hilbert F, Lindqvist R, Nauta M, Peixe L, Ru G, Simmons M, Skandamis P, Suffredini E, Cocconcelli PS, Fernández Escámez PS, Maradona MP, Querol A, Sijtsma L, Suarez JE, Sundh I, Vlak J, Barizzone F, Hempen M, Correia S, Herman L

(2022) Updated list of QPS-recommended biological agents for safety risk assessments carried out by EFSA

Fekri A, Torbati M, Yari Khosrowshahi A, Bagherpour Shamloo H, Azadmard-Damirchi S (2020) Functional effects of phytate-degrading, probiotic lactic acid bacteria and yeast strains isolated from Iranian traditional sourdough on the technological and nutritional properties of whole wheat bread. Food Chem 306:125620. https://doi.org/10.1016/j.foodchem.2019.125620

Fernandez-Pacheco Rodríguez P, Arévalo-Villena M, Zaparoli Rosa I, Briones Pérez A (2018) Selection of potential non-*Sacharomyces* probiotic yeasts from food origin by a step-by-step approach. Food Res Int 112:143–151. https://doi.org/10.1016/j.foodres.2018.06.008

Font de Valdez G, Gerez CL, Inés Torino M, Rollán G (2010) New Trends in Cereal-Based Products Using Lactic Acid Bacteria. In: Mozzi F, Raya RR, Vignolo G (eds) Biotechnology of Lactic Acid Bacteria: Novel Applications. John Wiley & Sons, Ltd. https://doi.org/10.1002/9780813820866.ch15

Food and Drug Administration Microorganisms & Microbial-Derived Ingredients Used in Food (Partial List) | FDA. https://www.fda.gov/food/generally-recognized-safe-gras/microorganisms-microbial-derived-ingredients-used-food-partial-list. Accessed 19 Sep 2022

Fraberger V, Unger C, Kummer C, Domig KJ (2020) Insights into microbial diversity of traditional Austrian sourdough. LWT Food Sci Technol 127:109358. https://doi.org/10.1016/j.lwt.2020.109358

Franco W, Pérez-Díaz IM, Connelly L, Diaz JT (2020) Isolation of exopolysaccharide-producing yeast and lactic acid bacteria from quinoa (Chenopodium quinoa) sourdough fermentation. Foods 9(3):337. https://doi.org/10.3390/foods9030337

Fu W, Zhao G, Liu J (2022) Effect of preparation methods on physiochemical and functional properties of yeast β-glucan. LWT Food Sci Technol 160:113284. https://doi.org/10.1016/j.lwt.2022.113284

Fujimoto A, Ito K, Narushima N, Miyamoto T (2019) Identification of lactic acid bacteria and yeasts, and characterization of food components of sourdoughs used in Japanese bakeries. J Biosci Bioeng 127(5):575–581. https://doi.org/10.1016/j.jbiosc.2018.10.014

Gänzle MG (2014) Enzymatic and bacterial conversions during sourdough fermentation. Food Microbiol 37:2–10. https://doi.org/10.1016/j.fm.2013.04.007

Gänzle MG (2015) Lactic metabolism revisited: metabolism of lactic acid bacteria in food fermentations and food spoilage. Curr Opin Food Sci 2:106–117. https://doi.org/10.1016/j.cofs.2015.03.001

Gaspar-Pintiliescu A, Oancea A, Cotarlet M et al (2020) Angiotensin-converting enzyme inhibition, antioxidant activity and cytotoxicity of bioactive peptides from fermented bovine colostrum. Int J Dairy Technol 73:108–116. https://doi.org/10.1111/1471-0307.12659

Gennari A, Simon R, de Andrade BC et al (2021) Production of beta-galactosidase fused to a cellulose-binding domain for application in sustainable industrial processes. Bioresour Technol 326:124747. https://doi.org/10.1016/j.biortech.2021.124747

Gientka I, Błażejak S, Stasiak-Różańska L, Chlebowska-Śmigiel A (2015) Exopolysaccharides from yeast: insight into optimal conditions for biosynthesis, chemical composition and functional properties – review. Acta Sci Pol Technol Aliment. https://doi.org/10.17306/J.AFS.2015.4.29

Gientka I, Bzducha-Wróbel A, Stasiak-Różańska L et al (2016) The exopolysaccharides biosynthesis by *Candida* yeast depends on carbon sources. Electron J Biotechnol 22:31–37. https://doi.org/10.1016/J.EJBT.2016.02.008

Gobbetti M, De Angelis M, Di Cagno R, Calasso M, Archetti G, Rizzello CG (2019) Novel insights on the functional/nutritional features of the sourdough fermentation. Int J Food Microbiol 302:103–113. https://doi.org/10.1016/j.ijfoodmicro.2018.05.018

Gottardi D, Siroli L, Vannini L et al (2021) Recovery and valorization of agri-food wastes and by-products using the non-conventional yeast *Yarrowia lipolytica*. Trends Food Sci Technol 115:74–86. https://doi.org/10.1016/j.tifs.2021.06.025

Gut AM, Vasiljevic T, Yeager T, Donkor ON (2022) Anti-salmonella properties of kefir yeast isolates: an in vitro screening for potential infection control. Saudi J Biol Sci 29:550–563. https://doi.org/10.1016/j.sjbs.2021.09.025

Hamidi M, Gholipour AR, Delattre C et al (2020) Production, characterization and biological activities of exopolysaccharides from a new cold-adapted yeast: *Rhodotorula mucilaginosa* sp. GUMS16. Int J Biol Macromol 151:268–277. https://doi.org/10.1016/J.IJBIOMAC.2020.02.206

Hamza Z, El-Hashash M, Aly S et al (2019) Preparation and characterization of yeast cell wall beta-glucan encapsulated humic acid nanoparticles as an enhanced aflatoxin B1 binder. Carbohydr Polym 203:185–192. https://doi.org/10.1016/j.carbpol.2018.08.047

Han M, Liao W-Y, Wu S-M, Gong X, Bai C (2020) Use of *Streptococcus thermophilus* for the in situ production of γ-aminobutyric acid-enriched fermented milk. J Dairy Sci 103(1):98–105. https://doi.org/10.3168/jds.2019-16856

Hapeta P, Kerkhoven EJ, Lazar Z (2020) Nitrogen as the major factor influencing gene expression in *Yarrowia lipolytica*. Biotechnol Rep 27:e00521. https://doi.org/10.1016/j.btre.2020.e00521

Harth H, Van Kerrebroeck S, De Vuyst L (2018) Impact of process conditions on the microbial community dynamics and metabolite production kinetics of teff sourdough fermentations under bakery and laboratory conditions. Food Sci Nutr 6(6):1438–1455. https://doi.org/10.1002/fsn3.690

Hazelwood LA, Daran J-M, van Maris AJA, Pronk JT, Dickinson JR (2008) The Ehrlich pathway for fusel alcohol production: a century of research on *Saccharomyces cerevisiae* metabolism. Appl Environ Microbiol 74(8):2259–2266. https://doi.org/10.1128/AEM.02625-07

Herman L, Chemaly M, Cocconcelli PS et al (2019) The qualified presumption of safety assessment and its role in EFSA risk evaluations: 15 years past. FEMS Microbiol Lett 366:260. https://doi.org/10.1093/FEMSLE/FNY260

Hosoglu MI, Guneser O, Yuceer YK (2018) Different bioengineering approaches on production of bioflavor compounds. In: Role of materials science in food bioengineering. Elsevier, pp 37–71. https://doi.org/10.1016/B978-0-12-811448-3.00002-4

Huys G, Daniel H-M, De Vuyst L (2013) Taxonomy and biodiversity of sourdough yeasts and lactic acid bacteria. In: Gobbetti M, Gänzle M (eds) Handbook on sourdough biotechnology. Springer, Boston, pp 105–154

Ilango S, Antony U (2021) Probiotic microorganisms from non-dairy traditional fermented foods. Trends Food Sci Technol 118:617–638. https://doi.org/10.1016/j.tifs.2021.05.034

Janek T, Mirończuk AM, Rymowicz W, Dobrowolski A (2020) High-yield expression of extracellular lipase from *Yarrowia lipolytica* and its interactions with lipopeptide biosurfactants: a biophysical approach. Arch Biochem Biophys 689:108475. https://doi.org/10.1016/j.abb.2020.108475

Jensen S, Oestdal H, Skibsted LH, Larsen E, Thybo AK (2011) Chemical changes in wheat pan bread during storage and how it affects the sensory perception of aroma, flavour, and taste. J Cereal Sci 53(2):259–268. https://doi.org/10.1016/j.jcs.2010.11.007

Jin J, Nguyen TTH, Humayun S, Park S, Oh H, Lim S, Mok I-K, Li Y, Pal K, Kim D (2021) Characteristics of sourdough bread fermented with *Pediococcus pentosaceus* and *Saccharomyces cerevisiae* and its bio-preservative effect against *Aspergillus flavus*. Food Chem 345:128787. https://doi.org/10.1016/j.foodchem.2020.128787

Johansson L, Nikulin J, Juvonen R, Krogerus K, Magalhães F, Mikkelson A, Nuppunen-Puputti M, Sohlberg E, de Francesco G, Perretti G, Gibson B (2021) Sourdough cultures as reservoirs of maltose-negative yeasts for low-alcohol beer brewing. Food Microbiol 94:103629. https://doi.org/10.1016/j.fm.2020.103629

Karaman K, Sagdic O, Durak MZ (2018) Use of phytase active yeasts and lactic acid bacteria isolated from sourdough in the production of whole wheat bread. LWT Food Sci Technol 91:557–567. https://doi.org/10.1016/j.lwt.2018.01.055

Karim A, Gerliani N, Aïder M (2020) *Kluyveromyces marxianus*: an emerging yeast cell factory for applications in food and biotechnology. Int J Food Microbiol 333:108818. https://doi.org/10.1016/j.ijfoodmicro.2020.108818

Kurniawati M, Nurliyani N, Budhijanto W, Widodo W (2022) Isolation and identification of lactose-degrading yeasts and characterisation of their fermentation-related ability to produce ethanol. Fermentation 8:183. https://doi.org/10.3390/FERMENTATION8040183

Kwak S, Robinson SJ, Lee JW et al (2022) Dissection and enhancement of prebiotic properties of yeast cell wall oligosaccharides through metabolic engineering. Biomaterials 282:121378. https://doi.org/10.1016/j.biomaterials.2022.121379

Lahue C, Madden A, Dunn R, Smukowski Heil C (2020) History and domestication of *Saccharomyces cerevisiae* in bread baking. Front Genet 11:584718. https://doi.org/10.3389/fgene.2020.584718

Laurent J, Timmermans E, Struyf N et al (2020) Variability in yeast invertase activity determines the extent of fructan hydrolysis during wheat dough fermentation and final FODMAP levels in bread. Int J Food Microbiol 326:108648. https://doi.org/10.1016/j.ijfoodmicro.2020.108648

Lhomme E, Lattanzi A, Dousset X, Minervini F, De Angelis M, Lacaze G, Onno B, Gobbetti M (2015) Lactic acid bacterium and yeast microbiotas of sixteen French traditional sourdoughs. Int J Food Microbiol 215:161–170. https://doi.org/10.1016/j.ijfoodmicro.2015.09.015

Lhomme E, Urien C, Legrand J, Dousset X, Onno B, Sicard D (2016) Sourdough microbial community dynamics: an analysis during French organic bread-making processes. Food Microbiol 53:41–50. https://doi.org/10.1016/j.fm.2014.11.014

Li Z, Li H, Bian K (2016) Microbiological characterization of traditional dough fermentation starter (Jiaozi) for steamed bread making by culture-dependent and culture-independent methods. Int J Food Microbiol 234:9–14. https://doi.org/10.1016/j.ijfoodmicro.2016.06.024

Li H-Y, Zhou D-D, Gan R-Y et al (2021a) Effects and mechanisms of probiotics, prebiotics, synbiotics, and postbiotics on metabolic diseases targeting gut microbiota: a narrative review. Nutrients 13(9):3211. https://doi.org/10.3390/NU13093211

Li S, Zhang Y, Yin P, Zhang K, Liu Y, Gao Y, Li Y, Wang T, Lu S, Li B (2021b) Probiotic potential of γ-aminobutyric acid (GABA)–producing yeast and its influence on the quality of cheese. J Dairy Sci 104(6):6559–6576. https://doi.org/10.3168/jds.2020-19845

Li Y, Wang T, Li S, Yin P, Sheng H, Wang T, Zhang Y, Zhang K, Wang Q, Lu S, Dong J, Li B (2022) Influence of GABA-producing yeasts on cheese quality, GABA content, and the volatilome. LWT Food Sci Technol 154:112766. https://doi.org/10.1016/j.lwt.2021.112766

Liu Y, Huang G (2018) The derivatization and antioxidant activities of yeast mannan. Int J Biol Macromol 107:755–761. https://doi.org/10.1016/j.ijbiomac.2017.09.055

Liu T, Li Y, Chen J, Sadiq FA, Zhang G, Li Y, He G (2016) Prevalence and diversity of lactic acid bacteria in Chinese traditional sourdough revealed by culture dependent and pyrosequencing approaches. LWT Food Sci Technol 68:91–97. https://doi.org/10.1016/j.lwt.2015.12.025

Liu T, Li Y, Sadiq FA, Yang H, Gu J, Yuan L, Lee YK, He G (2018) Predominant yeasts in Chinese traditional sourdough and their influence on aroma formation in Chinese steamed bread. Food Chem 242:404–411. https://doi.org/10.1016/j.foodchem.2017.09.081

Liu L, Xu J, Na R et al (2022) Purification, characterization and partial biological activities of exopolysaccharide produced by *Saccharomyces cerevisiae* Y3. Int J Biol Macromol 206:777–787. https://doi.org/10.1016/j.ijbiomac.2022.03.083

Lo S-C, Yang C-Y, Mathew DC, Huang C-C (2021) Growth and autolysis of the kefir yeast *Kluyveromyces marxianus* stimulated with lactate provides insights into symbiosis. Sci Rep 11(1):14552. https://doi.org/10.1038/s41598-021-94101-y

Lombardi SJ, De Leonardis A, Lustrato G, Testa B, Iorizzo M (2015) Yeast autolysis in sparkling wine aging: use of killer and sensitive *Saccharomyces cerevisiae* strains in co-culture. Recent Pat Biotechnol 9(3):223–230. https://doi.org/10.2174/1872208310666160414102015

Luo Y, Liu L, Yuan L et al (2022) The characteristics of patulin degradation by probiotic yeast – *Pichia guilliermondii* S15-8. Food Control 133:108627. https://doi.org/10.1016/j.foodcont.2021.108627

Ma S, Wang Z, Guo X, Wang F, Huang J, Sun B, Wang X (2021) Sourdough improves the quality of whole-wheat flour products: mechanisms and challenges—a review. Food Chem 360:130038. https://doi.org/10.1016/j.foodchem.2021.130038

Madzak C (2021) Yarrowia lipolytica strains and their biotechnological applications: how natural biodiversity and metabolic engineering could contribute to cell factories improvement. J Fungi 7:548. https://doi.org/10.3390/JOF7070548

Mancini A, Carafa I, Franciosi E, Nardin T, Bottari B, Larcher R, Tuohy KM (2019) In vitro probiotic characterization of high GABA producing strain *Lactobacilluas brevis* DSM 32386 isolated from traditional "wild" Alpine cheese. Ann Microbiol 69(13):1435–1443. https://doi.org/10.1007/s13213-019-01527-x

Martínez JM, Delso C, Aguilar D, Cebrián G, Álvarez I, Raso J (2018) Factors influencing autolysis of *Saccharomyces cerevisiae* cells induced by pulsed electric fields. Food Microbiol 73:67–72. https://doi.org/10.1016/j.fm.2017.12.008

Martínez MP, Magnoli AP, González Pereyra ML, Cavaglieri L (2019) Probiotic bacteria and yeasts adsorb aflatoxin M1 in milk and degrade it to less toxic AFM1-metabolites. Toxicon 172:1–7. https://doi.org/10.1016/j.toxicon.2019.10.001

Martínez-Avila O, Sánchez A, Font X, Barrena R (2018) Bioprocesses for 2-phenylethanol and 2-phenylethyl acetate production: current state and perspectives. Appl Microbiol Biotechnol 102(23):9991–10004. https://doi.org/10.1007/s00253-018-9384-8

Maurya DK, Kumar A, Chaurasiya U et al (2021) Modern era of microbial biotechnology: opportunities and future prospects. In: Microbiomes Plant Health. Academic Press, pp 317–343. https://doi.org/10.1016/B978-0-12-819715-8.00011-2

Menezes AGT, Ramos CL, Dias DR, Schwan RF (2018) Combination of probiotic yeast and lactic acid bacteria as starter culture to produce maize-based beverages. Food Res Int 111:187–197. https://doi.org/10.1016/j.foodres.2018.04.065

Merchán AV, Benito MJ, Galván AI, Seco R-M, de Herrera S (2020) Identification and selection of yeast with functional properties for future application in soft paste cheese. LWT- Food Sci Technol 124:109173. https://doi.org/10.1016/j.lwt.2020.109173

Mierzejewska J, Dąbkowska K, Chreptowicz K, Sokołowska A (2019) Hydrolyzed corn stover as a promising feedstock for 2-phenylethanol production by nonconventional yeast. J Chem Technol Biotechnol 94(3):777–784. https://doi.org/10.1002/jctb.5823

Minervini F, De Angelis M, Di Cagno R, Gobbetti M (2014) Ecological parameters influencing microbial diversity and stability of traditional sourdough. Int J Food Microbiol 171:136–146. https://doi.org/10.1016/j.ijfoodmicro.2013.11.021

Mitri S, Koubaa M, Maroun RG, Rossignol T, Nicaud J-M, Louka N (2022) Bioproduction of 2-Phenylethanol through yeast fermentation on synthetic media and on agro-industrial waste and by-products: a review. Foods 11(1):109. https://doi.org/10.3390/foods11010109

Morita K, Seike T, Ishii J et al (2022) Improvement of 2,3-butanediol production by dCas9 gene expression system in *Saccharomyces cerevisiae*. J Biosci Bioeng 133:208–212. https://doi.org/10.1016/j.jbiosc.2021.12.007

Nehmi VA, Murata GM, de Moraes RCM et al (2021) A novel supplement with yeast β-glucan, prebiotic, minerals and *Silybum marianum* synergistically modulates metabolic and inflammatory pathways and improves steatosis in obese mice. J Integr Med 19:439–450. https://doi.org/10.1016/j.joim.2021.05.002

Nguyen TL, Toh M, Lu Y et al (2022) Biovalorization of market surplus bread for development of probiotic-fermented potential functional beverages. Foods 11:250. https://doi.org/10.3390/FOODS11030250/S1

Ogunsakin AO, Vanajakshi V, Anu-Appaiah KA, Vijayendra SVN, Walde SG, Banwo K, Sanni AI, Prabhasankar P (2017) Evaluation of functionally important lactic acid bacteria and yeasts from Nigerian sorghum as starter cultures for gluten-free sourdough preparation. LWT Food Sci Technol 82:326–334. https://doi.org/10.1016/j.lwt.2017.04.048

Olojede AO, Sanni AI, Banwo K (2020) Rheological, textural and nutritional properties of gluten-free sourdough made with functionally important lactic acid bacteria and yeast from Nigerian sorghum. LWT Food Sci Technol 120:108875. https://doi.org/10.1016/j.lwt.2019.108875

Oro L, Feliziani E, Ciani M et al (2018) Volatile organic compounds from *Wickerhamomyces anomalus*, *Metschnikowia pulcherrima* and *Saccharomyces cerevisiae* inhibit growth of decay causing fungi and control postharvest diseases of strawberries. Int J Food Microbiol 265:18–22. https://doi.org/10.1016/J.IJFOODMICRO.2017.10.027

Oshiro M, Tanaka M, Momoda R, Zendo T, Nakayama J (2021) Mechanistic insight into yeast bloom in a lactic acid bacteria relaying-community in the start of sourdough microbiota evolution. Microbiol Spectr 9(2):e00662–e00621. https://doi.org/10.1128/Spectrum.00662-21

Păcularu-Burada B, Turturică M, Rocha JM, Bahrim GE (2021) Statistical approach to potentially enhance the postbiotication of gluten-free sourdough. Appl Sci 11:5306. https://doi.org/10.3390/app11115306

Palla M, Agnolucci M, Calzone A, Giovannetti M, Di Cagno R, Gobbetti M, Rizzello CG, Pontonio E (2019) Exploitation of autochthonous Tuscan sourdough yeasts as potential starters. Int J Food Microbiol 302:59–68. https://doi.org/10.1016/j.ijfoodmicro.2018.08.004

Papadimitriou K, Zoumpopoulou G, Georgalaki M, Alexandraki V, Kazou M, Anastasiou R, Tsakalidou E (2019) Sourdough bread. In: Galanakis CM (ed) Innovations in traditional foods, 1st edn. Woodhead Publishing, pp 127–158. https://doi.org/10.1016/B978-0-12-814887-7.00006-X

Parfene G, Horincar V, Tyagi AK et al (2013) Production of medium chain saturated fatty acids with enhanced antimicrobial activity from crude coconut fat by solid state cultivation of *Yarrowia lipolytica*. Food Chem 136:1345–1349. https://doi.org/10.1016/j.foodchem.2012.09.057

Park J, Seo JS, Kim S-A, Shin S-Y, Park JH, Han NS (2017) Microbial diversity of commercial Makgeolli and its influence on the organoleptic characteristics of Korean rice sourdough, Jeung-Pyun. J Microbiol Biotechnol 27(10):1736–1743. https://doi.org/10.4014/jmb.1708.08003

Park SJ, Kim DH, Kang HJ, Shin M, Yang S-Y, Yang J, Jung YH (2021) Enhanced production of γ-aminobutyric acid (GABA) using *Lactobacillus plantarum* EJ2014 with simple medium composition. LWT Food Sci Technol 137:110443. https://doi.org/10.1016/j.lwt.2020.110443

Peluzio M d CG, Martinez JA, Milagro FI (2021) Postbiotics: metabolites and mechanisms involved in microbiota-host interactions. Trends Food Sci Technol 108:11–26. https://doi.org/10.1016/j.tifs.2020.12.004

Pereira de Paula B, de Souza LH, Firmino L et al (2021) Technological features of *Saccharomyces cerevisiae* var. *boulardii* for potential probiotic wheat beer development. LWT Food Sci Technol 135:110233. https://doi.org/10.1016/j.lwt.2020.110233

Pétel C, Onno B, Prost C (2017) Sourdough volatile compounds and their contribution to bread: a review. Trends Food Sci Technol 59:105–123. https://doi.org/10.1016/j.tifs.2016.10.015

Phukon LC, Singh SP, Pandey A, Rai AK (2022) Microbial bioprocesses for production of nutraceuticals and functional foods. In: Rai AK, Pendey A, Soccol CR, Singh SP, Larroche C (eds) Current developments in biotechnology and bioengineering. Elsevier, pp 1–29. https://doi.org/10.1016/C2020-0-00498-6

Pico J, Bernal J, Gómez M (2015) Wheat bread aroma compounds in crumb and crust: a review. Food Res Int 75:200–215. https://doi.org/10.1016/j.foodres.2015.05.051

Pino A, Russo N, Solieri L, Sola L, Caggia C, Randazzo CL (2022) Microbial consortia involved in traditional Sicilian sourdough: characterization of lactic acid bacteria and yeast populations. Microorganisms 10(2):283. https://doi.org/10.3390/microorganisms10020283

Qadir F, Shariq M, Ahmed A, Sohail M (2018) Evaluation of a yeast co-culture for cellulase and xylanase production under solid state fermentation of sugarcane bagasse using multivariate approach. Ind Crop Prod 123:407–415. https://doi.org/10.1016/j.indcrop.2018.07.021

Rafieenia R, Atkinson E, Ledesma-Amaro R (2022) Division of labor for substrate utilization in natural and synthetic microbial communities. Curr Opin Biotechnol 75:102706. https://doi.org/10.1016/J.COPBIO.2022.102706

Rahbar Saadat Y, Yari Khosroushahi A, Movassaghpour AA et al (2020) Modulatory role of exopolysaccharides of *Kluyveromyces marxianus* and *Pichia kudriavzevii* as probiotic yeasts from dairy products in human colon cancer cells. J Funct Foods 64:103675. https://doi.org/10.1016/j.jff.2019.103675

Rahbar Saadat Y, Yari Khosroushahi A, Pourghassem Gargari B (2021) Yeast exopolysaccharides and their physiological functions. Folia Microbiol 66(2):171–182. https://doi.org/10.1007/s12223-021-00856-2

Rai AK, Pandey A, Sahoo D (2019) Biotechnological potential of yeasts in functional food industry. Trends Food Sci Technol 83:129–137. https://doi.org/10.1016/j.tifs.2018.11.016

Rakowska R, Sadowska A, Dybkowska E, Świderski F (2017) Spent yeasts as natural source of functional food additives. Ann Natl Inst Hyg 68:115–121. bwmeta1.element.agro-93584634-72d6-437c-8256-72d7560a5f69

Rengarajan S, Palanivel R (2020) High purity prebiotic isomalto-oligosaccharides production by cell associated transglucosidase of isolated strain *Debaryomyces hansenii* SCY204 and selective fermentation by *Saccharomyces cerevisiae* SYI065. Process Biochem 98:93–105. https://doi.org/10.1016/j.procbio.2020.07.024

Rengel dos Passos F, Lopes Maestre K, Florêncio da Silva B et al (2021) Production of a synbiotic composed of galacto-oligosaccharides and *Saccharomyces boulardii* using enzymatic-fermentative method. Food Chem 129486. https://doi.org/10.1016/j.foodchem.2021.129486

Rusinova-Videva S, Pavlova K, Panchev I, Georgieva K, Kuncheva M (2010) Effect of different factors on biosynthesis of exopolysaccharide from Antarctic yeast. Biotechnol Biotechnol Equip 24(1):507–511. https://doi.org/10.1080/13102818.2010.10817891

Sampaolesi S, Gamba RR, de Antoni GL, León Peláez ÁM (2019) Potentiality of yeasts obtained as beer fermentation residue to be used as probiotics. LWT Food Sci Technol 113:108251. https://doi.org/10.1016/j.lwt.2019.108251

Sampaolesi S, Briand LE, de Antoni G, León Peláez A (2022) The synthesis of soluble and volatile bioactive compounds by selected brewer's yeasts: Antagonistic effect against enteropathogenic bacteria and food spoiler – toxigenic *Aspergillus* sp. Food Chemistry: X 13:100193. https://doi.org/10.1016/J.FOCHX.2021.100193

Sen S, Mansell TJ (2020) Yeasts as probiotics: mechanisms, outcomes, and future potential. Fungal Genet Biol 137:103333. https://doi.org/10.1016/j.fgb.2020.103333

Shruthi B, Deepa N, Somashekaraiah R et al (2022) Exploring biotechnological and functional characteristics of probiotic yeasts: a review. Biotechnol Rep 34:e00716. https://doi.org/10.1016/J.BTRE.2022.E00716

Siepmann FB, Sousa de Almeida B, Waszczynskyj N, Spier MR (2019) Influence of temperature and of starter culture on biochemical characteristics and the aromatic compounds evolution on type II sourdough and wheat bread. LWT Food Sci Technol 108:199–206. https://doi.org/10.1016/J.LWT.2019.03.065

Sieuwerts S, Bron PA, Smid EJ (2018) Mutually stimulating interactions between lactic acid bacteria and *Saccharomyces cerevisiae* in sourdough fermentation. LWT Food Sci Technol 90:201–206. https://doi.org/10.1016/j.lwt.2017.12.022

Silva LC, Schmidt GB, Alves LGO et al (2020) Use of probiotic strains to produce beers by axenic or semi-separated co-culture system. Food Bioprod Process 124:408–418. https://doi.org/10.1016/j.fbp.2020.10.001

Singh A, Vishwakarma V, Singhal B (2018) Metabiotics: the functional metabolic signatures of probiotics: current state-of-art and future research priorities—metabiotics: probiotics effector molecules. Adv Biosci Biotechnol 09:147–189. https://doi.org/10.4236/abb.2018.94012

Sivieri K, de Oliveira SM, de Souza Marquez A et al (2022) Insights on β-glucan as a prebiotic coadjuvant in the treatment of diabetes mellitus: a review. Food Hydrocoll Health 2:100056. https://doi.org/10.1016/j.fhfh.2022.100056

Struyf N, Laurent J, Lefevere B, Verspreet J, Verstrepen KJ, Courtin CM (2017a) Establishing the relative importance of damaged starch and fructan as sources of fermentable sugars in wheat flour and whole meal bread dough fermentations. Food Chem 218:89–98. https://doi.org/10.1016/j.foodchem.2016.09.004

Struyf N, Laurent J, Verspreet J, Verstrepen KJ, Courtin CM (2017b) *Saccharomyces cerevisiae* and *Kluyveromyces marxianus* cocultures allow reduction of fermentable Oligo-, Di-, and Monosaccharides and Polyols levels in whole wheat bread. J Agric Food Chem 65(39):8704–8713. https://doi.org/10.1021/acs.jafc.7b02793

Sun X, Wang J, Li C, Zheng M, Zhang Q, Xiang W, Tang J (2022) The use of γ-Aminobutyric Acid-producing *Saccharomyces cerevisiae* SC125 for functional fermented beverage production from apple juice. Foods 11(9):1202. https://doi.org/10.3390/foods11091202

Swieca M, Kordowska-Wiater M, Pytka M et al (2019) Nutritional and pro-health quality of lentil and adzuki bean sprouts enriched with probiotic yeast *Saccharomyces cerevisiae* var. boulardii. LWT Food Sci Technol 100:220–226. https://doi.org/10.1016/j.lwt.2018.10.081

Syrokou MK, Themeli C, Paramithiotis S, Mataragas M, Bosnea L, Argyri AA, Chorianopoulos NG, Skandamis PN, Drosinos EH (2020) Microbial ecology of Greek wheat sourdoughs, identified by a culture-dependent and a culture-independent approach. Foods 9(11):1603. https://doi.org/10.3390/foods9111603

Syrokou MK, Tziompra S, Psychogiou E-E, Mpisti S-D, Paramithiotis S, Bosnea L, Mataragas M, Skandamis PN, Drosinos EH (2021) Technological and safety attributes of lactic acid bacteria and yeasts isolated from spontaneously fermented Greek wheat sourdoughs. Microorganisms 9(4):671. https://doi.org/10.3390/microorganisms9040671

Tao Y, García JF, Sun D-W (2014) Advances in wine aging technologies for enhancing wine quality and accelerating wine aging process. Crit Rev Food Sci Nutr 54(6):817–835. https://doi.org/10.1080/10408398.2011.609949

Terpou A, Rai AK (2022) Microbial transformation for improving food functionality. In: Rai AK, Pendey A, Soccol CR, Singh SP, Larroche C (eds). Elsevier, Current developments in biotechnology and bioengineering, pp 31–45. https://doi.org/10.1016/C2020-0-00498-6

Tian X, Yang P, Jiang W (2019) Effect of alkali treatment combined with high pressure on extraction efficiency of β-d-glucan from spent brewer's yeast. Waste and Biomass Valorization 10:1131–1140. https://doi.org/10.1007/s12649-017-0130-8

Tiefenbacher KF (2017) Chapter four – technology of minor ingredients for wafers and waffles. In: Tiefenbacher KF (ed) Wafer and Waffle. Academic Press, pp 227–311. https://doi.org/10.1016/C2015-0-04250-2

Tinzl-Malang SK, Rast P, Grattepanche F, Sych J, Lacroix C (2015) Exopolysaccharides from co-cultures of *Weissella confusa* 11GU-1 and *Propionibacterium freudenreichii* JS15 act synergistically on wheat dough and bread texture. Int J Food Microbiol 214:91–101. https://doi.org/10.1016/j.ijfoodmicro.2015.07.025

Urien C, Legrand J, Montalent P, Casaregola S, Sicard D (2019) Fungal species diversity in French bread sourdoughs made of organic wheat flour. Front Microbiol 10:201. https://doi.org/10.3389/fmicb.2019.00201

Valdez Castillo M, Pachapur VL, Brar SK et al (2022) Effect of the concentration of L-Phenylalanine and lactose on 2-Phenylethanol production by whey fermentation using the yeasts *Kluyveromyces marxianus* and *Debaryomyces hansenii* under co-culture mode. Bioresource Technol Rep 18:100994. https://doi.org/10.1016/J.BITEB.2022.100994

Van Kerrebroeck S, Bastos FCC, Harth H, De Vuyst L (2016) A low pH does not determine the community dynamics of spontaneously developed backslopped liquid wheat sourdoughs but does influence their metabolite kinetics. Int J Food Microbiol 239:54–64. https://doi.org/10.1016/j.ijfoodmicro.2016.07.019

Van Kerrebroeck S, Maes D, De Vuyst L (2017) Sourdoughs as a function of their species diversity and process conditions, a meta-analysis. Trends Food Sci Technol 68:152–159. https://doi.org/10.1016/j.tifs.2017.08.016

Van Wyk N, Kroukamp H, Pretorius IS (2018) The smell of synthetic biology: engineering strategies for aroma compound production in yeast. Fermentation 4(3):54. https://doi.org/10.3390/fermentation4030054

Vanden Braber NL, Díaz Vergara LI, Rossi YE et al (2020) Effect of microencapsulation in whey protein and water-soluble chitosan derivative on the viability of the probiotic *Kluyveromyces marxianus* VM004 during storage and in simulated gastrointestinal conditions. LWT Food Sci Technol 118:108844. https://doi.org/10.1016/j.lwt.2019.108844

Venturi M, Guerrini S, Vincenzini M (2012) Stable and non-competitive association of *Saccharomyces cerevisiae*, *Candida milleri* and *Lactobacillus sanfranciscensis* during manufacture of two traditional sourdough baked goods. Food Microbiol 31(1):107–115. https://doi.org/10.1016/j.fm.2012.02.011

Vrancken G, De Vuyst L, Van der Meulen R, Huys G, Vandamme P, Daniel H-M (2010) Yeast species composition differs between artisan bakery and spontaneous laboratory sourdoughs. FEMS Yeast Res 10(4):471–481. https://doi.org/10.1111/j.1567-1364.2010.00621.x

Wang J, Li M, Zheng F, Niu C, Liu C, Li Q, Sun J (2018) Cell wall polysaccharides: before and after autolysis of brewer's yeast. World J Microbiol Biotechnol 34(9):137. https://doi.org/10.1007/s11274-018-2508-6

Xu D, Yin Y, Ali B, Zhang Y, Guo L, Xu X (2019a) Isolation of yeast strains from Chinese liquor Daqu and its use in the wheat sourdough bread making. Food Biosci 31:100443. https://doi.org/10.1016/j.fbio.2019.100443

Xu D, Zhang Y, Tang K, Hu Y, Xu X, Gänzle MG (2019b) Effect of mixed cultures of yeast and lactobacilli on the quality of wheat sourdough bread. Front Microbiol 10:2113. https://doi.org/10.3389/fmicb.2019.02113

Yabaci Karaoglan S, Jung R, Gauthier M et al (2022) Maltose-negative yeast in non-alcoholic and low-alcoholic beer production. Fermentation 8:273. https://doi.org/10.3390/FERMENTATION8060273/S1

Yan B, Sadiq FA, Cai Y, Fan D, Chen W, Zhang H, Zhao J (2019a) Microbial diversity in traditional type I sourdough and jiaozi and its influence on volatiles in Chinese steamed bread. LWT Food Sci Technol 101:764–773. https://doi.org/10.1016/j.lwt.2018.12.004

Yan B, Sadiq FA, Cai Y, Fan D, Zhang H, Zhao J, Chen W (2019b) Identification of key aroma compounds in type i sourdough-based Chinese steamed bread: application of untargeted metabolomics analysisp. Int J Mol Sci 20(4):818. https://doi.org/10.3390/ijms20040818

Yang Q, Rutherfurd-Markwick K, Mutukumira AN (2021) Identification of dominant lactic acid bacteria and yeast in rice sourdough produced in New Zealand. Curr Res Food Sci 4:729–736. https://doi.org/10.1016/j.crfs.2021.10.002

Young S, Rai R, Nitin N (2020) Bioaccessibility of curcumin encapsulated in yeast cells and yeast cell wall particles. Food Chem 309:125700. https://doi.org/10.1016/j.foodchem.2019.125700

Yupanqui-Mendoza SL, Vaz de Arruda P, Castelo da Silva GM (2022) Statistical sequential optimization of process parameters for inulinase production by *Kluyveromyces marxianus* ATCC 36907 in solid-state fermentation using beer residue. Biocatal Agric Biotechnol 39:102252. https://doi.org/10.1016/j.bcab.2021.102252

Zhang Q, Sun Q, Tan X, Zhang S, Zeng L, Tang J, Xiang W (2020) Characterization of γ-aminobutyric acid (GABA)-producing *Saccharomyces cerevisiae* and coculture with *Lactobacillus plantarum* for mulberry beverage brewing. J Biosci Bioeng 129(4):447–453. https://doi.org/10.1016/j.jbiosc.2019.10.001

Zhang L, Yue Y, Wang X, Dai W, Piao C, Yu H (2022a) Optimization of fermentation for γ-aminobutyric acid (GABA) production by yeast *Kluyveromyces marxianus* C21 in okara (soybean residue). Bioprocess Biosyst Eng. https://doi.org/10.1007/s00449-022-02702-2

Zhang M, Che Y, Wu C (2022b) A novel exopolysaccharide produced by *Zygosaccharomyces rouxii* with cryoprotective and freeze-drying protective activities. Food Chem 392:133304. https://doi.org/10.1016/J.FOODCHEM.2022.133304

Zheng J, Wittouck S, Salvetti E et al (2020) A taxonomic note on the genus Lactobacillus: description of 23 novel genera, emended description of the genus Lactobacillus Beijerinck 1901, and union of Lactobacillaceae and Leuconostocaceae. Int J Syst Evol Microbiol 70:2782–2858. https://doi.org/10.1099/ijsem.0.004107

Zhou RY, Chua JY, Liu SQ (2022) Exploring the feasibility of biotransforming salted soy whey into a soy sauce-like condiment using wine yeast *Torulaspora delbrueckii* and soy sauce yeasts *Zygosaccharomyces rouxii* and *Candida versatilis* as single starter cultures. Food Res Int 156:111350. https://doi.org/10.1016/j.foodres.2022.111350

Żółkiewicz J, Marzec A, Ruszczyński M, Feleszko W (2020) Postbiotics—a step beyond pre- and probiotics. Nutrients 12:2189. https://doi.org/10.3390/nu12082189

Chapter 13
Health Promoting Features of Sourdough Lactic Acid Bacteria

S. Paramithiotis, L. Bosnea, and M. Mataragas

13.1 Introduction

Sourdough is a starter culture that is created by spontaneous fermentation of a mixture of cereal flour and water and is used in breadmaking through the backslopping regime. On the basis of their production characteristics, they have been distinguished into three types, designated I, II and III (De Vuyst et al. 2017). Spontaneously fermented ones belong to type I, with the ones sustained through daily refreshments (subtype Ib) being the most commonly occurring ones. Sourdough microecosystem consists principally of lactic acid bacteria and yeasts. Regarding the former, *Fructilactobacillus sanfranciscensis*, *Lactiplantibacillus plantarum*, *Levilactobacillus brevis*, *Limosilactobacillus fermentum* and *Companilactobacillus paralimentarius* are most frequently reported to dominate. As far as yeasts are concerned, *Saccharomyces cerevisiae* seems to dominate in the majority of the cases, accompanied by species such as *Kazachstania humilis*, *K. exigua*, *Torulaspora delbrueckii*, *Wickerhamomyces anomalus* and *Pichia kudriavzevii* (Syrokou et al. 2020).

The use of sourdough confers technological and functional benefits, most of which have been assigned to the metabolic activities of the microconsortium. Acidification is a key function of LAB; it enhances solubility of gluten and pentosans and modulates enzymatic activities. Especially regarding the latter, activation of endogenous or microbial phytases may lead to increased mineral bioavailability, while activation of endogenous or microbial proteolytic enzymes may modulate

S. Paramithiotis (✉)
Department of Biological Applications and Technology, University of Ioannina, Ioannina, Greece
e-mail: paramithiotis@uoi.gr

L. Bosnea · M. Mataragas
Department of Dairy Researchl, Institute of Technology of Agricultural Products, Hellenic Agricultural Organization "DEMETER", Ioannina, Greece

E. B. Ceresino et al. (eds.), *Sourdough Microbiota and Starter Cultures for Industry*, https://doi.org/10.1007/978-3-031-48604-3_13

celiac and non-celiac response and even lead to the decryption of bioactive peptides. In addition, production of bioactive compounds such as γ-aminobutyric acid, vitamins or phenolic compounds may increase the functional potential of the final product. Detection of these metabolic activities has been traditionally based on phenotype-based protocols. Indeed, a number of approaches are currently available for the qualitative and quantitative assessment of properties such as exopolysaccharide production (Smitinont et al. 1999), amylase (Bartkiene et al. 2017), protease (Di Cagno et al. 2002), lipase (Kamiloglu et al. 2020), and phytase activity (Anastasio et al. 2010) as well as the production of antimicrobial compounds (Syrokou et al. 2021). Advancements in the field of next generation sequencing enabled the detection of the genes that are related to the aforementioned activities to the genome of the isolates or metagenomic data. However, the results obtained by this approach may only be used as indicative and not conclusive; therefore, phenotypic assessment is still necessary. The aim of the present chapter is to collect all information related to the aforementioned activities of LAB and present them in a comprehensive and critical way.

13.2 Dough Acidification

Acidification is a very important function of sourdough LAB. The extent of acidification depends upon the mode of carbohydrate fermentation (homofermentative, heterofermentative), the presence of organic compounds that can be used as electron acceptors and the ability of the strains to utilize them. For example, utilization of fructose as an electron acceptor allows the production of additional ATP as well as regeneration of NAD^+ without ethanol formation being necessary. Instead, through the acetate kinase, additional acetic acid is formed (Gobbetti et al. 2005). Acidification may affect the solubility of structure forming components, such as gluten and pentosans, affecting dough softness, elasticity, and bread volume. Indeed, disaggregation and ionization of gluten proteins, which takes place at pH values away from its isoelectric point (6.2–6.4), enhances interaction with water and concomitantly affects the mechanical properties of the dough (Garg et al. 2019). On the other hand, the water holding capacity of pentosans, a property that is critical in rye breadmaking, is improved by acidification (Lindhauer and Dreisoerner 2003). In addition, acidification affects enzymatic activities and starch digestibility. The former will be further discussed in the following paragraphs. Regarding starch digestibility, the use of sourdough, especially with pH value lower than 4.0, is an effective way to reduce the glycemic index. Indeed, this has been reported in the case of bread made by wheat, rye and wholemeal barley. This effect has been assigned to a number of reasons, including the reduction of starch digestibility by lactic acid, the reduction of gelatinization degree and formation of resistant starch and the regulation of glucose metabolism by compounds liberated during sourdough fermentation (Liljeberg et al. 1995; Burton and Lightowler 2006; Katina et al. 2007; Nilsson et al. 2007; Gobbetti et al. 2014).

13.3 Enrichment with γ-Aminobutyric Acid

Gamma-aminobutyric acid (GABA) is an amino acid that holds very important physiological functions in plants, vertebrates, and microorganisms. In plants, it has been associated with their development and stress response (Kinnersley and Turano 2000; Podlesakova et al. 2019; Ramos-Ruiz et al. 2019), while in microorganisms, it directs physiological functions such as spore germination and resistance to acidic conditions (Foester and Foester 1973; Hao and Schmit 1993; Cotter and Hill 2003). In mammals, it is the major inhibitory neurotransmitter in the central nervous system; as such it is involved in the control of action potential through regulation of synaptic transmission (Ochoa-de la Paz et al. 2021). Therefore, enrichment of food commodities with GABA or *in situ* production through the use of GABA-producing microorganisms, has been extensively considered (Sun et al. 2022). However, it is worth mentioning that according to Hepsomali et al. (2020), a positive correlation between oral intake of GABA and improvement of the associated physiological functions, has not been established yet. Enrichment of a wide variety of lactic acid fermented food commodities with GABA, including dairy, soymilk, meat, fish, fruit, vegetable and sourdough products, through the use of GABA-producing starter cultures, has been assessed (Rizzello et al. 2008; Kim et al. 2009; Yu et al. 2017; Sanchart et al. 2017; El-Fattah et al. 2018; Lee et al. 2018; Venturi et al. 2019; Tanamool et al. 2020; Jin et al. 2021; Le et al. 2021; Moore et al. 2021; Wang et al. 2021; Galli et al. 2022; Woraratphoka et al. 2022).

Elucidation of the biochemical pathway through which GABA is produced, along with the genes that either encode for the necessary enzymes or play regulatory role is very important for two reasons; the first one refers to the capacity to evaluate the GABA-producing potential of microbial strains through genomic assessment and the second to revelation of opportunities for productivity improvement. In microorganisms, GABA production takes place principally via L-glutamate decarboxylation. This reaction is catalyzed by glutamic acid decarboxylase (GAD), in the presence of pyridoxal 5′-phosphate (PLP) as co-factor. L-glutamate may be present in the growth medium and transported into the cell through the glutamate/GABA antiporter (GadC) or result from the transamination of a-ketoglutarate and L-glutamine, a reaction catalyzed by glutamate synthase (GltS). The GABA that is produced may be degraded to succinic semialdehyde and subsequently to succinic acid, which enters the TCA cycle. GABA aminotransferase (GABA-AT) catalyzes the degradation of GABA to succinic semialdehyde and succinate semialdehyde dehydrogenase (SSADH) the conversion of succinic semialdehyde to succinic acid. Alternatively, GABA may be transported extracellularly, through the GadC. When GABA is present in the growth medium, it may also enter the cell through the GABA-specific transporter (GabP), degraded and enter the TCA cycle. GABA production through putrescine catabolism has also been reported (Kurihara et al. 2010; Schneider and Reitzer 2012). It is apparent that a set of functional genes is required for effective GABA production, the genetic and transcriptional organization of which has been studied to some extent. Occurrence of two homologous glutamate decarboxylases, namely GadA and GadB, has been reported (Lyu et al. 2018), which are encoded by *gadA* and *gadB*, respectively. Interestingly, genomic analysis of GABA-producing LAB revealed that *gadA*, when present, is always accompanied by *gadB* and occurrence of *gadB* is much more frequent. In addition, *gadB* is usually immediately downstream of *gadC*, the gene that encodes the glutamate/GABA antiporter. The transcription of the operon *gadC-gadB* is regulated by the positive transcriptional regulator GadR, which is encoded by *gadR* that is located immediately upstream of the *gadC*, forming the operon *gadRCB*. Genes *gltB* and *gltC*, which encode the two subunits of GltS, *gabT* that encodes the GABA-AT and *gabD* that encoded the SSADH are located elsewhere in the chromosome and no transcriptional correlation has been yet established with *gadRCB* (Su et al. 2011; Lyu et al. 2018; Cui et al. 2020).

Regarding the capacity of sourdough LAB to synthesize GABA, it has been characterized as a strain-dependent property and has been reported for strains belonging to the *C. farciminis*, *C. paralimentarius*, *Fr. sanfranciscensis*, *Furfurilactobacillus rossiae*, *Lp. paraplantarum*, *Lp. plantarum*, *Latilactobacillus graminis*, *Lt. curvatus*, *Leuconostoc mesenteroides*, *Leu. pseudomesenteroides*, *Leu. citreum*, *Lv. brevis*, *Lm. reuteri*, *Weissella cibaria* and *W. paramesenteroides* species (Pallin et al. 2016; Villegas et al. 2016; Demirbas et al. 2017; Venturi et al. 2019).

Occurrence of GABA in commercially available breads and artisanal sourdough breads has been reported (Diana et al. 2014). This is facilitated by the fact that baking results in the reduction of GABA concentration only in crust and not in crumb (Polak et al. 2021). In commercially available breads, it ranged between 1.57–3.95 mg/100 g while in artisanal sourdough breads between 2.01–8.84 mg/100 g. Apart from the already recognized importance of the biosynthetic capacity of the strains that participate in the sourdough microecosystem, the raw materials used seem to be equally important. Indeed, Coda et al. (2010a, b) reported that doughs made by pseudo-cereals and leguminous flours contained 7–78 mg/Kg GABA, without any prior incubation. The highest endogenous GABA concentration was reported at 40 and 78 mg/Kg for buckwheat and quinoa doughs, respectively. Fermentation of the doughs with *Lp. plantarum* C48 or *Lactococcus lactis* subsp. *lactis* PU1 resulted in the significant enhancement of GABA concentration. In the first case, the highest GABA concentration was reported at 654 and 643 mg/Kg, for amaranth and buckwheat sourdoughs, respectively, while in the latter case at 813 and 1031 mg/Kg, for amaranth and chickpea sourdoughs, respectively. The significance of the type of flour used for sourdough fermentation was also highlighted by Rizzello et al. (2008, 2010), Kim et al. (2019), Molfetta et al. (2021) and Polak et al. (2021). Enhancement of GABA accumulation through inoculation with GABA-producing strains was also studied by Bhanwar et al. (2013) and Venturi et al. (2019). In the first study *Lc. lactis* subsp. *lactis* was used for the preparation of 'Bhatura', a fried sourdough bread. *Vigna mungo* flour was used as source of glutamate resulting in the accumulation of 226.22 mg/100 g of GABA in the final product. Venturi et al. (2019) aimed to enhance GABA concentration using amaranth flour (20% of total flour weight) and strains with high GABA producing capacity, namely *Lv. brevis* A7 and *C. farciminis* A11. The highest GABA concentration, 39 mg/Kg, was noticed in the bread made with *C. farciminis* A11. However, sensory analysis revealed that addition of amaranth reduced the acceptability of the final product.

13.4 Improvement of Micronutrient Bioavailability

Phytic acid (myo-inositol hexakisphosphate) is the phosphorus warehouse of the cereal grains; around 50–80% of the total phosphorus is stored in that form. Phytic acid is also characterized by the ability to chelate multivalent metal ions, reducing, thus, their bioavailability. Hydrolysis of phytic acid by endogenous or microbial phytases has been proposed as a feasible solution to address this issue (Coulibaly et al. 2011). Indeed, Leenhardt et al. (2005) demonstrated that only a slight acidification (to pH 5.5) in the form of sourdough or addition of lactic acid, resulted in the decomposition of the 70% of the phytate of wholemeal wheat flour. In addition, the capacity of sourdough LAB to hydrolyze phytic acid has been assessed to some extent. Several strains belonging to *C. paralimentarius*, *Enterococcus casseliflavus*, *E. faecium*, *E. gallinarum*, *E. hermanniensis*, *E. hirae*, *E. mundtii*, *Lacticaseibacillus casei*, *La. paracasei*, *La. rhamnosus*, *Lc. lactis* subsp. *cremoris*, *Leu. citreum*, *Leu. mesenteroides*, *Lp. plantarum*, *Lp. pentosus*, *Lt. curvatus*, *Lt. graminis*, *Lt. sakei*, *Lv. brevis*, *Pediococcus acidilactici*, *P. pentosaceus*, *W. cibaria* and *W. confusa*, have been reported to possess such capacity (Carrizo et al. 2016; Ispirli et al. 2018; Karaman et al. 2018; Cakir et al. 2020; Fekria et al. 2020). Sourdough breadmaking has been repeatedly reported to result in phytate decomposition and increase of the cations amount. Lopez et al. (2001) compared the effect of yeast, sourdough and their combination to the phytate concentration of the resulting bread. Addition of sourdough resulted in the degradation of the 62% of phytate while the use of yeast to 38%. Phytate decomposition reached 90% and 40% when bran was incubated with the sourdough or the yeast, respectively. Phytate breakdown resulted in the increase of magnesium and phosphorus solubility (Lopez et al. 2001). Taking their study one step further, Lopez et al. (2003) fed Wistar rats with these breads and reported that sourdough bread was a better source of available minerals, especially iron, magnesium and zinc. Rodriguez-Ramiro et al. (2017) compared the effect on phytate content of wholemeal wheat bread of three baking processes, namely Chorleywood, conventional yeast and sourdough breadmaking as well as the impact on iron uptake by Caco-2 cells. The complete decomposition of phytate by the sourdough process and the 75% reduction of the concentration by Chorleywood and conventional yeast processes, was reported. However, no increase of ferritin formation, which was employed as a measure of iron uptake by Caco-2 cells, was observed. On the contrary, the bioaccessibility of exogenously added iron increased only after simulated gastrointestinal digestion of sourdough bread and not bread made by the other two processes. These results led to the conclusion that iron bioavailability increased after phytate degradation, but only upon combination of the sourdough bread with another iron source (Rodriguez-Ramiro et al. 2017). The effect of strain variability was highlighted by Karaman et al. (2018). In their study, the capacity of different strains to reduce phytate content during wholemeal wheat sourdough breadmaking was assessed; the reduction of phytate content reached 43.4% with the utilization of a *S. cerevisiae* and *P. pentosaceus* combination. The increase of the total amount of free calcium, zinc, iron and magnesium was also reported by De Angelis et al. (2019) in breads started with wholemeal wheat sourdoughs and made by either wholemeal soft or durum wheat flour. Finally, Fekria et al. (2020) reported that the utilization of a *Kluyveromyces marxianus* strain resulted in the reduction of phytate content of bread to 17.49 mg/5 g, which represents a significant reduction from the 51.49 mg/5 g of the flour.

13.5 Enrichment with Vitamins

The vitamin content of cereal grains depends upon cultivar, environmental conditions as well as agricultural practices (Garutti et al. 2022). The degree of refining seems to be the factor that plays decisive role regarding the amount of vitamins in the flour, since the majority of them are located in the germ and bran (Garg et al. 2021). Then, the amount of vitamins in the final product depends upon the biosynthetic capacity of the yeast and LAB (in the case of sourdough) strains, the fermentation time and baking conditions (Batifoulier et al. 2005; Gujska and Majewska 2005; Hjortmo et al. 2008; Capozzi et al. 2011; Mihhalevski et al. 2013).

In general, several LAB relevant to the sourdough microecosystem have been reported capable to produce vitamins of the B complex, among which *Fu. rossiae*, *Lm. reuteri*, *Lm, fermentum* and *Lp. plantarum* (Santos et al. 2008; Capozzi et al. 2011; de Angelis et al. 2014; Li et al. 2017; Tamene et al. 2019). However, application of such strains in sourdough breadmaking remains rather limited. Capozzi et al. (2011) isolated roseoflavin-resistant strains from sourdoughs and used two *Lp. plantarum* strains, together with baker's yeast, in standard breadmaking procedure. Accumulation of 6.81 μg/g riboflavin was reported, which was a significant improvement compared to the 2.41 μg/g that were accumulated when baker's yeast was used alone. Mihhalevski et al. (2013) studied the stability of B vitamins during rye sourdough breadmaking and reported that sourdough inoculated with *Lm. panis* N915 and incubated for 24 h contained more nicotinamide but less nicotinic and pantothenic acids compared to uninoculated rye flour. No statistically significant differences were observed in the amount of thiamine, riboflavin, pyridoxal and pyridoxine. Obviously, the effect of sourdough addition on the vitamin content of bread, is yet to be thoroughly explored.

13.6 Metabolism of Phenolic Compounds

The term phenolic compounds refers to a very large and diverse cluster of compounds that contain one or more phenol groups. They are plant secondary metabolites that serve a variety of important physiological functions, including signaling, pathogen inhibition etc. (Beckman 2000). Their classification has been the subject of intensive study and several classification schemes have been proposed (Singla et al. 2019). In cereals, they are located mainly in the bran and their amount depends upon the variety, environmental conditions and agricultural practices (Calinoiu and Vodnar 2018). They occur mostly in conjugated form or bound to cell wall polysaccharides.

A wide variety of health benefits have been attributed to the phenolic compounds of cereal grains. Their antioxidant capacity is probably the most studied one with a series of antioxidants being identified. For example, occurrence of phytosterols as well as tocopherol, caffeic, coumaric, ferulic, syringic and vanillic acid has been reported in wheat, while p-hydroxybenzoic, p-coumaric, caffeic, ferulic, protocatechuic, sinapic syringic and vanillic acids have been reported in rye (Nayak et al. 2015). In addition, their anti-inflammatory (Anson et al. 2011, 2012; Laddomada et al. 2015; Vitaglione et al. 2015), antimicrobial (Sanchez-Maldonado et al. 2011; Borges et al. 2012) and possible antidiabetic (Belobrajdic and Bird 2013) and anticancer capacity (Dai and Mumper 2010) has also been reported. However, since they mostly occur in conjugated form, their release may increase their bioavailability (Katina et al. 2012) and, possibly, enhance the associated health benefits. In addition, possible bioconversion, which may occur during sourdough fermentation, may generate volatile compounds that may affect the sensorial properties of bread (Rodríguez et al. 2009; Czerny and Schieberle 2002).

The fate of cereal phenolic compounds during sourdough breadmaking has been marginally studied. Research has focused on the metabolism of hydroxybenzoic and hydroxycinnamic acids, due to their relative abundance and technological importance (Gaenzle 2014). They may be decarboxylated, reduced, or both. In addition, esters of phenolic acids may be hydrolyzed through esterase activity (Rodríguez et al. 2009; Hole et al. 2012). So far, genomic screening of phenolic acid degradation capacity is based on the detection of *pdc*, which encodes for phenolic acid decarboxylase. Ripari et al. (2019), applied this screening approach and subsequently verified ferulic acid decomposition by *Lv. brevis*, *Lp. plantarum*, *La. manihotivorans* and *P. pentosaceus* strains. Then, strains with different phenolic acid decomposition capacity were used for whole wheat and rye malt sourdough fermentation. In whole wheat sourdough, ferulic acid was reported as the most abundant phenolic acid, occurring mainly in bound form, with syringic acid being the second most abundant, occurring both in free and conjugated form. In the case of rye malt sourdough, ferulic acid was also reported as the most abundant phenolic acid of all fractions while sinapic acid was very abundant among conjugated phenolic acids. In both cases, strains with decarboxylase and reductase activities (*Lp. plantarum* MAXIII and TMW1.460) resulted in degradation of the free ferulic acid, while use of esterase-positive strain (*Lv. hammesii* DSM 16381) resulted in the reduction of the bound ferulic acid concentration. The aforementioned complementarity was verified and seemed to be enhanced in co-fermentation experiments (Ripari et al. 2019). Konopka et al. (2014) prepared bread using white and whole meal wheat and rye flour, using either yeast or sourdough. Ferulic acid was the predominant phenolic compound, which was mostly in the bound form. The crumb and crust of the bread made with whole meal wheat or rye flour, contained more Folin-Ciocalteu reactive compounds and free ferulic acid that the respective made with white flour. Similarly, the crumb and crust of the bread made with the use of sourdough contained more Folin-Ciocalteu reactive compounds and free ferulic acid that the respective made with the use of baker's yeast. In all cases, the quantity of Folin-Ciocalteu reactive compounds and free ferulic acid in the crumb and crust of the bread made with rye was higher than the respective made with wheat, which was assigned to the initially higher content in rye flour. The latter was also verified by Skrajda-Brdak et al. (2019). Ravisankar et al. (2021) assessed the changes in flavonoids and extractable phenolic acids after spontaneous fermentation of wheat, sorghum and teff varieties. The flavonoid content of red and white wheat as well as white sorghum was qualitatively and quantitatively not affected by fermentation. Based on the compounds detected, this observation refers to the flavone aglycones or their di-C-linked glycosides. In addition, these compounds have been reported to withstand baking; therefore, their stability throughout sourdough breadmaking has been documented (Ravisankar et al. 2020). In this study, the partial hydrolysis of O-glycosides to their aglycones during rye sourdough fermentation was also reported. Significant changes were observed in the flavonoid content of lemon yellow sorghum as well as white and brown teff. More specifically, the significant reduction of the mixed O/C-linked flavone and flavone O-acyl glycosides, in white and brown teff sourdoughs was highlighted; the reduction of flavone O-acyl glycosides was more rapid compared to the respective of the mixed O/C-linked flavone. Significant reduction was also observed in the case of flavonoid-O-glycosides, which were the dominant class in lemon yellow sorghum. Regarding phenolic acids, the results obtained could only be explained through possible hydrolysis of cell walls accompanied by variable degree of free phenolic acids decomposition (Ravisankar et al. 2021). The importance of thermal treatment before fermentation of sorghum flour was highlighted by Ofosu et al. (2022). The authors reported that fermentation after wet or pressurized cooking resulted in different metabolic profile that in both cases enhanced ABTS and DPPH antioxidant capacity.

13.7 Proteolytic Activities

Regarding the proteolytic activities of the sourdough LAB, research has recently focused on the alleviation of disorders inflicted by gluten and on the decryption of bioactive peptides.

Gluten is the causative agent of a series of disorders that have been classified into three types: celiac disease, IgE-mediated allergy and non-celiac sensitivity (Cabanillas 2019).

Celiac disease is an immune-mediated disorder of the small intestine that is based on genetic predisposition but also significantly affected by environmental factors. In this condition, gluten consumed is only partially hydrolyzed by proteolytic enzymes of the digestive system; the resulting peptides contain T cell epitopes, triggering production of a variety of cytokines by innate and adaptive immune cells, which collectively results in epithelial damage (Sedda et al. 2020). Such is the case of the immunodominant 33-mer peptide from a2-gliadin that has been described as the most significant immunogenic sequence in gluten, since it contains three such epitopes (Schalk et al. 2017). Such epitopes have also been detected in γ-gliadin, ω-gliadin, hordein and secalin (Camarca et al. 2009, 2012; Tye-Din et al. 2010).

Gluten consumption may also trigger an allergenic inflammation through IgE-mediated release of inflammatory and vasoactive mediators upon recognition of specific allergenic epitopes (Sabenca et al. 2021). Such epitopes are prevalent in wheat and less common in rye and barley. Indeed, according to WHO/IUIS Allergen Nomenclature Sub-Committee, a total of 28 such proteins have been identified in wheat and only one and six in rye and barley, respectively.

Non-celiac gluten sensitivity includes any distress that is caused by gluten consumption and do not involve autoimmune or IgE-mediated responses. This situation has not drawn much scientific attention and therefore the molecular mechanisms implicated are yet to be fully elucidated.

The effect of sourdough utilization on the aforementioned conditions has been studied to some extent. It seems that the breakdown of wheat proteins is performed by both endogenous and microbial proteolytic activities. The endogenous proteolytic activity is largely dependent upon grain species and variety (Brijs et al. 1999) while the microbial proteolytic activity upon the microbial strain. Sourdough offers moisture and acidic pH, both of which activate aspartic proteases (Loponen et al. 2009). The latter increases prolamin solubility, which, along with the increased levels of thiol groups, generated through glutathione reductase activity, enhances their susceptibility to proteolysis (Gaenzle 2014). Di Cagno et al. (2004) reported the complete hydrolysis of the 33-mer peptide in wheat sourdough started with a mixture of living cells of *C. alimentarius* 15 M, *Lv. brevis* 14G, *Fr. sanfranciscensis* 7A and *Lentilactobacillus hilgardii* 51B and their cytoplasmic extracts. This sourdough was used in combination with oat, millet and buckwheat flours at technologically relevant proportion and the bread that was baked was tolerated by celiac disease patients. The bread made by incorporation of this sourdough also exhibited significant reduction of IgE-binding low molecular weight proteins and peptides (Rizzello et al. 2006). Similar results were obtained with the use of the VSL#3 preparation (De Angelis et al. 2007). More specifically, a sourdough was prepared with the use of the aforementioned microbial consortium, consisting of *Streptococcus thermophilus*, *Lp. plantarum*, *Lactobacillus acidophilus*, *La. casei*, *Lb. delbrueckii* spp. *bulgaricus*, *Bifidobacterium breve*, *B. longum* and *B. infantis*, which was subsequently freeze-dried and used for breadmaking, leading to significant degradation of wheat proteins, including some IgE-binding ones. The efficient hydrolysis of the 33-mer by a combination of the aforementioned strains with 6 *Fr. sanfranciscensis* strains and proteases by *Aspergillus oryzae* and *A. niger* was reported by Rizzello et al. (2007). The sourdough was then freeze-dried and combined with structural agents to form a bread loaf. The same approach was employed by De Angelis et al. (2010) for fermentation of durum wheat semolina, which was subsequently freeze-dried and used for pasta making. The pasta made by this procedure was sensorially superior to conventional gluten-free one.

Germination is a process that activates a variety of endogenous proteolytic enzymes. Indeed, occurrence of cysteine proteinase, serine proteinase, metallo proteinase and serine carboxypeptidases I-V, has been reported in germinated wheat grain (Gaenzle et al. 2008). The use of germinated seeds to create products suitable with patients with celiac disease has occurred but need further study (Loponen et al. 2009).

Bioactive peptides are short peptides that are liberated from the parent protein through the activity of endogenous, microbial or proteolytic enzymes of the digestive system. The term bioactive refers to the specific biological activity that is revealed upon their decryption, which, in most of the cases, is different from the respective of the parent protein. The most studied such activities are anti-diabetic, −oxidant, −microbial, −thrombotic, hypocholesterolemic, hypotensive, mineral binding, opioid and antiopioid ones. Occurrence of such peptides has also been reported in cereals. More specifically, Hu et al. (2011) reported that occurrence of the ACE inhibitory peptides LQP, LLP, VPP and IPP in wheat, rye and barley prolamins was indicated through bioinformatic analysis. Furthermore, their amount in sourdough was reported to depend upon the proteolytic capacity of the LAB strains as well as the endogenous proteolytic activity, which may be enhanced either by the use of germinated grains or exogenous addition of fungal proteases. Rizzello et al. (2017) reported the detection of 5 peptides with antioxidant capacity in sourdough made with quinoa flour and *Lp. plantarum* T0A10. Luti et al. (2020) reported the decryption of a series of bioactive peptides with antioxidant and/or anti-inflammatory activity from wheat flour, due to the proteolytic capacity of *C. farciminis* strains A11 and H3 and *Fr. sanfranciscensis* strain I4. Interestingly, the peptides were reported to retain their biological activity even after baking.

13.8 Conclusions and Future Perspectives

The health promoting features of sourdough LAB have attracted significant scientific attention over the last decades. As a result, our knowledge regarding their metabolic activities as well as the specific conditions under which they take place, has been enriched. At the same time, novel approaches including the use of genomic and metagenomic data, have improved our understanding of their metabolic capacity and activities at single culture and microecosystem level. However, several research gaps are still evident, directing future research activities. Among them, metabolism of phenolic compounds and alleviation of gluten adverse effects, are very likely to draw scientific attention in the near future. The functional potential of sourdough bread has been adequately highlighted. However, further research is still necessary in order to elucidate LAB metabolic activities within the specific sourdough microenvironment that would lead to further improvement of our knowledge regarding the biotic and abiotic factors that direct them, optimization of the production of these functional metabolites and possibly further improve the level of standardization, which would collectively lead to further enhancement of sourdough bread quality.

References

Anastasio M, Pepe O, Cirillo T et al (2010) Selection and use of phytate-degrading LAB to improve cereal-based products by mineral solubilization during dough fermentation. J Food Sci 75:M28–M35

Anson NM, Aura A-M, Selinheimo E et al (2011) Bioprocessing of wheat bran in whole wheat bread increases the bioavailability of phenolic acids in men and exerts antiinflammatory effects ex vivo. J Nutr 141:137–143

Anson NM, Hemery YM, Bast A et al (2012) Optimizing the bioactive potential of wheat bran by processing. Food Funct 3:362–375

Bartkiene E, Bartkevics V, Pugajeva I et al (2017) The contribution of *P. acidilactici*, *L. plantarum*, and *L. curvatus* starters and L-(+)-lactic acid to the acrylamide content and quality parameters of mixed rye—wheat bread. LWT-Food Sci Technol 80:43–50

Batifoulier F, Verny M-A, Chanliaud E et al (2005) Effect of different breadmaking methods on thiamine, riboflavin and pyridoxine contents of wheat bread. J Cereal Sci 42:101–108

Beckman CH (2000) Phenolic-storing cells: keys to programmed cell death and periderm formation in wilt disease resistance and in general defence responses in plants? Physiol Mol Plant Pathol 57:101–110

Belobrajdic DP, Bird AR (2013) The potential role of phytochemicals in wholegrain cereals for the prevention of type-2 diabetes. Nutr J 12:62

Bhanwar S, Bamnia M, Ghosh M et al (2013) Use of *Lactococcus lactis* to enrich sourdough bread with γ-aminobutyric acid. Int J Food Sci Nutr 64:77–81

Borges A, Saavedra MJ, Simões M (2012) The activity of ferulic and gallic acids in biofilm prevention and control of pathogenic bacteria. Biofouling 28:755–767

Brijs K, Bleukx W, Delcour JA (1999) Proteolytic activities in dormant rye (*Secale cereale* L.) grain. J Agric Food Chem 47:3572–3578

Burton P, Lightowler HJ (2006) Influence of bread volume on glycaemic response and satiety. Br J Nutr 96:877–882

Cabanillas B (2019) Gluten-related disorders: celiac disease, wheat allergy, and nonceliac gluten sensitivity. Crit Rev Food Sci Nutr 60:2606–2621

Cakir E, Arici M, Durak MZ (2020) Biodiversity and techno-functional properties of lactic acid bacteria in fermented hull-less barley sourdough. J Biosci Bioeng 130:450–456

Calinoiu LF, Vodnar DC (2018) Whole grains and phenolic acids: a review on bioactivity, functionality, health benefits and bioavailability. Nutrients 10:1615

Camarca A, Anderson RP, Mamone G et al (2009) Intestinal t cell responses to gluten peptides are largely heterogeneous: implications for a peptide-based therapy in celiac disease. J Immunol 182:4158–4166

Camarca A, Del Mastro A, Gianfrani C (2012) Repertoire of gluten peptides active in celiac disease patients: perspectives for translational therapeutic applications. Endocr Metab Immune Disord Drug Targets 12:207–219

Capozzi V, Menga V, Digesu AM et al (2011) Biotechnological production of vitamin B2-enriched bread and pasta. J Agric Food Chem 59:8013–8020

Carrizo SL, Montes de Oca CE, Laiño JE et al (2016) Ancestral Andean grain quinoa as source of lactic acid bacteria capable to degrade phytate and produce B-group vitamins. Food Res Int 89:488–494

Coda R, Rizzello CG, Gobbetti M (2010a) Use of sourdough fermentation and pseudo-cereals and leguminous flours for the making of a functional bread enriched of γ-aminobutyric acid (GABA). Int J Food Microbiol 137:236–245

Coda R, Di Cagno R, Edema MO et al (2010b) Exploitation of Acha (*Digitaria exiliis*) and Iburu (*Digitaria iburua*) flours: chemical characterization and their use for sourdough fermentation. Food Microbiol 27:1043–1050

Cotter PD, Hill C (2003) Surviving the acid test: responses of gram-positive bacteria to low pH. Microbiol Mol Biol Rev 67:429–453

Coulibaly A, Kouakou B, Chen J (2011) Phytic acid in cereal grains: structure, healthy or harmful ways to reduce phytic acid in cereal grains and their effects on nutritional quality. Am J Plant Nutr Fert Technol 1:1–22

Cui Y, Miao K, Niyaphorn S et al (2020) Production of gamma-aminobutyric acid from lactic acid bacteria: a systematic review. Int J Mol Sci 21:995

Czerny M, Schieberle P (2002) Important aroma compounds in freshly ground whole meal and white wheat flour – identification and quantitative changes during sourdough fermentation. J Agric Food Chem 50:6835–6840

Dai J, Mumper RJ (2010) Plant phenolics: extraction, analysis and their antioxidant and anticancer properties. Molecules 15:7313–7352

De Angelis M, Rizzello CG, Scala E et al (2007) Probiotic preparation has the capacity to hydrolyze wheat protein responsible for food allergy. J Food Prot 70:135–144

De Angelis M, Cassone A, Rizzello CG et al (2010) Mechanism of degradation of immunogenic gluten epitopes from *Triticum turgidum* L. var. *durum* by sourdough lactobacilli and fungal proteases. Appl Environ Microbiol 76:508–518

de Angelis M, Bottacini F, Fosso B et al (2014) *Lactobacillus rossiae*, a vitamin B12 producer, represents a metabolically versatile species within the genus *Lactobacillus*. PLoS One 9:e107232

De Angelis M, Minervini F, Siragusa S et al (2019) Wholemeal wheat flours drive the microbiome and functional features of wheat sourdoughs. Int J Food Microbiol 302:35–46

De Vuyst L, Van Kerrebroeck S, Leroy F (2017) Microbial ecology and process technology of sourdough fermentation. Adv Appl Microbiol 100:49–160

Demirbas F, Ispirli H, Kurnaz AA et al (2017) Antimicrobial and functional properties of lactic acid bacteria isolated from sourdoughs. LWT – Food Sci Technol 79:361–366

Di Cagno R, De Angelis M, Lavermicocca P et al (2002) Proteolysis by sourdough lactic acid bacteria: effects on wheat flour protein fractions and gliadin peptides involved in human cereal intolerance. Appl Environ Microbiol 68:623–633

Di Cagno R, De Angelis M, Auricchio S et al (2004) Sourdough bread made from wheat and nontoxic flours and started with selected lactobacilli is tolerated in celiac sprue patients. Appl Environ Microbiol 70:1088–1096

Diana M, Rafecas M, Quílez J (2014) Free amino acids, acrylamide and biogenic amines in gamma-aminobutyric acid enriched sourdough and commercial breads. J Cereal Sci 60:639–644

El-Fattah AA, Sakr S, El-Dieb S et al (2018) Developing functional yogurt rich in bioactive peptides and gamma-aminobutyric acid related to cardiovascular health. LWT – Food Sci Technol 98:390–397

Fekria A, Torbatia M, Khosrowshahib AY et al (2020) Functional effects of phytate-degrading, probiotic lactic acid bacteria and yeast strains isolated from Iranian traditional sourdough on the technological and nutritional properties of whole wheat bread. Food Chem 306:125620

Foester CW, Foester HF (1973) Glutamic acid decarboxylase in spores of *Bacillus megaterium* and its possible involvement in spore germination. J Bacteriol 114:1090–1098
Gaenzle MG (2014) Enzymatic and bacterial conversions during sourdough fermentation. Food Microbiol 37:2–10
Gaenzle MG, Loponen J, Gobbetti M (2008) Proteolysis in sourdough fermentations: mechanisms and potential for improved bread quality. Trends Food Sci Technol 19:513–521
Galli V, Venturi M, Mari E et al (2022) Gamma-aminobutyric acid (GABA) production in fermented milk by lactic acid bacteria isolated from spontaneous raw milk fermentation. Int Dairy J 127:105284
Garg S, Cran MJ, Mishra VK (2019) Effect of heating and acidic pH on characteristics of wheat gluten suspension. Int J Food Sci Technol 54:1892–1900
Garg M, Sharma A, Vats S et al (2021) Vitamins in cereals: a critical review of content, health effects, processing losses, bioaccessibility, fortification, and biofortification strategies for their improvement. Front Nutr 8:586815
Garutti M, Nevola G, Mazzeo R et al (2022) The impact of cereal grain composition on the health and disease outcomes. Front Nutr 9:888974
Gobbetti M, De Angelis M, Corsetti A et al (2005) Biochemistry and physiology of sourdough lactic acid bacteria. Trends Food Sci Technol 16:57–69
Gobbetti M, Rizzello CG, Di Cagno R et al (2014) How the sourdough may affect the functional features of leavened baked goods. Food Microbiol 37:30–40
Gujska E, Majewska K (2005) Effect of baking process on added folic acid and endogenous folates stability in wheat and rye breads. Plant Foods Hum Nutr 60:37–42
Hao R, Schmit JC (1993) Cloning of the gene for glutamate decarboxylase and its expression during conidiation in *Neurospora crassa*. Biochem J 293:735–738
Hepsomali P, Groeger JA, Nishihira J et al (2020) Effects of oral gamma-aminobutyric acid (GABA) administration on stress and sleep in humans: a systematic review. Front Neurosci 14:923
Hjortmo S, Patring J, Jastrebova J et al (2008) Biofortification of folates in white wheat bread by selection of yeast strain and process. Int J Food Microbiol 127:32–36
Hole AS, Rud I, Grimmer S et al (2012) Improved bioavailability of dietary phenolics in whole grain barley and oat groat following fermentation with probiotic *Lactobacillus acidophilus*, *Lactobacillus johnsonii*, and *Lactobacillus reuteri*. J Agric Food Chem 60:6369–6375
Hu Y, Stromeck A, Loponen J et al (2011) LCMS/ MS quantification of bioactive antiotensin I-converting enzyme inhibitory peptides in rye malt sourdoughs. J Agric Food Chem 59:11983–11989
Ispirli H, Demirbaş F, Yüzer MO et al (2018) Identification of lactic acid bacteria from spontaneous rye sourdough and determination of their functional characteristics. Food Biotechnol 32:222–235
Jin YH, Hong JH, Lee J-H et al (2021) Lactic acid fermented green tea with *Levilactobacillus brevis* capable of producing γ-aminobutyric acid. Fermentation 7:110
Kamiloglu A, Kaban G, Kaya M (2020) Technological properties of autochthonous *Lactobacillus plantarum* strains iso-lated from sucuk (Turkish dry-fermented sausage). Braz J Microbiol 51:1279–1287
Karaman K, Sagdic O, Durak MZ (2018) Use of phytase active yeasts and lactic acid bacteria isolated from sourdough in the production of whole wheat bread. LWT – Food Sci Technol 91:557–567
Katina K, Laitila A, Juvonen R et al (2007) Bran fermentation as a means to enhance technological properties and bioactivity of rye. Food Microbiol 24:175–186
Katina K, Juvonen R, Laitila A et al (2012) Fermented wheat bran as a functional ingredient in baking. Cereal Chem 89:126–134
Kim JY, Lee MY, Ji GE et al (2009) Production of γ-aminobutyric acid in black raspberry juice during fermentation by *Lactobacillus brevis* GABA100. Int J Food Microbiol 130:12–16
Kim S-Y, Kim KJ, Chung H-C et al (2019) Use of rice bran for preparation of GABA (γ-aminobutyric acid)-rich sourdough. Food Sci Technol Res 25:399–404

Kinnersley AM, Turano FJ (2000) Gamma aminobutyric acid (GABA) and plant responses to stress. Crit Rev Plant Sci 19:479–509

Konopka I, Tańska M, Faron A et al (2014) Release of free ferulic acid and changes in antioxidant properties during the wheat and rye bread making process. Food Sci Biotechnol 23:831–840

Kurihara S, Kato K, Asada K et al (2010) A putrescine-inducible pathway comprising PuuE-YneI in which γ-aminobutyrate is degraded into succinate in *Escherichia coli* K-12. J Bacteriol 192:4582–4591

Laddomada B, Durante M, Minervini F et al (2015) Phytochemical composition and anti-inflammatory activity of extracts from the whole-meal flour of Italian durum wheat cultivars. Int J Mol Sci 16:3512–3527

Le PH, Parmentier N, Le TT et al (2021) Evaluation of using a combination of enzymatic hydrolysis and lactic acid fermentation for γ-aminobutyric acid production from soymilk. LWT – Food Sci Technol 142:111044

Lee KW, Shim JM, Yao Z et al (2018) Properties of kimchi fermented with GABA-producing lactic acid bacteria as a starter. J Microbiol Biotechnol 28:534–541

Leenhardt F, Levrat-Verny M-A, Chanliaud E et al (2005) Moderate decrease of pH by sourdough fermentation is sufficient to reduce phytate content of whole wheat flour through endogenous phytase activity. J Agric Food Chem 53:98–102

Li P, Gu Q, Yang L et al (2017) Characterization of extracellular vitamin B12 producing *Lactobacillus plantarum* strains and assessment of the probiotic potentials. Food Chem 234:494–501

Liljeberg H, Lönner C, Björck I (1995) Sourdough fermentation or addition of organic acids or corresponding salts to bread improves nutritional properties of starch in healthy humans. J Nutr 125:1503–1511

Lindhauer MG, Dreisoerner J (2003). Rye. In: Caballero B, Trugo L, Finglas PM (eds) Encyclopedia of food sciences and nutrition, 2nd edn. Academic Press, FL, USA, pp. 5027–5031

Lopez HW, Krespine V, Guy C et al (2001) Prolonged fermentation of whole wheat sourdough reduces phytate level and increases soluble magnesium. J Agric Food Chem 49:2657–2662

Lopez HW, Duclos V, Coudray C et al (2003) Making bread with sourdough improves mineral bioavailability from reconstituted whole wheat flour in rats. Nutrition 19:524–530

Loponen J, Kanerva P, Zhang C et al (2009) Prolamin hydrolysis and pentosane solubilization in germinated-rye sourdoughs determined by chromatography and immunological methods. J Agric Food Chem 57:746–753

Luti S, Mazzoli L, Ramazzotti M et al (2020) Antioxidant and anti-inflammatory properties of sourdoughs containing selected Lactobacilli strains are retained in bread. Food Chem 322:126710

Lyu C, Zhao W, Peng C et al (2018) Exploring the contributions of two glutamate decarboxylase isozymes in *Lactobacillus brevis* to acid resistance and γ-aminobutyric acid production. Microb Cell Factories 17:180

Mihhalevski A, Nisamedtinov I, Hälvin K et al (2013) Stability of B-complex vitamins and dietary fiber during rye sourdough bread production. J Cereal Sci 57:30–38

Molfetta M, Celano G, Minervini F (2021) Functional, nutritional, and sensory quality of mixed flours-based breads as compared to durum wheat semolina-based breads. Foods 10:1613

Moore JF, DuVivier R, Johanningsmeier SD (2021) Formation of γ-aminobutyric acid (GABA) during the natural lactic acid fermentation of cucumber. J Food Compos Anal 96:103711

Nayak B, Liu RH, Tang J (2015) Effect of processing on phenolic antioxidants of fruits, vegetables, and grains—a review. Crit Rev Food Sci Nutr 55:887–918

Nilsson M, Holst JJ, Björck IME (2007) Metabolic effects of amino acid mixtures and whey protein in healthy subjects: studies using glucose-equivalent drinks. Am J Clin Nutr 85:996–1004

Ochoa-de la Paz L, Gulias-Canizo R, Ruiz-Leyja EA et al (2021) The role of GABA neurotransmitter in the human central nervous system, physiology, and pathophysiology. Rev Mex de Neurocienc 22:67–76

Ofosu FK, Elahi F, Daliri EB-M et al (2022) Impact of thermal treatment and fermentation by lactic acid bacteria on sorghum metabolite changes, their antioxidant and antidiabetic activities. Food Biosci 45:101502

Pallin A, Agback P, Jonsson H et al (2016) Evaluation of growth, metabolism and production of potentially bioactive components during fermentation of barley with *Lactobacillus reuteri*. Food Microbiol 57:159–171

Podlesakova K, Ugena L, Spichal L et al (2019) Phytohormones and polyamines regulate plant stress responses by altering GABA pathway. New Biotechnol 48:53–65

Polak T, Mejaš R, Jamnik P et al (2021) Accumulation and transformation of biogenic amines and gamma-aminobutyric acid (GABA) in chickpea sourdough. Foods 10:2840

Ramos-Ruiz R, Martinez F, Knauf-Beiter G (2019) The effects of GABA in plants. Cogent Food Agric 5:1670553

Ravisankar S, Queiroz VAV, Awika JM (2020) Rye flavonoids – structural profile of the flavones in diverse varieties and effect of fermentation and heat on their structure and antioxidant properties. Food Chem 324:126871

Ravisankar S, Dizlek H, Awika JM (2021) Changes in extractable phenolic profile during natural fermentation of wheat, sorghum and teff. Food Res Int 145:110426

Ripari V, Bai Y, Gaenzle MG (2019) Metabolism of phenolic acids in whole wheat and rye malt sourdoughs. Food Microbiol 77:43–51

Rizzello CG, De Angelis M,·Coda R et al (2006) Use of selected sourdough lactic acid bacteria to hydrolyze wheat and rye proteins responsible for cereal allergy. Eur Food Res Technol 223:405–411

Rizzello CG, De Angelis M, Di Cagno R et al (2007) Highly efficient gluten degradation by lactobacilli and fungal proteases during food processing: new perspectives for celiac disease. Appl Environ Microbiol 73:4499–4507

Rizzello CG, Cassone A, Di Cagno R et al (2008) Synthesis of angiotensin I-converting enzyme (ACE)-inhibitory peptides and γ-aminobutyric acid (GABA) during sourdough fermentation by selected lactic acid bacteria. J Agric Food Chem 56:6936–6943

Rizzello CG, Nionelli L, Coda R et al (2010) Effect of sourdough fermentation on stabilisation, and chemical and nutritional characteristics of wheat germ. Food Chem 119:1079–1089

Rizzello CG, Lorusso A, Russo V et al (2017) Improving the antioxidant properties of quinoa flour through fermentation with selected autochthonous lactic acid bacteria. Int J Food Microbiol 241:252–326

Rodríguez H, Curiel JA, Landete JM et al (2009) Food phenolics and lactic acid bacteria. Int J Food Microbiol 132:79–90

Rodriguez-Ramiro I, Brearley CA, Bruggraber SFA et al (2017) Assessment of iron bioavailability from different bread making processes using an in vitro intestinal cell model. Food Chem 228:91–98

Sabenca C, Ribeiro M, Sousa T et al (2021) Wheat/gluten-related disorders and gluten-free diet misconceptions: a review. Foods 10:1765

Sanchart C, Rattanaporn O, Haltrich D et al (2017) Enhancement of gamma-aminobutyric acid (GABA) levels using an autochthonous *Lactobacillus futsaii* CS3 as starter culture in Thai fermented shrimp (Kung-Som). World J Microbiol Biotechnol 33:152

Sanchez-Maldonado AF, Schieber A, Gänzle MG (2011) Structure–function relationships of the antibacterial activity of phenolic acids and their metabolism by lactic acid bacteria. J Appl Microbiol 111:1176–1184

Santos F, Vera JL, van der Hejden R et al (2008) The complete coenzyme B12 biosynthesis gene cluster of *Lactobacillus reuteri* CRL 1098. Microbiology 154:81–93

Schalk K, Lang C, Wieser H et al (2017) Quantitation of the immunodominant 33-mer peptide from a-gliadin in wheat flours by liquid chromatography tandem mass spectrometry. Sci Rep 7(1):45092

Schneider BL, Reitzer L (2012) Pathway and enzyme redundancy in putrescine catabolism in *Escherichia coli*. J Bacteriol 194:4080–4088

Sedda S, Dinallo V, Marafini I et al (2020) mTOR sustains inflammatory response in celiac disease. Sci Rep 10:10798

Singla RK, Dubey AK, Garg A et al (2019) Natural polyphenols: chemical classification, definition of classes, subcategories, and structures. J AOAC Int 102:1397–1400
Skrajda-Brdak M, Konopka I, Tańska M et al (2019) Changes in the content of free phenolic acids and antioxidative capacity of wholemeal bread in relation to cereal species and fermentation type. Eur Food Res Technol 245:2247–2256
Smitinont T, Tansakul C, Keeratipibul S et al (1999) Exopolysaccharide-producing lactic acid bacteria strains from traditional fermented food: isolation, identification and exopolysaccharide characterization. Int J Food Microbiol 15:105–111
Su MS, Schlicht S, Gänzle MG (2011) Contribution of glutamate decarboxylase in *Lactobacillus reuteri* to acid resistance and persistence in sourdough fermentation. Microb Cell Factories 10:S8
Sun Y, Mehmood A, Battino M et al (2022) Enrichment of gamma-aminobutyric acid in foods: from conventional methods to innovative technologies. Food Res Int 162:111801
Syrokou MK, Themeli C, Paramithiotis S et al (2020) Microbial ecology of Greek wheat sourdoughs identified by culture-dependent and culture-independent approach. Foods 9:1603
Syrokou MK, Tziompra S, Psychogiou E-E et al (2021) Technological and safety attributes of lactic acid bacteria and yeasts isolated from spontaneously fermented Greek wheat sourdoughs. Microorganisms 9:671
Tamene A, Baye K, Kariluoto S et al (2019) *Lactobacillus plantarum* P2R3FA isolated from traditional cereal-based fermented food increase folate status in deficient rats. Nutrients 11:2819
Tanamool V, Hongsachart P, Soemphol W (2020) Screening and characterisation of gamma-aminobutyric acid (GABA) producing lactic acid bacteria isolated from Thai fermented fish (Plaa-som) in Nong Khai and its application in Thai fermented vegetables (Som-pak). Food Sci Technol (Campinas) 40:483–490
Tye-Din JA, Stewart JA, Dromey JA et al (2010) Comprehensive, quantitative mapping of T cell epitopes in gluten in celiac disease. Sci Transl Med 2(41):41ra51
Venturi M, Galli V, Pini N et al (2019) Use of selected lactobacilli to increase γ-aminobutyric acid (GABA) content in sourdough bread enriched with amaranth flour. Foods 8:218
Villegas JM, Brown L, de Giori GS et al (2016) Optimization of batch culture conditions for GABA production by *Lactobacillus brevis* CRL 1942, isolated from quinoa sourdough. LWT – Food Sci Technol 67:22–26
Vitaglione P, Mennella I, Ferracane R et al (2015) Whole-grain wheat consumption reduces inflammation in a randomized uncontrolled trial on overweight and obese subjects with unhealthy dietary and lifestyle behaviors: role of polyphenols bound to cereal dietary fiber. Am J Clin Nutr 101:251–261
Wang D, Wang Y, Lan H et al (2021) Enhanced production of γ-aminobutyric acid in litchi juice fermented by *Lactobacillus plantarum* HU-C2W. Food Biosci 42:101155
Woraratphoka J, Innok S, Soisungnoen P et al (2022) γ-Aminobutyric acid production and antioxidant activities in fresh cheese by *Lactobacillus plantarum* L10-11. Food Sci Technol (Campinas) 42:e03121
Yu H-H, Choi JH, Kang KM et al (2017) Potential of a lactic acid bacterial starter culture with gamma-aminobutyric acid (GABA) activity for production of fermented sausage. Food Sci Biotechnol 26:1333–1341

Chapter 14
Rope and Mould Spoilage in Breadmaking – Risk Analysis

M. Piotrowska, A. Nowak, and A. Koziróg

14.1 Introduction

Bread has been the basis of human nutrition for centuries. Bread perfectly satisfies hunger, is easily divisible, has a relatively long shelf life and a unique taste and aroma. Bakery products are a source of quite a large dose of energy, and also contain many nutrients: proteins, fats and fat-soluble vitamins, B vitamins and minerals. The global market for bread and bakery products grew from USD 226.25 billion in 2022 to USD 241.69 billion in 2023, an increase of 6.8%. It is estimated that due to the growing world population, in 2027 this market will amount to USD 304.89 billion.

The production of bread and bakery products is widespread all over the world. Nevertheless, bread is the most wasted product in Europe. The global production of bread in 2016 amounted to about 130 million tons, but losses of bakery products worldwide amounted to about 10 million tons (Demirci et al. 2016). According to Goryńska-Goldmann et al. (2020) losses occur mainly in the production section and during the transport of final products (Goryńska-Goldmann et al. 2020). For this reason, in order to effectively reduce the risk of losses, it is necessary to know what causes them using risk analysis for this purpose.

Risk analysis is a systematic approach to the study and assessment of public health and food safety hazards. It is used to formulate, implement and communicate risk management decisions and consists of three interrelated elements – risk assessment, risk management and risk communication. In 1963 Food and Agriculture Organization (FAO) and World Health Organization (WHO) prepared the Codex

M. Piotrowska (✉) · A. Nowak · A. Koziróg
Faculty of Biotechnology and Food Sciences, Institute of Fermentation Technology and Microbiology, Lodz University of Technology, Łódź, Poland
e-mail: malgorzata.piotrowska@p.lodz.pl; agnieszka.nowak@p.lodz.pl; anna.kozirog@p.lodz.pl

E. B. Ceresino et al. (eds.), *Sourdough Microbiota and Starter Cultures for Industry*, https://doi.org/10.1007/978-3-031-48604-3_14

which contains international food standards and guidelines under the joint FAO/WHO Food Standards Programme. Since then, it has been systematically replenished and its main goal is to protect the health of consumers and ensure fair practices in global food trade (WHO 2011).

Risk in food production is the danger of negative health effects and the severity of such effects as a result of the hazard. Elements of the risk assessment under food law include:

- hazard identification – identification of potential biological, chemical and physical hazards and their impact on human health. Basic data sources – epidemiological, microbiological and clinical studies.
- hazard characterization – conducting a qualitative analysis (determining the importance and probability of a hazard) and quantitative (determining the dose of the consumed product).
- exposure assessment – determining the size of microbiological and chemical hazards in food (then it is possible to determine the size of the dose for consumption of these factors in a given population).
- risk characterization – collecting prior information and on this basis determining possible adverse health effects for a given population (Mayes and Mortimore 2009; Dury et al. 2019; Koutsoumanis et al. 2021).

According to WHO (2022) data, almost 600 million people in the world fall ill per year after eating contaminated food, 420,000 of whom die (WHO 2022). The level of food poisoning caused by the consumption of bakery products is low compared to other types of food. However, microorganisms (bacteria, yeast, moulds) can cause baked goods to spoil, which in turn leads to food safety problems.

14.2 Bakery Products Contamination

Contaminated bakery products contribute not only to human poisoning, but also cause huge food waste and economic losses for producers and consumers. Microbial spoilage is often the main factor limiting the shelf life of high and intermediate moisture bakery products ($a_w > 0.6$) and is also a major cause of economic loss in the baking industry. In small and medium-sized bakeries, the waste pose to 1.5% of the produced mass of bakery products, in bigger bakeries it is even 5% (Saranraj and Sivasakthivelan 2016; Melini and Melini 2018; Axel et al. 2017; Melikoglu and Webb 2013).

In bakery products, the overgrowth of microorganisms can occur at various stages of production, starting with grain harvested from the field, through processing, packaging, and storage. Identification of hazards in the case of bakery products does not only apply to the raw materials used for baking (Fig. 14.1), but also to the entire process in accordance with the "from farm to fork" principle.

During the production of semi-finished products and dough, microbiological hazards may result from e.g.:

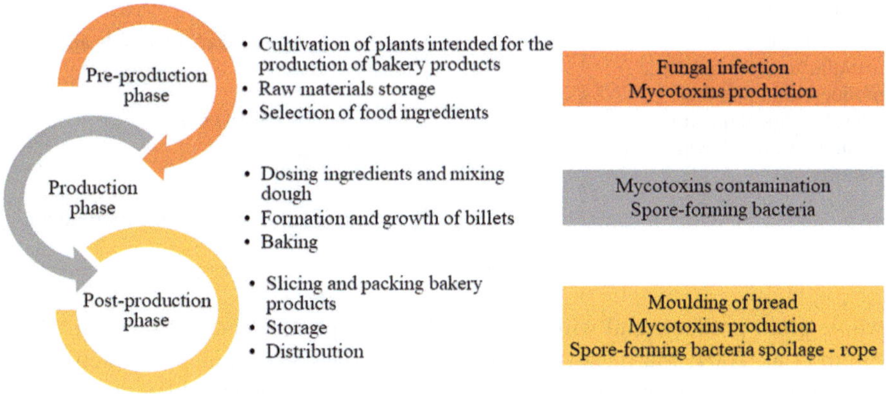

Fig. 14.1 Bread production stages, taking into account areas with the highest potential risk of microbiological infections. (Source: Authors' own elaboration)

- failure to comply with the principle of stronger leavening of the dough when using flour infected with bacteria of the *Bacillus* genus,
- the use of bread with signs of mould coming from returns, for the preparation of acids and baker's sours,
- incorrect hygiene rules – use of boards for proving pieces with traces of dried dough, proving baskets with traces of mould,
- incorrect conditions in the billet growth chambers,
- poor microbiological quality of the air.

The implementation of the HACCP system (Hazard Analysis and Critical Control Points) and the related risks analysis in the entire food chain and the systematic raising of awareness of people related to it (from the farmer to the seller) is an important element in the production of high-quality food. The health quality is particularly important in the case of food items consumed commonly, and regularly throughout the entire lifetime, and bread is one of them (Garcia et al. 2019; Miśniakiewicz 2011; Czerwinska and Piotrowski 2009).

The risk of microbiological contamination of bakery products depends on their type and stage of production as well as the addition of preservatives. The chemical composition of raw materials and other ingredients of the dough, high water content (approx. 40%) and pH of about 6 favour the growth of microorganisms, but the high temperature during baking eliminates most of them. However, fungal metabolites and spore-forming bacteria may remain in the bread (Özçakmak 2019; Yibar et al. 2012). Therefore, the microbial contamination is mainly secondary in the post-baking stage. The most susceptible to contamination is white bread and wholegrain bread due to its fairly low pH, around 5.6, and high water activity of 0.96–0.98. The growth of microorganisms will be prevented by the reduced water activity (aw) or low pH. Such parameters are characteristic for some bakery products, such as e.g. sourdough bread, pH = 4.2–4.6; cookies and crackers aw = 0.2–0.3. Currently, the process of bread spoilage is favoured by the increase in consumer demand for

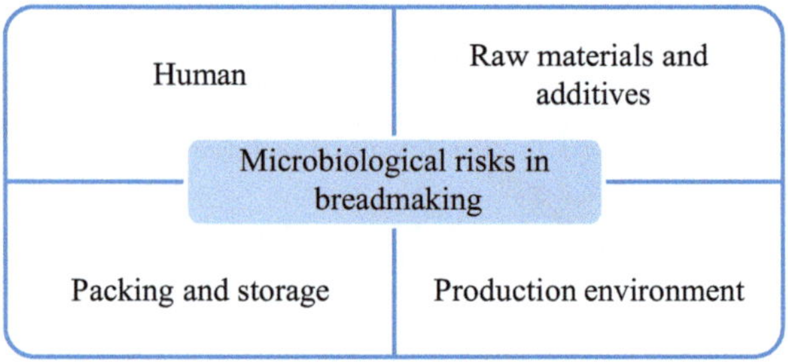

Fig. 14.2 The main factors in the manufacturing of bakery products affecting their microbiological quality. (Source: Authors' own elaboration)

whole-grain products and without the addition of preservatives (Axel et al. 2017; Smith et al. 2004; Martinez-Anaya et al. 1990). The development of microorganisms reduces the bakery products shelf life and leads to economic losses both for the producers and consumers due to the need to utilization of contaminated products and the loss of the manufacturer's good brand.

The scale of contamination in food is reflected in the number of notifications issued by the Rapid Alert System for Food and Feed (RASFF), which is a platform for exchanging information about foods and feeds that may pose a hazard to consumers in Europe. In the last three years, the RASFF system has registered few reports of microbiological contamination of bread. There were 13 reports defined as non-serious regarding visible mould symptoms, mainly in pita bread (about 40%), also in kebab-roll bread, hot dog bread, Napoli, Turkish, naan and ginger bread. There have been no reports of contamination of bread by mycotoxins or spore-forming bacteria causing rope defect.

For the production of bread and its derivatives, grains and flour from various plants are used. The three most important cereals grown in the world are wheat, corn and rice. Wheat and rye are most often used to make bread, but barley, oats and gluten-free plants such as quinoa or sorghum are also used in bakery products. The grains themselves at the harvest stage are often contaminated with microorganisms, the number of which, depending on the storage and processing conditions, may increase or decrease. The source of microbiological contamination, and consequently spoilage of bakery products may also be other ingredients used in production (Pereira et al. 2020; Hernandez Nopsa et al. 2015).

Also the baking process itself can be a critical point, despite the fact that it takes place at high temperature, which causes the majority of microorganisms to die. Both underbaked bread and its improper cooling may contribute to the development of microbial infections. In such conditions, not only spore-forming bacteria will be able to develop, but during the cooling process secondary microbiological infections (bacteria, moulds) may appear coming from personnel, equipment and the production environment – including air. Figure 14.2 shows the main factors that affect microbiological quality throughout the production chain.

Not without significance is the post-production phase, which includes bread packaging, i.e. cutting and packaging. These activities should take place in specially designated rooms, and the devices should be carefully cleaned of crumbs and subjected to systematic washing and disinfection processes. Our own research shows that this is an important critical point, especially in the production of hot-dog bread, where pitting machines are used.

The production environment has a large impact on the microbiological quality in the production process, which translates into the final product. All food contact surfaces, equipment and materials in the food processing area must comply with good hygiene practices, should be easy to clean, corrosion resistant, suitable for contact with food, and must not absorb water. The aforementioned microbiological quality of the air is also important, i.e. proper natural or mechanical ventilation systems, properly distributed and maintained ventilation ducts (Garcia et al. 2019; Miśniakiewicz 2011; Czerwinska and Piotrowski 2009).

The last, but often the most important link in the correct bread production process is human. Regardless of whether bakeries use manual or mechanical bread forming, compliance with the rules of production hygiene within the workplace and in the entire production facility is extremely important. Despite monitoring the quality of raw materials and the course of the technological process, it is the human factor that ultimately determines the quality of bread.

In this chapter the attention was paid to spore-forming bacteria and fungi – two main groups of microorganisms responsible for rope defect and "mouldy bread" spoilage defects in bread bakery products as well as mycotoxins contamination.

14.3 Mould Spoilage

The most common group of microorganisms which cause spoilage in breadmaking are fungi. Due to the high temperature during baking, mold spores are killed at this stage. Therefore, mould contamination in bakery products is associated with secondary contamination during the slicing, packaging and distribution stages. Mycotoxins are not destroyed during baking and are therefore a problem throughout the production process.

The source of moulds in the dough is mainly the raw material (flour, seeds, yeasts) as well as air, water, equipment or employees. The air and surfaces inside the bakery can be contaminated with fungi from dry ingredients, mainly flour, which spread easily through the air and fall onto surfaces during the weighing and mixing steps. Moulds can grow rapidly in a humid atmosphere (relative humidity above 90%). Contamination is also favoured by high humidity that results from the condensation of water vapor inside the package of insufficiently cooled bread. In addition, sliced bread and packed in airtight packaging is particularly exposed to the growth of mould coming from the surface of the slicer (Melini and Melini 2018). Therefore, it is good practice to separate the flour handling areas from the areas where the finished bread is being prepared for distribution (Cauvain 2015).

The main sources of bakery environment contamination by fungi are the raw materials used in production, flour and grains. During cultivation cereals can be infected by plant pathogens also called field fungi, include genera such as *Alternaria, Fusarium, Cladosporium, Helminthosporium, Aureobasidium, Epicoccum,* and

Botrytis species (Oliveira et al. 2014). During grains storage, pre- and post-processing the microbiota is expanded to include saprophytic species from *Aspergillus*, *Penicillium*, *Eurotium*, *Rhizopus* and *Mucor* genera (Hernandez Nopsa et al. 2015). They are include also toxigenic species. Their development leads to contamination of the crops with mycotoxins, such as aflatoxin B1, ochratoxin A, citrinin, *Fusarium* toxins (fumonisin, zearalenone, and deoxynivalenol), alternariol, and tenauzonic acid.

The number of yeasts and moulds found on various grains was 10^6–10^7 cfu/g and depends on the kind of commodities, location, environmental factors, e.g. temperature, pH, water activity, nutrients availability, atmosphere composition, presence of antimicrobial compounds, mechanical damage, the handling and storage conditions after harvest. Poor hygiene, excessive temperature and moisture during harvesting, storage and handling may contribute increasing of number of contaminated microorganisms. The cereals could cause spoilage during storage, if the moisture of the commodity reaches levels above 12% (Tournas and Niazi 2018). The fungus growth in the grain is the main cause of primary flour contamination. Weidenbörner et al. (2000) isolated from wheat flour (whole meal and type 405) 51 fungal species belonging to 14 different genera. The most common isolated moulds belong to the genera *Aspergillus* and *Penicillium*. The most frequently encountered species were *Aspergillus candidus*, *A.flavus*, *A.versicolor*, *A.niger*, *A.penicillioides*, *Penicillium aurantiogriseum*, *P.verrucosum*, *P.viridicatum*, *P.griseofulvum*, *P.brevicompactum*, *P.citrinum*, *Scopulariopsis brevicaulis*, *Cladosporium cladosporioides*, *Aureobasidium pullulans*, *Alternaria alternata*, moreover xerophilic species such as *Wallemia sebi*, *Eurotium chevalieri*, *E.herbariorum* and *E.amstelodami*. From the identified fungi 93.3% belong to the group of toxigenic moulds (Weidenbörner et al. 2000; Pitt and Hocking 2009a).

Other studies have shown contamination of self-rising, unbleached and whole wheat flours at around the level 10^3 CFU/g regardless of the type of flour. The authors isolated 20 mould species belonging to the genus *Alternaria*, *Aspergillus*, *Cladosporium*, *Eurotium*, *Fusarium*, and *Penicillium*. Strains of *Aspergillus* *A.candidus*, *A.flavus*, *A.terreus*, *A.tritici*, *A.versicolor*, *Eurotium chevalierii*, *Penicillium* spp., *Fusarium* spp. were present in the largest number of samples (Tournas and Niazi 2018; Pitt and Hocking 2009a).

The process of baking bread destroys the spores of fungi or wild yeasts contaminating the flour. Hence the growth of moulds in the finished bread is the result of post-processing contamination during cooling, slicing, packaging and storage, mainly from the mould spores present in the atmosphere of the bakery, parts of equipment and packaging. The air spoilage by fungi in bakery depends on hygienic status of facilities, equipment, furniture, utensils, raw material, ingredients, and packaging. In a study conducted by Garcia et al. (2019) in 20 bakeries in Brazil, it was shown that the air contamination by fungi in the cooling area was at level below 10^3 CFU/m^3 in 45% of bakeries, in 45% between 1.0×10^3 and 5.0×10^3 CFU/m^3, and in 10% was above 2.4×10^4 CFU/m^3 (Valle Garcia et al. 2019). The air contamination levels in Canadian bakeries were similar at the rate of 50–2.0×10^3 CFU/m^3 (Legan 1993).

Microbiological spoilage by yeasts and moulds consists on visible growth, production of mycotoxins and formation of off-flavours, which might be produced even before fungal outgrowth is visible (Melini and Melini 2018). The bread spoilage commonly known as "bread mould" is mainly caused by fungi from *Rhizopus nigricans, R.stolonifer, Aspergillus niger* or *Penicillium expansum, P.roqueforti, P.corylophilum, P.solitum, P.paneum, P.brevicompactum*. This defect in bread manifested by the development of mycelium and spore deposit on its surface visible to the naked eye as a black, green, grey or red spots (Table 14.1, Fig. 14.3). In addition, fungi of the genus *Cladosporium, Mucor*, or *Neurospora* with black, grey, or pink lesions on the surface of the bread have been observed (Garcia et al. 2019; Cauvain 2015; Nielsen and Rios 2000). It was found that wheat bread left in plastic bags at 22 °C after 5–6 days was covered with an intense mould growth. In case of sourdough bread, the time which the mould symptoms appear is longer. It was observed that part of loaves stored at 25 °C and 70% relative air humidity became mouldy after 7 days (Legan 1993).

Bread defects can also be caused by yeasts showing mycelial or pseudomycelial growth. In this case, they are called "chalky bread" and its symptoms are white flour-like spots appearing on the crumb and bread crust (Fig. 14.4), especially in par-baked MAP packaged bread. The organisms responsible for the early spoilage of bread in the form of white spreading colonies similar to chalk dust are *Hyphopichia burtonii* and *Endomyces fibuliger*. Their growth in bread is very fast and they are

Table 14.1 Bread defects caused by fungi

Spoilage agents	Spoilage symptoms
Penicillium expansum, P. roqueforti, P. paneum, P. brevicompactum	Blue, green, flat spots, fast growing, slowly spreading with loose conidia
Aspergillus niger	Black, white or yellowish mycelium, fluffy, fast spreading
Aspergillus flavus	Olive green colony
Aspergillus candidus	Cream spots
Eurotium herbariorum	Green, yellow spots
Chrysonila sitophila	Pink, reddish, fluffy and fast spreading in bread stored in high humidity or wrapped while still warm, growth very rapidly
Rhizopus nigricans	"Black bread mould" grey/black, white cottony mycelium and black sporangia, very fluffy and fast spreading
Cladosporium sp.	Dark olive green spots, flat, spreads slowly, often present on damp bakery walls
Mucor sp.	Fluffy and fast spreading colony, grey turns into black
Monascus ruber	Red spots
Endomyces fibuliger	Chalky-like spot
Hypopichia burtonii	Chalky-like spot
Pichia anomala	Ethyl acetate production, flavour defect

Source: Authors' own elaboration based on Garcia et al. (2019), Cauvain (2015), Pitt and Hocking (2009a, b), and Legan (1993)

Fig. 14.3 "Mouldy bread" contaminated by *Penicillium* sp. (Source: Authors)

Fig. 14.4 Chalky mould on bread. (Source: Authors)

resistant to preservatives and disinfectants (Deschuyffeleer et al. 2011; Pitt and Hocking 2009b). Modified atmosphere packaging (MAP) can extend the shelf life of bakery products, but this method is ineffective against chalky moulds spoilage. The effect of secondary contamination by yeasts, including fermentative *Saccharomyces cerevisiae*, is an "alcoholic" or "estery" off-odour (Cauvain 2015).

Fungi are responsible for changes in sensory characteristics of the raw materials and the bread too, such as taste and odour due to produce volatiles. Moulds can produce a range of volatile odour compounds form alcohols, esters, aldehydes, ketones, hydrocarbons, monoterpenes, sesquiterpenes and furanes group, including 3-octanone, 1-octen-3-ol, 2-methyl-1-propanol, geosmine, 2-methoxy-3-isopropylpyrazine, which are responsible for an earthy-musty off-odour and affected the quality of raw materials even when present in very small amounts (Buśko et al. 2010).

Moulds produce a vast number of enzymes: lipases, proteases, carbohydrases (amylases, galactosidases, glucosidases, etc.) which are break down bread ingredients leading to its spoilage. They also affected on grain or flour polysaccharides, such as the starch and pentosans, which play an important role in breadmaking. Wang et al. (2005) showed that contamination of wheat grains by *Fusarium culmorum* causes starch grains damage, changes in their size and shape, which is one of the factors affecting the quality of bread crumb. In addition, under the influence of mould enzymes a significant increase in soluble dietary fiber and a slight decrease in insoluble dietary fiber and pentoses were observed (Wang et al. 2005).

In order to protect the bread against the development of undesirable moulds, chemical preservatives such as propionic, sorbic or acetic acids are used, but they are mostly not accepted by consumers. Therefore, it proposes the use of natural biopreservatives. The use of sourdough fermented with strains of lactic acid bacteria (LAB) is a safe alternative (Bartkiene et al. 2022). The antifungal effect of sourdough microorganisms is associated with competition for specific nutrients (e.g. vitamins), modification of the pH value and production of metabolites with antifungal properties. The use of sourdough for the production of bread significantly extends its shelf life, which was shown in the studies of Garofalo et al. (2012). Sourdough bread artificially contaminated with *Aspergillus japonicus* fermented with *Lactobacillus rossiae* and *L.paralimentarius* had a shelf life of 32 days, which was related to the production of lactic and phenyllactic acids, moreover the occurrence of a series of peptides corresponding to wheat α-gliadin proteolysis fragments (Garofalo et al. 2012). According to Lavermicocca et al. (2000) *Lactiplantibacillus plantarum* isolated from sourdough possess antifungal activity against *Penicillium roqueforti*, *P.expansum*, *Aspergillus niger*, *A.flavus*, and *Fusarium graminearum*. Two novel compounds responsible for this phenomenon such as phenyllactic acid (PLA) and 4-hydroxy-phenyllactic acid (OH-PLA) were identified (Lavermicocca et al. 2000).

The antimicrobial metabolites include low-molecular-mass compounds, bacteriocins, organic acids (lactic, acetic, caproic, formic, propionic, and butyric) (Hassan et al. 2016). Lactic acid bacteria, in addition to those listed above, produce other compounds with antifungal properties, such as 2-hydroxyisocaproic, 2-hydroxy-4-methypentanoic, hydroferulic acid, methylhydantoin, mevalonolactone, δ-dodecalactone, cinnamic acid derivatives, reuterin/acrolein and cyclic dipeptides. It was shown that sourdough bread fermented by *L.amylovorus* was characterized by 6 days longer mould-free time compared to the control bread (Axel et al. 2016). Various mechanisms of the antifungal activity of these metabolites are described, including the cell membrane destabilization, enzyme inhibition, proton gradient interference or creation of reactive oxygen species (Siedler et al. 2019).

14.4 Mycotoxins Contamination

Bread spoilage by moulds causes great not only economic losses for bakeries. More serious, however, is the risk to public health directly related to the consumption of bread, which can be contaminated with mycotoxins produced by some mould species. The high temperature during baking kills moulds, but the mycotoxins do not decompose if the temperature does not exceed 200 °C, which makes them an "invisible contaminant" of bread (Saranraj and Sivasakthivelan 2016; Garcia et al. 2019). Mycotoxin contamination mainly comes from raw materials that have developed moulds during vegetation in the field or during storage. The formation of mycotoxins by fungi contaminating bread is of less importance due to the rejection of such a product by consumers. Reducing the risk of contamination of bakery products by mycotoxins includes the proper selection of raw materials (flour) that meet quality requirements and testing of the finished product before it is released for distribution.

Mycotoxins are secondary metabolites of fungi that contaminate both plant raw materials and products of plant and animal origin. They can be produced, primarily in cereals in the field, during harvest, storage, transport, and processing of crops mainly by fungi belonging to *Aspergillus*, *Penicillium* and *Fusarium* genera. Mycotoxins diffuse into grain and can be found in all grind fractions, flour, and, due to their thermo-resistant properties, also in bread (Palumbo et al. 2020).

Currently, more than 400 compounds are identified as mycotoxins. Most scientific attention has been focused on only a few, due to their frequency and toxic properties. The most important mycotoxins in cereal products are aflatoxins (AFs), ochratoxin A (OTA), citrinin (CIT), moreover *Fusarium* toxins represented by fumonisins (FUM), deoxynivalenol (DON), zearalenone (ZEN), T-2 and HT-2 toxins (T2, HT-2), *Alternaria* toxins, such as alternariol (AOH), alternariol methyl ether (AME), tenuazonic acid (TeA), and *Claviceps* ergot alkaloids (Richard 2007; Murphy et al. 2006). The result of long-term exposure to mycotoxins may be poisoning called mycotoxicosis, which may be acute or chronic. Mycotoxins have numerous negative effects on human, causing liver and kidney damage, as well as interfering with the functioning of the digestive tract and the immune system. They may exhibit carcinogenic, mutagenic, cytotoxic, teratogenic, neurotoxic, or estrogenic properties (Richard 2007). The co-occurrence of different mycotoxin is common in food and can result in antagonistic, additive, or synergistic effects on consumers health (Alassane-Kpembi et al. 2017).

The problem of recent years is the so-called masked mycotoxins, which are derivative compounds formed as a result of plant or microorganisms activity or during technological processes. They are not detected during routine analyses, but they pose a threat to the health of consumers (Zhang et al. 2020; Tan et al. 2023).

One of the most dangerous mycotoxins is aflatoxin B1 classified by The International Agency for Research on Cancer (IARC) as a human carcinogen (1A group). This toxin can induce liver tumours, Reye's syndrome and encephalopathy. Aflatoxins (B1, B2, G1 and G2) are produced by *Aspergillus flavus* and *Aspergillus parasiticus* mainly in subtropical climate regions. The most contaminated products are different kind of nuts, grains and spices which are often bread ingredients (Bhardwaj et al. 2023).

Ochratoxin A is a widespread mycotoxin which is produced mainly by species from *Aspergillus* and *Penicillium* genera: *Aspergillus ochraceus, A.melleus, A.westerdijkiae, A.carbonarius, Penicillium verrucosum* and *P.nordicum.* This group of mycotoxin contaminates plant and animal derived food, e.g. grains and grain products, spices, fruits. OTA was classified a 2B cancer compound as possibly carcinogenic compounds for human. Ochratoxin A is carcinogenic, genotoxic, teratogenic, immuno-toxic and nephrotoxic agent (Li et al. 2021; Ganesan et al. 2022).

In the moderate climate regions, including Central Europe, cereals and derived products are mainly contaminated with mycotoxins produced by pathogenic *Fusarium* species. Among cultivated cereals, wheat is very often susceptible to the development of these pathogens, which not only reduce crop yields but also produce mycotoxins such as deoxynivalenol, zearalenone, nivalenol and, to a lesser extent, HT-2 toxin and T-2 toxin. Contamination of cereals by several recently revealed *Fusarium* toxins, represented mainly by enniathins and of minor importance beauvericin, fusaproliferin and moniliformin has been documented also (Vaclavikova et al. 2013; Malachova et al. 2011).

Deoxynivalenol (DON, vomitoxin) is a type-B trichothecene mycotoxin produced by several field fungi, including *Fusarium graminearum, F.sambucinum* and *F.incarnatum-equiseti.* DON is very frequently occurring toxin, and is found worldwide, particularly in cereal crops such as wheat, barley, oats, maize, and their products like flour, bread and germ. DON presents a wide range of toxic influence on human, resulting in their lack of appetite, nausea, diarrhea, hemorrhages and anemia (Ganesan et al. 2022).

Zearalenone (ZEN) is a non-steroidal oestrogenic mycotoxin produced by many species of the genus *Fusarium*, mainly *F.graminearum, F.culmorum, F.cerealis, F.equiseti, F.crookwellense, F.semitectum, F.verticillioides, F.sporotrichioides, F.oxysporum* and *F.acuminatum.* Zearalenone has been detected in different cereals, such as wheat, barley, maize, rye, rice, sesame seed, flour, as well as in grain-based products such as grains for human consumption, baked goods, pasta, breakfast cereals and bread. Zearalenone has a strong estrogenic, anabolic and immunotoxic effects. ZEN is responsible for the hyperestrogenic syndrome (Ropejko and Twarużek 2021; Hueza et al. 2014).

Other toxins belonging to trichothecene type A produced by *Fusarium* are T-2 and HT-2 toxin produced in temperate region of Europe, America and Asia by different species, mainly by *F.equiseti, F.sporotrichioides* and *F.poae.* They can be found in various cereal crops (wheat, maize, barley, oats, and rye), processed grains (cereal-based foodstuffs, flour, bread). Although the detected concentrations of trichothecenes type A are generally low, however, due to their high toxicity, they should be included in the analysis of consumer exposure to mycotoxins.

T-2 and HT-2 toxins have a very strong toxic effect on the liver, kidneys, nervous, immune and reproductive systems. They could decrease cell viability, inhibit cell proliferation, induce oxidative stress and cause mitochondria damage, and induce oxidative stress. Intoxication manifests by vomiting, anorexia, feed refusal, stomach necrosis, alimentary toxic aleukia, and skin irritation (Mousavi Khaneghah et al. 2020; Janik et al. 2021; Mahato et al. 2022).

To protect public health in the European Union (EU), the maximum levels of the most toxicologically important mycotoxins permitted in foodstuffs are established by the regulations of the European Union Commission (Table 14.2). It has

Table 14.2 European Union mycotoxin legislation – cereal and cereal products

Mycotoxin	Commodity	Maximum level [μg/kg]	References
Aflatoxin B1	All cereals and all products derived from cereals, including processes cereal products, except maize and rice	2	European Commission (2010)
Ochratoxin A	Unprocessed cereals	5	European Commission (2022)
	All products, derived/processed from unprocessed cereals. Cereals placed on the market for the final consumer	3	
	Bakery wares:		
	Products not containing oilseeds, nuts or dried fruit	2	
	Products containing at least 20% dried vine fruit and/or dried figs	4	
	Other products containing oilseeds, nuts and/or dried fruit	3	
Zearalenone	Unprocessed cereals other than maize	100	European Commission (2007)
	Cereals intended for direct human consumption, cereal flour	75	
	Bread (including small bakery wares)	50	
Deoxynivalenol	Unprocessed cereals (excluding durum wheat, oats and maize)	1250	
	Unprocessed durum wheat and oats	1750	
	Cereals intended for direct human consumption, cereal flour	750	
	Bread (including small bakery wares)	500	
T-2 and HT-2 toxin	Unprocessed cereals:		European Commission (2013)
	Barley	200	
	Oats	1000	
	Wheat, rye and other	100	
	Cereal grains for direct human consumption:		
	Oats	200	
	Other	50	
	Cereal products for human consumption:		
	Cereal milling products (excluding oat and maize products)	50	
	Bread (including small bakery wares)	25	

Source: Authors' own elaboration based on literature

established maximum levels of aflatoxins, ochratoxin A, and a *Fusarium* toxins (fumonisin, deoxynivalenol, zearalenone, T-2 and HT-2) in different food products, including unprocessed cereal grains and cereal grains for direct consumption, and cereal products such as bread, flour or breakfast products.

Recently, Wan et al. (2020) reviewed the contamination of cereal products, including flour and bread, in individual countries in 2009–2018 (Wan et al. 2020). Table 14.3 shows the contamination levels of this type of products, based on available data.

Results of Saladino et al. (2017) showed that among the tested bread loaves from Spanish supermarkets, 20% were contaminated with aflatoxins, of which about 10% exceeded the limits for the total amount of aflatoxins. Contamination by zaralenone was observed in as many as 96% of the samples and 30% of them contained ZEA above the limits set by the EU. Enniathins were also present in 65% of the samples

Table 14.3 Occurrence of mycotoxin in flour and bread

Product	Country	Mycotoxin	Contamination level [μg/kg]	References
Bread	Spain	Aflatoxin B1, B2, G1	0.5–7.1	Saladino et al. (2017)
		ZEN	27–905	
		Enniathins	0.2–54	
Flour	UK	DON	63–284	Scudamore et al. (2009)
Wholemeal flour			51–232	
White bread			31–186	
Wholemeal bread			28–125	
Wheat	Italy	DON	9.6–99.6	Juan et al. (2013)
		HT-2	6.78–60.1	
		T-2	7.14–17.8	
		ZEN	2.35–27.15	
Bread	Portugal	OTA	0.03–2.09	Paíga et al. (2013)
Wheat flour		ZEN	13.0–22.0	Cunha and Fernandes (2010)
Wheat flour		DON	78.9–325.8	Trombete et al. (2016)
Whole rye flour	Germany	DON	26.0–32.0	Gottschalk et al. (2009)
Whole wheat flour		DON	39.0–434.0	Gottschalk et al. (2009)
Bread	Undefined	T-2	3.28–5.42	Mousavi Khaneghah et al. (2020)
		HT-2	11.54–27.06	
		DON	37.50–44.40	
Rye bread	Czech	T-2/HT2 sum	0.3–4.2	Ostry et al. (2020)
High fibre bread	Belgium	DON	10.0–138.0	de Boevre et al. (2013)

Source: Authors' own elaboration based on literature

but the level of which is not regulated by the EU (Saladino et al. 2017). The presence of deoksynivalenol was found in all tested samples of white and wholegrain flour and bread from UK, but none of them exceeded the EU limits (Scudamore et al. 2009).

Mycotoxin are stable during bread baking. Few data show that the mycotoxin content is slightly reduced during baking due to transformation to other, often undefined products of unknown toxicity. A big risk for consumers is the fact that modified forms of mycotoxins called "masked mycotoxins" present in flour can be reverted to their original forms during breadmaking processes (Tan et al. 2023; Zhang et al. 2019).

It have been shown that deoxynivalenol, and zearalenone and nivalenol contaminating flour, mostly transferred to the bread. Only the DON content was reduced during bread baking by 35% and 39% in white and wholegrain bread, respectively (Scudamore et al. 2009). Studies of Vaclavikova et al. (2013) have shown that the level of ennianthines after bread baking decreased to 30% of their initial concentration in the flour (Vaclavikova et al. 2013).

In order to limit the risk of consumer exposure to mycotoxins, it is necessary to take appropriate measures. These are mainly preventive actions involving strategies at various stages of production. With regard to bakery, the appropriate quality of the raw material – flour and additives – is important, both in terms of contamination with moulds and mycotoxins. This is influenced by agrotechnical treatments in the field, storage and processing conditions of cereals, and in the case of the final product, appropriate hygienic conditions for cutting, packing and storing bread. During sourdough fermentation, fate of mycotoxins are also observed as a result of the interaction of lactic acid bacteria and yeasts with toxins. However, the literature data on this subject varies depending on the authors. According to Giménez et al. (2014) during the breadmaking process, DON was reduced by 2.1% after fermentation and dropped by 7.1% after baking, reaching a maximum reduction of 19.8% in the crust as compared with a decrease of 5.6% in the crumb (Giménez et al. 2014). A higher reduction in the amount of DON was obtained during the production of French bread and Vienna bread from flour naturally contaminated with DON at 150 μg/kg according to the Argentinean bread-processing technology. A maximum reduction above 40% was observed in the dough fermented at 50 °C (Samar et al. 2001). During the fermentation of the dough containing artificially contaminated flour with ochratoxin A, the toxin level decreased in the range of 29.8–33.5%, while after baking it did not change (Valle-Algarra et al. 2009). According to Banu et al. (2014) during the sourdough fermentation with selected lactic acid bacteria (*Lactiplantibacillus plantarum, Levilactobacillus brevis*, and *Lactobacillus helveticus*) the significant decrease (58.6–66.5%) of deoxynivalenol content of flour was observed. The spontaneous fermentation induced a much lower decrease from 26.2% to 29.1% (Banu et al. 2014). It was found that during the sourdough fermentation by mixed population of lactic acid bacteria *Lactiplantibacillus plantarum, Levilactobacillus brevis*, *Fructilactobacillus sanfranciscensis* and *Saccharomyces cerevisiae* yeasts the OTA removed degree was elevated from 33% after 10 h to 68% after 40 h (Piotrowska and Żakowska 2000). Different results obtained Milani and

Heidari (2017). It was shown that the greatest loss of ochratoxin A in artificially contaminated wheat flour during bread making using compressed yeast (total destruction 83.3%), and the lowest in sourdough bread (only 18.8%) (Milani and Heidari 2017). Various mechanisms for removing toxins from the environment are known. The two basic ones are adsorption to the cell wall components of lactic acid bacteria (peptidoglycan) and yeast (glucomannans, β-D-glucans or mannooligosaccharides) and biotransformation to less toxic or non-toxic compounds (Piotrowska 2021). The efficiency of adsorption to the cell wall is related to the surface properties – hydrophobicity/hydrophilicity and electrostatic properties. However, the stability of the cell-bound toxin during bread baking and digestion in gastrointestinal conditions is questionable. Studies have shown that the binding process is reversible, so even if the toxin was adsorbed during the fermentation of the sourdough, it may be desorbed in the digestive tract and the consumer is exposed to its original form (Haskard et al. 2001; Piotrowska 2014). Biotransformation of a mycotoxin to another compound gives a better perspective and takes place by the reduction of ketone carbonyl groups, modification of phenolic hydroxyl groups, the hydrolysis of lactone rings, or the creation of connections with glutathione, deamination or decarboxylation. Glutathione, deamination or decarboxylation. Toxicity of the biotransformation product must be well understood (Piotrowska 2021).

14.5 Bacterial Spoilage

A less common but still significant problem, especially when bread is stored at higher temperatures, is contamination with bacteria belonging to *Bacillus* genus. The source of the presence of bacteria of this type in bread is primarily flour and other raw materials used in the bread production process. Their additional source can be the production environment – air, equipment surface and employees (Viljoen and von Holy 1997). Bread defect caused by spore-forming *Bacillus* species is called rope spoilage. The first symptom of overgrowth of *Bacillus* spp. in the crumb of bread is the characteristic fruity-ester aroma. It is often described as the smell of ripe melon or pineapple (Rumeus and Turtoi 2016). Next comes the colour changes and the crumb becomes soft and sticky and in more advanced stages can almost liquify (Pereira et al. 2020).

Cereal grain used for the production of flour is very often contaminated with *Bacillus* spp. spores originating from the soil and plant surface. The degree of flour contamination depends not only on the hygiene of the milling process, but also on the type of product obtained. The number of spores in milled products is lowest when the husk particles are reduced to a minimum. Because wholemeal flour contains more outer layers of grain, its microbiological load is much higher compared to white flours. Pereira et al. (2020) found that the number of mesophilic aerobic bacterial spores in white flours ranged from 10 to 10^2 CFU/g. In the case of wholegrain flour, there are on average ten times more spores (Pereira et al. 2020). The authors identified five *Bacillus* species: *B.licheniformis, B.sonorensis, B.cereus,*

B.subtilis, B.pumilus and *B.polymyxa.* Most of the isolates belonged to the species *B.licheniformis.* On the other hand, Sorokulova et al. (2003) tested eleven samples of wheat flour. In this case, the *Bacillus* count was higher and ranged from 1.1×10^4 to 8.3×10^4 CFU/g. The dominant species isolated from the flour samples were *Bacillus subtilis* and *B. licheniformis.* The authors did not study wholegrain flours (Sorokulova et al. 2003). Berghofer et al. (2003) found that the origin of flour, bran and wheat germ with spores of *Bacillus* sp. ranged from 10^2 to 10^5 CFU/g. They got the lowest mean values for the flour. The authors did not identify the isolated bacteria (Berghofer et al. 2003). Baker's yeast used for raising the dough is also mentioned as a potential source of spore-forming bacteria (Pacher et al. 2022).

The consequence of the presence of *Bacillus* sp. in raw materials and production environment is their presence in bread dough. Although the baking temperature is high in the centre of the crumb, it only reaches 97 to 100 °C for a few minutes. The kinetics of temperature changes in the centre of the bread crumb was studied by Chhanwal and Anandharamakrishnan (2014). They studied changes in temperature in the centre of a 9.8 cm loaf. For the first three minutes the temperature did not change, then it increased linearly for about 20 minutes. In the final stage of baking, the activation of the evaporation-condensation mechanism and the addition of latent heat during the phase transition keeps the temperature of the bread crumb below 100 °C. Under these temperature conditions, heat-resistant spores survive (Chhanwal and Anandharamakrishnan 2014).

The decimal reduction time of *Bacillus* spores most often isolated from bread is presented in Table 14.4. Even in the case of a low level of *Bacillus* spores in the flour used in the production process (below 10^2 CFU/g), the multiplication of these bacteria in bread within two days reaches the level of 10^7 CFU/g. Many researchers have been concerned with identifying bacteria of the genus *Bacillus* that cause bread rope spoilage. Identification by biochemical methods has usually indicated that spoilage of bread is primarily caused by *Bacillus subtilis. Bacillus licheniformis, B.pumilus* and *B.cereus* were also isolated (Thompson et al. 1998; Rosenkvist 1995). The use of molecular techniques allowed to revise the knowledge about *Bacillus* sp., responsible for the rope spoilage. An interesting study was conducted by Valerio et al. (2012). The authors investigated the diversity of spore-forming bacteria isolated from raw materials and bread using molecular methods along with the FT-NIR spectroscopy. They found that the dominant species in both the raw materials and the bread is *Bacillus amyloliquefaciens. B.cereus., B.subtilis* and *B.licheniformis* have also been isolated from bread (Valerio et al. 2012). The characteristics of bacteria of the genus *Bacillus* most often isolated from bread are presented in the Table 14.4.

Heat-resistant spores of bacteria belonging to *Bacillus* genera can cause rope spoilage which is bread crumb disease and even lead to food-born poisoning. *Bacillus cereus* is the cause of two kinds of foodborne diseases, an emetic (vomiting) intoxication due to the ingestion of a toxin (cereulide) pre-formed in the food and a diarrhoeal infection due to the ingestion of bacterial cells/spores which produce enterotoxins in the small intestine. Other *Bacillus* spp., *B. amyloliquefaciens*, B. *licheniformis*, B. *pumilus* have more rarely been identified as agents of foodborne

Table 14.4 Characteristic of rope spoilage *Bacillus*

Species	***Bacillus amyloliquefaciens***
Morphology	Gram-positive bacilli, often occurring in chains, strictly aerobic, motile Spores – ellipsoidal or cylindrical, located centrally or subcentrally in unswollen sporangia Colony – smooth, shiny with irregular margin, white to cream
Growth temperature	Minimum 15 °C; maximum 50 °C (there are strains capable of growing at 58 °C), optimal in the range of 30–40 °C
pH	Minimum 5; maximum 9, optimum 6–7
a_w	Minimum a_w = 0.93–0.97;
Sources	Soil, Contamination of bread, UHT-treated milk. Commonly found in a variety of raw materials and food products, including flour, cocoa, herbs, spices, soup powders, milk and powdered milk
Spoilage potential	Capable of breaking down starch and proteins
Virulence	GRAS in the USA (FDA), recognized as a probiotic used for animal feed supplementation Group 1 (according to Leibniz Institut DSMZ-Deutsche Sammlung von Mikroorganismen und Zellkulturen GmbH) Few reports on the isolation of *B. amyloliquefaciens* from immunocompromised patients. Some strains are capable of producing enterotoxin.
Spore temperature sensivity	D_{100} = 23–44 min, z = 12,8 °C
References	Pacher et al. (2022), Lee et al. (2017), Berendsen et al. (2016), André et al. (2017), and Vos et al. (2011)
Species	***Bacillus subtilis***
Morphology	Gram-positive spore-forming bacteria, aerobic, motile,. Spores ellipsoidal to cylindrical located centrally or subterminally, in unswollen sporangia Colonies variable white, through cream to brown, opaque, moist, smooth or wrinkled
Growth temperature	Minimum 5–20 °C, optimum 28–30 °C, maximum 45–55 °C
pH	5.5–8.5
a_w	Minimum a_w = 0.93–0.96
Source	Soil, dust, plant surface, organic matter
Spoilage potential	Capable of breaking down starch and proteins
Virulence	GRAS in the USA (FDA) Biological agents at workstations group 2A (allergenic effect) Group 1 (according to Leibniz Institut DSMZ-Deutsche Sammlung von Mikroorganismen und Zellkulturen GmbH)
Spore temperature sensivity	$D_{100=}$14 min

(continued)

Table 14.4 (continued)

References	Pacher et al. (2022), Condon et al. (1989), Condon and Sala (1992), Serp et al. (2002), and Boone et al. (2001)
Species	***Bacillus cereus***
Morphology	Gram-positive bacilli, facultatively anaerobic, motile Spores – ellipsoidal sometimes cylindrical, located subterminally in unswollen sporangia Colonies 2–7 mm diameter, variable may be smooth or undulating, usually dull but may be shiny. Colony color is usually white to cream, some strains are capable of forming a pinkish-brown pigment.
Growth temperature	Minimum 10–20 °C; optimum 37 °C maximum 40–45 °C,
pH	Minimum 4.9; maximum 9.3
a_w	Minimum a_w = 0.91–0.93;
Sources	*B. cereus* is widely distributed in natural environments. The primary habitat of *B. cereus* is believed to be soil. It is often found in foods such as meat, poultry, milk, cereals, starch raw materials, herbs and spices.
Spoilage potential	It produces various extracellular proteolytic and lipolytic enzymes, which may be responsible for the deterioration of the organoleptic quality and shortening the stability of food products, especially dairy products (e.g. the formation of the so-called sweet curd). It also has the ability to hydrolyze starch.
Virulence	Group 2 (wg Leibniz Institut DSMZ-Deutsche Sammlung von Mikroorganismen und Zellkulturen GmbH) Hemolysin BL (HBL), as well as cytolysin K (CytK), non-hemolytic enterotoxin (Nhe), are considered to be the basic etiological factor of toxicoinfection caused by *B. cereus* sensu stricto. The emetic form of food poisoningis the result of intoxication with cereulide – an emetic toxin
Spore temperature sensivity	D_{100} < 10 min, z = 7,7 °C
References	Pacher et al. (2022), Vos et al. (2011), Bartoszewicz and Czyżewska (2017), Soni et al. (2016), and Berthold-Pluta and Garbowska (2019)
Species	***Bacillus pumilus***
Morphology	Gram-positive rods, motile, Spores cylindrical to ellipsoidal located centrally, paracentrally and subterminally in unswollen sporangia Colonies wrinkled and irregular, unpigmented with an even margin
Growth temperature	Minimum 10 °C, optimum 30–37 °C, maximum 50 °C
pH	Minimum 4.5; maximum 9.5
a_w	Minimum a_w = 0.94
Sources	Soil and many other environments Isolated from meat dishes, eggs, bread and canned tomato juice
Spoilage potential	Capable of breaking down starch and proteins

(continued)

Table 14.4 (continued)

Virulence	Group 1 (wg Leibniz Institut DSMZ-Deutsche Sammlung von Mikroorganismen und Zellkulturen GmbH) Causes mastitis in cattle, opportunistic infections in humans, some strains are toxinogenic – can form a cereulide-like toxin
Spore temperature sensivity	D_{100} = 56 min
References	Pacher et al. (2022), Boone et al. (2001), Suominen et al. (2001), Satomi et al. (2006), Janštová and Lukášová (2001), and Gibbiel and Abd-et al. (1973)
Species	***Bacillus licheniformis***
Morfology	Gram-positive bacilli, relatively anaerobic, motile Spores ellipsoidal or cylindrical located centrally or subcentrally Colonies wrinkled, dull, adherent to the medium, cream to brown
Growth temperature	Minimum 15 °C; maximum 50–55 °C,
pH	Minimum 5.7; maximum 10
a_w	Minimum a_w = 0.93–0.98
Source	Widely distributed in soil and many other environments, including milk and other foods,
Spoilage potential	Capable of breaking down starch and proteins
Virulence	GRAS in the USA (FDA), recognized as a probiotic Group 1 (wg Leibniz Institut DSMZ-Deutsche Sammlung von Mikroorganismen und Zellkulturen GmbH) Opportunistic human and animal pathogen, can cause foodborne poisoning, synthesizes thermolabile emetic toxins similar to *B. cereus* cereulide
Spore temperature sensivity	D_{100} = 56 min
References	Pacher et al. (2022), Boone et al. (2001), Janštová and Lukášová (2001), Taylor et al. (2005), Palop (1996), Montville and Sapers (1981), Khyami-Horani (1996), and He et al. (2006)

Source: Authors' own elaboration based on literature

diseases characterized by diarrhoea and/or vomiting. The infectious dose of *Bacillus cereus* ranges from 10^3 to 10^8 cells (Panel EB 2005). Currently, there are no legal requirements in the EU regarding the abundance limits of pathogenic *Bacillus* sp. in bread. Guidelines from the Health Protection Agency may be taken into account when developing the criteria (Table 14.5) (Health Protection Agency 2009).

Due to the fact that ropiness spoilage is a common disease of bread, it is important to properly protect the product. There are many methods of extending the shelf life of bread, e.g. the addition of preservatives and natural antimicrobial, modified atmosphere packaging, active packaging systems (Smith et al. 2004; Pattison et al. 2003). The method used for years is the addition of baker's sourdough to the formulation. Typically, sourdough contains two main ingredients, flour and water, which, during microbial bioconversion, develop sour characteristics. Sourdough

Table 14.5 Guidance on the interpretation of results for enumeration of pathogenic *Bacillus* in ready-to-eat foods placed on the market (Health Protection Agency 2009)

Microorganism	Number [CFU/g]	Risk category	Interpretation
Bacillus cereus and other pathogenic *Bacillus* spp.	$>10^5$	High	Unsatisfactory: Potentially injurious to health and/or unfit for human consumption
	10^3–$\leq 10^5$	Moderate	Borderline
	$<10^3$	Low	Satisfactory

source: authors' own elaboration based on literature

microbiota is dominated by LAB, mainly belonging to the genus *Lactobacillus*, and yeasts, mainly belonging to the genus *Saccharomyces* or *Candida.* The predominant microflora and their characteristics in fermented cereals are related to many factors, including geographical location, processing, handling (Bartkiene et al. 2022).

The antimicrobial activity of many metabolites of lactic acid bacteria is well known. Organic acids (lactic and acetic acid), hydrogen peroxide, ethanol, diacetyl, acetaldehyde, acetoine, carbon dioxide, reuterin, reutericyclin and bacteriocins, are examples of antimicrobial agents produced by LAB. Just lowering the pH of the environment significantly affects the growth dynamics of rope-spoilage *Bacillus* sp. The Fig. 14.5 shows the prediction of the growth of *Bacillus licheniformis, B.cereus and B.subtilis,* typical species responsible for bread spoilage in conditions corresponding to sourdough bread (pH 4.9, and in 0.96) and whole wheat bread (pH 5.7, and in 0.96). Predictions were made using the ComBase database.

As mentioned earlier, not only acidic LAB metabolites are responsible for their antagonistic activity. This is confirmed by the research of Katina et al. (2002). The authors stated that rope spoilage of wheat bread was inhibited by adding 20–30 g of sourdough/100 g of wheat dough if the sourdough was fermented with *Lactiplantibacillus plantarum*, *Pediococcus pentosaceus* or *Levilactobacillus brevis* and the pH values of sourdoughs were adjusted below 4.0. They did not observe the same effect if they added an equivalent amount of lactic acid to the bread dough (Katina et al. 2002). Inhibition of the proliferation of *Bacillus* sp. in bread dough under the influence of *Lactobacillus acidophilus* and *Bifidobacterium lactis* was also observed by Elsanhoty et al. (2017) (Elsanhoty et al. 2017). Pepe et al. (2003) showed that the addition of *L plantarum* and *L. mesenteroides* to the starter culture extended the shelf life of bread by 1 to 5 days (Pepe et al. 2003). In conclusion, there is a lot of data confirming that the addition of sourdough to bread prevents rope-spoilage.

As mentioned earlier, lactic acid bacteria form many compounds with antimicrobial activity. One of them are bacteriocins. A review of the literature indicates that the bacteriocins and bacteriocin-like inhibitory substances produced by LAB on sourdough are heat-stable peptides belonging mainly to class II, but also to class I (Table 14.6). Class II bacteriocins are small molecules (<10 kDa) that do not contain lanthionine. They are divided into four main subclasses. Subclass IIa includes pediocin-like bacteriocins. The molecules grouped in this subclass are characterized

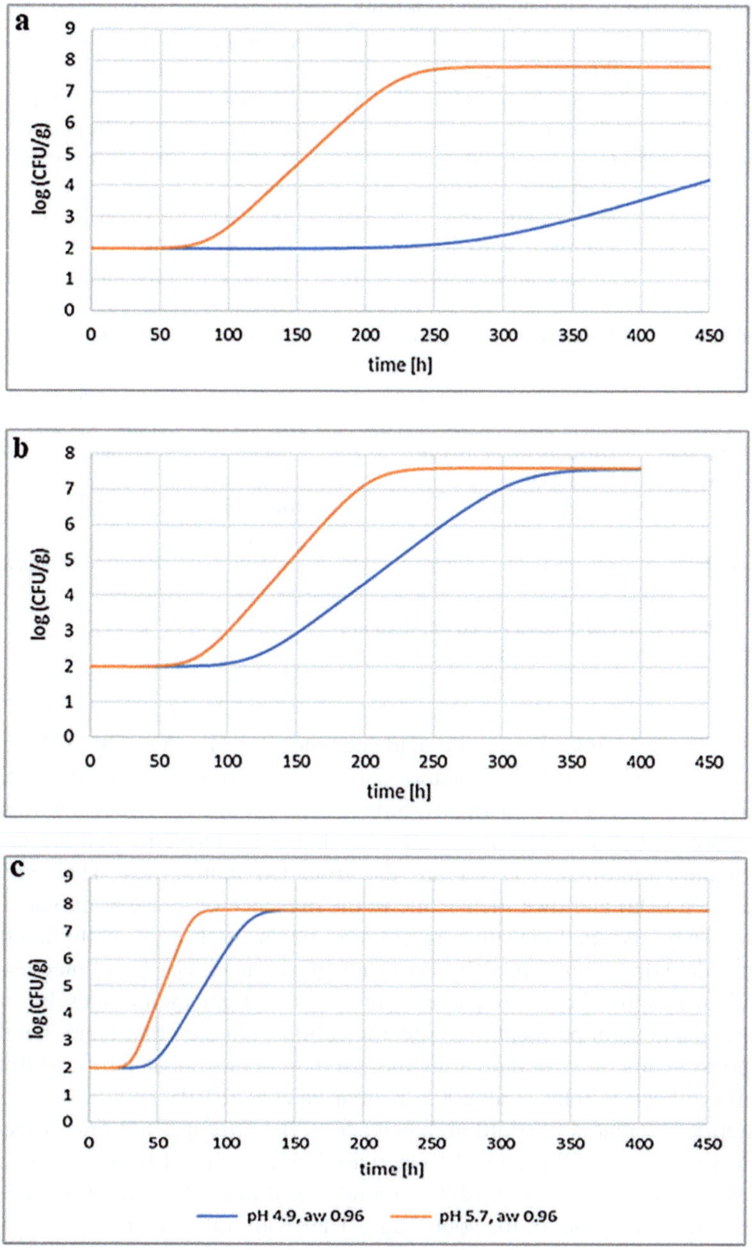

Fig. 14.5 Prediction of *Bacillus licheniformis* (**a**), *B. cereus* (**b**) and *B. subtilis* (**c**) growth at 20 °C. (Source: Authors' own elaboration based on ComBase database)

Table 14.6 Bacteriocins with anti-*Bacillus* activity

Bacteriocin	Producer	Class	Inhibited *Bacillus*	References
Plantaricin ST31	*Lactiplantibacillus plantarum*	II	*Bacillus subtilis*	Mokoena (2017) and Todorov et al. (1999)
Plantaricin GZ1–27	*Lactiplantibacillus plantarum*	II	*Bacillus cereus*	Du et al. (2018)
BLIS C57[a]	*Fructilactobacillus sanfranciscensis*	nd[b]	*Bacillus subtilis*	Owusu-Kwarteng et al. (2013)
Sakacin 806	*Latilactobacillus sakei*	II	*Bacillus subtilis*	Narbutaite et al. (2008)
Pediocin 808, 809, 810,	*Pediococcus pentosaceus*	II	*Bacillus subtilis*	Narbutaite et al. (2008)
Pediocin Ac807	*Pediococcus acidilactici*	II	*Bacillus subtilis*	Narbutaite et al. (2008)
BLIS identical to lacticin 3147	*Lactococus lactis* ssp. *lactis* M30	I	*Bacillus subtilis* *Bacillus cereus*	Settanni et al. (2005)
Acidocin LF221 A and B	*Lactobacillus acidophilus*	II	*Bacillus cereus*	Bogovič-Matijašic et al. (1998)

Source: Authors' own elaboration based on literature
[a]Bacteriocin-like inhibitory substance
[b]No data

by the N-terminal consensus amino acid sequence Tyr-Gly-Asn-Gly-Val-Xaa-Cys. Subclass IIb consists of binary bacteriocins that require two separate peptides for full activity, which act synergistically. Subclass IIc grouped molecules with a circular structure. Subclass IId contains all other linear bacteriocins that are not post-translationally modified. Class I are the so-called lantibiotics. They are small peptides (< 5 kDa) and owe their name to the presence of a rare amino acid – lanthionine. They also include other rare amino acids, such as α-methyllanthionine (MeLan), dehydroalanine, or dehydroaminobutyric acid. The basic mechanism of bactericidal action of class II bacteriocins act on target membranes by creating pores that make them permeable to specific cations. On the other hand, lantibiotics are responsible not only for damaging the cell membrane, but also for disturbing the cell wall synthesis (Cintas et al. 2001; Kumariya et al. 2019).

Bacteriocins created in sourdough can act on *Bacillus* spp. not only directly. An example would be BLIS identical to lacticin 3147, produced by *Lactococus lactis* ssp. *lactis* M30. Settanni et al. (2005) found that it may affect *Bacillus* due to the change in the microbiota of the sourdough. The proliferation of *Fructilactobacillus sanfranciscensis* is enhanced in the presence of this BLIS. Bacteria of this species are capable of forming anti-*Bacillus* bacteriocins (Settanni et al. 2005).

The compound with antimicrobial activity, created by *Limosilactobacillus reuteri* present in sourdough, is also reutericycline. It is structurally but not functionally related to tetramic acids. Reutericyclin activity was tested against a broad range of gram-positive and gram-negative bacteria as well as yeasts and fungi. Gänzle et al. (2000) studied the sensitivity of *B.cereus* and *B.subtilis* to this compound. All 10 investigated strains (including 6 isolated from bread) were susceptible and the MIC ranged from 0.13 to 0.3 mg/l (Gänzle et al. 2000). Reutericyclin acts as a proton-ionophore, translocating protons across the cytoplasmic membrane and consequently dissipating the transmembrane ΔpH. The mode of action of reutericyclin is essentially comparable to that of weak organic acids, e.g. acetic acid and sorbic acid (Gänzle 2004).

In conclusion, there is a lot of data confirming that the addition of sourdough to bread prevents rope-spoilage.

14.6 Conclusion

A review of the literature indicates that bread may be contaminated not only with spoilage microorganisms but also with pathogens. It may also contain dangerous metabolites of microbial origin, e.g. mycotoxins. When considering the spoilage of this product, connection with Regulation EC/178/2002 extending the definition of unsafe foodstuffs to encompass all those which are unfit for human consumption should also be taken into account. It is therefore important to maintain appropriate care for microbiological safety throughout the entire production and distribution chain. A good way to protect the final product is to add sourdough to the bread dough.

References

Alassane-Kpembi I, Schatzmayr G, Taranu I et al (2017) Mycotoxins co-contamination: methodological aspects and biological relevance of combined toxicity studies. Crit Rev Food Sci Nutr 57:3489–3507. https://doi.org/10.1080/10408398.2016.1140632

André S, Vallaeys T, Planchon S (2017) Spore-forming bacteria responsible for food spoilage. Res Microbiol 168:379–387. https://doi.org/10.1016/j.resmic.2016.10.003

Axel C, Brosnan B, Zannini E et al (2016) Antifungal activities of three different Lactobacillus species and their production of antifungal carboxylic acids in wheat sourdough. Appl Microbiol Biotechnol 100:1701–1711. https://doi.org/10.1007/s00253-015-7051-x

Axel C, Zannini E, Arendt EK (2017) Mold spoilage of bread and its biopreservation: a review of current strategies for bread shelf life extension. Crit Rev Food Sci Nutr 57:3528–3542. https://doi.org/10.1080/10408398.2016.1147417

Banu I, Dragoi L, Aprodu I (2014) From wheat to sourdough bread: a laboratory scale study on the fate of deoxynivalenol content. Qual Assur Saf Crop Foods 6:53–60. https://doi.org/10.3920/QAS2012.0226

Bartkiene E, Özogul F, Rocha JM (2022) Bread sourdough lactic acid bacteria—technological, antimicrobial, toxin-degrading, immune system-, and faecal microbiota-modelling biological agents for the preparation of food, nutraceuticals and feed. Foods 11:452. https://doi.org/10.3390/foods11030452

Bartoszewicz M, Czyżewska U (2017) Taksonomia, wirulencja i cykle życiowe *Bacillus cereus* sensu lato. Postępy Mikrobiologii 56:440–450

Berendsen EM, Koning RA, Boekhorst J et al (2016) High-level heat resistance of spores of *Bacillus amyloliquefaciens* and *Bacillus licheniformis* results from the presence of a spoVA operon in a Tn1546 transposon. Front Microbiol 7. https://doi.org/10.3389/fmicb.2016.01912

Berghofer LK, Hocking AD, Miskelly D, Jansson E (2003) Microbiology of wheat and flour milling in Australia. Int J Food Microbiol 85:137–149. https://doi.org/10.1016/S0168-1605(02)00507-X

Berthold-Pluta P, Garbowska S (2019) Prevalence and toxicity characterization of Bacillus cereus in food products from Poland. Foods 8:269. https://doi.org/10.3390/foods8070269

Bhardwaj K, Meneely JP, Haughey SA et al (2023) Risk assessments for the dietary intake aflatoxins in food: a systematic review (2016–2022). Food Control:109687. https://doi.org/10.1016/j.foodcont.2023.109687

Bogovič-Matijašic B, Rogelj I, Nes IF, Holo H (1998) Isolation and characterization of two bacteriocins of *Lactobacillus acidophilus* LF221. Appl Microbiol Biotechnol 49:606–612. https://doi.org/10.1007/s002530051221

Boone D, Castenholz R, Garrity G (2001) Bergey's Manual® of Systematic Bacteriology. Springer

Buśko M, Jeleń H, Góral T et al (2010) Volatile metabolites in various cereal grains. Food Additiv Contamin: Part A 27:1574–1581. https://doi.org/10.1080/19440049.2010.506600

Cauvain S (2015) Bread spoilage and staling. In: Technology of breadmaking. Springer, Cham, pp 279–302

Chhanwal N, Anandharamakrishnan C (2014) Temperature- and moisture-based modeling for prediction of starch gelatinization and crumb softness during bread-baking process. J Texture Stud 45:462–476. https://doi.org/10.1111/jtxs.12097

Cintas LM, Casaus MP, Herranz C et al (2001) Review: bacteriocins of lactic acid bacteria. Food Sci Technol Int 7:281–305. https://doi.org/10.1106/R8DE-P6HU-CLXP-5RYT

Condon S, Sala FJ (1992) Heat resistance of *Bacillus subtilis* in buffer and foods of different pH. J Food Prot 55:605–608. https://doi.org/10.4315/0362-028X-55.8.605

Condon S, Lopez P, Oria R, Sala FJ (1989) Thermal death determination: design and evaluation of a thermoresistometer. J Food Sci 54:451–457. https://doi.org/10.1111/j.1365-2621.1989.tb03104.x

Cunha SC, Fernandes JO (2010) Development and validation of a method based on a QuEChERS procedure and heart-cutting GC-MS for determination of five mycotoxins in cereal products. J Sep Sci 33:600–609. https://doi.org/10.1002/jssc.200900695

Czerwinska E, Piotrowski W (2009) Technological aspects of bread backing with assessment of critical points in microbiological contamination. Rocznik Ochrona Srodowiska 11:449–464

de Boevre M, Jacxsens L, Lachat C et al (2013) Human exposure to mycotoxins and their masked forms through cereal-based foods in Belgium. Toxicol Lett 218:281–292. https://doi.org/10.1016/j.toxlet.2013.02.016

Demirci AS, Palabiyik I, Gumus T (2016) Bread wastage and recycling of waste bread by producing biotechnological products. J Biotechnol 231:S13. https://doi.org/10.1016/j.jbiotec.2016.05.071

Deschuyffeleer N, Audenaert K, Samapundo S et al (2011) Identification and characterization of yeasts causing chalk mould defects on par-baked bread. Food Microbiol 28:1019–1027. https://doi.org/10.1016/j.fm.2011.02.002

Du H, Yang J, Lu X et al (2018) Purification, characterization, and mode of action of plantaricin GZ1-27, a novel Bacteriocin against *Bacillus cereus*. J Agric Food Chem 66:4716–4724. https://doi.org/10.1021/acs.jafc.8b01124

Dury S, Bendjebbar P, Hainzelin E et al (2019) Food systems at risk. New trends and challenges. FAO/CIRAD, Rome

Elsanhoty RM, Ghonamy AG, El-Adly NA, Fawzy Ramadan M (2017) Impact of lactic acid bacteria and *Bifidobacterium* on the survival of *Bacillus subtilus* during fermentation of wheat sourdough. J Food Process Preserv 41:e13086. https://doi.org/10.1111/jfpp.13086

European Commission (2007) Commission Regulation (EC) No 1126/2007 of 28 September 2007 amending Regulation (EC) No 1881/2006 setting maximum levels for certain contaminants in foodstuffs as regards Fusarium toxins in maize and maize products. Off J Eur Union 255:14–17

European Commission (2010) Commission Regulation (EU) No 165/2010 of 26 February 2010 amending Regulation (EC) No 1881/2006 setting maximum levels for certain contaminants in foodstuffs as regards aflatoxins. Off J Eur Union 50:8–12

European Commission (2013) Commission recommendation of 27 March 2013 on the presence of T-2 and HT-2 toxin in cereals and cereal products. Off J Eur Union 91:12–15

European Commission (2022) Commission regulation (EU) 2022/1370 of 5 August 2022 amending Regulation (EC) No 1881/2006 as regards maximum levels of ochratoxin A in certain foodstuffs. Off J Eur Union 206:11–14

Ganesan AR, Mohan K, Karthick Rajan D et al (2022) Distribution, toxicity, interactive effects, and detection of ochratoxin and deoxynivalenol in food: a review. Food Chem 378:131978. https://doi.org/10.1016/j.foodchem.2021.131978

Gänzle MG (2004) Reutericyclin: biological activity, mode of action, and potential applications. Appl Microbiol Biotechnol 64:326–332. https://doi.org/10.1007/s00253-003-1536-8

Gänzle MG, Höltzel A, Walter J et al (2000) Characterization of reutericyclin produced by *Lactobacillus reuteri* LTH2584. Appl Environ Microbiol 66:4325–4333. https://doi.org/10.1128/AEM.66.10.4325-4333.2000

Garcia MV, Bernardi AO, Copetti MV (2019) The fungal problem in bread production: insights of causes, consequences, and control methods. Curr Opin Food Sci 29:1–6. https://doi.org/10.1016/j.cofs.2019.06.010

Garofalo C, Zannini E, Aquilanti L et al (2012) Selection of sourdough lactobacilli with antifungal activity for use as biopreservatives in bakery products. J Agric Food Chem 60:7719–7728. https://doi.org/10.1021/jf301173u

Gibbiel AY, Abd-El Al ATH (1973) Measurement of heat resistance parameters for spores isolated from canned products. J App Bacteriol 36:321–327. https://doi.org/10.1111/j.1365-2672.1973.tb04109.x

Giménez I, Blesa J, Herrera M, Ariño A (2014) Effects of bread making and wheat germ addition on the natural deoxynivalenol content in bread. Toxins (Basel) 6:394–401. https://doi.org/10.3390/toxins6010394

Goryńska-Goldmann E, Gazdecki M, Rejman K et al (2020) How to prevent bread losses in the baking and confectionery industry?—measurement, causes, management and prevention. Agriculture 11:19. https://doi.org/10.3390/agriculture11010019

Gottschalk C, Barthel J, Engelhardt G et al (2009) Simultaneous determination of type A, B and D trichothecenes and their occurrence in cereals and cereal products. Food Additiv Contamin: Part A 26:1273–1289. https://doi.org/10.1080/02652030903013260

Haskard CA, El-Nezami HS, Kankaanpää PE et al (2001) Surface binding of aflatoxin B_1 by lactic acid bacteria. Appl Environ Microbiol 67:3086–3091. https://doi.org/10.1128/AEM.67.7.3086-3091.2001

Hassan YI, Zhou T, Bullerman LB (2016) Sourdough lactic acid bacteria as antifungal and mycotoxin-controlling agents. Food Sci Technol Int 22:79–90. https://doi.org/10.1177/1082013214565722

He L, Chen W, Liu Y (2006) Production and partial characterization of bacteriocin-like pepitdes by *Bacillus licheniformis* ZJU12. Microbiol Res 161:321–326. https://doi.org/10.1016/j.micres.2005.12.002

Health Protection Agency (2009) Guidelines for assessing the microbiological safety of ready-to-eat foods placed on the market

Hernandez Nopsa JF, Daglish GJ, Hagstrum DW et al (2015) Ecological networks in stored grain: key postharvest nodes for emerging pests, pathogens, and mycotoxins. Bioscience 65:985–1002. https://doi.org/10.1093/biosci/biv122

Hueza I, Raspantini P, Raspantini L et al (2014) Zearalenone, an estrogenic mycotoxin, is an immunotoxic compound. Toxins (Basel) 6:1080–1095. https://doi.org/10.3390/toxins6031080

Janik E, Niemcewicz M, Podogrocki M et al (2021) T-2 toxin—the most toxic trichothecene mycotoxin: metabolism, toxicity, and decontamination strategies. Molecules 26:6868. https://doi.org/10.3390/molecules26226868

Janštová B, Lukášová J (2001) Heat resistance of *Bacillus* spp. spores isolated from cow's mMilk and farm environment. Acta Vet Brno 70:179–184. https://doi.org/10.2754/avb200170020179

Juan C, Ritieni A, Mañes J (2013) Occurrence of *Fusarium* mycotoxins in Italian cereal and cereal products from organic farming. Food Chem 141:1747–1755. https://doi.org/10.1016/j.foodchem.2013.04.061

Katina K, Sauri M, Alakomi H-L, Mattila-Sandholm T (2002) Potential of lactic acid bacteria to inhibit rope spoilage in wheat sourdough bread. LWT – Food Sci Technol 35:38–45. https://doi.org/10.1006/fstl.2001.0808

Khyami-Horani H (1996) Thermotolerant strain of *Bacillus licheniformis* producing lipase. World J Microbiol Biotechnol 12:399–401. https://doi.org/10.1007/BF00340219

Koutsoumanis K, Tsaloumi S, Aspridou Z et al (2021) Application of quantitative microbiological risk assessment (QMRA) to food spoilage: principles and methodology. Trends Food Sci Technol 114:189–197. https://doi.org/10.1016/j.tifs.2021.05.011

Kumariya R, Garsa AK, Rajput YS et al (2019) Bacteriocins: classification, synthesis, mechanism of action and resistance development in food spoilage causing bacteria. Microb Pathog 128:171–177. https://doi.org/10.1016/j.micpath.2019.01.002

Lavermicocca P, Valerio F, Evidente A et al (2000) Purification and characterization of novel antifungal compounds from the sourdough *Lactobacillus plantarum* strain 21B. Appl Environ Microbiol 66:4084–4090. https://doi.org/10.1128/AEM.66.9.4084-4090.2000

Lee A, Cheng K-C, Liu J-R (2017) Isolation and characterization of a *Bacillus amyloliquefaciens* strain with zearalenone removal ability and its probiotic potential. PLoS One 12:e0182220. https://doi.org/10.1371/journal.pone.0182220

Legan JD (1993) Mould spoilage of bread: the problem and some solutions. Int Biodeterior Biodegrad 32:33–53. https://doi.org/10.1016/0964-8305(93)90038-4

Li X, Ma W, Ma Z et al (2021) The occurrence and contamination level of ochratoxin a in plant and animal-derived food commodities. Molecules 26:6928. https://doi.org/10.3390/molecules26226928

Mahato DK, Pandhi S, Kamle M et al (2022) Trichothecenes in food and feed: occurrence, impact on human health and their detection and management strategies. Toxicon 208:62–77. https://doi.org/10.1016/j.toxicon.2022.01.011

Malachova A, Dzuman Z, Veprikova Z et al (2011) Deoxynivalenol, deoxynivalenol-3-glucoside, and enniatins: the major mycotoxins found in cereal-based products on the Czech market. J Agric Food Chem 59:12990–12997. https://doi.org/10.1021/jf203391x

Martinez-Anaya MA, Pitarch B, Bayarri P, Benedito de Barber C (1990) Microflora of the sourdoughs of wheat flour bread. X. Interactions between yeasts and lactic acid bacteria in wheat doughs and their effects on bread quality. Cereal Chem 67:85–91

Mayes A, Mortimore S (2009) The effective implementation of HACCP systems in food processing. In: Foodborne Pathogens. Elsevier, pp 481–517

Melikoglu M, Webb C (2013) Use of waste bread to produce fermentation products. In: Food industry wastes. Elsevier, pp 63–76

Melini V, Melini F (2018) Strategies to extend bread and gf bread shelf-life: from sourdough to antimicrobial active packaging and nanotechnology. Fermentation 4:9. https://doi.org/10.3390/fermentation4010009

Milani J, Heidari S (2017) Stability of ochratoxin A during bread making process. J Food Saf 37:e12283. https://doi.org/10.1111/jfs.12283

Miśniakiewicz M (2011) Factors determining the quality and safety of bakery products. Zeszyty Naukowe/Uniwersytet Ekonomiczny w Krakowie:35–52

Mokoena MP (2017) Lactic acid bacteria and their bacteriocins: classification, biosynthesis and applications against uropathogens: a mini-review. Molecules 22:1255. https://doi.org/10.3390/molecules22081255

Montville TJ, Sapers GM (1981) Thermal resistance of spores from ph, elevating strains of *Bacillus licheniformis*. J Food Sci 46:1710–1712. https://doi.org/10.1111/j.1365-2621.1981.tb04469.x

Mousavi Khaneghah A, Farhadi A, Nematollahi A et al (2020) A systematic review and meta-analysis to investigate the concentration and prevalence of trichothecenes in the cereal-based food. Trends Food Sci Technol 102:193–202. https://doi.org/10.1016/j.tifs.2020.05.026

Murphy PA, Hendrich S, Landgren C, Bryant CM (2006) Food mycotoxins: an update. J Food Sci 71:R51–R65. https://doi.org/10.1111/j.1750-3841.2006.00052.x

Narbutaite V, Fernandez A, Horn N et al (2008) Influence of baking enzymes on antimicrobial activity of five bacteriocin-like inhibitory substances produced by lactic acid bacteria isolated from Lithuanian sourdoughs. Lett Appl Microbiol 47:555–560. https://doi.org/10.1111/j.1472-765X.2008.02466.x

Nielsen PV, Rios R (2000) Inhibition of fungal growth on bread by volatile components from spices and herbs, and the possible application in active packaging, with special emphasis on mustard essential oil. Int J Food Microbiol 60:219–229. https://doi.org/10.1016/S0168-1605(00)00343-3

Oliveira PM, Zannini E, Arendt EK (2014) Cereal fungal infection, mycotoxins, and lactic acid bacteria mediated bioprotection: from crop farming to cereal products. Food Microbiol 37:78–95. https://doi.org/10.1016/j.fm.2013.06.003

Ostry V, Dofkova M, Blahova J et al (2020) Dietary exposure assessment of sum deoxynivalenol forms, sum T-2/HT-2 toxins and zearalenone from cereal-based foods and beer. Food Chem Toxicol 139:111280. https://doi.org/10.1016/j.fct.2020.111280

Owusu-Kwarteng J, Tano-Debrah K, Akabanda F et al (2013) Partial characterization of bacteriocins produced by *Lactobacillus reuteri* 2-20B and *Pediococcus acidilactici* 0-11A isolated from fura, a millet-based fermented food in Ghana. J Food Res 2:50. https://doi.org/10.5539/jfr.v2n1p50

Özçakmak S (2019) A model of hazard and risk analysis for bread production and the awareness of food safety. Qual Assur Saf Crop Foods 11:719–724. https://doi.org/10.3920/QAS2019.1592

Pacher N, Burtscher J, Johler S et al (2022) Ropiness in bread—a re-emerging spoilage phenomenon. Foods 11:3021. https://doi.org/10.3390/foods11193021

Paíga P, Morais S, Oliva-Teles T et al (2013) Determination of ochratoxin A in bread: evaluation of microwave-assisted extraction using an orthogonal composite design coupled with response surface methodology. Food Bioprocess Technol 6:2466–2477. https://doi.org/10.1007/s11947-012-0876-8

Palop A (1996) Influence of pH on heat resistance of *Bacillus licheniformis* in buffer and homogenised foods. Int J Food Microbiol 29:1–10. https://doi.org/10.1016/0168-1605(95)00010-0

Palumbo R, Crisci A, Venâncio A et al (2020) Occurrence and co-occurrence of mycotoxins in cereal-based feed and food. Microorganisms 8:74. https://doi.org/10.3390/microorganisms8010074

Panel EB (2005) Opinion of the scientific panel on biological hazards on *Bacillus cereus* and other Bacillus spp. in foodstuffs. EFSA J 175:1–48

Pattison TL, Lindsay D, von Holy A (2003) In vitro growth response of bread-spoilage *Bacillus* strains to selected natural antimicrobials. J Basic Microbiol 43:341–347. https://doi.org/10.1002/jobm.200390037

Pepe O, Blaiotta G, Moschetti G et al (2003) Rope-producing strains of *Bacillus* spp. from wheat bread and strategy for their control by lactic acid bacteria. Appl Environ Microbiol 69:2321–2329. https://doi.org/10.1128/AEM.69.4.2321-2329.2003

Pereira APM, Stradiotto GC, Freire L et al (2020) Occurrence and enumeration of rope-producing spore forming bacteria in flour and their spoilage potential in different bread formulations. LWT 133:110108. https://doi.org/10.1016/j.lwt.2020.110108

Piotrowska M (2014) The adsorption of ochratoxin A by *Lactobacillus* species. Toxins (Basel) 6:2826–2839. https://doi.org/10.3390/toxins6092826

Piotrowska M (2021) Microbiological decontamination of mycotoxins: opportunities and limitations. Toxins (Basel) 13:819. https://doi.org/10.3390/toxins13110819

Piotrowska M, Żakowska Z (2000) The biodegradation of ochratoxin A in food products by lactic acid bacteria and baker's yeast. In: Food biotechnology. Elsevier, pp 307–310

Pitt JI, Hocking AD (2009a) Spoilage of stored, processed and preserved foods. In: Fungi and food spoilage. Springer, Boston, pp 401–421

Pitt JI, Hocking AD (2009b) Primary keys and miscellaneous fungi. In: Fungi and food spoilage. Springer, Boston, pp 53–143

Richard JL (2007) Some major mycotoxins and their mycotoxicoses—an overview. Int J Food Microbiol 119:3–10. https://doi.org/10.1016/j.ijfoodmicro.2007.07.019

Ropejko K, Twarużek M (2021) Zearalenone and its metabolites—general overview, occurrence, and toxicity. Toxins (Basel) 13:35. https://doi.org/10.3390/toxins13010035

Rosenkvist H (1995) Contamination profiles and characterisation of *Bacillus* species in wheat bread and raw materials for bread production. Int J Food Microbiol 26:353–363. https://doi.org/10.1016/0168-1605(94)00147-X

Rumeus I, Turtoi M (2016) The influence of the classical cooling process on the occurrence of the rope spoilage in bread. In: Modern technologies in the food industry. pp 82–87

Saladino F, Quiles JM, Mañes J et al (2017) Dietary exposure to mycotoxins through the consumption of commercial bread loaf in Valencia, Spain. LWT 75:697–701. https://doi.org/10.1016/j.lwt.2016.10.029

Samar MM, Neira MS, Resnik SL, Pacin A (2001) Effect of fermentation on naturally occurring deoxynivalenol (DON) in Argentinean bread processing technology. Food Addit Contam 18:1004–1010. https://doi.org/10.1080/02652030110051284

Saranraj P, Sivasakthivelan P (2016) Microorganisms involved in spoilage of bread and its control measures. In: Bread and its fortification: nutrition and health benefits. Taylor & Francis, pp 132–149

Satomi M, la Duc MT, Venkateswaran K (2006) *Bacillus safensis* sp. nov., isolated from spacecraft and assembly-facility surfaces. Int J Syst Evol Microbiol 56:1735–1740. https://doi.org/10.1099/ijs.0.64189-0

Scudamore KA, Hazel CM, Patel S, Scriven F (2009) Deoxynivalenol and other *Fusarium* mycotoxins in bread, cake, and biscuits produced from UK-grown wheat under commercial and pilot scale conditions. Food Additiv Contamin: Part A 26:1191–1198. https://doi.org/10.1080/02652030902919426

Serp D, von Stockar U, Marison IW (2002) Immobilized bacterial spores for use as bioindicators in the validation of thermal sterilization processes. J Food Prot 65:1134–1141. https://doi.org/10.4315/0362-028X-65.7.1134

Settanni L, Massitti O, Van Sinderen D, Corsetti A (2005) In situ activity of a bacteriocin-producing Lactococcus lactis strain. Influence on the interactions between lactic acid bacteria during sourdough fermentation. J Appl Microbiol 99:670–681. https://doi.org/10.1111/j.1365-2672.2005.02647.x

Siedler S, Balti R, Neves AR (2019) Bioprotective mechanisms of lactic acid bacteria against fungal spoilage of food. Curr Opin Biotechnol 56:138–146. https://doi.org/10.1016/j.copbio.2018.11.015

Smith JP, Daifas DP, El-Khoury W et al (2004) Shelf life and safety concerns of bakery products—a review. Crit Rev Food Sci Nutr 44:19–55. https://doi.org/10.1080/10408690490263774

Soni A, Oey I, Silcock P, Bremer P (2016) *Bacillus* spores in the food industry: a review on resistance and response to novel inactivation technologies. Compr Rev Food Sci Food Saf 15:1139–1148. https://doi.org/10.1111/1541-4337.12231

Sorokulova IB, Reva ON, Smirnov VV et al (2003) Genetic diversity and involvement in bread spoilage of Bacillus strains isolated from flour and ropy bread. Lett Appl Microbiol 37:169–173. https://doi.org/10.1046/j.1472-765X.2003.01372.x

Suominen I, Andersson MA, Andersson MC et al (2001) Toxic *Bacillus pumilus* from indoor air, recycled paper pulp, Norway spruce, food poisoning outbreaks and clinical samples. Syst Appl Microbiol 24:267–276. https://doi.org/10.1078/0723-2020-00025

Tan H, Zhou H, Guo T et al (2023) Recent advances on formation, transformation, occurrence, and analytical strategy of modified mycotoxins in cereals and their products. Food Chem 405:134752. https://doi.org/10.1016/j.foodchem.2022.134752

Taylor JMW, Sutherland AD, Aidoo KE, Logan NA (2005) Heat-stable toxin production by strains of *Bacillus cereus*, *Bacillus firmus*, *Bacillus megaterium*, *Bacillus simplex* and *Bacillus licheniformis*. FEMS Microbiol Lett 242:313–317. https://doi.org/10.1016/j.femsle.2004.11.022

Thompson JM, Waites WM, Dodd CER (1998) Detection of rope spoilage in bread caused by *Bacillus* species. J Appl Microbiol 85:481–486. https://doi.org/10.1046/j.1365-2672.1998.853512.x

Todorov S, Onno B, Sorokine O et al (1999) Detection and characterization of a novel antibacterial substance produced by Lactobacillus plantarum ST 31 isolated from sourdough. Int J Food Microbiol 48:167–177. https://doi.org/10.1016/S0168-1605(99)00048-3

Tournas VH, Niazi NS (2018) Potentially toxigenic fungi from selected grains and grain products. J Food Saf 38. https://doi.org/10.1111/jfs.12422

Trombete F, Barros A, Vieira M et al (2016) Simultaneous determination of deoxynivalenol, deoxynivalenol-3-glucoside and nivalenol in wheat grains by HPLC-PDA with immunoaffinity column cleanup. Food Anal Methods 9:2579–2586. https://doi.org/10.1007/s12161-016-0450-5

Vaclavikova M, Malachova A, Veprikova Z et al (2013) 'Emerging' mycotoxins in cereals processing chains: changes of enniatins during beer and bread making. Food Chem 136:750–757. https://doi.org/10.1016/j.foodchem.2012.08.031

Valerio F, de Bellis P, di Biase M et al (2012) Diversity of spore-forming bacteria and identification of Bacillus amyloliquefaciens as a species frequently associated with the ropy spoilage of bread. Int J Food Microbiol 156:278–285. https://doi.org/10.1016/j.ijfoodmicro.2012.04.005

Valle Garcia M, Sonnenstrahl Bregão A, Parussolo G et al (2019) Incidence of spoilage fungi in the air of bakeries with different hygienic status. Int J Food Microbiol 290:254–261. https://doi.org/10.1016/j.ijfoodmicro.2018.10.022

Valle-Algarra FM, Mateo EM, Medina Á et al (2009) Changes in ochratoxin A and type B trichothecenes contained in wheat flour during dough fermentation and bread-baking. Food Additiv Contamin: Part A 26:896–906. https://doi.org/10.1080/02652030902788938

Viljoen CR, von Holy A (1997) Microbial populations associated with commercial bread production. J Basic Microbiol 37:439–444. https://doi.org/10.1002/jobm.3620370612

Vos P, Garrity G, Jones D et al (2011) Bergey's manual of systematic bacteriology: Volume 3: The Firmicutes. Springer

Wan J, Chen B, Rao J (2020) Occurrence and preventive strategies to control mycotoxins in cereal-based food. Compr Rev Food Sci Food Saf 19:928–953. https://doi.org/10.1111/1541-4337.12546

Wang J, Pawelzik E, Weinert J, Wolf GA (2005) Impact of *Fusarium culmorum* on the Polysaccharides of Wheat Flour. J Agric Food Chem 53:5818–5823. https://doi.org/10.1021/jf050525g

Weidenbörner M, Wieczorek C, Appel S, Kunz B (2000) Whole wheat and white wheat flour—the mycobiota and potential mycotoxins. Food Microbiol 17:103–107. https://doi.org/10.1006/fmic.1999.0279

WHO (2011) FAO/WHO guide for application of risk analysis principles and procedures during food safety emergencies. Food and Agriculture Organization of the United Nations

WHO Food safety fact sheet. https://www.who.int/news-room/fact-sheets/detail/food-safety. Accessed 3 Sept 2022

Yibar A, Cetinkaya F, Soyutemiz GE (2012) Detection of rope-producing bacillus in bread and ıdentification of ısolates to species level by vitek 2 system. J Biol Environ Sci 6:243–248

Zhang Y, Pei F, Fang Y et al (2019) Comparison of concentration and health risks of 9 *Fusarium* mycotoxins in commercial whole wheat flour and refined wheat flour by multi-IAC-HPLC. Food Chem 275:763–769. https://doi.org/10.1016/j.foodchem.2018.09.127

Zhang Z, Nie D, Fan K et al (2020) A systematic review of plant-conjugated masked mycotoxins: occurrence, toxicology, and metabolism. Crit Rev Food Sci Nutr 60:1523–1537. https://doi.org/10.1080/10408398.2019.1578944

Chapter 15
Impact of Sourdough Microbiota on FODMAPs and ATI Content in Bakery Products

Sabrina Geisslitz and Katharina Anne Scherf

15.1 Introduction

Wheat and rye products are important staple foods for human nutrition in many countries worldwide. They provide energy and 15–20% of the required daily intake of macro- and micronutrients including digestible carbohydrates, dietary fiber, protein, vitamins and minerals (Shiferaw et al. 2013). Especially wholegrain products are associated with reduced risks of type 2 diabetes and cardiovascular diseases, improved weight management and other health benefits (Chen et al. 2016; Reynolds et al. 2020). These undoubtedly positive aspects have been overshadowed recently by countless popular media reports conveying the overall message that the consumption of wheat products leads to a variety of adverse health effects and contributes to chronic diseases (Brouns et al. 2013). It is true that wheat-related disorders including celiac disease, wheat allergy and non-celiac wheat sensitivity (NCWS) affect about 1%, 0.5% and 0.5–15% of the population, respectively (Nwaru et al. 2014; Bradauskiene et al. 2021; Cárdenas-Torres et al. 2021). While individuals suffering from celiac disease and wheat allergy need to strictly eliminate wheat, rye and barley products from their diet, a wheat-reduced diet appears to be sufficient to alleviate symptoms of NCWS and irritable bowel syndrome (IBS) (Zevallos et al. 2017; Dionne et al. 2018). There is a lot of anecdotal evidence suggesting that sourdough products are better tolerated compared to yeast-leavened breads with short dough fermentation times, especially by those suffering from NCWS and IBS. The main reason is that the wide variety of yeasts and lactic acid bacteria (LAB) present in sourdoughs use many dough constituents, including starch, sugars and proteins, as substrates for their metabolic activity (Loponen and Gänzle 2018). In this way,

S. Geisslitz (✉) · K. A. Scherf
Department of Bioactive and Functional Food Chemistry, Institute of Applied Biosciences, Karlsruhe Institute of Technology (KIT), Karlsruhe, Germany
e-mail: sabrina.geisslitz@kit.edu; katharina.scherf@kit.edu; https://bioactivefc.iab.kit.edu

E. B. Ceresino et al. (eds.), *Sourdough Microbiota and Starter Cultures for Industry*, https://doi.org/10.1007/978-3-031-48604-3_15

specific constituents known to cause wheat-related disorders, such as fermentable oligosaccharides, disaccharides, monosaccharides and polyols (FODMAPs), amylase/trypsin-inhibitors (ATIs) and gluten proteins, are more or less extensively degraded, contributing to improved digestibility and tolerability of sourdough products. At the same time, dietary fibre, vitamins and minerals largely remain intact to provide their well-known health benefits.

This chapter will provide an overview on the occurrence and molecular structures of FODMAPs and ATIs, the inhibitory activity and immunoreactivity of ATIs, the positive dietary effects of FODMAPs, IBS, the low FODMAP diet as well as the effect of sourdough fermentation on FODMAPs and ATIs.

15.2 FODMAPs

15.2.1 Molecular Structures

FODMAPs is an acronym for **f**ermentable **o**ligosaccharides, **d**isaccharides, **m**onosaccharides **a**nd **p**olyols. The term was first coined by Gibson and Shepherd (2005) and comprises polyols, fructose, fructooligosaccharides (FOS), fructans, lactose, galactooligosaccharides and others including polydextrose that are poorly absorbed in the proximal small intestine, but rapidly fermented in the distal small intestine and proximal colon (Fig. 15.1).

Polyols refer to sugar alcohols including maltitol, mannitol, sorbitol, xylitol and erythritol. They occur naturally in apples, pears and stone fruits or are commonly added to foods as humectants and low calorie sugar substitutes. While the polyol content of wheat or rye wholemeal flours is negligible (0.05 g/100 g), yeast-leavened breads may contain glycerol and wheat sourdough breads contained 0.21 g/100 g of polyols (sum of mannitol, sorbitol and xylitol). The presence of mannitol depends on the presence of certain heterofermentative LAB that possess mannitol-dehydrogenase and are capable of reducing fructose to mannitol (Ispiryan et al. 2020).

Fructose is present in foods either in its free form as a monosaccharide or as a building block of disaccharides (e.g., sucrose or isomaltulose), FOS or fructans. The absorption efficiency of fructose in the human small intestine is about 85% that of glucose as long as both sugars are present in equimolar ratios, because glucose facilitates fructose absorption via the high capacity GLUT-2 transporter (Rumessen and Gudmand-Høyer 1986). In absence of glucose, fructose is absorbed via a low capacity facilitated diffusion across the small intestinal epithelium via GLUT-5 transporters. Fructose malabsorption universally occurs in case of high excess fructose levels, while about 30% of the population always have limited ability to absorb free fructose (Rumessen and Gudmand-Høyer 1986).

FOS consist of 2–10 fructose monomers (degree of polymerization [DP] 2–10) linked via β-(2→1)-glycosidic bonds with one α-D-glucose unit at the end of the

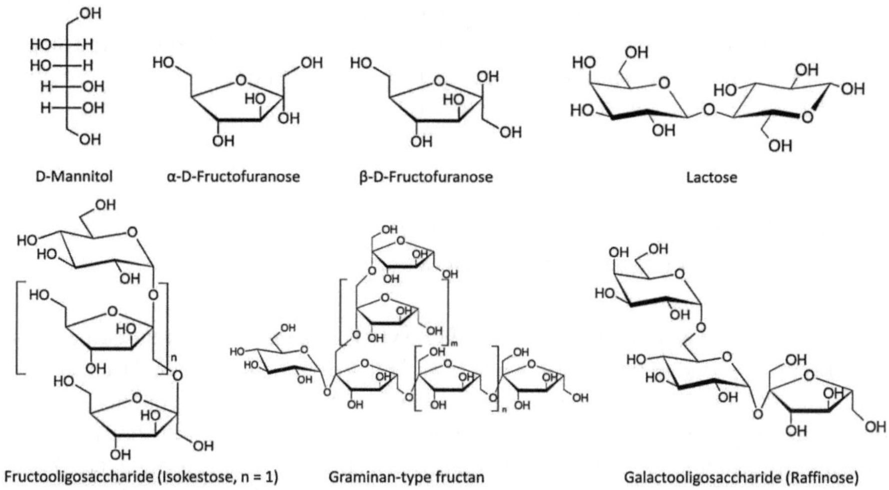

Fig. 15.1 Molecular structures of the most common fermentable oligosaccharides, disaccharides, monosaccharides, and polyols (FODMAPs) in cereal-based products, including D-mannitol as polyol; fructose in its α-D-fructofuranose and β-D-fructofuranose form, respectively, as monosaccharide; lactose (O-β-D-galactopyranosyl-(1→4)-D-glucopyranose) as disaccharide; isokestose (n = 1, O-α-D-glucopyranosyl-(1→2)-O-β-D-fructofuranosyl-(1→2)-β-D-fructofuranoside) as fructooligosaccharide; fructan (graminan-type); and raffinose (O-α-D-galactopyranosyl-(1→6)-O-α-D-glucopyranosyl-(1→2)-β-D-fructofuranoside) as galactooligosaccharide

chain. Important FOS are kestose (2 fructose units), nystose (3 fructose units) and 1-fructofuranosyl-nystose (4 fructose units) that are commonly used as prebiotics in foods. Fructans are FOS with DP >10–50. They are divided into inulin-type fructans with linear β-(2→1)-D-fructofuranose units, levan-type fructans with linear β-(2→6)-D-fructofuranose units and graminan-type fructans with both β-(2→1)-D- and β-(2→6)-D-fructofuranose units present in branched form (Lewis 1993). Wheat and rye predominantly contain graminan-type fructans (Grausgruber et al. 2020). The average DP is 5–7, but may also be less or up to 19. Fructans are present in the endosperm (1–2%), the germ (≈2%) and mainly in the bran and shorts (3–4%) (Haskå et al. 2008).

Lactose, with its systematic name β-D-galactopyranosyl-(1→4)-D-glucose, composes 2–8% of milk (by weight) and is the main FODMAP in dairy-based products. Although not naturally present in grain-based raw materials, grain-based products may still contain significant quantities of lactose depending on the ingredients used in the recipe (Ispiryan et al. 2020).

Galactooligosaccharides, also called raffinose family oligosaccharides (RFO) or α-galactosides, are α-(1→6) linked galactosyl derivatives of sucrose, mainly found in legumes. The most common RFO are raffinose (α-D-galactopyranosyl-(1→6)-α-D-glucopyranosyl-(1→2)-β-D-fructofuranoside) with one galactose unit, as well as stachyose and verbascose with two and three galactose units, respectively. RFO are only minor components of wheat- and rye-based raw materials (≈0.1 g/100 g), except for wheat bran (0.4 g/100 g), whereas barley (0.6 g/100 g) and oats (0.3 g/100 g) contain higher amounts of raffinose and stachyose. For comparison, lentils have 2.5 g/100 g of raffinose and stachyose plus 1.4 g/100 g of verbascose (Ispiryan et al. 2020).

15.2.2 FODMAP Content in Cereals and Cereal-Based Products

Table 15.1 provides an overview of the FODMAP content in different cereals and cereal-based products. In general, fructans are most abundant, followed by fructose, while the RFO raffinose and especially stachyose are minor components. Polyols such as mannitol are present in trace amounts in some cereal-based products, but sourdough-fermented bread may contain levels as high as 1.5%. As expected due to the enrichment of fructans in the outer layers of the grain (Haskå et al. 2008), wheat bran has the highest fructan content (3.4%) and white flour the lowest (1.2%), with wholemeal flour in between (1.7%).

Wheat starch has no detectable FODMAPs whereas vital gluten contains 0.2% of fructose and 0.6% of fructans, which corresponds to about 30% of the amount present in the flour. Whole rye grains have the highest fructan content (4.3–5.0%) compared to wheat and barley. Comparative analyses of wholegrain flours from wheat, spelt, emmer and einkorn show that einkorn has the highest fructan content on average (1.6–3.6%) followed by spelt (1.4–1.8%). Wheat and emmer have a comparable fructan content (0.8–1.4%), with two exceptions in the wheat varieties, that have up to 2% of fructans. The FODMAP content in the flour is one factor that determines the final content in baked or otherwise processed cereal-based products. The second and more important factor is the procedure of dough making and fermentation, as discussed in detail below (see Sect. 15.2.6). As a third factor, the final baking step may contribute to reducing fructan levels by 10–20% (Fretzdorff and Welge 2003), while pasta making can cause a loss of about 50% of fructans (Gélinas et al. 2016).

In general, the fructan content of yeast-leavened wheat breads is about 60–80% lower compared to the flour, mainly depending on the length of the proofing time, with longer times leading to greater fructan degradation (Longin et al. 2020; Schmidt and Sciurba 2021). As fructose is the degradation product of fructans, fructose levels in wheat breads may be two to three times higher, similar, or about 50% lower compared to the flour (Schmidt and Sciurba 2021), depending on how well fructose is used as a substrate by different yeast or LAB strains. The use of yeast and sourdough causes up to 90% fructan degradation and may be, therefore, more efficient compared to yeast fermentation alone. On the other hand, sourdough-fermented wheat breads contain 0.2–0.5% of mannitol, which is not present in the flour and formed by certain heterofermentative LAB. Similar effects are observed for rye breads except that the overall FODMAP content of rye breads is about five to seven times higher compared to wheat breads, even after sourdough fermentation (Schmidt and Sciurba 2021).

Table 15.1 FODMAP content [g/100 g] of cereals and cereal products

Sample	Fructose	Mannitol	Raffinose	Stachyose	Fructans	Reference
Grain, whole wheat	0.02–0.15	nd	0.12–0.78	0.01–0.20	0.66–3.20	Grausgruber et al. (2020)
Grain, whole wheat	0.24	–	–	–	1.69	Pitsch et al. (2021)
Grain, whole spelt	0.15	–	–	–	1.53	Pitsch et al. (2021)
Grain, whole rye, n = 13	–	–	–	–	4.6–6.6	Karppinen et al. (2003)
Grain, whole rye	–	–	0.1–0.7	Tr	4.3–5.0	Loponen and Gänzle (2018)
Grain, whole rye	0.25	–	–	–	3.36	Pitsch et al. (2021)
Grain, whole barley	0.05	nd	0.56	R&S sum	1.38	Ispiryan et al. (2020)
Flour, wheat, type 550	0.11	nd	0.06	–	1.19	Schmidt and Sciurba (2021)
Flour, wheat, wholemeal	0.20	nd	0.19	–	1.69	Schmidt and Sciurba (2021)
Flour, wheat, wholegrain, n = 21	tr – 0.10	–	–	–	0.85–2.00	Longin et al. (2020)
Flour, spelt, wholegrain, n = 9	–	–	–	–	1.35–1.76	Longin et al. (2020)
Flour, emmer, wholegrain, n = 9	–	–	–	–	0.80–1.40	Longin et al. (2020)
Flour, einkorn, wholegrain, n = 9	–	–	–	–	1.56–3.61	Longin et al. (2020)
Flour, rye, type 1150	1.20	nd	0.06	–	3.24	Schmidt and Sciurba (2021)
Flour, rye, wholemeal	1.50	nd	0.05	–	3.35	Schmidt and Sciurba (2021)
Bread, wheat, white	0.26	tr	0.20	nd	0.68	Biesiekierski et al. (2011)
Bread, wheat, white, n = 2	0.10–0.22	nd	nd	–	0.27–0.34	Schmidt and Sciurba (2021)
Bread, wheat, white, sourdough, n = 2	0.05–0.12	0.18–0.19	nd	–	0.22–0.24	Schmidt and Sciurba (2021)
Bread, wheat, white, ciabatta, n = 2	0.26–0.33	nd	nd	–	0.44–0.48	Schmidt and Sciurba (2021)
Bread, wheat, wholegrain, n = 3	0.19–0.27	nd/tr	0.19–0.38	0.31–0.36	0.48–0.69	Biesiekierski et al. (2011)
Bread, wheat, wholegrain, n = 2	0.11–0.65	–	–	–	0.10–0.30	Longin et al. (2020)
Bread, wheat, wholemeal	2.28	0.03	nd	–	0.55	Schmidt and Sciurba (2021)

(continued)

Table 15.1 (continued)

Sample	Fructose	Mannitol	Raffinose	Stachyose	Fructans	Reference
Bread, wheat, wholemeal, sourdough	1.75	0.52	nd	–	0.45	Schmidt and Sciurba (2021)
Bread, spelt, n = 2	0.13–0.16	nd	tr – 0.26	nd	0.14–0.20	Biesiekierski et al. (2011)
Bread, rye, n = 5	–	–	–	–	2.1–2.8	Karppinen et al. (2003)
Bread, rye, n = 2	0.38	nd	0.24–0.27	nd	1.05–1.42	Biesiekierski et al. (2011)
Bread, rye, sourdough	0.11	0.16	0.33	nd	1.07	Biesiekierski et al. (2011)
Bread, rye, type 1150, sourdough, n = 3	1.03–1.27	0.93–1.16	nd	–	1.70–1.76	Schmidt and Sciurba (2021)
Bread, rye, wholemeal, sourdough, n = 3	0.65–0.94	1.21–1.53	nd	–	1.91–1.94	Schmidt and Sciurba (2021)
Pasta, wheat	nd	nd	nd	nd	0.34	Biesiekierski et al. (2011)
Vital gluten, wheat	0.18	nd	0.03	R&S sum	0.60	Ispiryan et al. (2020)
Starch, wheat	nd	nd	nd	nd	nd	Ispiryan et al. (2020)
Bran, wheat	0.15	0.04	0.41	R&S sum	3.40	Ispiryan et al. (2020)

Values given on dry matter basis (Grausgruber et al. 2020; Ispiryan et al. 2020; Longin et al. 2020; Schmidt and Sciurba 2021); values given on basis of eaten weight (Biesiekierski et al. 2011)
– not analysed, *nd* not detected, *tr* traces, *R&S sum* given as sum of raffinose and stachyose

15.2.3 Positive Dietary Effects

FODMAPs, especially inulin-type fructans, are recognized as one form of dietary fiber, because they are largely resistant to breakdown in the human stomach and small intestine. They reach the colon where they are selectively fermented by intestinal microorganisms. By definition, prebiotics are substances that benefit the host's health by selectively targeting and stimulating the growth and/or activity(ies) of one or a limited number of beneficial host microorganisms (Gibson et al. 2017). Many FODMAPs including FOS, RFO, other galactooligosaccharides, xylooligosaccharides, polyols and possibly fructose exert prebiotic effects by promoting specific types of bacteria such as *Bifidobacteria*, *Faecalibacterium prausnitzii* and *Akkermansia municiphila* that are thought to improve the host's health (Verhoog et al. 2019). Dietary intervention in healthy subjects and IBS patients with a high FODMAP dose (5.5 g/day), habitual dietary intake (3.9 g/day) and a low FODMAP dose (1.6 g/day) demonstrated that the prebiotic effect appeared to be dose-dependent in this range (Halmos et al. 2015).

It is unclear whether very high doses (15–20 g/day) have additional positive effects, considering that these doses already lead to increased fermentation in the colon and related symptoms (Wilson et al. 2019). The main physiological outcomes of prebiotics are improved gut microbiota composition, improved intestinal functions (stool bulking, consistency and regularity), increased mineral absorption leading to improved bone mineral density, modulation of gastrointestinal peptide production, energy metabolism and satiety, regulation or modulation of immune functions, improved intestinal barrier function and reduced risk of intestinal infections. Further positive effects are tentatively being discussed including reduced risk of obesity, type 2 diabetes, metabolic syndrome, intestinal inflammation and colon cancer (for a detailed review, please see Roberfroid et al. (2010)).

These positive physiological effects of prebiotics are generally accepted based on metabolic and immunological markers, but evidence on health-related outcomes is often missing. One confounding factor in many human dietary intervention studies is that background FODMAP levels in the normal diet are inadequately taken into account. If the diet already contains moderate to high FODMAP levels, supplementation with FODMAPs is unlikely to have an additional benefit (Gibson et al. 2020). Almost all dietary interventions so far have been carried out with inulin or FOS, so that there is little knowledge whether graminan-type fructans from cereal-based products behave in a similar way.

15.2.4 Irritable Bowel Syndrome (IBS)

IBS has a prevalence of about 5–10% of the general population and is the most common reason to see a gastroenterologist (Sperber et al. 2017). Women are affected more frequently than men as well as individuals below the age of 50 (Palsson et al. 2020). As a chronic disorder of gut-brain interaction, IBS is characterized by recurrent symptoms of abdominal pain at least once per week on average, related to defecation and associated with a change in stool frequency and/or a change in stool appearance (Ford et al. 2017; Lacy et al. 2021). Other commonly reported symptoms include excessive gas, distension, bloating and urgency to defecate (Muir et al. 2019).

IBS does not cause mortality, but morbidity and has a pronounced negative effect on the patients' quality of life. In addition to being a significant burden to healthcare systems, IBS decreases work productivity and social integration and worsens psychosocial factors including anxiety, or depression, and vice versa (Koloski et al. 2012). Due to lack of biomarkers for IBS, a symptom-based diagnosis is made based on the Rome IV criteria that consider abdominal pain together with altered stool form or frequency for at least six months. Four IBS subgroups are commonly recognized according to the predominant stool pattern: (i) IBS with diarrhea, (ii) IBS with constipation, (iii) IBS with mixed stool pattern, and (iv) IBS unclassified for patients who meet criteria for IBS, but do not fall into either of the subgroups (i–iii) (Ford et al. 2020). Acute bacterial, viral or protozoal enteric infection increases the risk of developing post-infection IBS and this is the case for about 10% of IBS patients (Klem et al. 2017). Other factors affecting symptom severity in IBS can be peripheral (food, mucosal inflammation, abdominal surgery, menstruation) or central (stress, somatization, anxiety, depression, poor coping skills, poor social support, abuse). These associations are supported by the pathophysiological model to explain IBS. It links a certain genetic predisposition to gastrointestinal infections, psychological factors or adverse events early in life that trigger modulations of the enteric nervous system, which controls sensory, gastrointestinal motor, mucosal barrier and secretory responses (Ringel et al. 2001). Genome-wide association studies show that mutations in the sucrose-isomaltase gene and genes encoding for ion channels, e.g., the voltage-gated sodium channel NaV1.5, may play a role in IBS (Beyder et al. 2014; Henström et al. 2018).

The majority of IBS patients have increased sensitivity of the gut to different stimuli (e.g., slight pressure) and altered gut motility affecting contractions and transit times (Kellow and Phillips 1987; Posserud et al. 2007).

IBS symptoms are typically associated with foods and, therefore, diet and intestinal microbiota are likely involved. FODMAPs increase small intestinal water content and colonic volume to the same degree in IBS patients and in healthy individuals, but IBS patients respond with increased symptoms due to higher gut sensitivity (Major et al. 2017). The gut microbiome of IBS patients shows differences to that of healthy individuals, but the implications of these changes are not clear yet, because it remains to be established whether they are cause or effect (Pittayanon et al. 2019). Other pathophysiological alterations frequently associated with IBS are idiopathic bile acid diarrhea, increased gastrointestinal permeability and immune activation (Ford et al. 2020).

Treatment of IBS is directed towards the predominant symptom, but also very difficult, because most therapies lead to an improvement in only 25–30% of patients (Ford et al. 2020). Therefore, it is important that IBS patients are fully informed of available treatment options to make their own choice together with the physician. Overall, options include first-line medical therapies (e.g., laxatives, antidiarrheal, antispasmodics), second-line medical therapies (e.g., antidepressants, neuromodulators, medications targeting the central nervous system), psychological therapies (e.g., cognitive behavioral therapy, gut-directed hypnotherapy) and lifestyle changes (e.g., intake of probiotics, exercise, increased intake of soluble dietary fiber, low FODMAP diet). The best treatment option(s) need(s) to be determined individually for each patient, because the efficacy of each option is unclear in many cases, or the treatment is effective, but may cause other adverse effects. For current detailed recommendations on therapies for IBS, please refer to Lacy et al. (2021). A gluten-free diet is not recommended to reduce IBS symptoms (Dionne et al. 2018).

15.2.5 Low FODMAP Diet

The large majority of IBS patients (63%) perceive their symptoms to be related to certain foods, including fatty foods, coffee, alcohol, spicy foods and those containing high concentrations of FODMAPs, such as dairy products, pulses, onion, cabbage, garlic and wheat (Simrén et al. 2001). FODMAPs are osmotically active and poorly absorbed in the small intestine, which leads to water secretion into the small intestine and subsequent distension and abdominal symptoms. Further, they are rapidly degraded by microorganisms in the colon causing gas production, again associated with distension, pain and bloating. Both effects are likely to exacerbate IBS symptoms and support the rationale of restricting dietary intake of FODMAPs. Despite limitations in study design, number of participants and/or study completion or response, the low FODMAP diet led to an improvement of overall gastrointestinal symptoms of adult IBS patients (Nanayakkara et al. 2016; Dionne et al. 2018). In their randomized, controlled, single-blind, cross-over trial, Halmos et al. (2014) found that 70% of IBS patients had lower overall gastrointestinal symptom scores (reduced by almost 50%) while on the low FODMAP diet, compared to a standard Australian diet. Healthy controls had minimal or no symptoms without any changes.

Cut-off values have been defined for each FODMAP considering serving size and FODMAP content of typical foods based on a comparison of foods that are generally well tolerated to those that trigger symptoms in IBS. The total intake of FODMAPs should not exceed 0.5 g per single meal. Individual cut-off values are 0.3 g/serving for oligosaccharides (total fructans and RFO) from grain products, 1 g/serving for lactose, 0.15 g/serving for excess fructose and 0.4 g/serving for total polyols, or 0.2 g/serving for sorbitol or mannitol (Varney et al. 2017). Two slices of wheat bread (49 g), for example, are classified as high FODMAP, because the content of oligosaccharides (0.43 g/serving) exceeds the cut-off, even if all other FODMAPs are below the respective cut-off values. A comprehensive FODMAP food database and app is available from Monash University (Melbourne, Australia). The app provides information to users based on a traffic light system to identify low FODMAP choices (green light) and avoid medium or high FODMAP choices (yellow or red light) (https://monashfodmap.com/).

Based on all available studies so far, a limited trial of a low FODMAP diet is conditionally recommended to improve overall symptoms, although the quality of evidence is graded as very low (Lacy et al. 2021). While the low FODMAP diet appears to be safe overall, over-restriction of FODMAPs causes unintended effects such as micronutrient deficiencies or alterations of the gut microbiota (Halmos et al. 2015; Staudacher et al. 2020). Expert dietary counselling is thus essential to maintain adequate nutrient intake and diet diversity (Staudacher et al. 2020). The diet should follow three stages: (i) substitution of foods regularly eaten with low FODMAP alternatives, (ii) gradual reintroduction of regular foods containing individual FODMAPs while assessing symptoms, and (iii) personalization of the diet for long-term adherence to specifically avoid only the causative foods. Further well-designed and -controlled long-term studies are needed to evaluate efficacy, adherence, and potential adverse effects of the low FODMAP diet for IBS management.

15.2.6 Effect of Sourdough Fermentation on FODMAPs

Sourdough fermentation is one of the oldest and most widely used strategies to lower the FODMAP content in cereal-based products. In general, processing technology is more important than the cereal flour with respect to the final FODMAP content in the product as consumed (Pejcz et al. 2019; Grausgruber et al. 2020). Milling can be used as a first step to influence FODMAP content, because it is highest in the bran (3.7%) and the shorts (3.6%), followed by the germ (2.1%), the middling flour (2.0%) and the refined flour (1.5%) (Haskå et al. 2008). Although baking contributes to further FODMAP reduction, the main degradation occurs during dough mixing and fermentation through the combined actions of endogenous enzymes from wheat flour, yeast and, in case of sourdough, LAB (Nyyssölä et al. 2020).

Straight dough processes use baker's yeast (*Saccharomyces cerevisiae*) as the only microorganism for fermentation of sugars to carbon dioxide to achieve dough leavening. Typical fermentation times are 2 h or less. Sourdoughs contain LAB in addition to yeast, which broadens the diversity of metabolic activities. Type I sourdoughs undergo fermentation at 15–30 °C with one to three propagation steps. Common procedures use fermentation of about 10% of the flour for >12 h, then 20–30% of the flour for >6 h and all of the flour for 2–3 h. The predominant microorganisms are *Fructilactobacillus sanfranciscensis*, *Candida milleri* and *S. cerevisiae*, along with *Levilactobacillus brevis*, *Lactiplantibacillus plantarum* and *Limosilactobacillus reuteri*. Type II sourdoughs are fermented at about 40 °C and include *Lactobacillus delbrueckii*, *L. reuteri*, *Limosilactobacillus pontis* and *Limosilactobacillus panis* as main microorganisms (Loponen and Gänzle 2018).

Starch, especially damaged starch, is partially degraded to maltose and glucose by wheat or rye flour amylase and glucoamylase, respectively (Fig. 15.2). Sucrose is rapidly cleaved to glucose and fructose by yeast invertase. LAB use sucrose phosphorylase or sucrose-1-phosphate hydrolase to metabolize sucrose. The subsequent metabolism of glucose and fructose follows the Emden-Meyerhoff pathway in homofermentative LAB or the phosphoketolase pathway in heterofermentative LAB, some of which also reduce fructose to mannitol via mannitol-dehydrogenase (for details on LAB metabolic pathways, please see Gänzle (2015)). Fructans undergo partial hydrolysis catalyzed by yeast invertase to release fructose and fructans with a lower DP than in the flour. LAB contribute to fructan degradation through fructosidases or phosphofructosidases, but they can only use fructans with a DP of four or less. FOS can be degraded by homofermentative LAB, but only by some heterofermentative LAB. Other LAB metabolic pathways lead to the formation of isomalto-oligosaccharides (polymeric glucans) from sucrose via

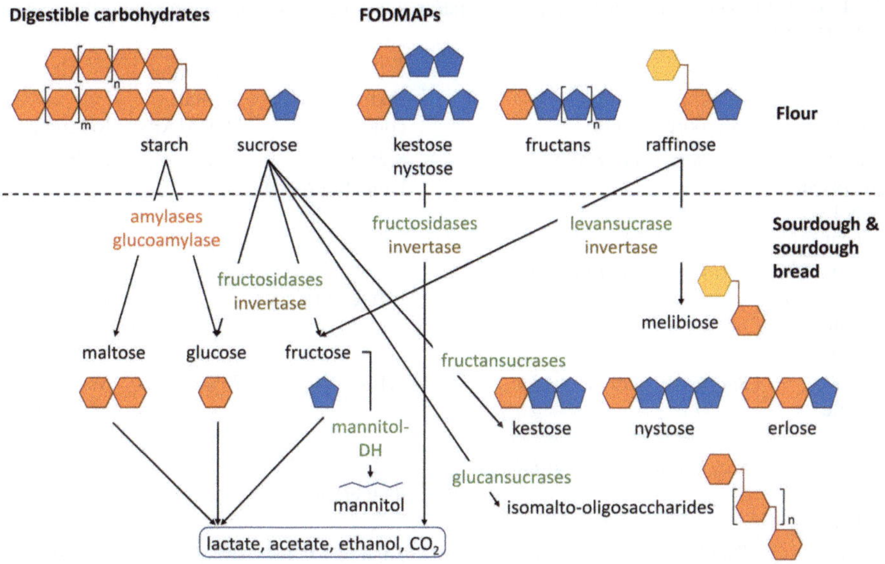

Fig. 15.2 Metabolism of carbohydrates, including fermentable oligosaccharides, disaccharides, monosaccharides, and polyols (FODMAPs) in wheat and rye sourdoughs. Amylases and glucoamylase are from the flour, fructosidases, levansucrase, fructansucrases and glucansucrases from lactic acid bacteria, mannitol-DH (dehydrogenase) from heterofermentative lactic acid bacteria, and invertase from yeast. Glucose is represented by orange hexagons, galactose by yellow hexagons and fructose by blue pentagons. (Figure modified from Loponen and Gänzle (2018))

glucansucrases or to the formation of levan or inulin and FOS from sucrose via fructansucrases. Relevant concentrations of these indigestible oligosaccharides are only formed when sucrose is added to sourdough. Raffinose is converted to melibiose and fructose or fructan by LAB levansucrase, followed by melibiose hydrolysis to galactose and glucose by α-galactosidase. Alternatively, raffinose is cleaved to galactose and sucrose by α-galactosidase. Yeast invertase is also capable of hydrolyzing raffinose to fructose and melibiose. In sum, a wide variety of different metabolic pathways contribute to yeast- and LAB-mediated degradation of FODMAPs, even if minor concentrations of new FODMAPs may be formed (Loponen and Gänzle (2018) and references cited therein). Supplementation with exogenous fructan-degrading enzymes such as invertase or inulinase may be useful, especially for degradation of fructans with DP >4 (Nyyssölä et al. 2020).

Optimized degradation of FODMAPs during (sour)dough fermentation requires a careful selection of microorganisms, especially those capable of utilizing fructans and mannitol. Table 15.2 provides an overview of recent studies that utilized different microorganisms, either alone or in combination, and various fermentation conditions to reduce total FODMAPs in breads made of wholemeal or refined wheat or rye flours. In general, prolonged fermentation times lead to more substantial reductions of FODMAPs (Ziegler et al. 2016; Schmidt and Sciurba 2021). Dough fermentation for about 2 h at 30 °C with *S. cerevisiae* alone already decreased

Table 15.2 Overview of studies using yeast and/or sourdough fermentation to reduce the content of FODMAPs

Product	Microorganism(s)	Fermentation conditions	Reduction of total FODMAPs[a]	Reference
Flour suspension, whole wheat	*Saccharomyces cerevisiae*	8.25 h, temperature n/a	83%	Fraberger et al. (2018)
Flour suspension, whole wheat	*Torulaspora delbrueckii*	8.25 h, temperature n/a	ca. 75%	Fraberger et al. (2018)
Bread, whole wheat	*Saccharomyces cerevisiae*	2.1 h at 30 °C	78%	Verspreet et al. (2013)
Bread, whole wheat	*Saccharomyces cerevisiae*	2.1 h, temperature n/a	47%	Knez et al. (2014)
Bread, whole wheat	*Saccharomyces cerevisiae*	1.0 h, 2.5 h and 4.5 h at 30 °C	0%, 67% and 90%	Ziegler et al. (2016)
Bread, whole wheat	*Saccharomyces cerevisiae*	2.1 h at 30 °C	56%	Struyf et al. (2017)
Bread, whole wheat	*Saccharomyces cerevisiae* + *Kluyveromyces marxianus* NCYC587	2.1 h at 30 °C	>90%	Struyf et al. (2017)
Bread, whole wheat	*Kluyveromyces marxianus* CBS6014	2.1 h at 30 °C	>90%	Struyf et al. (2018)
Bread, whole wheat	*Saccharomyces cerevisiae* (bioethanol strain 1, bread strain 4)	2.1 h at 30 °C	78% to 80%	Laurent et al. (2020)
Bread, whole wheat	*Saccharomyces cerevisiae* + *Kluyveromyces marxianus* CBS6014	1.2 h at 37 °C	>80%	Laurent et al. (2021)
Bread, whole wheat	*Saccharomyces cerevisiae* + *Lactiplantibacillus plantarum* 20,174 (*Lp*)	Stage I: *Lp* only for 24 h at 30 °C, stage II with dough (1 + 1): 2.5 h at 30 °C	56%	Pejcz et al. (2019)
Bread, whole wheat	*Saccharomyces cerevisiae*	1.8 h at RT and 24.3 h (1 h at RT, 17 h at 7 °C, 6.3 h at RT)	>65%	Longin et al. (2020)

(continued)

Table 15.2 (continued)

Product	Microorganism(s)	Fermentation conditions	Reduction of total FODMAPs[a]	Reference
Bread, wheat	Sourdough (not specified)	>12 h, temperature n/a	74%[b]	Laatikainen et al. (2017)
Bread, wheat	Sourdough (native from wheat flour, not specified)	Initially cultured at 21 °C and 30 °C; then for 10 h at 25 °C	69% and 75%	Menezes et al. (2019)
Bread, refined wheat	*Saccharomyces cerevisiae*	2.1 h, temperature n/a	62%	Knez et al. (2014)
Bread, refined wheat	*Saccharomyces cerevisiae*	3 h at 35 °C	68%	Gélinas et al. (2016)
Bread, refined wheat	*Saccharomyces cerevisiae* + *Lactiplantibacillus plantarum* 20174 (*Lp*)	Stage I: *Lp* only for 24 h at 30 °C, stage II with dough (1 + 1): 2.5 h at 30 °C	60%	Pejcz et al. (2019)
Bread, refined wheat	*Saccharomyces cerevisiae* + *Kluyveromyces marxianus* CBS6014	1.2 h at 37 °C	70% to 80%	Laurent et al. (2021)
Bread, refined wheat	*Saccharomyces cerevisiae* + *Weissella minor* 4451 + *Companilactobacillus farciminis* 4841 + *Levilactobacillus brevis* 4901	6 h at 30 °C	93%[2]	Menezes et al. (2021)
Bread, refined wheat	*Saccharomyces cerevisiae*	1.5 h and 1.8 h at 32 °C	58% and 71%	Schmidt and Sciurba (2021)
Bread, refined wheat	*Saccharomyces cerevisiae* + *Fructilactobacillus sanfranciscensis* + *Candida milleri*	Pre-dough (16 h, 26 °C), with dough (1 + 8.5) 1.5 h and 1.8 h at 32 °C	58% and 62%	Schmidt and Sciurba (2021)
Bread, whole spelt	*Saccharomyces cerevisiae*	1.0 h, 2.5 h and 4.5 h at 30 °C	6%, 71% and 77%	Ziegler et al. (2016)
Dough, durum	*Apilactobacillus kunkeei* B23I + *Fructobacillus fructosus* MBIII5, single or mixed	24 h, 30 °C	75% to 81%	Acín Albiac et al. (2020)
Bread, whole emmer	*Saccharomyces cerevisiae*	3 h at 35 °C	66%	Gélinas et al. (2016)

(continued)

Table 15.2 (continued)

Product	Microorganism(s)	Fermentation conditions	Reduction of total FODMAPs[a]	Reference
Bread, whole Khorasan wheat	*Saccharomyces cerevisiae*	3 h at 35 °C	98%	Gélinas et al. (2016)
Bread, wheat/rye mix	*Saccharomyces cerevisiae* + sourdough extract (not specified)	2.1 h, temperature n/a	40%	Knez et al. (2014)
Bread, whole rye	*Saccharomyces cerevisiae* + sourdough (not specified)	n/a	62%	Andersson et al. (2009)
Bread, whole rye	*Saccharomyces cerevisiae* + *Lactiplantibacillus plantarum* (*Lp*)	Stage I: *Lp* only for 24 h at 30 °C, stage II with dough (1 + 1): 3 h at 30 °C	16%	Pejcz et al. (2020)
Bread, whole rye	*Saccharomyces cerevisiae* + *Kluyveromyces marxianus* CBS6014	0.7 h at 37 °C	Ca. 90%	Laurent et al. (2021)
Bread, whole rye	*Saccharomyces cerevisiae*	1.5 h at 32 °C	6%	Schmidt and Sciurba (2021)
Bread, rye type 1150	*Saccharomyces cerevisiae* + *Fructilactobacillus sanfranciscensis* + *Candida milleri*	Pre-dough (16 h, 26 °C), with dough (1 + 1.2)	20%	Schmidt and Sciurba (2021)

n/a not available, *RT* room temperature
[a]Content in the final product relative to the flour; [b]content relative to a yeast-leavened bread

FODMAPs in breads made of wholemeal wheat flour by at least 47% (Knez et al. 2014) and up to 80% (Laurent et al. 2020). Similar reductions of up to 71% were achieved in breads produced from refined wheat flour (Schmidt and Sciurba 2021). FODMAPs were also degraded by 71% in whole spelt breads (Ziegler et al. 2016), by 66% in whole emmer breads (Gélinas et al. 2016) and even by 98% in whole Khorasan wheat breads (Gélinas et al. 2016) using *S. cerevisiae* and fermentation conditions from 2.5 to 3 h at 30 °C or 35 °C. In contrast, fermentation with *S. cerevisiae* for 1.5 h at 32 °C was not sufficient to effectively degrade FODMAPs in wholemeal rye breads (Schmidt and Sciurba 2021).

The comparison of 27 different *S. cerevisiae* strains from different industrial applications for making bread, bioethanol, beer or other alcoholic beverages showed a wide range of invertase activities towards sucrose or FOS. Two strains, one bioethanol and one bread strain, had high activity and specificity for FOS, resulting in faster and more extensive fructan hydrolysis during breadmaking. Appropriate selection of *S. cerevisiae* strains thus offers the opportunity to produce low FODMAP wheat breads (Laurent et al. 2020).

It is also important to note that the relative distribution of FODMAPs changes during dough fermentation (Ziegler et al. 2016; Struyf et al. 2017). After 1 h, more than 50% of the fructans present in wheat flour were degraded in bread, but high percentages of excess fructose from sucrose and fructan hydrolysis were detected, so that the total FODMAP level remained unchanged. Significant reductions of FODMAPs by 67% and 90% occurred after 2.5 h and 4.5 h, respectively, after substantial degradation of excess fructose and further reduction of fructans (Ziegler et al. 2016).

Compared to *S. cerevisiae*, the yeast *Kluyveromyces marxianus* is beneficial for FODMAP reduction, because it secretes inulinase, which efficiently degrades lactose and especially fructans. Wholemeal wheat doughs with *K. marxianus* thus showed faster fructan degradation compared to those with *S. cerevisiae*, and the resulting breads had less than 10% of the original fructan content. One disadvantage of using *K. marxianus* for breadmaking is its inability to metabolize maltose, because this results in low CO_2 production during dough leavening. Therefore, combining *K. marxianus* and *S. cerevisiae* as coculture was the most favorable approach to achieve high loaf volume and low FODMAP content (Struyf et al. 2017). This strategy reduced FODMAP levels by up to >90% not only in wholemeal and white wheat breads (Struyf et al. 2017; Laurent et al. 2021), but also in whole rye breads (Laurent et al. 2021) (Table 15.2).

Sourdough fermentation combining *S. cerevisiae* with single LAB (e.g., *L. plantarum*) further achieved FODMAP reductions of 56% in wholemeal wheat bread and of 60% in white wheat bread (Pejcz et al. 2019). The addition of mixtures of different LAB (e.g., *Weissella minor*, *Companilactobacillus farciminis* and *L. brevis*) degraded up to 93% of FODMAPs compared to a yeast-leavened bread (Menezes et al. 2021). Fructophilic LAB that preferentially consume fructans and fructose are promising candidates to reduce FODMAPs in wheat dough more efficiently than conventional LAB. *Apilactobacillus kunkeei* B23I demonstrated high fructose utilization, whereas *Fructobacillus fructosus* MBIII5 also metabolized mannitol as a substrate and produced lactic and acetic acids. Liquid dough and conditions with sufficient oxygen were favorable and there may be synergistic interactions with *S. cerevisiae* and *K. marxianus* that can be explored (Acín Albiac et al. 2020).

In conclusion, yeast and/or sourdough fermentation is essential for FODMAP reduction in wheat or rye breads. Maximal reduction can be achieved by appropriately selecting the grain raw material, by combining yeasts and LAB with high FODMAP-degrading activities and by determining the optimal conditions for fermentation and sourdough propagation.

Two randomized, double-blind controlled crossover studies evaluated whether low FODMAP rye breads (0.3–0.4 g of fructans/100 g) could be beneficial compared to regular rye breads (1.1–1.2 g of fructans/100 g) for IBS patients. While the low FODMAP rye bread reduced colonic fermentation as well as individual symptoms including flatulence, abdominal pain, cramps and stomach rumbling compared to regular rye bread, overall IBS symptom and quality of life scores showed no significant differences (Laatikainen et al. 2016). Comparable results were obtained by Pirkola et al. (2018), indicating that a replacement of regular by low FODMAP rye bread alone was not sufficient to relieve all IBS symptoms, but at least reduced some symptoms. In contrast, a double-blind controlled pilot study found no significant differences in IBS symptoms when yeast-fermented (0.23 g of fructans/100 g) and sourdough wheat breads (0.06 g of fructans/100 g) were compared (Laatikainen et al. 2017). Further studies with more participants and careful consideration of placebo and nocebo effects are needed to help clarify whether low FODMAP wheat or rye breads are indeed suitable for inclusion into the diet of subjects with IBS.

15.3 Amylase/Trypsin-Inhibitors (ATIs)

15.3.1 Occurrence and Molecular Structures

The name of this group of proteins already describes their biological function: ATIs inhibit amylase and trypsin. Cereal ATIs inhibit amylase and trypsin from different (non-wheat) sources, such as mammals, insects and mites (Silano et al. 1975; Buonocore et al. 1977; Priya et al. 2013; Cuccioloni et al. 2016) and as consequence ATIs play a role in the response to damage and infection (van Loon et al. 2006). ATIs are generally present in cereals including wheat, rye and barley, but there are also proteins that have inhibitory activity in, e.g., pulses (Shi et al. 2017).

ATIs correspond to about 4% of all wheat proteins (Altenbach et al. 2011; Call et al. 2020; Geisslitz et al. 2020) and belong to the albumin/globulin fraction according to their solubility in water and diluted salt solutions. In contrast to the gluten proteins that make up about 80% of wheat proteins, ATIs have a more balanced amino acid composition. Thus, they are supposed to play an important role as storage proteins for seedling growth besides their protective properties (Dupont and Altenbach 2003).

Wheat ATIs are classified in three groups (Table 15.3): (i) the wheat monomeric amylase inhibitor (WMAI), (ii) the wheat dimeric amylase inhibitors (WDAI) and (iii) the wheat tetrameric amylase inhibitors (WTAI) (Altenbach et al. 2011). The first two groups mainly have inhibitory activity against amylase and the last group has activity against both amylase and trypsin. The monomeric 0.28 belongs to the WMAI and the homodimeric 0.19 and 0.53 to the WDAI. These ATIs are named according to their electrophoretic mobility. In the WTAI group, the chloroform and methanol (CM) soluble, heterotetrameric CM1, CM2, CM3, CM16 and CM17 are summarized. Wheat ATIs are encoded on several chromosomes of the A, B and D genomes in wheat (Altenbach et al. 2011; Juhász et al. 2018). For example, 0.19 and 0.53 are encoded on 3B and 3D, CM3 on 4B and 4D, and CM16 on 2A, 4B and 4D. In wheat, there are 35 ATI genes (Juhász et al. 2018) and at least 33 isoforms

Table 15.3 Overview of ATI-types according to their aggregation state with number of amino acids and molecular mass according to the UniProtKB database (The UniProt Consortium 2017) and content in wheat flour (Geisslitz et al. 2020)

Aggregation state	Name	UniProtKB accession	Number of amino acids	Molecular mass (Da)	% total flour protein
Monomeric	0.28	P01083	123	13,326	0.4
Dimeric	0.19	P01085	124	13,337	0.9
	0.53	P0184	124	13,185	0.2
Tetrameric	CM1	P16850	120	13,086	0.2
	CM2	P16851	120	13,035	0.3
	CM3	P17314	143	15,832	0.7
	CM16	P16159	119	13,437	0.4
	CM17	Q41540	119	13,431	0.5
Total ATI content					3.6

already known (Altenbach et al. 2011; Bose et al. 2020) showing the complex genetic architecture of ATIs (El Hassouni et al. 2021).

With the exception of 0.53 (nine cysteine residues), all wheat ATIs have ten highly conserved cysteine residues at the same position within the amino acid sequence (Wang et al. 2010; Bose et al. 2020). The nine to ten cysteine residues form four to five intramolecular disulfide bonds resulting in a compact three-dimensional structure (Oda et al. 1997). This compact structure is the reason for the resistance to heat and their high pH stability between 3.0 and 11.7 (Buonocore et al. 1977). In addition to the conserved pattern of cysteine residues, wheat ATIs have a very high homology in their amino acid sequence (Junker et al. 2012; Møller and Svensson 2022). Even though ATIs from barley and rye are less well-characterized compared to wheat until now, all cereal ATIs show high homology in their sequence (Møller and Svensson 2022). Except for CM3, wheat ATIs comprise 119 to 124 amino acids and have a molecular mass between 13.0 kDa and 13.5 kDa. CM3 has more amino acids (143) and a higher molecular mass (15.8 kDa) than the other ATIs.

In addition to the ATIs summarized in Table 15.3, there are also other proteins included in the family of ATIs, which have inhibitory activity against amylase and subtilisin (WASI), chymotrypsin (WCI) or only against trypsin (WTI and CMX1/3). These minor ATIs have lower contents in wheat flour in comparison to the major ATIs (Geisslitz et al. 2020).

15.3.2 Inhibitory Activity and Immunoreactivity of ATIs

Cereal research on ATIs has increased over the last decades, as ATIs have many interesting properties. Besides the presence and the amount, two different activities of ATIs are described: (i) inhibitory activity: ATIs inhibit amylase and trypsin of different sources; (ii) immunoreactivity: ATIs trigger wheat-related disorders.

15.3.2.1 Inhibitory Activity

The tertiary structure is essential for the inhibitory activity of ATIs: after reduction of the disulfide bonds, ATIs lose their inhibitory activity (Buonocore et al. 1977). The inhibitory activity against amylase and trypsin from different sources depends on the respective ATI-type. The monomeric 0.28 has a higher inhibitory activity against amylases from insect pests (e.g., mealworm) than against human salivary amylase, but the dimeric inhibitors 0.19 and 0.53 have similar inhibitory activities against these sources (Feng et al. 1996). When comparing the inhibitory activity of monomeric and dimeric ATIs, it is even more complicated. The monomeric 0.28 has a higher activity against mealworm amylase than 0.19 and 0.53, but 0.19 and 0.53 have a higher activity against Colorado potato beetle amylase than 0.28. Other sources of insect amylases were equally inhibited (Gutierrez et al. 1990). The inhibitory activity of the CM-types is not only dependent on the tertiary structure, but also on the quaternary structure. In general, three types (e.g., CM2, CM3 and CM16 or CM1, CM3 and CM17) are necessary for ATIs to inhibit amylase, whereas each tetramer contains two CM3 units (Gomez et al. 1989). Furthermore, the CM-types are more active against amylase from mealworm than against that in human saliva (Gomez et al. 1989). In addition to the differential activity against different amylase sources, various wheat cultivars inhibit amylases to a different extent (Priya et al. 2013). Some cultivars show similar inhibitory activity against insect and human amylase, but interestingly, there are cultivars that inhibit amylases from insects more than those from human saliva and pancreas. Thus, these cultivars have a high resistance against storage pests, but might not influence human digestion (Priya et al. 2013).

Processing such as baking and cooking has an effect on the inhibitory activity against human amylase. Gélinas et al. (2018) showed that there was no detectable inhibitory activity against human amylase in commercially available cereal foods including cookies, crackers, pretzels or cakes. Cooking of spaghettini led to a remarkable decrease of inhibitory activity against human amylase. Furthermore, fermentation with and without yeast did not decrease the inhibitory activity, but there was no inhibitory activity detectable after baking for 34 min.

The majority of studies investigated only the inhibitory activity against trypsin from wheat extracts, but not from purified ATI-types (reviewed by Geisslitz et al. (2021)). The inhibitory activity against trypsin is influenced by genotype, environment (e.g., precipitation), location and interactions of the aforementioned factors (Priya et al. 2013; Call et al. 2020). Furthermore, the inhibitory activity depends on the source of trypsin, i.e., from mammals or insects.

The effect of the inhibitory activity against amylase and trypsin on human digestion is not absolutely clear. ATIs slow the release rate of glucose from starch during digestion and this can lead to modestly reduced glycemia and insulinemia (Lankisch et al. 1998). Furthermore, it might be conceivable that the inhibition of trypsin may result in increased levels of non-digested peptides and proteins, but currently there is no evidence for this (Geisslitz et al. 2021).

15.3.2.2 Immunoreactivity in the Context of NCWS

ATIs have the ability to activate the toll-like receptor 4 (TLR4)-MD2-CD14 complex on myeloid cells, monocytes, macrophages, and dendritic cells with a subsequent release of pro-inflammatory cytokines (Junker et al. 2012; Zevallos et al. 2017). This activation of innate immunity may cause NCWS, characterized by gastrointestinal symptoms including bloating, abdominal pain and diarrhea, as well as extraintestinal symptoms such as headache, tiredness or depression (reviewed by Cárdenas-Torres et al. (2021)). Furthermore, the activation may cause the worsening of preexisting inflammations, e.g., non-alcoholic fatty liver, metabolic syndrome and allergic airway inflammation (Zevallos et al. 2018; Ashfaq-Khan et al. 2019). In contrast to other wheat-related disorders, NCWS is less studied and there are still many open questions. The main reasons for this are the overlapping symptoms of IBS and NCWS and the lack of specific serological markers for NCWS, although potential biomarkers are already suggested (Cárdenas-Torres et al. 2021). The prevalence of NCWS is estimated to be between 0.49% and 14.9% and to be higher in women than in men (reviewed by Cárdenas-Torres et al. (2021)). In contrast to celiac disease, a wheat-reduced diet seems to be sufficient to treat NCWS (Zevallos et al. 2017). More details about NCWS are summarized in the reviews by Schuppan et al. (2015), Dale et al. (2019) and Cárdenas-Torres et al. (2021).

The number of studies investigating the NCWS immunoreactivity of wheat is very limited until now and only one study analyzed different cereal flours (Zevallos et al. 2017). Not only wheat flour possesses a high NCWS immunoreactivity, but also barley and rye are characterized by a high NCWS immunoreactivity. Contrary to that, other wheat species (emmer, einkorn and spelt), pseudocereals (buckwheat, millet, quinoa) and other cereals (oats and maize) have a lower NCWS immunoreactivity than wheat. Furthermore, the NCWS immunoreactivity depends on the source of the wheat flour: Some wheat flours showed a 50% higher NCWS immunoreactivity than the reference wheat flour. Baking and cooking reduced the NCWS immunoreactivity of wheat flour between 25% and 50%, showing that processing has an effect. Nevertheless, more studies are required that compare more flours of different species from different growing conditions.

15.3.2.3 Immunoreactivity in the Context of Celiac Disease and Wheat Allergy

Besides NCWS, the other two major wheat-related disorders are celiac disease (autoimmunogenic) and food allergy to wheat and cereals (allergic) (reviewed by Scherf et al. (2015)). The worldwide prevalence of celiac disease was recently estimated to be 1.25% (Bradauskiene et al. 2021) and about 0.5% of the population are affected by wheat allergy (Nwaru et al. 2014). Countless studies demonstrated that gluten proteins are the most important causative factors for these harmful immune responses, but it has also been shown that ATIs may either intensify or trigger these wheat-related disorders.

Serum antibodies of the immunoglobulin (Ig) classes A and G directed against gluten proteins are commonly detected in celiac disease patients. It has been shown that 60% of 14 celiac disease patients also had IgG against ATIs (Huebener et al. 2015). However, the role of ATIs in celiac disease pathogenesis is still not clear.

IgE antibodies play a central role in wheat allergy. Several studies showed that ATIs are recognized by IgE antibodies. The recombinant ATIs 0.28, CM3 and CM16 were active as allergens because they showed immunoreactivity with IgE from wheat allergic patients (Tundo et al. 2018). In another study, all patients' sera with atopic dermatitis showed positive IgE reactions with the ATIs 0.28, CM2 and CM16 and more than 90% of the patients reacted with 0.53 and CM17 (Šotkovský et al. 2011). Two studies with 101 and 94 bakers with wheat allergy revealed that 0.19, CM1, CM2, CM3 and CM17 resulted in positive IgE tests (Sander et al. 2015; Sander et al. 2016). Last, sera from 50 patients with wheat allergy showed an IgE-binding for 0.19, CM3, CM16 and CM17 (Šotkovský et al. 2008). Baking had an effect on the IgE-binding of ATIs, because dough showed IgE-binding, but crumb and crust did not (Simonato et al. 2001). In contrast, cooking of pasta did not reduce the IgE-binding of ATIs (Pastorello et al. 2007).

15.3.3 *Effect of Sourdough Fermentation on ATIs*

Several studies already showed that processing including cooking and baking has an effect on the inhibitory activity and immunoreactivity of ATIs. Sourdough fermentation is one promising tool to reduce ATI contents and activities.

15.3.3.1 Analytical Challenges and Requirements

The analysis of the effects of sourdough fermentation on ATIs requires robust, precise and correct methods. The extraction, detection and quantitation of ATIs are already well established in flours (Geisslitz et al. 2020; Sielaff et al. 2021), but the analysis of dough and bread is confronted with challenges due to the more complex matrix (reviewed by Geisslitz et al. (2021)). During dough resting, disulfide bonds are cleaved and newly formed, proteins are cross-linked or may be partially degraded by yeast and LAB and finally proteins are modified by the Maillard reaction, among other various reactions during baking. All these modifications lead to a decreased extractability of proteins (Lambrecht et al. 2015) and thus, also of ATIs. To compensate these challenges, two different controls containing no LAB are required to evaluate the effect of LAB on ATIs: the first control contains only yeast and in the second control, dough acidification is obtained by chemicals (e.g., lactic acid and acetic acid in a molar ratio of 4:1) until a pH of about 4 is reached (Rizzello et al. 2006; Won et al. 2021).

After the extraction of ATIs as completely as possible, various methods are suitable to monitor the degradation of ATIs. Sodium dodecyl sulfate-polyacrylamide gel electrophoresis (SDS-PAGE) and high-performance liquid chromatography (HPLC) are well established for the characterization of intact ATIs from dough and bread and these methods mainly determine the molecular mass of proteins. Immunoblots with sensitized patients' sera (IgG or IgE) are performed after SDS-PAGE to determine the immunoreactivity of different molecular mass ranges. The NCWS immunoreactivity and pro-inflammatory activity is analyzed by cell-based assays (Zevallos et al. 2017). The inhibitory activity against amylase and trypsin is determined by enzymatic assays (Gélinas and Gagnon 2018; Call et al. 2019). In contrast to SDS-PAGE and HPLC, enzymatic and cell-based assays are not compatible with high concentrations of chaotropic salts, reducing agents and organic solvents and it has always been questioned, if a decrease in activity is due to a decrease of ATIs or due to a low extractability (Won et al. 2021).

The number of studies combining all methods and analyzing the effect of sourdough on ATIs are very limited until now. However, some sourdough studies did not have the focus on ATIs, but on gluten proteins; nevertheless, those allow conclusions on the effect of sourdough on ATIs based on the published results.

15.3.3.2 Degradation of ATIs at Protein Level

Sourdough fermentation indeed has an effect on ATIs – the intensity of bands corresponding to ATIs (13.0–16.0 kDa) decreased in SDS-PAGE analysis. An exemplary SDS-PAGE gel is shown in Fig. 15.3, in which yeast-fermented dough at different conditions (lanes 1–5) is compared to chemically acidified dough (lanes 6

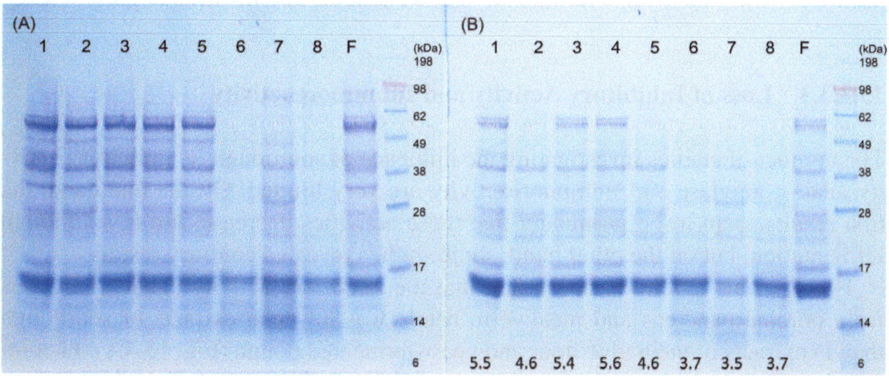

Fig. 15.3 SDS-PAGE under reducing condition of the albumin/globulin extract of doughs and sourdoughs (**a**) before baking and (**b**) after baking. The pH value after fermentation is indicated below in (**b**). 1–5, yeast-fermented dough: 1, at 30 °C for 1 h; 2, at 30 °C for 17 h; 3, at 4 °C for 17 h; 4, with glutathione at 30 °C for 1 h; 5, with glutathione at 30 °C for 17 h; 6, chemically acidified dough, pH 3.5, at 30 °C for 17 h; 7, wheat sourdough at 30 °C for 17 h; 8, chemically acidified dough, pH 3.5, with glutathione at 30 °C for 17 h; F, wheat flour. (Figure taken unmodified from Huang et al. (2020), available under a Creative Commons Attribution 4.0 International License: https://creativecommons.org/licenses/by/4.0/)

and 8) and sourdough fermentation (lane 7). Sourdough fermentation had a lower degrading effect before baking (Fig. 15.3A) than after baking (Fig. 15.3B). Especially when comparing yeast-fermented dough with sourdough fermentation, the latter had a remarkable effect on the degradation of ATIs and even a more pronounced effect than chemically acidified dough (Huang et al. 2020). This confirms that not only the lower pH value is responsible for ATI degradation.

Several studies performing SDS-PAGE gels confirmed that sourdough fermentation degrades ATIs more effectively than yeast fermentation and chemical acidification (Table 15.4). However, there are also studies that are contrary to this statement (Thiele et al. 2004; Fu et al. 2021). Furthermore, some studies summarized in Table 15.3 showed limitations in either the study design or in the interpretation for the degradation of ATIs. Finally, if different LAB are used in one study, some LAB had a higher degrading effect than others (Fu et al. 2020b) and even some LAB did not degrade ATIs at all (Di Cagno et al. 2002).

As an alternative method to SDS-PAGE, protein profiles and protein degradation are monitored by HPLC and UV-detection. One study investigated the potential of different LAB to degrade ATIs (Fraberger et al. 2020). In the model system, the LAB were cultivated on agar that contained an ATI isolate. After cultivation, the remaining ATIs were extracted from the medium and analyzed by HPLC-UV. The degradation capacity ranged from 52% to 85% and was dependent on the LAB. *Pediococcus pentosaceus* had the highest potential to degrade ATIs compared to the other nine tested LAB. Furthermore, there seemed to be a relationship between ATI degradation and concentration of amino acids, but this was only true for *P. pentosaceus*, but not for the other LAB.

In summary, there are indications that sourdough fermentation degrades ATIs more efficiently than yeast fermentation and chemical acidification at protein level.

15.3.3.3 Loss of Inhibitory Activity and Immunoreactivity

The number of studies investigating the influence of sourdough on inhibitory activity against amylase and immunoreactivity are very limited. On the one hand, the first results seem to be promising that these activities decrease during sourdough fermentation, but on the other hand, some studies revealed contrary results.

Huang et al. (2020) demonstrated that the inhibitory activity against amylases from porcine pancreas and mealworm beetle was lower in extracts of sourdough than in extracts of the native flour and yeast-fermented dough (Fig. 15.4A). In combination with the increase of free thiols in sourdough (sourdough had a higher content of free thiols than yeast-fermented dough), the authors postulate that the disulfide bonds of the ATIs are reduced to free thiols, which is the reason for the decreased inhibitory activity against amylase. The reduction of disulfide bonds opens the compact protein structure of ATIs and this leads to higher availability to proteolytic degradation (Junker et al. 2012; Huang et al. 2020). In contrast to Huang et al. (2020), Diowksz et al. (2020) revealed a higher inhibitory activity against amylase from porcine pancreas in extracts from sourdough bread compared to

Table 15.4 Overview of ATI degradation by different starter cultures and different fermentation times determined by SDS-PAGE. The ATI band refers to the band around 15 kDa

Starter culture[a]	Duration of fermentation	Findings	Limitations in interpretation or study design	Reference
Lactiplantibacillus plantarum	0, 2, 4, 6, 8, 10 h	Degradation of the ATI band after 6, 8 and 10 h compared to 0, 2 and 4 h in sourdough, but not in yeast-fermented dough	Not all ATIs visible due to inappropriate gel	Yin et al. (2015)
Enterococcus faecium *Lacticaseibacillus paracasei* *Lactiplantibacillus plantarum* *Pediococcus pentosaceus* *Pediococcus acidilactici* *Latilactobacillus sakei* *Torulaspora delbrueckii* *Pichia anomala* *Issatchenkia orientalis*	0, 24 h	Degradation of the ATI band after 24 h compared to 0 h Highest degradation efficiency: *Pediococcus acidilactici* *Pediococcus pentosaceus* *Latilactobacillus sakei*	The band intensity of the control (without LAB) also decreased.	Fu et al. (2020b)
Lentibacillus alimentarius *Levilactobacillus brevis* *Fructilactobacillus sanfranciscensis* *Lentilactobacillus hilgardii*	4, 8 h	Degradation of ATI spots depending on the LAB used: degradation more than 50%, but *L. hilgardii* had no effect	The study focus was not on ATI and identification of ATI spots was done by comparing with Dupont et al. (2011)	Di Cagno et al. (2002)
Rye sourdough starter (not specified)	1, 17 h	See Fig. 15.3 Most effective ATI degradation in chemically acidified dough, but no degradation in sourdough Degradation of ATIs in breads made of sourdough, but presence of ATIs in the controls	Use of commercially available sourdough starter	Huang et al. (2020)

(continued)

Table 15.4 (continued)

Starter culture[a]	Duration of fermentation	Findings	Limitations in interpretation or study design	Reference
Mixture of *Lentibacillus alimentarius* *Levilactobacillus brevis* *Fructilactobacillus sanfranciscensis* *Lentilactobacillus hilgardii*	24 h	Higher degree of degradation of the ATI band in albumin/ globulin extracts of sourdough compared to chemically acidified dough	No single LAB was used	Rizzello et al. (2006)
Mixture of *Lactiplantibacillus plantarum* *Levilactobacillus zymae* *Limosilactobacillus fermentum* Addition of Flavourzyme	12, 24 h	More effective degradation of the ATI band after 24 h than after 12 h in almost all samples, but most effective degradation (nearly 100%) in the combination of LAB and Flavourzyme	High amounts of Flavourzyme were added (1 g per 10 g wheat protein) to degrade proteins and also very effective degradation using Flavourzyme without LAB	Mickowska et al. (2019)
VSL#3 mixture containing *Streptococcus thermophilus* *Lactiplantibacillus plantarum* *Lactobacillus acidophilus* *Lacticaseibacillus casei* *Lactobacillus delbrueckii* spp. *Bulgaricus* *Bifidobacterium breve* *Bifidobacterium longum* *Bifidobacterium infantis*	24 h	More effective degradation of the ATI band in sourdough compared to chemically acidified dough	Not clear, if VSL#3 is usually used for sourdoughs and only the mixture of all microorganisms is responsible for the degradation of wheat proteins	De Angelis et al. (2007)
Combination of *Pediococcus acidilactici* *Lactobacillus delbrueckii*	0, 6, 12, 24 h	The ATI band of albumin/globulin extracts did not decrease, but higher molecular mass proteins degraded between 12 h and 24 h. This degradation was not observed in the acidified control.	It is in contrast to other studies that the ATI band did not decrease	Fu et al. (2021)

(continued)

Table 15.4 (continued)

Starter culture[a]	Duration of fermentation	Findings	Limitations in interpretation or study design	Reference
Fructilactobacillus sanfranciscensis Limosilactobacillus pontis	0, 6, 24 h	Lower intensity of the ATI band in sourdough after 0 and 6 h compared to 24 h, but similar intensity of the ATI band in 24 h sourdough and chemically acidified dough	No clear explanation why intensity of the ATI band increased from 6 to 24 h	Thiele et al. (2004)

[a]no additional information of the LAB is provided here; if different strains of the same LAB were used multiple times, the LAB is only listed once

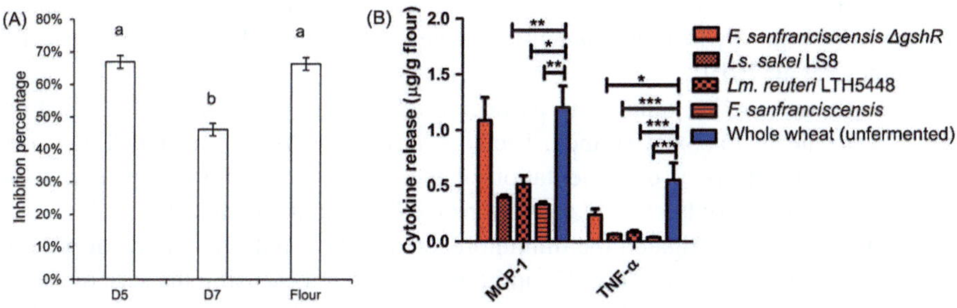

Fig. 15.4 (**A**) Inhibitory activity of the albumin/globulin extract from yeast-fermented dough (D5), sourdough (D7) and flour. Different letters indicate significant differences, one-way analysis of variance (ANOVA) with Tukey's honest significance test, $p < 0.05$. (**B**) ATI immunoreactivity of sourdough fermented by two accessions of *Fructilactobacillus sanfranciscensis* and one accession each of *Latilactobacillus sakei* and *Limosilactobacillus reuteri*. Quantitation of monocyte chemoattractant protein-1 (MCP-1) and tumour necrosis factor alpha (TNF-α) by ELISA. Asterisks indicate significant differences, one-way ANOVA with Dunnett's post-hoc test, *$p < 0.05$; **$p < 0.01$; ***$p < 0.001$. (**A**) of the figure is an excerpt with slight modification from a figure from Huang et al. (2020) and (**B**) is taken unmodified from Huang et al. (2020), both available under a Creative Commons Attribution 4.0 International License: https://creativecommons.org/licenses/by/4.0/

yeast-fermented bread. Overall, the study methodology showed limitations, especially because the naturally present wheat amylase was not inactivated by heat as it is usually done and the inhibitory activity in yeast-fermented bread was 0%. Furthermore, no correlation to wheat-related disorders, but correlations to possible health benefits (e.g., for diabetics) were made. Further studies are required to elucidate the effect of sourdough on the inhibitory activity against (human) amylases.

Only a limited number of research groups are currently able to analyze the NCWS immunoreactivity. In this experiment, ATI extracts are added to TLR4 expressing THP-1 human monocytic cells and the release of monocyte chemoattractant protein-1 (MCP-1) and tumor necrosis factor alpha (TNF-α) into the culture medium is quantitated by enzyme-linked immunosorbent assay (ELISA). The only study (Huang et al. 2020) comparing flour and sourdough showed that sourdough fermentation reduced the NCWS immunoreactivity in sourdough compared to the native flour (Fig. 15.4B). Unfortunately, no yeast-fermented control was included in this experiment. Nevertheless, the reduction of NCWS immunoreactivity was dependent on the LAB and *Latilactobacillus sakei* and *F. sanfranciscensis* were most effective.

Another study screened several LAB for their ability to degrade ATIs in a model system using a bioassay. The LAB were cultivated on agar that contained purified ATI. The LAB were sorted in two groups: (i) *Lactobacillus* strains with a high hydrolytic capacity against ATI (*Ligilactobacillus salivarius*, *Limosilactobacillus mucosae*, *Lacticaseibacillus rhamnosus*) and (ii) *Lactobacillus* strains with a low hydrolytic capacity (*Limosilactobacillus fermentum* and *L. reuteri*). The immunoreactivity (TNF-α) was the same for the flour extract and the extract that was incubated with the LAB with low hydrolytic capacity, but the extract containing the LAB with high hydrolytic capacity showed only half the immunoreactivity (Caminero et al. 2019).

Even though the immunoreactivity decreased in these two studies, the effect on NCWS still needs to be determined. The high variability during bread making using sourdough and the poor documentation of the effect of sourdough in the diet of NCSW patients are only exemplary reasons for this (Huang et al. 2020).

Many studies investigated the immunoreactivity (IgG and IgE) of wheat protein extracts by combining SDS-PAGE and immunoblots and by using the sera of sensitized patients with different kinds of wheat-related disorders. Most of these studies focused on gluten proteins, but it is possible to draw conclusions on ATIs, if the ATIs are visible at around 15 kDa in the SDS-PAGE and if the control (e.g., flour, yeast-fermented or chemically acidified dough) has an immunoreactivity at 15 kDa, but the sourdough either less or none. Some studies indicate that sourdough indeed reduces the immunoreactivity.

Immunoblots with goat anti-human IgG showed that bands around 15 kDa decreased in sourdoughs made with *Pediococcus acidilactici*, *P. pentosaceus* and *L. sakei* after 12 h compared to 0 h (Fu et al. 2020a). Four other LAB did not decrease the immunoreactivity. However, it has to be stated that this study has some limitations for the transfer to ATIs, since the protein marker in the immunoblots was not visible very well and differed in the blots made for 0 h and 12 h. Another study of the same group revealed that the immunoreactivity with goat anti-human IgG was the same for 15 kDa proteins in the albumin/globulin extracts of chemically acidified dough and sourdough (Fu et al. 2021). This study had limitations again, because different pooled sera were used for one sourdough than for the control.

Immunoblots with sera of patients with food allergy (IgE-dependent sensitization) showed that albumin/globulin extracts of chemically acidified dough had a higher immunoreactivity than extracts from sourdough at the band around 15 kDa (Rizzello et al. 2006). This study is in agreement with De Angelis et al. (2007) who revealed that proteins at the 15 kDa band had a higher immunoreactivity (immunoblots with IgE-dependent sensitization) in albumin/globulin extracts of chemically acidified dough compared to sourdough that was started with a mixture of eight probiotic LAB and *Bifidobacteria*. The band at 15 kDa was identified as 0.19 by mass spectrometry, which proved that 0.19 was degraded in the sourdough.

Sourdough fermentation can reduce the inhibitory activity and immunoreactivity of ATIs, but the effects on human health and on the wheat-related disorders NCWS, wheat allergy and celiac disease are still mostly unknown.

15.4 Conclusion

Taken together, a rapidly increasing number of studies shows that sourdough fermentation of wheat and rye flours is indeed suitable to degrade FODMAPs, ATIs and gluten proteins more or less extensively. Although this reduction of potentially immunoreactive constituents likely contributes to better tolerability of sourdough products compared to products with only short fermentation times, it is still unclear whether this really translates to clinically relevant benefits for NCWS and/or IBS patients. The few dietary intervention studies comparing sourdough to non-sourdough products so far found either limited or insignificant effects in IBS patients. One confounding factor in all studies so far is that sourdough fermentation simultaneously reduces the content of FODMAPs, ATIs, gluten and other potentially immunoreactive flour constituents, even if the extent is variable and depends on the specific microorganisms. Studies focusing on FODMAPs usually did not consider ATIs and gluten. There is only one pilot study reporting the content of fructans and gluten and at least the predominant form of ATIs assessed by SDS-PAGE. It found no difference in tolerability between the sourdough and the yeast-fermented wheat breads (Laatikainen et al. 2017). This calls for well-designed randomized, double-blind placebo-controlled crossover studies that consider potential confounders by working with (i) well-diagnosed and well-characterized NCWS or IBS patients, (ii) sufficient number of patients to achieve statistical power, (iii) sufficient study duration, (iv) well-characterized wheat or rye products considering FODMAPs, ATIs and gluten and (v) careful control of placebo and nocebo effects.

Furthermore, it is unclear whether branched graminan-type fructans from wheat and rye flours behave in a similar way to the linear inulin or FOS studied so far. With the huge microbial and associated metabolic diversity in sourdoughs, there is still great potential to uncover specific strains capable of efficiently degrading FODMAPs, ATIs and gluten. Maximal reduction can be achieved by using grain raw materials with already inherently low concentrations of potentially harmful constituents, by combining yeasts and LAB with high and specific metabolic activities and by determining the optimal conditions for fermentation and sourdough propagation. Careful disentanglement of the current confounders will pave the way towards grain-based foods with high nutritional and sensory value and better tolerability for individuals suffering from NCWS or IBS.

Acknowledgements This work is related to COST Action 18101 SOURDOMICS – Sourdough biotechnology network towards novel, healthier and sustainable food and bioprocesses (https://sourdomics.com; https://www.cost.eu/actions/CA18101. Accessed on 30 June, 2022).

References

Acín Albiac M, Di Cagno R, Filannino P, Cantatore V, Gobbetti M (2020) How fructophilic lactic acid bacteria may reduce the FODMAPs content in wheat-derived baked goods: a proof of concept. Microb Cell Factories 19(1):182. https://doi.org/10.1186/s12934-020-01438-6

Altenbach SB, Vensel WH, Dupont FM (2011) The spectrum of low molecular weight α-amylase/protease inhibitor genes expressed in the US bread wheat cultivar Butte 86. BMC Res Notes 4:242. https://doi.org/10.1186/1756-0500-4-242

Andersson R, Fransson G, Tietjen M, Åman P (2009) Content and molecular-weight distribution of dietary fiber components in whole-grain rye flour and bread. J Agri Food Chem 57(5):2004–2008. https://doi.org/10.1021/jf801280f

Ashfaq-Khan M, Aslam M, Qureshi MA, Senkowski MS, Yen-Weng S, Strand S, Kim YO, Pickert G, Schattenberg JM, Schuppan D (2019) Dietary wheat amylase trypsin inhibitors promote features of murine non-alcoholic fatty liver disease. Sci Rep 9(1):17463. https://doi.org/10.1038/s41598-019-53323-x

Beyder A, Mazzone A, Strege PR, Tester DJ, Saito YA, Bernard CE, Enders FT, Ek WE, Schmidt PT, Dlugosz A, Lindberg G, Karling P, Ohlsson B, Gazouli M, Nardone G, Cuomo R, Usai–Satta P, Galeazzi F, Neri M, Portincasa P, Bellini M, Barbara G, Camilleri M, Locke GR, Talley NJ, D'Amato M, Ackerman MJ, Farrugia G (2014) Loss-of-function of the voltage-gated sodium channel NaV1.5 (Channelopathies) in patients with irritable bowel syndrome. Gastroenterology 146(7):1659–1668. https://doi.org/10.1053/j.gastro.2014.02.054

Biesiekierski JR, Rosella O, Rose R, Liels K, Barrett JS, Shepherd SJ, Gibson PR, Muir JG (2011) Quantification of fructans, galacto-oligosacharides and other short-chain carbohydrates in processed grains and cereals. J Hum Nutr Diet 24(2):154–176. https://doi.org/10.1111/j.1365-277X.2010.01139.x

Bose U, Juhász A, Broadbent JA, Byrne K, Howitt CA, Colgrave ML (2020) Identification and quantitation of amylase trypsin inhibitors across cultivars representing the diversity of bread wheat. J Proteome Res 19(5):2136–2148. https://doi.org/10.1021/acs.jproteome.0c00059

Bradauskiene V, Vaiciulyte-Funk L, Martinaitiene D, Andruskiene J, Verma AK, Lima JPM, Serin Y, Catassi C (2021) Wheat consumption and prevalence of celiac disease: correlation from a multilevel analysis. Crit Rev Food Sci Nutr. https://doi.org/10.1080/10408398.2021.1939650

Brouns FJPH, van Buul VJ, Shewry PR (2013) Does wheat make us fat and sick? J Cereal Sci 58(2):209–215. https://doi.org/10.1016/j.jcs.2013.06.002

Buonocore V, Petrucci T, Silano V (1977) Wheat protein inhibitors of α-amylase. Phytochemistry 16(7):811–820. https://doi.org/10.1016/S0031-9422(00)86672-8

Call L, Reiter E, Wenger-Oehn G, Strnad I, Grausgruber H, Schoenlechner R, D'Amico S (2019) Development of an enzymatic assay for the quantitative determination of trypsin inhibitory activity in wheat. Food Chem 299:125038. https://doi.org/10.1016/j.foodchem.2019.125038

Call L, Kapeller M, Grausgruber H, Reiter E, Schoenlechner R, D'Amico S (2020) Effects of species and breeding on wheat protein composition. J Cereal Sci 93:102974. https://doi.org/10.1016/j.jcs.2020.102974

Caminero A, McCarville JL, Zevallos VF, Pigrau M, Yu XB, Jury J, Galipeau HJ, Clarizio AV, Casqueiro J, Murray JA, Collins SM, Alaedini A, Bercik P, Schuppan D, Verdu EF (2019) *Lactobacilli* degrade wheat amylase trypsin inhibitors to reduce intestinal dysfunction induced by immunogenic wheat proteins. Gastroenterology 156(8):2266–2280. https://doi.org/10.1053/j.gastro.2019.02.028

Cárdenas-Torres FI, Cabrera-Chávez F, Figueroa-Salcido OG, Ontiveros N (2021) Non-celiac gluten sensitivity: an update. Medicina 57(6):526. https://doi.org/10.3390/medicina57060526
Chen G-C, Tong X, Xu J-Y, Han S-F, Wan Z-X, Qin J-B, Qin L-Q (2016) Whole-grain intake and total, cardiovascular, and cancer mortality: a systematic review and meta-analysis of prospective studies. Am J Clin Nutr 104(1):164–172. https://doi.org/10.3945/ajcn.115.122432
Cuccioloni M, Mozzicafreddo M, Ali I, Bonfili L, Cecarini V, Eleuteri AM, Angeletti M (2016) Interaction between wheat alpha-amylase/trypsin bi-functional inhibitor and mammalian digestive enzymes: kinetic, equilibrium and structural characterization of binding. Food Chem 213:571–578. https://doi.org/10.1016/j.foodchem.2016.07.020
Dale HF, Biesiekierski JR, Lied GA (2019) Non-coeliac gluten sensitivity and the spectrum of gluten-related disorders: an updated overview. Nutr Res Rev 32(1):28–37. https://doi.org/10.1017/S095442241800015X
De Angelis M, Rizzello CG, Scala E, De Simone C, Farris GA, Turrini F, Gobbetti M (2007) Probiotic preparation has the capacity to hydrolyze proteins responsible for wheat allergy. J Food Prot 70(1):135–144. https://doi.org/10.4315/0362-028x-70.1.135
Di Cagno R, De Angelis M, Lavermicocca P, De Vincenzi M, Giovannini C, Faccia M, Gobbetti M (2002) Proteolysis by sourdough lactic acid bacteria: effects on wheat flour protein fractions and gliadin peptides involved in human cereal intolerance. Appl Environ Microbiol 68(2):623–633. https://doi.org/10.1128/AEM.68.2.623-633.2002
Dionne J, Ford AC, Yuan Y, Chey WD, Lacy BE, Saito YA, Quigley EMM, Moayyedi P (2018) A systematic review and meta-analysis evaluating the efficacy of a gluten-free diet and a low FODMAPs diet in treating symptoms of irritable bowel syndrome. Am J Gastroenterol 113(9):1290–1300. https://doi.org/10.1038/s41395-018-0195-4
Diowksz A, Malik A, Jaśniewska A, Leszczyńska J (2020) The inhibition of amylase and ACE enzyme and the reduction of immunoreactivity of sourdough bread. Foods 9(5):656. https://doi.org/10.3390/foods9050656
Dupont FM, Altenbach SB (2003) Molecular and biochemical impacts of environmental factors on wheat grain development and protein synthesis. J Cereal Sci 38(2):133–146. https://doi.org/10.1016/S0733-5210(03)00030-4
Dupont FM, Vensel WH, Tanaka CK, Hurkman WJ, Altenbach SB (2011) Deciphering the complexities of the wheat flour proteome using quantitative two-dimensional electrophoresis, three proteases and tandem mass spectrometry. Proteome Sci 9:10. https://doi.org/10.1186/1477-5956-9-10
El Hassouni K, Sielaff M, Curella V, Neerukonda M, Leiser W, Würschum T, Schuppan D, Tenzer S, Longin CFH (2021) Genetic architecture underlying the expression of eight α-amylase trypsin inhibitors. Theor Appl Genet 134(10):3427–3441. https://doi.org/10.1007/s00122-021-03906-y
Feng GH, Richardson M, Chen MS, Kramer KJ, Morgan TD, Reeck GR (1996) α-Amylase inhibitors from wheat: amino acid sequences and patterns of inhibition of insect and human α-amylases. Insect Biochem Mol Biol 26(5):419–426. https://doi.org/10.1016/0965-1748(95)00087-9
Ford AC, Lacy BE, Talley NJ (2017) Irritable bowel syndrome. N Engl J Med 376(26):2566–2578. https://doi.org/10.1056/NEJMra1607547
Ford AC, Sperber AD, Corsetti M, Camilleri M (2020) Irritable bowel syndrome. Lancet 396(10263):1675–1688. https://doi.org/10.1016/S0140-6736(20)31548-8
Fraberger V, Call L-M, Domig KJ, D'Amico S (2018) Applicability of yeast fermentation to reduce fructans and other FODMAPs. Nutrients 10(9):1247. https://doi.org/10.3390/nu10091247
Fraberger V, Ladurner M, Nemec A, Grunwald-Gruber C, Call LM, Hochegger R, Domig KJ, D'Amico S (2020) Insights into the potential of sourdough-related lactic acid bacteria to degrade proteins in wheat. Microorganisms 8(11):1689. https://doi.org/10.3390/microorganisms8111689
Fretzdorff B, Welge N (2003) Abbau von getreideeigenen Fructanen während der Herstellung von Roggenvollkornbrot. Getreide, Mehl und Brot 57:147–151. (in German)

Fu W, Rao H, Tian Y, Xue W (2020a) Bacterial composition in sourdoughs from different regions in China and the microbial potential to reduce wheat allergens. LWT 117:108669. https://doi.org/10.1016/j.lwt.2019.108669

Fu W, Xue W, Liu C, Tian Y, Zhang K, Zhu Z (2020b) Screening of lactic acid bacteria and yeasts from sourdough as starter cultures for reduced allergenicity wheat products. Foods 9(6):751. https://doi.org/10.3390/foods9060751

Fu W, Liu C, Meng X, Tao S, Xue W (2021) Co-culture fermentation of *Pediococcus acidilactici* XZ31 and yeast for enhanced degradation of wheat allergens. Int J Food Microbiol 347:109190. https://doi.org/10.1016/j.ijfoodmicro.2021.109190

Gänzle MG (2015) Lactic metabolism revisited: metabolism of lactic acid bacteria in food fermentations and food spoilage. Curr Opin Food Sci 2:106–117. https://doi.org/10.1016/j.cofs.2015.03.001

Geisslitz S, Longin CFH, Koehler P, Scherf KA (2020) Comparative quantitative LC–MS/MS analysis of 13 amylase/trypsin inhibitors in ancient and modern Triticum species. Sci Rep 10(1):14570. https://doi.org/10.1038/s41598-020-71413-z

Geisslitz S, Shewry P, Brouns F, America AHP, Caio GPI, Daly M, D'Amico S, De Giorgio R, Gilissen L, Grausgruber H, Huang X, Jonkers D, Keszthelyi D, Larré C, Masci S, Mills C, Møller MS, Sorrells ME, Svensson B, Zevallos VF, Weegels PL (2021) Wheat ATIs: characteristics and role in human disease. Front Nutr 8:667370. https://doi.org/10.3389/fnut.2021.667370

Gélinas P, Gagnon F (2018) Inhibitory activity towards human α-amylase in wheat flour and gluten. Int J Food Sci Technol 53(2):467–474. https://doi.org/10.1111/ijfs.13605

Gélinas P, McKinnon C, Gagnon F (2016) Fructans, water-soluble fibre and fermentable sugars in bread and pasta made with ancient and modern wheat. Int J Food Sci Technol 51(3):555–564. https://doi.org/10.1111/ijfs.13022

Gélinas P, McKinnon C, Gagnon F (2018) Inhibitory activity towards human α-amylase in cereal foods. LWT 93:268–273. https://doi.org/10.1016/j.lwt.2018.03.049

Gibson PR, Shepherd SJ (2005) Personal view: food for thought – western lifestyle and susceptibility to Crohn's disease. The FODMAP hypothesis. Aliment Pharmacol Ther 21(12):1399–1409. https://doi.org/10.1111/j.1365-2036.2005.02506.x

Gibson GR, Hutkins R, Sanders ME, Prescott SL, Reimer RA, Salminen SJ, Scott K, Stanton C, Swanson KS, Cani PD, Verbeke K, Reid G (2017) Expert consensus document: The International Scientific Association for Probiotics and Prebiotics (ISAPP) consensus statement on the definition and scope of prebiotics. Nat Rev Gastroenterol Hepatol 14(8):491–502. https://doi.org/10.1038/nrgastro.2017.75

Gibson PR, Halmos EP, Muir JG (2020) Review article: FODMAPS, prebiotics and gut health-the FODMAP hypothesis revisited. Aliment Pharmacol Ther 52(2):233–246. https://doi.org/10.1111/apt.15818

Gomez L, Sanchez-Monge R, Garcia-Olmedo F, Salcedo G (1989) Wheat tetrameric inhibitors of insect α-amylases: Alloploid heterosis at the molecular level. Proc Natl Acad Sci 86(9):3242–3246. https://doi.org/10.1073/pnas.86.9.3242

Grausgruber H, Lovegrove A, Shewry P, Békés F (2020) FODMAPs in wheat. In: Igrejas G, Ikeda TM, Guzmán C (eds) Wheat quality for improving processing and human health. Springer, Cham, pp 517–534. https://doi.org/10.1007/978-3-030-34163-3_21

Gutierrez C, Sanchez-Monge R, Gomez L, Ruiz-Tapiador M, Castañera P, Salcedo G (1990) α-Amylase activities of agricultural insect pests are specifically affected by different inhibitor preparations from wheat and barley endosperms. Plant Sci 72(1):37–44. https://doi.org/10.1016/0168-9452(90)90184-P

Halmos EP, Power VA, Shepherd SJ, Gibson PR, Muir JG (2014) A diet low in FODMAPs reduces symptoms of irritable bowel syndrome. Gastroenterology 146(1):67–75.e65. https://doi.org/10.1053/j.gastro.2013.09.046

Halmos EP, Christophersen CT, Bird AR, Shepherd SJ, Gibson PR, Muir JG (2015) Diets that differ in their FODMAP content alter the colonic luminal microenvironment. Gut 64(1):93–100. https://doi.org/10.1136/gutjnl-2014-307264

Haskå L, Nyman M, Andersson R (2008) Distribution and characterisation of fructan in wheat milling fractions. J Cereal Sci 48(3):768–774. https://doi.org/10.1016/j.jcs.2008.05.002

Henström M, Diekmann L, Bonfiglio F, Hadizadeh F, Kuech E-M, von Köckritz-Blickwede M, Thingholm LB, Zheng T, Assadi G, Dierks C, Heine M, Philipp U, Distl O, Money ME, Belheouane M, Heinsen F-A, Rafter J, Nardone G, Cuomo R, Usai-Satta P, Galeazzi F, Neri M, Walter S, Simrén M, Karling P, Ohlsson B, Schmidt PT, Lindberg G, Dlugosz A, Agreus L, Andreasson A, Mayer E, Baines JF, Engstrand L, Portincasa P, Bellini M, Stanghellini V, Barbara G, Chang L, Camilleri M, Franke A, Naim HY, Amato M (2018) Functional variants in the sucrase–isomaltase gene associate with increased risk of irritable bowel syndrome. Gut 67(2):263–270. https://doi.org/10.1136/gutjnl-2016-312456

Huang X, Schuppan D, Rojas Tovar LE, Zevallos VF, Loponen J, Gänzle M (2020) Sourdough fermentation degrades wheat alpha-amylase/trypsin inhibitor (ATI) and reduces pro-inflammatory activity. Foods 9(7):943. https://doi.org/10.3390/foods9070943

Huebener S, Tanaka CK, Uhde M, Zone JJ, Vensel WH, Kasarda DD, Beams L, Briani C, Green PH, Altenbach SB, Alaedini A (2015) Specific nongluten proteins of wheat are novel target antigens in celiac disease humoral response. J Proteome Res 14(1):503–511. https://doi.org/10.1021/pr500809b

Ispiryan L, Zannini E, Arendt EK (2020) Characterization of the FODMAP-profile in cereal-product ingredients. J Cereal Sci 92:102916. https://doi.org/10.1016/j.jcs.2020.102916

Juhász A, Belova T, Florides CG, Maulis C, Fischer I, Gell G, Birinyi Z, Ong J, Keeble-Gagnère G, Maharajan A, Ma W, Gibson P, Jia J, Lang D, Mayer KFX, Spannagl M, Tye-Din JA, Appels R, Olsen O-A (2018) Genome mapping of seed-borne allergens and immunoresponsive proteins in wheat. Sci Adv 4(8):eaar8602. https://doi.org/10.1126/sciadv.aar8602

Junker Y, Zeissig S, Kim SJ, Barisani D, Wieser H, Leffler DA, Zevallos V, Libermann TA, Dillon S, Freitag TL, Kelly CP, Schuppan D (2012) Wheat amylase trypsin inhibitors drive intestinal inflammation via activation of toll-like receptor 4. J Exp Med 209(13):2395–2408. https://doi.org/10.1084/jem.20102660

Karppinen S, Myllymäki O, Forssell P, Poutanen K (2003) Fructan content of rye and rye products. Cereal Chem 80(2):168–171. https://doi.org/10.1094/CCHEM.2003.80.2.168

Kellow JE, Phillips SF (1987) Altered small bowel motility in irritable bowel syndrome is correlated with symptoms. Gastroenterology 92(6):1885–1893. https://doi.org/10.1016/0016-5085(87)90620-2

Klem F, Wadhwa A, Prokop LJ, Sundt WJ, Farrugia G, Camilleri M, Singh S, Grover M (2017) Prevalence, risk factors, and outcomes of irritable bowel syndrome after infectious enteritis: a systematic review and meta-analysis. Gastroenterology 152(5):1042–1054.e1041. https://doi.org/10.1053/j.gastro.2016.12.039

Knez M, Abbott C, Stangoulis JCR (2014) Changes in the content of fructans and arabinoxylans during baking processes of leavened and unleavened breads. Eur Food Res Technol 239(5):803–811. https://doi.org/10.1007/s00217-014-2273-1

Koloski NA, Jones M, Kalantar J, Weltman M, Zaguirre J, Talley NJ (2012) The brain–gut pathway in functional gastrointestinal disorders is bidirectional: a 12-year prospective population-based study. Gut 61(9):1284–1290. https://doi.org/10.1136/gutjnl-2011-300474

Laatikainen R, Koskenpato J, Hongisto S-M, Loponen J, Poussa T, Hillilä M, Korpela R (2016) Randomised clinical trial: low-FODMAP rye bread vs. regular rye bread to relieve the symptoms of irritable bowel syndrome. Aliment Pharmacol Ther 44(5):460–470. https://doi.org/10.1111/apt.13726

Laatikainen R, Koskenpato J, Hongisto SM, Loponen J, Poussa T, Huang X, Sontag-Strohm T, Salmenkari H, Korpela R (2017) Pilot tudy: comparison of sourdough wheat bread and yeast-fermented wheat bread in individuals with wheat sensitivity and irritable bowel syndrome. Nutrients 9(11):1215. https://doi.org/10.3390/nu9111215

Lacy BE, Pimentel M, Brenner DM, Chey WD, Keefer LA, Long MD, Moshiree B (2021) ACG clinical guideline: management of irritable bowel syndrome. Am J Gastroenterol 116(1):17–44. https://doi.org/10.14309/ajg.0000000000001036

Lambrecht MA, Rombouts I, Van Kelst L, Delcour JA (2015) Impact of extraction and elution media on non-size effects in size exclusion chromatography of proteins. J Chromatogr A 1415:100–107. https://doi.org/10.1016/j.chroma.2015.08.070

Lankisch M, Layer P, Rizza RA, DiMagno EP (1998) Acute postprandial gastrointestinal and metabolic effects of wheat amylase inhibitor (WAI) in normal, obese, and diabetic humans. Pancreas 17(2):176–181. https://doi.org/10.1097/00006676-199808000-00011

Laurent J, Timmermans E, Struyf N, Verstrepen KJ, Courtin CM (2020) Variability in yeast invertase activity determines the extent of fructan hydrolysis during wheat dough fermentation and final FODMAP levels in bread. Int J Food Microbiol 326:108648. https://doi.org/10.1016/j.ijfoodmicro.2020.108648

Laurent J, Struyf N, Bautil A, Bakeeva A, Chmielarz M, Lyly M, Herrera-Malaver B, Passoth V, Verstrepen KJ, Courtin CM (2021) The potential of *Kluyveromyces marxianus* to produce low-FODMAP straight-dough and sourdough bread: a pilot-scale study. Food Bioprocess Technol 14(10):1920–1935. https://doi.org/10.1007/s11947-021-02681-4

Lewis DH (1993) Nomenclature and diagrammatic representation of oligomeric fructans—a paper for discussion. New Phytol 124(4):583–594. https://doi.org/10.1111/j.1469-8137.1993.tb03848.x

Longin CFH, Beck H, Gütler A, Gütler H, Heilig W, Zimmermann J, Bischoff SC, Würschum T (2020) Influence of wheat variety and dough preparation on FODMAP content in yeast-leavened wheat breads. J Cereal Sci 95:103021. https://doi.org/10.1016/j.jcs.2020.103021

Loponen J, Gänzle MG (2018) Use of sourdough in low FODMAP baking. Foods 7(7):96. https://doi.org/10.3390/foods7070096

Major G, Pritchard S, Murray K, Alappadan JP, Hoad CL, Marciani L, Gowland P, Spiller R (2017) Colon hypersensitivity to distension, rather than excessive gas production, produces carbohydrate-related symptoms in individuals with irritable bowel syndrome. Gastroenterology 152(1):124–133.e122. https://doi.org/10.1053/j.gastro.2016.09.062

Menezes LAA, Molognoni L, de Sá Ploêncio LA, Costa FBM, Daguer H, Dea Lindner JD (2019) Use of sourdough fermentation to reducing FODMAPs in breads. Eur Food Res Technol 245(6):1183–1195. https://doi.org/10.1007/s00217-019-03239-7

Menezes LAA, De Marco I, Neves Oliveira dos Santos N, Costa Nunes C, Leite Cartabiano CE, Molognoni L, Pereira GVM, Daguer H, De Dea Lindner J (2021) Reducing FODMAPs and improving bread quality using type II sourdough with selected starter cultures. Int J Food Sci Nutr 72(7):912–922. https://doi.org/10.1080/09637486.2021.1892603

Mickowska B, Romanova K, Urminska D (2019) Reduction of immunoreactivity of rye and wheat prolamins by lactobacilli and Flavourzyme proteolysis during sourdough fermentation – a way to obtain low-gluten bread. J Food Nutr Res 58(2):153–166

Møller MS, Svensson B (2022) Structure, function and protein engineering of cereal-type inhibitors acting on amylolytic enzymes. Front Mol Biosci 9:868568. https://doi.org/10.3389/fmolb.2022.868568

Muir JG, Varney JE, Ajamian M, Gibson PR (2019) Gluten-free and low-FODMAP sourdoughs for patients with coeliac disease and irritable bowel syndrome: a clinical perspective. Int J Food Microbiol 290:237–246. https://doi.org/10.1016/j.ijfoodmicro.2018.10.016

Nanayakkara WS, Skidmore PM, O'Brien L, Wilkinson TJ, Gearry RB (2016) Efficacy of the low FODMAP diet for treating irritable bowel syndrome: the evidence to date. Clin Exp Gastroenterol 9:131–142. https://doi.org/10.2147/ceg.S86798

Nwaru B, Hickstein L, Panesar S, Roberts G, Muraro A, Sheikh A, Allergy EF, Group AG (2014) Prevalence of common food allergies in Europe: a systematic review and meta-analysis. Allergy 69(8):992–1007. https://doi.org/10.1111/all.12423

Nyyssölä A, Ellilä S, Nordlund E, Poutanen K (2020) Reduction of FODMAP content by bioprocessing. Trends Food Sci Technol 99:257–272. https://doi.org/10.1016/j.tifs.2020.03.004

Oda Y, Matsunaga T, Fukuyama K, Miyazaki T, Morimoto T (1997) Tertiary and quaternary structures of 0.19 α-amylase inhibitor from wheat kernel determined by x-ray analysis at 2.06 Å resolution. Biochemistry 36(44):13503–13511. https://doi.org/10.1021/bi971307m

Palsson OS, Whitehead W, Törnblom H, Sperber AD, Simren M (2020) Prevalence of Rome IV functional bowel disorders among adults in the United States, Canada, and the United Kingdom. Gastroenterology 158(5):1262–1273.e1263. https://doi.org/10.1053/j.gastro.2019.12.021

Pastorello EA, Farioli L, Conti A, Pravettoni V, Bonomi S, Iametti S, Fortunato D, Scibilia J, Bindslev-Jensen C, Ballmer-Weber B, Robino AM, Ortolani C (2007) Wheat IgE-mediated food allergy in European patients: α-Amylase inhibitors, lipid transfer proteins and low-molecular-weight glutenins. Int Arch Allergy Immunol 144(1):10–22. https://doi.org/10.1159/000102609

Pejcz E, Spychaj R, Gil Z (2019) Technological methods for reducing the content of fructan in wheat bread. Foods 8(12):663. https://doi.org/10.3390/foods8120663

Pejcz E, Spychaj R, Gil Z (2020) Technological methods for reducing the content of fructan in rye bread. Eur Food Res Technol 246(9):1839–1846. https://doi.org/10.1007/s00217-020-03537-5

Pirkola L, Laatikainen R, Loponen J, Hongisto SM, Hillilä M, Nuora A, Yang B, Linderborg KM, Freese R (2018) Low-FODMAP vs regular rye bread in irritable bowel syndrome: randomized SmartPill(®) study. World J Gastroenterol 24(11):1259–1268. https://doi.org/10.3748/wjg.v24.i11.1259

Pitsch J, Sandner G, Huemer J, Huemer M, Huemer S, Weghuber J (2021) FODMAP fingerprinting of bakery products and aourdoughs: quantitative assessment and content reduction through fermentation. Foods 10(4):894. https://doi.org/10.3390/foods10040894

Pittayanon R, Lau JT, Yuan Y, Leontiadis GI, Tse F, Surette M, Moayyedi P (2019) Gut microbiota in patients with irritable bowel syndrome—a aystematic review. Gastroenterology 157(1):97–108. https://doi.org/10.1053/j.gastro.2019.03.049

Posserud I, Syrous A, Lindström L, Tack J, Abrahamsson H, Simrén M (2007) Altered rectal perception in irritable bowel syndrome is associated with symptom severity. Gastroenterology 133(4):1113–1123. https://doi.org/10.1053/j.gastro.2007.07.024

Priya S, Kumar S, Kaur N, Gupta A (2013) Specificity of α-amylase and trypsin inhibitor proteins in wheat against insect pests. New Zeal J Crop Hort Sci 41(1):49–56. https://doi.org/10.1080/01140671.2012.722112

Reynolds AN, Akerman AP, Mann J (2020) Dietary fibre and whole grains in diabetes management: systematic review and meta-analyses. PLoS Med 17(3):e1003053. https://doi.org/10.1371/journal.pmed.1003053

Ringel Y, Sperber AD, Drossman DA (2001) Irritable bowel syndrome. Annu Rev Med 52(1):319–338. https://doi.org/10.1146/annurev.med.52.1.319

Rizzello CG, De Angelis M, Coda R, Gobbetti M (2006) Use of selected sourdough lactic acid bacteria to hydrolyze wheat and rye proteins responsible for cereal allergy. Eur Food Res Technol 223(3):405–411. https://doi.org/10.1007/s00217-005-0220-x

Roberfroid M, Gibson GR, Hoyles L, McCartney AL, Rastall R, Rowland I, Wolvers D, Watzl B, Szajewska H, Stahl B, Guarner F, Respondek F, Whelan K, Coxam V, Davicco M-J, Léotoing L, Wittrant Y, Delzenne NM, Cani PD, Neyrinck AM, Meheust A (2010) Prebiotic effects: metabolic and health benefits. Br J Nutr 104(S2):S1–S63. https://doi.org/10.1017/S0007114510003363

Rumessen JJ, Gudmand-Høyer E (1986) Absorption capacity of fructose in healthy adults. Comparison with sucrose and its constituent monosaccharides. Gut 27(10):1161–1168. https://doi.org/10.1136/gut.27.10.1161

Sander I, Rihs H-P, Doekes G, Quirce S, Krop E, Rozynek P, van Kampen V, Merget R, Meurer U, Bruening T, Raulf M (2015) Component-resolved diagnosis of baker's allergy based on specific IgE to recombinant wheat flour proteins. J Allergy Clin Immunol 135(6):1529–1537. https://doi.org/10.1016/j.jaci.2014.11.021

Sander I, Rihs H-P, Brüning T, Raulf M (2016) A further wheat allergen for baker's asthma: Tri a 40. J Allergy Clin Immunol 137(4):1286. https://doi.org/10.1016/j.jaci.2015.11.026

Scherf KA, Koehler P, Wieser H (2015) Gluten and wheat sensitivities – an overview. J Cereal Sci 67:2–11. https://doi.org/10.1016/j.jcs.2015.07.008

Schmidt M, Sciurba E (2021) Determination of FODMAP contents of common wheat and rye breads and the effects of processing on the final contents. Eur Food Res Technol 247(2):395–410. https://doi.org/10.1007/s00217-020-03633-6

Schuppan D, Pickert G, Ashfaq-Khan M, Zevallos V (2015) Non-celiac wheat sensitivity: differential diagnosis, triggers and implications. Best Pract Res Clin Gastroenterol 29(3):469–476. https://doi.org/10.1016/j.bpg.2015.04.002

Shi L, Mu K, Arntfield SD, Nickerson MT (2017) Changes in levels of enzyme inhibitors during soaking and cooking for pulses available in Canada. J Food Sci Technol 54(4):1014–1022. https://doi.org/10.1007/s13197-017-2519-6

Shiferaw B, Smale M, Braun H-J, Duveiller E, Reynolds M, Muricho G (2013) Crops that feed the world 10. Past successes and future challenges to the role played by wheat in global food security. Food Secur 5(3):291–317. https://doi.org/10.1007/s12571-013-0263-y

Sielaff M, Curella V, Neerukonda M, Afzal M, El Hassouni K, Distler U, Schuppan D, Longin CFH, Tenzer S (2021) Hybrid QconCAT-based targeted absolute and data-independent acquisition-based label-free quantification enables in-depth proteomic characterization of wheat amylase/trypsin inhibitor extracts. J Proteome Res 20(3):1544–1557. https://doi.org/10.1021/acs.jproteome.0c00752

Silano V, Furia M, Gianfreda L, Macri A, Palescandolo R, Rab A, Scardi V, Stella E, Valfre F (1975) Inhibition of amylases from different origins by albumins from the wheat kernel. Biochim Biophys Acta 391(1):170–178. https://doi.org/10.1016/0005-2744(75)90163-1

Simonato B, Pasini G, Giannattasio M, Peruffo AD, De Lazzari F, Curioni A (2001) Food allergy to wheat products: the effect of bread baking and in vitro digestion on wheat allergenic proteins. A study with bread dough, crumb, and crust. J Agric Food Chem 49(11):5668–5673. https://doi.org/10.1021/jf0104984

Simrén M, Månsson A, Langkilde AM, Svedlund J, Abrahamsson H, Bengtsson U, Björnsson ES (2001) Food-related gastrointestinal symptoms in the irritable bowel syndrome. Digestion 63(2):108–115. https://doi.org/10.1159/000051878

Šotkovský P, Hubálek M, Hernychová L, Novák P, Havranová M, Šetinová I, Kitanovičová A, Fuchs M, Stulík J, Tučková L (2008) Proteomic analysis of wheat proteins recognized by IgE antibodies of allergic patients. Proteomics 8(8):1677–1691. https://doi.org/10.1002/pmic.200700347

Šotkovský P, Sklenář J, Halada P, Cinová J, Šetinová I, Kainarová A, Goliáš J, Pavlásková K, Honzová S, Tučková L (2011) A new approach to the isolation and characterization of wheat flour allergens. Clin Exp Allergy: J Br Soc Allergy Clin Immunol 41(7):1031–1043. https://doi.org/10.1111/j.1365-2222.2011.03766.x

Sperber AD, Dumitrascu D, Fukudo S, Gerson C, Ghoshal UC, Gwee KA, Hungin APS, Kang J-Y, Minhu C, Schmulson M, Bolotin A, Friger M, Freud T, Whitehead W (2017) The global prevalence of IBS in adults remains elusive due to the heterogeneity of studies: a Rome Foundation working team literature review. Gut 66(6):1075–1082. https://doi.org/10.1136/gutjnl-2015-311240

Staudacher HM, Ralph FSE, Irving PM, Whelan K, Lomer MCE (2020) Nutrient intake, diet quality, and diet diversity in irritable bowel syndrome and the impact of the low FODMAP diet. J Acad Nutr Diet 120(4):535–547. https://doi.org/10.1016/j.jand.2019.01.017

Struyf N, Laurent J, Verspreet J, Verstrepen KJ, Courtin CM (2017) *Saccharomyces cerevisiae* and *Kluyveromyces marxianus* cocultures allow reduction of fermentable oligo-, di-, and monosaccharides and polyols levels in whole wheat bread. J Agri Food Chem 65(39):8704–8713. https://doi.org/10.1021/acs.jafc.7b02793

Struyf N, Vandewiele H, Herrera-Malaver B, Verspreet J, Verstrepen KJ, Courtin CM (2018) *Kluyveromyces marxianus* yeast enables the production of low FODMAP whole wheat breads. Food Microbiol 76:135–145. https://doi.org/10.1016/j.fm.2018.04.014

The UniProt Consortium (2017) UniProt: the universal protein knowledgebase. Nucleic Acids Res 45(Database issue):D115–D119. https://doi.org/10.1093/nar/gkh131

Thiele C, Grassl S, Gänzle M (2004) Gluten hydrolysis and depolymerization during sourdough fermentation. J Agri Food Chem 52(5):1307–1314. https://doi.org/10.1021/jf034470z

Tundo S, Lupi R, Lafond M, Giardina T, Larre C, Denery-Papini S, Morisset M, Kalunke R, Sestili F, Masci S (2018) Wheat ATI CM3, CM16 and 0.28 allergens produced in *Pichia pastoris* display a different eliciting potential in food allergy to wheat. Plants 7(4):101. https://doi.org/10.3390/plants7040101

van Loon LC, Rep M, Pieterse CM (2006) Significance of inducible defense-related proteins in infected plants. Annu Rev Phytopathol 44:135–162. https://doi.org/10.1146/annurev.phyto.44.070505.143425

Varney J, Barrett J, Scarlata K, Catsos P, Gibson PR, Muir JG (2017) FODMAPs: food composition, defining cutoff values and international application. J Gastroenterol Hepatol 32:53–61. https://doi.org/10.1111/jgh.13698

Verhoog S, Taneri PE, Roa Díaz ZM, Marques-Vidal P, Troup JP, Bally L, Franco OH, Glisic M, Muka T (2019) Dietary factors and modulation of bacteria strains of *Akkermansia muciniphila* and *Faecalibacterium prausnitzii*: a systematic review. Nutrients 11(7):1565. https://doi.org/10.3390/nu11071565

Verspreet J, Hemdane S, Dornez E, Cuyvers S, Delcour JA, Courtin CM (2013) Maximizing the concentrations of wheat grain fructans in bread by exploring strategies to prevent their yeast (*Saccharomyces cerevisiae*)-mediated degradation. J Agri Food Chem 61(6):1397–1404. https://doi.org/10.1021/jf3050846

Wang J-R, Pu Z-E, Lan X-J, Baum BR, Yan Z-H, Zheng Y-L, Wei Y-M (2010) Phylogenetic analysis of the dimeric alpha-amylase inhibitor sequences from an orthologous region in 21 different genomes of the tribe Triticeae (Poaceae). Biochem Syst Ecol 38(4):708–714. https://doi.org/10.1016/j.bse.2010.05.004

Wilson B, Rossi M, Dimidi E, Whelan K (2019) Prebiotics in irritable bowel syndrome and other functional bowel disorders in adults: a systematic review and meta-analysis of randomized controlled trials. Am J Clin Nutr 109(4):1098–1111. https://doi.org/10.1093/ajcn/nqy376

Won S, Curtis J, Gänzle M (2021) LC-MS/MS quantitation of α-amylase/trypsin inhibitor CM3 and glutathione during wheat sourdough breadmaking. J Appl Microbiol 00:1–10. https://doi.org/10.1111/jam.15346

Yin Y, Wang J, Yang S, Feng J, Jia F, Zhang C (2015) Protein degradation in wheat sourdough fermentation with *lactobacillus plantarum* M616. Interdiscip Sci 7(2):205–210. https://doi.org/10.1007/s12539-015-0262-0

Zevallos VF, Raker V, Tenzer S, Jimenez-Calvente C, Ashfaq-Khan M, Ruessel N, Pickert G, Schild H, Steinbrink K, Schuppan D (2017) Nutritional wheat amylase-trypsin inhibitors promote intestinal inflammation via activation of myeloid cells. Gastroenterology 152(5):1100–1113.e1112. https://doi.org/10.1053/j.gastro.2016.12.006

Zevallos VF, Raker VK, Maxeiner J, Scholtes P, Steinbrink K, Schuppan D (2018) Dietary wheat amylase trypsin inhibitors exacerbate murine allergic airway inflammation. Eur J Nutr. https://doi.org/10.1007/s00394-018-1681-6

Ziegler JU, Steiner D, Longin CFH, Würschum T, Schweiggert RM, Carle R (2016) Wheat and the irritable bowel syndrome – FODMAP levels of modern and ancient species and their retention during bread making. J Funct Foods 25:257–266. https://doi.org/10.1016/j.jff.2016.05.019

Chapter 16
Recent Advances in the Use of Sourdough Fermentation to Improve the Quality of Gluten-Free Bakery Products

Anıl Alper and Aylin Altan

16.1 Introduction

Today, gluten-free bakery products are characterized by a lower nutritional content and poorer taste than their traditional counterparts. The studies conducted in recent years have focused on alternative approaches that can mimic or substitute the gluten protein network to enhance the quality of gluten-free foods and increase their acceptability by the consumer. Some new biotechnological approaches are sourdough, in situ EPS production, and gluten detoxification (Bender and Schönlechner 2020; Moroni et al. 2009). Sourdough biotechnology is low-cost and ecological, having great potential in the gluten-free food market (Ramos et al. 2021). Sourdough, which has many advantages over commercial baker's yeast, is used in the modern biotechnology of bakery products as a natural fermenting agent to produce foods with superior properties such as high sensory quality and a characteristic taste (Corsetti and Settanni 2007). In sourdough technology, metabolites formed by using LAB and yeasts, alone or in combination, improve the taste, texture, flavor, shelf life, and nutritional content of foods, especially gluten-free ones. In addition, using sourdough with gluten-free flours increases the amount of certain bioactive compounds and reduces the amount of anti-nutritional compounds such as phytic acid (Dingeo et al. 2020). Furthermore, extracellular polysaccharides (commonly known as exopolysaccharides) produced by LAB can be an alternative to conventional thickeners to alter stability, structure, and viscosity of gluten-free bakery products. As EPS can replace hydrocolloids used as food additives, the sourdough application can constitute a cost-effective approach to improve the rheology of gluten-free doughs (Chochkov et al. 2022). In addition to EPS, the importance of sourdough LAB also lies in ensuring the hydrolysis of toxic metabolites in flour to detoxify

A. Alper · A. Altan (✉)
Department of Food Engineering, Mersin University, Mersin, Turkey
e-mail: aaltan@mersin.edu.tr

E. B. Ceresino et al. (eds.), *Sourdough Microbiota and Starter Cultures for Industry*, https://doi.org/10.1007/978-3-031-48604-3_16

gluten (Deora et al. 2014). Long-term *in vivo* trials with celiac patients proved that experimental bakery products made with hydrolyzed wheat flour via sourdough fermentation and fungal proteases were safe (Di Cagno et al. 2010; Greco et al. 2011). The gluten-free bread made with durum wheat semolina rendered gluten-free by fermentation with sourdough LAB and fungal proteases showed better nutritional properties than commercial gluten free products, and similar sensory properties to those of gluten containing bread (Rizzello et al. 2016). Various kinds of cereal, pseudocereals, and legume flours are widely used to produce gluten-free products, such as gluten-free bread, gluten-free biscuit, and gluten-free pizza (Arendt et al. 2002). In addition, microbiological studies on gluten-free sourdoughs indicate that these gluten-free flours are an essential source for new, competitive yeast and LAB strains that can be selected as starters for stable gluten-free sourdoughs (Moroni et al. 2010). Therefore, this chapter covers recent applications of sourdough fermentation for improving production and quality of gluten-free products.

16.2 Sourdough

Cereal fermentations are one of the essential biotechnological processes used since ancient Egypt to produce bread and beer through the interaction of LAB and yeasts (Poutanen et al. 2009). On the other hand, spontaneous sourdough fermentation is among the oldest cereal fermentation methods, which is based on fermenting the dough to aerate the bread (Chavan and Chavan 2011). Sourdough fermentation, which is the fermentation and acidification process of moist cereal flour by the bacteria and yeasts in the microbiota of sourdough, characterized by the micro-interactions between microorganisms and the environment (Edema et al. 2013). The microbiota of cereal and flour is important for grain-based foods, including sourdough bread (Papadimitriou et al. 2019). Sourdough is produced from spontaneously fermenting dough and renewed frequently at certain intervals under controlled conditions. Sourdough biotechnology gets its name from the acidification occurring in the dough due to microorganisms that release amino acids and organic acids during fermentation (Cappa et al. 2016). The first factor distinguishing sourdough processes from other dough processes is the presence of LAB and their interaction with yeasts. Another factor is the long fermentation period, which ensures the activation of endogenous enzymes (Gänzle 2014). The main metabolic activities from the sourdough stage to the final bread are the increase in acidity (LAB), fermentation (LAB and yeasts), and aroma formation (yeasts and LAB). This process results in a dough with a low pH (approximately 4.0) and sourdough bread with the desired taste and texture (De Vuyst et al. 2009). Sourdough can be produced in three different ways (Hansen 2012):

(i) Through spontaneous fermentation,
(ii) By adding a little ripe sourdough,
(iii) By adding starter culture.

Sourdough production through spontaneous fermentation has been used for years. However, using selected starter cultures has been gradually increasing due to their properties enhancing the nutritional and sensory quality, especially in cases where specific sensory, nutritional, and technological standards are required in traditional bakery products. The term starter culture denotes a microbial preparation incorporated into the raw material to obtain a fermented product by maintaining and accelerating the fermentation (Coda et al. 2014). There are two different forms of sourdough products on the market, which are used in both bakeries and industry: ready-to-use sourdoughs and sourdoughs containing live microorganisms. Sourdoughs containing live microorganisms are sold in liquid, stiff, crumbly, or pasty forms and are used as the starter culture in fermented bakery products. Some sourdoughs keep microorganisms alive and are dried in powder form using spray drying, freeze-drying, or fluidized bed drying technologies. On the other hand, ready-to-use sourdoughs are fermented products where microorganisms are inactivated to obtain a product with a long shelf life. The fermentation process of these sourdoughs is completed before they are used to produce bakery products (Brandt 2019).

The advantages of using sourdough in the production of bakery products can be listed as follows (Plessas 2021):

(i) They delay the staling of foods.
(ii) They increase the resistance of foods against mold and spoilage.
(iii) They enable foods with cleaner contents and fewer additives.
(iv) They improve nutritional properties.
(v) They contribute to decreasing starch digestibility and producing foods with a low glycemic index.
(vi) They provide improved texture, flavor, and organoleptic properties.

Sourdough fermentation is the most important fermentation method that has proven itself in improving the aroma, taste, nutritional value, texture, and shelf life of gluten-containing bakery products such as wheat and rye bread (Moroni et al. 2009). Therefore, it can be applied to improve product quality in the production of common gluten-free foods available on the market (Nionelli and Rizzello 2016). Commercially available gluten-free bread is characterized by poor crust and crumb properties, poor quality, flavor and mouthfeel. Furthermore, gluten-free bakery products, which are mainly starch-based, go stale fast and have a low nutritional content (Moroni et al. 2009). In addition, many additives used to improve the quality of these products are costly and do not meet the consumer's demand for natural products (Ramos et al. 2021). Using sourdough in gluten-free breadmaking meets the consumer demands for an additive-free, preservative-free, and clean-label product by preventing the use of expensive chemical additives and preservatives (Carbó et al. 2020). In this context, sourdough is both an economic and ecological process, and it can improve the flavor, texture, and nutritional value of gluten-free bakery products due to metabolic activities of microorganisms in sourdoughs such as lactic acid fermentation, EPS production and synthesis of antimicrobial compounds (Moroni et al. 2009).

16.3 Use of Sourdough in Gluten-Free Bakery Products

16.3.1 Gluten Structuring and Dietary Effect

Gluten is the storage protein of wheat and is a mixture of many associated and distinct proteins, gliadin and glutenin being in the first place. Similar storage proteins are also found in other cereals (e.g., avenin in oat, hordein in barley, and secalin in rye), and they are collectively called gluten (Biesiekierski 2017). The technological significance of gluten in bakery products is related to the presence and properties of gluten proteins (Arendt et al. 2008). Gliadin is associated with the viscous properties of dough, while glutenin fraction is associated with elastic properties of dough (Arendt et al. 2008; Shewry 1992). The balance of viscous and elastic properties is important in determining the suitability of flours for different uses such as breadmaking, biscuits and pasta (Shewry and Tatham 1997). The contribution of the monomeric gliadins to viscosity is due to intramolecular interactions via non-covalent bonds such as hydrogen bonds and hydrophobic bonds. The polymeric glutenins are composed of low molecular weight and high molecular weight glutenin subunits (Wieser 2007). These glutenin subunits are able to polymerize via disulfide bonds and resulting macromolecular polymer network contributes to the elasticity and strength of the wheat dough. The viscoelastic properties of dough and the quality of final product depend on the proper mixture of both fractions (Shewry and Tatham 1997; Wieser 2007). The water binding properties of gluten are related to the high glutamine and hydroxyl amino acids content (~10%) of gluten. During dough formation, cysteine and cysteine residues undergo sulfhydryl-disulfide interchange reaction, resulting in extensive polymerization of gluten protein (Arendt et al. 2008).

Gluten is a crucial component in bread production, which is called "structural protein" in breadmaking because many other ingredients are needed to produce quality bread without it (Arendt et al. 2002; Gallagher et al. 2004). Also, gluten contributes to crumb texture, consumer acceptability, and appearance of many bakery products. However, there are two main disorders related to gluten. These are celiac disease and non-celiac gluten sensitivity (NCGS) (Green et al. 2015). Celiac disease is a chronic autoimmune condition triggered by gluten consumption in genetically predisposed individuals (Bascuñán et al. 2017; Niewinski 2008; Saturni et al. 2010). On the other hand, NCGS, which should not be confused with celiac disease or wheat allergy, is a health issue occurring upon consuming gluten-containing foods and characterized by symptoms that are both intestinal and non-intestinal. (Barbaro et al. 2018; Lebwohl et al. 2015). Moreover, wheat allergy also differs from NCGS or celiac disease as it is an immunoglobulin E (IgE)-mediated hypersensitivity reaction whose symptoms occur minutes to hours after wheat consumption (through inhalation of wheat flour or ingestion of wheat products) (Green et al. 2015).

Celiac disease is treated through a dietary regime consisting of daily gluten intake of <20 mg for the entire lifetime to avoid the risk of further comorbidities and recurrence of symptoms (Scherf et al. 2018). A gluten-free diet is a diet type that does not contain proteins derived from cereals such as wheat, barley, and rye and is used to treat diseases related to gluten sensitivity, such as celiac disease (Ciclitira et al. 2005; Lundin et al. 2022). People who follow a gluten-free diet should consume foods made from gluten-free raw materials (Kaim and Wilk 2022). But it is difficult to follow a gluten-free diet throughout life. Gluten-free products are not widely available and are less tasty and more expensive than gluten-containing products (Zannini et al. 2012). Moreover, gluten free diet can lead to nutritional deficiencies in terms of calcium, iron, magnesium, zinc, fiber, vitamin D, and B vitamins (Kaim and Wilk 2022; Zannini et al. 2012). For this reason, when choosing gluten-free foods in the diet, it is necessary to examine the macronutrient profile, namely low fiber, high saturated fat content and micronutrient content. Pseudocereals (buckwheat, quinoa, amaranth and chia) have great potential for inclusion on their own or as ingredients used in the food industry to develop gluten-free food due to its richness in fiber, B, complex vitamins, unsaturated fat, proteins, minerals such as magnesium, iron, calcium, and other bioactive substances (Caeiro et al. 2022).

16.3.2 Challenges Faced During Processing of Gluten-Free Bakery Products

The absence of gluten results in bakery products that have a short shelf life, poor texture, less flavor, and low loaf volume, and therefore fail to meet consumer expectations (Cappelli et al. 2020). Producing high-quality gluten-free bakery products is therefore among the biggest challenges for bakers and food scientists in the gluten-free food industry (Arendt et al. 2008). Gluten-free doughs have a firm structure and are sticky and difficult to process compared to wheat doughs. For this reason, more energy is needed to process the dough, and mixing machines are often preferred for the kneading process (Houben et al. 2012). Due to the differences in flour compositions, breads made from gluten-free flour have poorer quality than those made from white wheat flour, even though they are superior in nutritional content (Hager et al. 2012). There are three crucial differences between gluten-free bread and wheat bread production processes. First, gluten-free bread formulations contain ingredients, such as starch, protein sources, and hydrocolloids, that improve the gas retention capability of dough and the volume and structure of dough and bread. Secondly, relatively more water is added to gluten-free bread during its production. Finally, gluten-free bread has shorter proofing, mixing, and cooking periods (Cappelli et al. 2020). As a result, the absence of gluten in the base flour leads to radically different processing requirements as well as significant differences in product characteristics such as texture and flavor (Dar 2013). Even though gluten plays a role in setting the structure of a cookie, gluten-free cookies are relatively easy to formulate compared to other baked products in which gluten is multifunctional (Casper and Atwell 2014). Gluten provides higher functionality in cakes, muffins, and similar baked goods, where it supports desirable cell structure development during the baking process. Additionally, it helps produce desirable sensory characteristics, including chewiness and moistness, as well as improves shelf stability. Removing of gluten results in lack of volume, crumbliness, dryness, and gumminess. Gluten-free doughs and batters can differ significantly in viscosity and elasticity compared to wheat dough and batters, leading to difficulties in mixing and machinability. Replacing gluten in gluten-free bread formulation presents a major technological challenge because gluten plays a key role in the dough development properties as well as development of cell structure and volume (Dar 2013).

Consumer expectations encourage the food industry to constantly modify and improve the techniques and formulations applied in gluten-free food production (El Khoury et al. 2018). One of these techniques is sourdough biotechnology, which is applied to eliminate the problems encountered in the production of common gluten-free bakery products that meet consumers' expectations, being cost-effective, environmentally friendly, and additive-free (Bender and Schönlechner 2020; Calasso et al. 2012; Ramos et al. 2021). Antimicrobial compounds, some volatiles, EPS and lactic acid produced during fermentation have positive effects on the quality of gluten-free products (Bender and Schönlechner 2020).

16.3.3 Effect of Sourdough Fermentation on Alternative Flours Processability and Product Quality

Many alternative flours are used to produce sourdough, which is popular due to the increasing demand for various gluten-free sourdough products. These are flours of pseudocereal products such as quinoa, buckwheat, amaranth, and chia (Carbó et al. 2020; Coda et al. 2010; Franco et al. 2020; Jagelaviciute and Cizeikiene 2021; Jekle et al. 2010; Różyło et al. 2015; Sterr et al. 2009), flours of legumes such as soy and chickpea (Coda et al. 2010), flours of cereals, e.g., sorghum, millet, maize, rice, and teff (Adepehin 2020; Chiş et al. 2020; Mert et al. 2014; Nami et al. 2019; Olojede et al. 2020; Różyło et al. 2016; Schober et al. 2007; Vogelmann et al. 2009; Yang et al. 2021), and chestnut flour (Aguilar et al. 2016; Aponte et al. 2013). Mygdalia et al. (2022) investigated the effect of gluten-free sourdough based on rice and a fermented chickpea extract and added water level on the quality and shelf-life of gluten-free rice-based breads. The loaf specific volume increased with increasing the amount of water added to the formulation. The inclusion of sourdough in the formulation at water level of 92% resulted in higher loaf volume in bread compared to the control bread. A fine macrostructure with small pore cells was reported in the crumb with the incorporation of sourdough. Jagelaviciute and Cizeikiene (2021) found an improvement in organoleptic and structural properties of gluten-free maize/rice bread by applying non-traditional seed flour (hemp, chia and quinoa) sourdough. Seed flour was fermented with *Fructilactobacillus sanfranciscensis* (former *Lactobacillus (L.) sanfranciscensis)* W2. Bread containing fermented seed flour was reported to show higher porosity, acidity and a slower rate of staling as compared to bread prepared with unfermented seed flour. In a different study, chia and flaxseed sourdoughs made by inoculating *Weissella (W.) cibaria* CH28, and *Lactiplantibacillus plantarum (*former *L. plantarum)* FUA3171 together with *Limosilactobacillus fermentum (*former *L. fermentum)* FUA3165 were used to prepare sorghum gluten-free bread (Maidana et al. 2020a). Sorghum bread containing 30% and 40% of chia sourdoughs fermented with *W. cibaria* exhibited higher specific volume compared to sorghum bread without sourdough (0%). This positive effect was attributed to the effect of arabinoxylans in oilseeds on dough's rheology, which slows down the rate of CO_2 diffusion from the dough. The inclusion of 40% chia sourdough in breads showed the highest sensory scores with respect to taste, texture and overall appearance. Olojede et al. (2020) investigated the effect of sourdough fermentation on the rheological properties of sourdoughs prepared from sorghum flour using LAB (*Pediococcus (P.) pentosaceus* SA8, *W. confusa* SD8, *P. pentosaceus* LD7) and yeast (*Saccharomyces (S.) cerevisiae* YC1). The sourdough formulation containing 5% sucrose led to EPS synthesis in EPS-positive isolates (*W. confusa* SD8, *P. pentosaceus* LD7), which increased the elastic component (G') of the batter compared with the EPS-negative starter (*P. pentosaceus* SA8). The textural properties of sorghum bread such as cohesiveness, springiness, gumminess and chewiness increased with sourdough addition. This was evaluated as beneficial effects due to a crumbly and brittle texture of gluten-free breads. In a study of Dingeo et al. (2020), type-I sourdough of teff prepared through back-slopping was used to fortify gluten-free teff muffins at three inoculation levels (5%, 10% and 15%). Muffins containing sourdough at 10% and 15% substitution were shown to have lower firmness than that of the control. The extended shelf life of muffins at higher sourdough contents was attributed to the release of organic acids (acetic, pheyllactic) and antimicrobial compounds during sourdough fermentation.

The amount of gluten-free sourdough used in the formulation of gluten-free bakery products varies depending on the type of flour used and is determined by sensory analyses. The most common ratio of gluten-free sourdough in the production of bakery products is about 20%, as it gives the best results in the dough and final product (Ramos et al. 2021). In a study investigating the effect of adding different amounts of sourdough on the gluten-free chestnut-rice dough's rheological characteristics, it was observed that as the amount of added sourdough increased, the acidity of the dough and bread also increased. The study reported that the ideal ratio of sourdough added was 20%, and adding a higher amount of sourdough affected the gluten-free bread's firmness and volume negatively (Mert et al. 2014). In another study investigating the changes in gluten-free breads' textural characteristics caused by fresh and freeze-dried rice sourdough addition, the breads with the highest sensory assessment rating were reported to be gluten-free breads containing 10% or 20% rice sourdough (Różyło et al. 2016). Schober et al. (2007) examined the quality of gluten-free sorghum bread. Figure 16.1 shows how the gluten-free crumb structure and bread volume are affected by adding sourdough. As shown in Fig. 16.1, upon adding sourdough, the hole in the center of the crumb disappeared, and the bread volume improved significantly compared to the bread without sourdough.

Sourdough fermentation increases water absorption and improves gas retention and dough strength by bringing about changes in starch granules, such as swelling

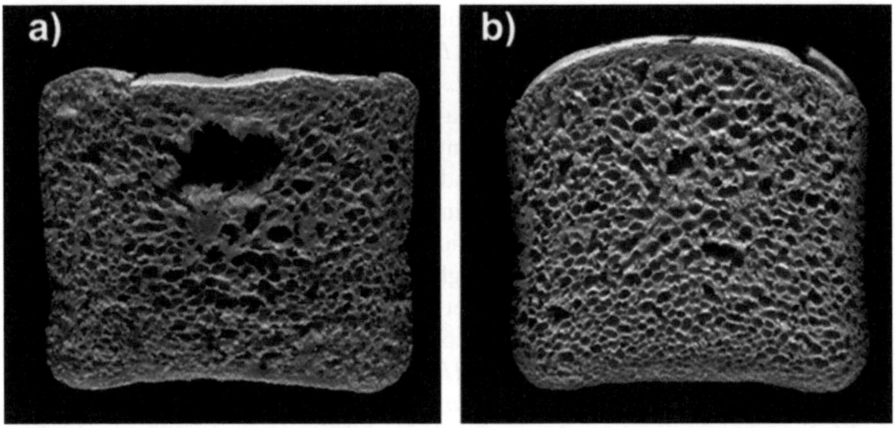

Fig. 16.1 Slice of gluten-free sorghum bread (the flour is a mixture of 70% sorghum, 30% potato starch, and 2% hydroxypropyl methylcellulose (HPMC)); (**a**) without sourdough, (**b**) with sourdough. (Reprinted from Schober et al. 2007. Copyright with permission from American Chemical Society)

and leaching, thus enhancing the quality of gluten-free dough and bread (Edema et al. 2013). Using sourdough in gluten-free breadmaking increases the dough elasticity (Nami et al. 2019; Wolter et al. 2014b) and bread volume (Falade et al. 2014; Różyło et al. 2015), decreases the crumb hardness (Novotni et al. 2012) and improves the bread porosity (Bender et al. 2018; Wolter et al. 2014b). On the other hand, compared to chemically acidified and non-acidified doughs, LAB-fermented sourdough improves bread quality by delaying bread staling in gluten-free breadmaking (Moore et al. 2007, 2008). In addition, sourdough adds a different aroma to gluten-free bread and increases enzymatic activity (Houben et al. 2012). Along with the improvements in its quality, sourdough also ensures the microbiological stability of the bread (Hüttner et al. 2010). Texture improvement is essential, especially to meet the consumer demand for gluten-free bakery products with good textures. Using sourdough in these products contributes to their textural improvement (Katina et al. 2005; Mariotti et al. 2017). In addition, sourdough used in bakery products can be regarded as an adjuvant that increases the healing rate of inflammation occurring in the intestines of celiac patients during the early phases of the gluten-free diet (Calasso et al. 2012).

16.4 Appropriateness of Starter Cultures in Gluten-Free Bakery Products

Since not every microorganism can adapt equally to the same gluten-free raw material, it is very important to determine the appropriate starter cultures in producing gluten-free sourdough. In addition, it is important to select suitable lactic acid bacterial strains to maximize the quality of gluten-free sourdough bread and to ensure the sourdough microbiota stability (Bender and Schönlechner 2020; Bender et al. 2018). In this context, Bender et al. (2018) investigated the compatibility of seven *Lactobacillus* spp. (*Limosilactobacillus fermentum (*former *L. fermentum), Levilactobacillus hammesii (*former *L. hammesii), Companilactobacillus paralimentarius (*former *L. paralimentarius), Lactiplantibacillus pentosus (*former *L. pentosus), Lactiplantibacillus plantarum (*former *L. plantarum), Fructilactobacillus sanfranciscensis (*former *L. sanfranciscensis)* LMG, DSM) strains and a commercial starter culture in buckwheat or millet sourdoughs, and their individual performance to improve bread quality. The stability of sourdough prepared using selected strains (*Levilactobacillus hammesii*, *Companilactobacillus paralimentarius*, *Lactiplantibacillus pentosus*, *Fructilactobacillus sanfranciscensis* DSM) were determined during back-slopping. Among LAB species, the lowest firmness and the highest specific volume were reported for *Fructilactobacillus sanfranciscensis* DSM strain in both buchwheat and millet breads. The crumb firmness was significantly improved with *Companilactobacillus paralimentarius* and *Lactiplantibacillus pentosus* for buckwheat bread, and *Levilactobacillus hammesii* for millet bread compared with the commercial starter culture. This study showed that buckwheat sourdough exhibited stable properties during 10 days of back-slopping, but the millet sourdough did not reach maturity during such a period. The results of this study highlighted that the quality of gluten-free breads could be improved by using the appropriate LAB strains selected. LAB strains usually isolated from sourdoughs of amaranth, teff, rice and maize are *Limosilactobacillus fermentum (*former *L. fermentum), Lactiplantibacillus plantarum* (former *L. plantarum*) and *Companilactobacillus paralimentarius* (former *L. paralimentarius*) (Bender and Schönlechner 2020; Moroni et al. 2009). Păcularu-Burada et al. (2020) investigated the isolation and selection of new LAB strains from different sources to eventually be used as starter cultures in gluten-free sourdough fermentations. Among LAB strains, the genera of *Lactobacillus*, *Leuconostoc*, *Pediococcus*, and *Streptococcus* were selected according to their acidification and EPS production capacities in the fermentation of chickpea, quinoa, and buckwheat flour extracts, and antimicrobial activity against food spoilage molds and bacteria. *Lactobacillus* spp. isolated from soil, chickpea seeds and whey, and *Leuconostoc* spp. isolated from sesame seeds exhibited valuable biotechnological characteristics in terms of acidification properties, antimicrobial properties, and EPS biosynthesis. The lactic acid bacteria (*Leuconostoc* spp.) isolated from sesame seeds produced high amounts of EPS and antimicrobial substances, while the lactic acid bacteria (*Lactobacillus* spp.) isolated from whey showed the best acidification potential. The study showed that these strains adapted to the gluten-free environment. One of the key parameters in sourdough fermentation is total titratable acidity because it reveals the production of organic acids and other valuable antimicrobial compounds that play an important role in the sourdough and bread industry in terms of final taste, aroma and shelf life extension (Păcularu-Burada et al. 2020). EPS produced by certain LAB strains enhance the rheological properties of batter and improve the texture and shelf-life of bread (Bender and Schönlechner 2020). Alibašić et al. (2020) studied the effect of the adition of buckwheat sourdough prepared with *Lactiplantibacillus plantarum* (former *L. plantarum*) on the quality of crackers. Sensory scores showed that crackers with 10% buckwheat sourdough had higher taste and crispness scores compared with wheat gluten-free crackers. Caponio et al. (2022) aimed to improve nutritional profile of gluten-free breadsticks through olive cake and sourdough fermentation using two selected LAB and a yeast. Among several LAB and yeast strains screened based on their acidification and growing capacity, *Lactiplantibacillus plantarum*, *Lactiplantibacillus plantarum/P. pentosaceus* and *Kluyveromyces marxianus* cultures have shown to be adapted to the plant raw material with increased cell density and acidity. The combination of olive cake addition and fermentation increased antioxidant activity and total phenolic content, but reduced predicted glycaemic index. Edema and Sanni (2008) investigated the functional properties (e.g., fermentation, acidification, antimicrobial compound generation) of LAB and yeasts isolated from spontaneously fermenting maize sourdough to be used as starter cultures in maize breadmaking where the sourdough technology is used. In another study, Jekle et al. (2010) characterized the effects of *Lactiplantibacillus plantarum, Companilactobacillus paralimentarius (former L. paralimentarius)*, and *L. helveticus* strains, which are promising in amaranth sourdough, including their effects on the sensory properties of the bread and the effects of their fermentation parameters on fermentation. The study reported that in addition to increasing the variety of quality bakery products, using amaranth in bakery products via sourdough fermentation is a promising procedure to benefit from this raw material as it is difficult to process.

Industrial applications of sourdough require long fermentation periods. Considering this disadvantage, which is also encountered in the production of gluten-free sourdough, some alternatives have been studied, and several solutions have been offered (Różyło et al. 2015). In the food industry, starter cultures, which are solo or mixed cultures of microorganism strains with certain properties, are used in fresh, freeze-dried, or frozen forms (Edema and Sanni 2008). Różyło et al. (2015) reported that freeze-dried buckwheat sourdough could be used directly in production, thus eliminating the long fermentation periods, which is a big problem in sourdough production.

16.4.1 Microbial Diversity of Gluten-Free Sourdoughs

There is an increasing interest in bread and other bakery products made with pseudocereals to produce different products with new sensory properties suitable for consumption by celiac patients and other consumers who apply this as a lifestyle as it evokes ecological, historical, civic, ethnical, health or cultural based interest (Foschia et al. 2016). For this purpose, using different flours in sourdough production facilitates producing new types of bread with high nutritional value and generates characteristic flavors (Sterr et al. 2009). In a study conducted with various gluten-containing and gluten-free flours, the competitiveness of microorganisms was reported to depend on the quality and variety of the substrate, even if each substrate was fermented with the same starters and under the same conditions (Vogelmann et al. 2009). In addition, by using different raw materials to make sourdough, different species are identified other than the LAB types known as endemic for sourdough, which shows that the raw material is effective in the microbiota of sourdough (Hüttner et al. 2010). While making sourdough, in addition to the use of different substrates, process parameters such as fermentation temperature and time, dough yield (is the dough consistency formulated by the mass ratio between dough and flour, multiplying by 100), oxygen, and the number of sourdough reproduction steps, as well as the cereal products' endogenous factors such as carbohydrates, minerals, nitrogen sources, free fatty acids, and lipids also significantly affect the sourdough microbiota (Corsetti and Settanni 2007). The microbial diversity of various gluten-free sourdoughs is shown in Table 16.1. *Limosilactobacillus fermentum* (former *L. fermentum*) has been reported frequently in the fermentation of gluten-free flours, either being spontaneous or by starter culture. As shown in Table 16.1, although species belonging to the genus *Lactobacillus* are usually predominant in gluten-free sourdoughs' LAB population, *P. pentosaceus* belonging to the genus *Pediococcus* is also frequently encountered. Moreover, in the studies where yeast species are identified, *S. cerevisiae* has been observed to be predominant in gluten-free sourdoughs (Meroth et al. 2004; Vogelmann et al. 2009; Yang et al. 2021). Moroni et al. (2011) produced four different gluten-free sourdoughs with spontaneous biota of the buckwheat or teff flours under two technological conditions (A and B) and then identified LAB and yeasts in the doughs (Table 16.1). Type A sourdough was developed at 25 °C by back-slopping with 10% ripe sourdough every 12 hours, with a dough yield of 200. Type B sourdough was developed at 35 °C by back-slopping with 10% ripe sourdough every 24 hours, with a dough yield of 275. In the study, the absence of *Kazachstania barnetti* in type B buckwheat sourdough was reported likely to have arisen from the selected higher incubation temperature and/or the dominant LAB present in the sourdough. In addition, it was reported that the *Limosilactobacillus fermentum* strain was identified in high temperature fermentation (type B) of both buckwheat and teff sourdoughs, while this strain was not found in low-temperature fermentation (type A).

Cereals (e.g., millet, rice, and oat), pseudocereals (e.g., amaranth, quinoa, buckwheat) and legumes (e.g., chickpea), which are valuable sources of dietary fiber, protein, and fat, are included in the formulations of gluten-free food products. The functional properties of these products can be improved by a sourdough process initiated through sourdough biota, which also improves the final fermented products' nutritional, textural and sensory properties (Păcularu-Burada et al. 2020). Table 16.1 shows the microbial diversity of gluten-free sourdoughs containing dominant LAB and yeast, which are prepared with cereals, pseudocereals, and legumes. As seen in Table 16.1, the dominant microbiota of the same type of flour varies across different countries. These results also show that the supply of flour from different countries is also effective on the microbiota of gluten-free sourdoughs, in addition to the endogenous factors of grain products. In the study of Meroth et al. (2004), different dominant strains were detected in rice sourdoughs fermented spontaneously or by starter culture. It was reported that *Lactiplantibacillus plantarum*

Table 16.1 Microbial diversity of gluten-free sourdoughs

Raw material	Country	Dominant Microbiota	Starter(St.)/ Spontaneous(Sp.)	References
Amaranth	Argentina	*Lactiplantibacillus plantarum, Enterococcus (E.) faecium, Pediococcus (P.) acidilactici*	Sp.	Ruiz Rodríguez et al. (2016b)
	Germany (St. Josefsgut, SG)	*P. pentosaceus, Lactiplantibacillus plantarum*	Sp.	Sterr et al. (2009)
	India (Alnatura®, AI)	*P. pentosaceus, Lactiplantibacillus plantarum*		
	Mexico (Allos, AM)	*P. pentosaceus, Lactiplantibacillus plantarum, Latilactobacillus sakei, E. spp.*		
	Peru (Alnatura®, AP, and Neuform, NP)	NP: *P. pentosaceus, Lactiplantibacillus plantarum, Companilactobacillus paralimentarius* AP: *P. Pentosaceus, Lactiplantibacillus plantarum, Latilactobacillus sakei, E. spp.*		
	–	*Companilactobacillus paralimentarius, Lactobacillus helveticus, Lactiplantibacillus plantarum* *Saccharomyces (S.) cerevisiae*	St.	Vogelmann et al. (2009)
Buckwheat	–	*Companilactobacillus paralimentarius, Lactiplantibacillus plantarum, Limosilactobacillus pontis*	St.	Vogelmann et al. (2009)
	Holland	For type A[a]: *Lactiplantibacillus plantarum, Weissella (W.) cibaria, P. pentosaceus, Leuconostoc holzapfelii, Latilactobacillus graminis, Latilactobacillus sakei* *Kazachstania barnetti* For type B[a]: *Limosilactobacillus fermentum, L. gallinarum, Lactiplantibacillus plantarum, L. crispatus, Limosilactobacillus vaginalis*	Sp.	Moroni et al. (2011)
Chia	Argentina	*E. faecium, E. mundtii, W. cibaria, Lacticaseibacillus rhamnosus*	Sp.	Maidana et al. (2020b)

(continued)

Table 16.1 (continued)

Raw material	Country	Dominant Microbiota	Starter(St.)/ Spontaneous(Sp.)	References
Chickpea	Spain	*Lactococcus garvieae, W. cibaria, W. paramesenteroides, E. durans, E. mundtii, P. pentosaceus*	Sp.	Sáez et al. (2018)
Maize	–	*Limosilactobacillus fermentum, Companilactobacillus paralimentarius, Limosilactobacillus pontis S. cerevisiae*	St.	Vogelmann et al. (2009)
Millet	–	*Limosilactobacillus fermentum, L. helvetivus, Limosilactobacillus pontis S. cerevisiae*	St.	Vogelmann et al. (2009)
	Kenya	For finger millet: *Bacillus subtilis, P. acidilactici, Candida glabrata, Brevibacillus borstelensis* For pearl millet: *Bacillus subtilis, Acinetobacter baumannii, P. pentosaceus*	Sp.	Adepehin (2020)
	Nigeria	For finger millet: *Lactiplantibacillus plantarum, Lactiplantibacillus pentosus, P. pentosaseus S. Cerevisiae and Candida milleri* For pearl millet: *Lactiplantibacillus plantarum, Lactiplantibacillus pentosus, P. Pentosaseus S. Cerevisiae*	Sp.	Akinola and Osundahunsi (2017)
Oat	Ireland	*Leuconostoc (Leu.) argentinum, P. pentosaceus, W. cibaria, Loigolactobacillus coryniformis*	Sp.	Hüttner et al. (2010)
Quinoa	–	*Limosilactobacillus fermentum, Companilactobacillus paralimentarius, Lactiplantibacillus plantarum S. cerevisiae, Issatchenkia orientalis*	St.	Vogelmann et al. (2009)
	Argentina	*Lactiplantibacillus plantarum, Levilactobacillus brevis*	Sp.	Ruiz Rodríguez et al. (2016a)

(continued)

Table 16.1 (continued)

Raw material	Country	Dominant Microbiota	Starter(St.)/ Spontaneous(Sp.)	References
Rice	New Zealand	*Lactiplantibacillus paraplantarum, L. fermentarum* *S. cerevisiae*	Sp.	Yang et al. (2021)
	–	*Limosilactobacillus fermentum, L. helvetivus, Lactiplantibacillus plantarum, Limosilactobacillus pontis* *S. cerevisiae*	St.	Vogelmann et al. (2009)
	–	*Levilactobacillus spicheri, Lacticaseibacillus paracasei, Companilactobacillus paralimentarius* *S. cerevisiae*	Sp.	Meroth et al. (2004)
	–	*Limosilactobacillus fermentum, L. gallinarum, Companilactobacillus kimchii, Lactiplantibacillus plantarum, Limosilactobacillus pontis* *Candida krusei, S. cerevisiae*	St.	Meroth et al. (2004)
Sorghum	South-West Nigeria	*P. pentosaceus, W. confusa,*	Sp.	Ogunsakin et al. (2017)
	Sudan	*E. faecalis, Lactococcus lactis, Limosilactobacillus fermentum, Limosilactobacillus reuteri*	Sp.	Hamad et al. (1997)
	Kenya	*Bacillus subtilis, Bacillus licheniformis, Bacillus megaterium, Bacillus coagulans, P. pentosaceus*	Sp.	Adepehin (2020)
Tartary buckwheat	China	*W. cibaria, Lactococcus taiwanensis*	Sp.	Zhou et al. (2022)
Teff	Holland	For type A[a]: *P. pentosaceus, Leuconostoc holzapfelii* *S. cerevisiae* For type B[a]: *Limosilactobacillus vaginalis, Limosilactobacillus fermentum, L. gallinarum, Limosilactobacillus pontis* *Candida glabrata*	Sp.	Moroni et al. (2011)

[a]Type A and type B: Sourdoughs fermented under different conditions

(former *L. plantarum*) obtained from the rice flour competed with the starter LAB and remained dominant during the entire fermentation process, although it could not be detected in spontaneous sourdough fermentation. Because the process parameters highly affected the microbiota. Unlike other studies in Table 16.1, the study by

Zhou et al. (2022) provided little information about the application of *Lactococcus taiwanensis*, the noticeable species they identified. Its application in sourdough is even rarer. In the study by Zhou et al. (2022), it was reported that the dominant bacterial genera were *Weissella* and *Lactococcus* in Tartary buckwheat, while *Penicillium, Aspergillus,* and *Wickerhamomyces* were the dominant fungal genera. Before applying LAB strains as starter cultures in food fermentation, specific characteristics related to the final product quality are needed to be analyzed. In this context, acidifying, amylolytic, and proteolytic activities of microorganisms can be examined as they contribute to the safety, protection, nutritional and organoleptic properties of the product (Sáez et al. 2018). Accordingly, Sáez et al. (2018) identified LAB from chickpea sourdough and investigated their techno-functional properties to obtain autochthonous cultures to develop new fermented legume products. As a result, *W. cibaria* (acidifying capacity), *W. paramesenteroides* (proteolytic activity), *Lactococcus garviae* (α-galactosidase and proteolytic activities), *P. pentosaceus* (gallate decarboxylase and peptidase activities) and *E. durans* (antimicrobial potential) were selected and identified as new functional starter cultures for value-added fermented legume products. The microbial diversity in sourdough enables the formation of different metabolites, thus offering unique flavors in sourdoughs (Zhou et al. 2022). In addition, microbiological studies on gluten-free sourdoughs indicate that gluten-free flours are an essential source for new, competitive yeast and LAB strains that can be selected as starters for stable gluten-free sourdoughs (Moroni et al. 2010).

16.5 Hydrocolloids and Their Natural Alternatives: Exopolysaccharides (EPS)

Hydrocolloids are used as food additives due to their effects on viscosity and texture, which are two fundamental properties in food production (Milani and Malcki 2012). Hydrocolloids can exhibit properties similar to the viscoelastic properties of gluten in gluten-containing doughs, such as wheat bread dough. They can improve the dough's gas retention and development through a weak gel network structure or by increasing the dough's viscosity. For this reason, they are added to gluten-free doughs as gluten substitutes, and when included in the production of gluten-free bakery products, they improve the viscoelastic characteristics of gluten-free doughs (Lazaridou and Biliaderis 2009). Hydrocolloids like guar gum, hydroxypropylmethylcellulose, and xanthan gum are incorporated in the production of gluten-free bakery products for desired texture, volume, and shelf life that the consumer may demand (Arendt et al. 2011). On the other hand, in situ EPS production is a natural alternative to hydrocolloids added to gluten-containing and gluten-free bakery products as additives. EPSs generated by LAB using sourdough technology have been determined to develop the quality properties, such as staling rate, loaf volume, and shelf life, of gluten-containing breads, wheat bread in particular (Lynch et al.

2018; Tamani et al. 2013). In addition, EPS formed during sourdough fermentation improves the textural properties of gluten-free bread (Galle et al. 2012; Valerio et al. 2020). Furthermore, in situ EPS generation during fermentation in gluten-free legume-based sourdoughs represents a potential approach to overcome the poor technological properties (poor textural and sensorial characteristics) of legume flours (Galli et al. 2020).

EPSs are microbial metabolites secreted by bacteria and microalgae with high molecular weight. EPS has applications in different fields, such as the pharmaceutical and food industries, due to its properties (e.g., stabilizing, gelling, thickening, and bioactive compound) (Abedfar et al. 2020). Especially LAB, bifidobacteria, dairy product propionibacteria, and strains of food microorganisms generally recognized as safe (GRAS) are used in the EPS food industry due to their ability to produce large amounts of EPS (De Vuyst et al. 2001). EPSs are divided into two groups depending on their composition: homopolysaccharides (HoPS) and heteropolysaccharides (HePS). HoPS consist of a single monosaccharide type (D-glucopyranose or D-fructofuranose) and are called glucans or fructans. Glucans and fructans are divided into subclasses depending on the predominant glucosidic bond. Glucans are dextrans [α-(1 $\rightarrow$ 6) linkages], alternans [alternating α-(1 $\rightarrow$ 6) and α-(1 $\rightarrow$ 3) linkages], reuterans [α-(1 $\rightarrow$ 4)] and mutants [α-(1 $\rightarrow$ 3)]. On the other hand, the fructans are inulins [β-(2 $\rightarrow$ 1) linkages] and levans [β-(2 $\rightarrow$ 6) linkages] (Lynch et al. 2018). HePS consist of branched or unbranched repeating subunits, three to eight monosaccharides (e.g., glucose, fructose, galactose, and rhamnose) (De Vuyst et al. 2001).

In a study emphasizing that using pseudocereals in gluten-free sourdough production supports EPS generation, it was reported that enriching EPS content by stimulating certain characteristics of LAB strains through the optimization of the dough yield, flour type, and sucrose amount parameters made pseudocereals suitable to produce liquid sourdough with a short fermentation period, which could be used as a bread improver (Valerio et al. 2020). Some of the sourdough LAB can produce EPS from sucrose. Therefore, using these microorganisms to produce bakery products can provide an opportunity to replace additives (Tieking et al. 2003). In addition to LAB, acetic acid bacteria can also produce high molecular weight EPS. Ua-Arak et al. (2016) reported that the capability of acetic acid bacteria strains in molasses-buckwheat sourdoughs producing a high amount of levans in situ was promising concerning using sourdoughs with acetic acid bacteria as a natural content to enhance the gluten-free bakery products' quality.

EPS improves fermented foods' taste perception, mouthfeel, texture, and stability (Abedfar et al. 2020). Studies aiming at determining the starter cultures that produce EPS in sourdoughs prepared with various gluten-free flours in order to make gluten-free breads with stable sensory and textural features have been increasing in recent years. Table 16.2 shows EPS producing microorganisms in gluten-free sourdoughs and their effects on doughs or products. Franco et al. (2020) characterized sourdough prepared using quinoa flour and investigated the EPS production of an isolated bacteria. In the study, *P. pentosaceous* was determined to have the potential to be an EPS-producing starter culture for sourdough prepared with quinoa

Table 16.2 Exopolysaccharide (EPS) producing microorganisms in gluten-free sourdoughs and their effects on doughs or products

EPS producing microorganisms	Gluten-free flours	Effects	References
Weissella (W.) cibaria MG1, Limosilactobacillus reuteri VIP, *Limosilactobacillus reuteri Y2*	Sorghum	All three strains improved the nutritional properties of gluten-free bread by producing oligosaccharides during the fermentation of gluten-free sourdough. Also, EPS formed during fermentation made the crumb of the gluten-free sorghum bread softer. Dextran was produced by *W. cibaria* MG1, reuteran was produced by *Limosilactobacillus reuteri* VIP, and fructan was produced by *Limosilactobacillus reuteri* Y2.	Galle et al. (2012)
Ligilactobacillus animalis, TMW 1.971	Buckwheat core	In order to obtain high EPS from gluten-free sourdoughs, dough yield, flour type, sucrose amount, cell count, and sucrose fed-batch parameters were optimized in the preparation of sourdoughs. The amounts of EPS obtained by optimized sourdoughs (0.3–0.7%) can be used to replace hydrocolloids.	Rühmkorf et al. (2012)
Limosilactobacillus reuteri TMW 1.106	Quinoa		
Latilactobacillus curvatus TMW 1.624	Buckwheat		
W. cibaria MG1	Buckwheat, oat, quinoa, teff	Dextran produced in fermentation improved dough rheology by showing a softening effect in gluten-free sourdoughs.	Wolter et al. (2014c)
Gluconobacter albidus TMW 2.1191, *Kozakia* baliensis NBRC 16680, *Neoasaia chiangmaiensis* NBRC 101099	Buckwheat	Three strains of acetic acid bacteria induced pH reduction in gluten-free sourdough fermentation with high in situ produced levan, predominant growth in pathogenic pollutants and organic acid production.	Ua-Arak et al. (2016)
Pediococcus (P.) pentosaceus SA8, *W. confusa* SD8, *P. pentosaceus* LD7	Sorghum	They showed the highest proteolytic activity, EPS production, and acidification property.	Ogunsakin et al. (2017)
P. pentosaceous	Quinoa	It showed a versatile sugar utilization profile and an ability to produce EPS from fructose, lactose, starch, and fructose and to rapidly acidify the quinoa fermentation liquor.	Franco et al. (2020)
W. confusa Ck15	Chickpea	Viscosity increased in doughs with in-situ dextran, and gluten-free chickpea sourdough had the highest EPS percentage.	Galli et al. (2020)

flour. Rühmkorf et al. (2012) applied in situ EPS optimization to investigate the EPS yields in gluten-free sourdough with *Latilactobacillus curvatus* (former *L. curvatus*) TMW 1.624, *Limosilactobacillus reuteri* (former *L. reuteri*) TMW 1.106, and *Ligilactobacillus animalis* (former *L. animalis*) TMW 1.971. Maximum yields were obtained in buckwheat core doughs with *Ligilactobacillus animalis* TMW 1.971, in quinoa sourdough with *Limosilactobacillus reuteri* TMW 1.106, and in buckwheat sourdough with *Latilactobacillus curvatus* TMW 1.624. Wolter et al. (2014c) investigated the EPS production in the sourdoughs prepared from gluten-free flours using *W. cibaria*, comparing the results with those prepared from wheat flour. While EPS was not observed in wheat sourdoughs, high EPS production was observed in sourdoughs prepared with gluten-free flours. In the study, high EPS production was associated with fermentable sugars, high protein amount, and mineral content in gluten-free flour.

16.6 Effect of Using Sourdough on Aroma of Gluten-Free Bakery Products

The taste and smell of cereal-based bakery products and bread are the main characteristics that consumers carefully evaluate to determine the quality of products (Corsetti and Settanni 2007). The taste of bakery products made from fermented dough is affected by the starter type as well as raw materials, sourdough fermentation, proofing, and baking. Although most aroma compounds form during baking, sourdough fermentation is essential for favored taste (Gobbetti 1998). When the protein hydrolysis is stimulated during the sourdough fermentation, the flavor and texture of the products are improved, and the bioactive peptide content and free amino acid amount increase (Păucean et al. 2019).

Sourdough fermentation is a promising method for improving the aroma profile of gluten-free breads where breads made from various gluten-free flours exhibit an undesirable aroma to consumers (Campo et al. 2016). While yeasts produce CO_2 in sourdough fermentation, the source of acidification of the dough is lactic acid and acetic acid produced by the LAB. The ratio between acetic acid and lactic acid is essential as it affects the aroma and texture of the bread (Bender and Schönlechner 2020). The aroma compounds depend on the microbiota of sourdough and the type of flour. Sourdough breads with a long and multi-staged fermentation period lasting 12 to 24 hours contain highly volatile compounds and are appreciated more in sensory tests (Aponte et al. 2013). Aponte et al. (2013) determined microbial flora in different chestnut flour-based sourdoughs and investigated the effect of fermentation on the formation of volatile compounds. The study reported that in sourdoughs produced completely with chestnut flour or with chestnut flour mixtures, 59 volatile compounds were identified in the aroma characterization, including alcohols, aldehydes, ketones, esters, and acids. The fact that the volatile metabolites are diverse this much contributes to the increased aroma and taste of fermented bakery products or bread. Nissen et al. (2020a) identified the production of volatile organic compounds in gluten-free bread fermented using sourdough and enriched with hemp seed flour, assessing its association with the product's health properties and aroma using a metabolic approach. A broader spectrum of bioactive volatile organic compounds, an increased concentration of microbial compounds, and a typical aroma profile were reported in the study compared to standard products. In another study, the gluten-free sourdough breads' general acceptability in the sensory assessment was reported to be higher than those produced with baker's yeast. This was attributed to the high amount of alcohol produced in sourdough during fermentation, which was identified in the aroma and taste of bread (Banwo et al. 2021).

Literature has reported that using LAB as a starter culture in sourdough fermentation contributes to making gluten-free bakery products with a high amino acid content and improved sensory properties (Chiş et al. 2019; Franco et al. 2021). In a study comparing gluten-free sourdoughs inoculated with a LAB strain with non-inoculated ones, differences in amino acid content released by proteolysis and the volatile derivatives in the final product (gluten-free muffin) were investigated. The study reported significant increases in lysine, histidine, leucine, isoleucine, and γ-aminobutyric acid (GABA) in gluten-free muffins produced with inoculated gluten-free sourdough. In addition, it was reported that substantial indicators of aroma compounds such as 2-methylbutanal, 3-methylbutanal, limonene, and 2.3-pentanedione were found in different and high concentrations, which were associated with the sensory test results in inoculated sourdough gluten-free muffins and considered to provide a pleasant aroma and taste to the final product (Chiş et al. 2019).

16.7 Improving Nutritional Properties of Gluten-Free Bakery Products Through Use of Sourdough

The leading desirable properties of an ideal bread are being a good source of vitamins, protein, minerals, antioxidants, and fiber, as well as having a low glycemic index. However, gluten-free breads have a high glycemic index and a low nutritional value and quality (Novotni et al. 2012). LAB in sourdough fermentation can affect the nutritional value of sourdough bakery products by acting on the number of bioactive compounds and increasing the shelf life and aroma (Chiş et al. 2018). Together with improving sensory properties of gluten-free, fiber-rich, and wholegrain foods, sourdough can increase mineral bioavailability and modulate bioaccessibility and bioactive compound levels. On the other hand, short-term cooking processes lead to vitamin losses. However, sourdough fermentation with long fermentation periods also preserves the vitamin content (Poutanen et al. 2009). In addition, using sourdough in gluten-free bakery products can provide a low glycemic index by delaying the digestibility of starch (Poutanen et al. 2009; Rinaldi et al. 2017).

Edema et al. (2013) investigated the effects of sourdough fermentation on starch properties and dough quality of one white sorghum species, two black and two white fonio species. A slight swelling was detected with scanning electron microscopy (SEM) in the starch granules due to sourdough fermentation. It was reported that this change was probably due to the impact of endogenous amylases from sourdough microorganisms causing increased water absorption and limited starch hydrolysis. Dingeo et al. (2020) used Type-1 sourdough and teff flour to produce gluten-free muffins with nutritional benefits. It was reported that using gluten-free teff flour in the production of muffins caused an increase in fiber and protein content, and sourdough LAB decreased phytic acids and increased total free amino acids. Furthermore, it was indicated that muffins showed high in vitro antioxidant and anti-mold properties contributing to increased shelf life. Wolter et al. (2014a) investigated the effects of sourdough to decrease the estimated glycemic index values of gluten-free breads obtained through fermentation of buckwheat, sorghum, quinoa, and teff flour using *Lactiplantibacillus plantarum* (former *L. plantarum*) FST1.7 and *W. cibaria* MG1. The study reported that the lactic acid bacterial strain and the type of flour used should be taken into account for breads with a low glycemic index in gluten-free bakery product formulations containing sourdough.

Gluten-free flours' fermentation by LAB allows producing functional gluten-free bakery products enriched with bioactive compounds (Naqash et al. 2017). Coda et al. (2010) investigated the fermentation of cereal flours (durum wheat, common wheat, spelt, rye, millet, rice, and oat), pseudocereal flours (amaranth, quinoa, and buckwheat), and legume flours (soy and chickpea) with GABA-producing strains. The study reported that when total phenols, *in vitro* starch hydrolysis, and antioxidant activity were examined, the nutritional properties of gluten-free sourdough breads were higher than those of gluten-free breads without sourdough and wheat breads. In addition, a study combining sourdough with partial baking and freezing process, a new approach in gluten-free breadmaking, reported that sourdough would help improve the glycemic index and quality of partially baked gluten-free bread (Novotni et al. 2012). In unrefined cereals, seeds, legumes, tubers, and nuts, phosphorus is found in the form of phytic acid. Since phytic acid is negatively charged, it forms complexes with proteins and positively charged ions, preventing the absorption of these compounds. The interaction of phytic acid with various minerals and proteins is one of the main factors that reduce the nutritional content of legume seeds and cereal grains. Although phytate hydrolysis occurs in traditional breadmaking with baker's yeast, sourdough also contains a high amount of bacteria that can degrade phytic acid (Lopez et al. 2000). The role of LAB in phytate degradation is to render the environment appropriate for endogenous cereal phytase activity by lowering its pH value (Reale et al. 2007). In addition, fermentation with LAB increases the bioavailability of micro and macro components due to phytate degradation (Diowksz and Kordialik-Bogacka 2014).

Spontaneous fermentation of sourdough is considered to be able to create differences in the quality of product. For this reason, substituting starter cultures for natural microorganisms in fermentation is considered likely to increase the repeatability of fermentation and make the final product more reliable and stable. In addition, in sourdough fermentation with selected starter cultures, the starter cultures enrich the nutritional content of gluten-free flour. Olojede et al. (2020) produced sourdough with sorghum flour using functional yeast and LAB as starter culture and investigated their effects on sourdough's textural, nutritional, and rheological characteristics. In the study, *P. pentosaceus* LD7, *P. pentosaceus* SA8, *W. confusa* SD8, and *S. cerevisiae* YC1, used as starter cultures in sorghum sourdough, were observed to improve the dough's textural, nutritional, and rheological properties. Therefore, the functionality of the selected cultures was confirmed.

Table 16.3 presents some recent studies showing sourdough's effects on the nutritional properties of gluten-free foods. Chiş et al. (2020) investigated the effect

Table 16.3 Effects of using sourdough on the nutritional properties of gluten-free foods

Raw material(s)/Starter culture(s)	Effects	References
Rice wholemeal flour/*Levilactobacillus spicheri* DSM 15429	Antioxidant activity (↑) Total phenols (↑) Amino-acids, lactic acid, and minerals (↑)	Chiş et al. (2020)
Common wheat, durum wheat, rye, spelt, oat, buckwheat, rice, amaranth, millet, chickpea, soy, and quinoa flours/ *Lactiplantibacillus plantarum* C48 and *Lactococcus lactis* subsp. *lactis* PU1	Free amino acids and GABA (↑) Phenolic compounds and antioxidant activity (↑) *In vitro* starch hydrolysis (↓)	Coda et al. (2010)
Teff flour	Phytic acid (↓) Total free amino acids (↑)	Dingeo et al. (2020)
Wholemeal rice, yellow pea, and wholemeal millet flours/ *Limosilactobacillus reuteri* DSM 20016, *Limosilactobacillus fermentum* DSM 20052 or *Levilactobacillus brevis* DSM 20054	Antioxidant capacity (↑) Phenolic content (↑)	Drakula et al. (2021)
Rice, psyllium, quinoa, chickpea, chestnut, red lentil, pea, flaxseed, faba, lupin, teff, corn, and chia flours/*Weissella (W.) cibaria* P9	Protein content and *in vitro* Protein digestibility (↑), Good textural properties, Fat and sugar content (↓)	Montemurro et al. (2021)
Corn and rice flours, cricket powder/*Lactiplantibacillus plantarum* 98a, *L. sanfranciscensis* Bb12 and *Saccharomyces (S.) cerevisiae* LBS	Antioxidant content (↑)	Nissen et al. (2020b)
Sourghum and corn flours/ *Pediococcus (P.) pentosaceus* SA8, *W. confusa* SD8, *P. Pentosaceus* LD7 and *S. cerevisiae* YC1	Total dietary fiber (↑)	Olojede et al. (2020)
Buckwheat, quinoa, sorghum and teff flours/*W. cibaria* MG1, *Lactiplantibacillus plantarum* FST1.7	Glycemic index (↓)	Wolter et al. (2014a)

of rice flour fermentation with *Levilactobacillus spicheri* (former *L. spicheri*) DSM 15429 strain and spontaneous rice flour fermentation on the gluten-free baked goods' nutritional properties. In the study, the gluten-free muffins made from rice dough fermented with *Levilactobacillus spicheri* DSM 15429 were reported to have a higher nutritional content than those produced through spontaneous fermentation. Incorporating rice sourdough fermented by LAB was determined to improve the mineral bioavailability, amino acids, antioxidant activity, and total phenols of the cooked product. It was also reported that calcium, magnesium, and potassium constituted the mineral content of rice flour, and phytic acid degraded during fermentation due to decreased pH, increasing the mineral values. In addition, the fact that the product contained lactic acid was observed to increase the shelf life of the muffins. Montemurro et al. (2021) incorporated 10% quinoa flour as a protein source and 6% hydrated chia, flaxseed, and psyllium as structuring agents in the rice and maize flour to produce "clean-labeled" gluten-free bread. Type-2 sourdough, produced from chestnut flour using *W. cibaria* P9 strain, was selected as a sugar source for in situ EPS production, and added to the mixture to constitute 30% of the dough. In the study, "clean-labeled" gluten-free bread was reported to have a high protein content and *in vitro* protein digestibility, good textural properties, and low fat and sugar amount. In addition, sourdough was reported to give the bread an acidic taste and the bread's crust and crumb a dark color. Drakula et al. (2021) examined the effects of substituting yellow pea flour for wholemeal rice flour and using different starter cultures in sourdough fermentation on gluten-free bread's antioxidant capacity and phenolic profile. Sourdough fermentation was observed to have a complementary effect as it increased bread's antioxidant capacity and phenolic content depending on the starter cultures applied. As for the study results, yellow pea flour was added to gluten-free bread to ensure high antioxidant capacity, while the importance of using sourdough fermentation with selected starter culture was highlighted. In order to enrich the nutritional content of foods, proteins are produced from vegetable and animal sources, but the stocks do not meet the market demands. In this regard, insect proteins are explored as an economical and practical option to replace proteins derived from animals such as poultry. In a study in which cricket flour was used to produce gluten-free sourdough bread with high protein content and suitable for consumption by those with celiac disease, cricket bread was reported to have a significantly high protein and antioxidant content (Nissen et al. 2020b).

16.8 Effect of Sourdough on Shelf Life of Gluten-Free Bakery Products

Both microbial spoilage and stale behavior determine the shelf life of bread. Gluten-free breads are more susceptible to spoilage due to their high water activity. For this reason, gluten-free bread's shelf life is lower than wheat bread (Axel et al. 2015; Hager et al. 2012). Moreover, as gluten-free bakery products do not include gluten and are basically starch-based, their staling is faster than gluten-free bakery products (Moore et al. 2008). Preventing starch retrogradation and thus increasing the shelf life of gluten-free bakery products and bread can be achieved in several ways (Šmídová and Rysová 2022):

(i) Addition of enzyme preparations
(ii) Using hydrocolloids
(iii) Applying sourdough fermentation
(iv) Convenient packaging method

Organic acids released in sourdough fermentation increase the shelf life of bread as they prevent rapid staling by affecting starch retrogradation. In addition to the organic acids, other metabolites, including ethanol, hydrogen peroxide, carbon dioxide, diacetyl, reuterin, and fatty acids, also contribute to preserving bread in sourdough fermentation (Bender and Schönlechner 2020). In gluten-free sourdough breadmaking, LAB, which can enzymatically hydrolyze starch, provide the opportunity to control the bread's staling (Moroni et al. 2009). Furthermore, using sourdough and EPS in situ produced by microorganisms also increases shelf life of gluten-free bread (Galle et al. 2012). The role of sourdough as a biological preservative in increasing the shelf life of gluten-free bakery products also largely satisfies consumers' demand for natural and high-quality food (Dingeo et al. 2020). Aguilar et al. (2016) investigated spontaneously fermented sourdough made from chestnut flour and its effect on chestnut flour- and cornstarch-based gluten-free bakery products. In the study, chestnut flour sourdough was reported to reduce the crumb hardness, increase the gluten-free bread's specific volume, and enable producing breads with a paler crust. In addition, Chestnut flour sourdough was determined to improve the bread's shelf life by providing softer crumbs after storage than chestnut flour without sourdough. Dubrovskaya et al. (2018) reported that gluten-free sourdough and rowan powder improved the gluten-free bread's sensory characteristics, and sourdough with an excessive amount of LAB, acetic acid, and lactic acid slowed down the onset of mold and bacteria formation which in turn form spores on bread during storage. Nami et al. (2019) examined the effects of sourdough fermented with LAB on the quality of gluten-free bread produced from pearl millet flour. Sourdough was reported to increase the elasticity of pearl millet dough by reducing its hardness. In addition, the study emphasized that gluten-free sourdough bread retained its moisture content better than traditional bread, and the addition of sourdough delayed mold formation for a more extended period. Jagelaviciute and Cizeikiene (2021) evaluated the effect of sourdough made with non-traditional chia, quinoa, and hemp flours on the quality of gluten-free maize/rice bread. In the study, the sourdough bread made from non-traditional flours fermented with *Fructilactobacillus sanfranciscensis* (former *L. sanfranciscensis*) W2 was determined to have a higher acidity, slower staling, and higher porosity than breads made from non-traditional flours without fermenting. One of the reasons for bread spoilage is fungal spores. In addition to the poor appearance of mold formation on the bread surface, fungi are responsible for the generation of mycotoxins, bad taste, and other allergenic components (Axel et al. 2015). Therefore, the effects of antifungal sourdough LAB on the shelf life of gluten-free bread have been investigated to extend the shelf life of bread. Accordingly, the preservative properties of carboxylic acids in the fermentation of gluten-free flours using LAB sourdough have been reported to contribute, even a little, to the antifungal activity of gluten-free bread (Axel et al. 2016).

16.9 Detoxification of Gluten in Wheat-Based Foods with Sourdough

When preparing pasta and bakery products, modified wheat flours are used to reduce or inhibit the immune toxicity of gluten (Rai et al. 2018). The pathogenic factor causing celiac disease has been traced down to the gluten's gliadin fraction. Peptides that gluten releases cause intolerance in people with a genetic predisposition (Blanco et al. 2011). In order to reduce the immunoreactivity caused by celiac disease, it is necessary to convert large gluten peptides into amino acid residue peptides. This is achieved by using genetically modified wheat, exogenous enzyme treatment, and LAB/sourdough. (Rai et al. 2018).

Celiac patients who need to follow a gluten-free diet may experience nutritional deficiencies due to gluten-free foods whose nutritional, textural, and sensory properties are relatively inadequate compared to gluten-containing foods. Therefore, using originally gluten-containing barley, wheat, and rye flours by removing their gluten component will contribute to both healthy nutrition and producing foods with better texture, aroma, taste, and flavor (Scherf et al. 2018). In a study comparing the bread produced from durum wheat semolina made gluten-free through sourdough biotechnology with the commercial gluten-free bread, the gluten-free product processed with wheat semolina was reported to have a high free amino acid content, low starch hydrolysis index, and high protein digestibility. Also, the technological properties of the gluten-free bread processed with wheat semolina were determined to be better compared to the commercial gluten-free bakery products on the market (Rizzello et al. 2016).

Peptidases of fungal, vegetable, animal, and bacterial origin can hydrolyze gluten into harmless peptides (Scherf et al. 2018). In a study aiming at producing wheat bread made solely from wheat flour, which can be tolerable by celiac patients (with a concentration of <20 ppm gluten), mixtures of fungal proteases and selected sourdough LAB were used to eliminate the wheat flour's toxicity through prolonged fermentation. In the study, *in vitro* assays conducted with 12 celiac patients were based on the proliferation of peripheral blood mononuclear cells (PBMCs) drawn from patients and the production of gamma interferon by intestinal T-cell lines (iTCLs) and PBMCs. The assays were used to determine the protein toxicity of pepsin-trypsin digests obtained from fermented wheat sourdough. The gluten's residual concentration in the sourdough, as determined by competitive enzyme-linked immunosorbent assay (ELISA) and R5-based sandwich, was found to be 12 ppm (Rizzello et al. 2007).

In a study in which LAB strains were screened for acidification rate and proteolytic activity in gluten-free flours, gluten concentrations were compared in sourdough gluten-free bread and bread started only with baker's yeast. In order to simulate gluten contamination, gluten was added to formulations before fermentation. The study emphasized that the gluten concentration decreased below 20 ppm in sourdough gluten-free bread alone, which was achieved using selected sourdough cultures in gluten-free breadmaking (Di Cagno et al. 2008). Di Cagno et al. (2010) reported that combining lactobacilli peptidases and fungal proteases known as bakery improvers degraded gluten (<10 ppm) in wheat flour through sourdough fermentation. In this context, eight celiac patients were given sweet bakery products containing 10 grams of native gluten equivalent daily for 60 days, and their serology, hematology, and intestinal permeability analyses were performed. The study emphasized that the bakery product fermented with wheat flour with completely degraded gluten was demonstrated not to be toxic for celiac patients. Nevertheless, it was stated that it should not be recommended for celiac patients without formal studies being conducted. Greco et al. (2011) *in vivo* assessed the toxicity of a bakery product made for celiac patients from wheat flour using sourdough lactobacilli and fungal proteases, by administering it daily to the patients for 60 days. In the study, the hydrolysis of gluten (<10 ppm) was ensured through the activity of peptidases present in the cytoplasm of LAB in sourdough fermentation. In addition, fungal proteases used in bakery to alter the resistance and elasticity of gluten were added to break down gluten completely. The results showed that baked wheat products were not toxic for celiac patients. As can be seen from the studies conducted, sourdough's role in detoxifying gluten is to ensure the hydrolysis of toxic metabolites included in flour with LAB (Deora et al. 2014). In this context, the biotechnological process involving the combination of LAB and fungal enzymes and resulting in gluten's complete degradation (<20 ppm gluten concentration) in wheat, rye, barley, and oat flours has been patented and industrialized (Giuliani et al. 2010). The produced gluten-free wheat flour has come into use in making various gluten-free bakery products with improved sensory and nutritional properties. In this context, in a study where gluten-free pasta was characterized and produced from wheat flour rendered gluten-free by selected sourdough LAB and fungal proteases, the properties of gluten-free pasta were compared to both gluten-free and gluten-containing commercial pasta. The gluten-free pasta was determined to show a fast water intake and shorter optimum cooking time than other pasta. In addition, it was reported that gluten-free pasta had a high in vitro protein digestibility, and its essential amino acid profile, nutritional index, biological value, and chemical scores were relatively high due to the proteolysis occurring in sourdough fermentation (Curiel et al. 2014).

16.10 Future Trends

The use of yeast and LAB fermentation in the gluten-free food and beverage industry represents a powerful economic/ecological process enabling the production of quality and nutritious foods demanded by consumers. Furthermore, in order to meet the increasing demands of the world's growing population for traditional products, it is necessary to develop industrial-scale production of home-fermented products and thus increase the variety of value-added products. Fermentation of cereals, pseudocereals, and legumes is a crucial biotechnological process in producing various foods for different needs worldwide. Therefore, for cereal, pseudocereal, and legume-based products to be commercialized, raw material resources must be abundant and easily accessible. In this context, sowing and harvesting of these products are vital. In addition, studying regionally consumed products such as black chickpea and brown rice with high nutritional values and generalizing their use worldwide will also contribute to the increase in the diversity of gluten-free bakery products.

The gluten-free product market needs to be developed for celiac patients, patients with gluten sensitivity, and consumers who prefer to follow a gluten-free diet. Studies on sourdough and gluten-free products reveal that sourdough is generally used in gluten-free bread and sometimes muffin production. However, the potential of sourdough in gluten-free healthy snacks, gluten-free crackers, gluten-free biscuits, and ready-to-use powder cooking products is huge. In addition, the increasing variety of sourdough gluten-free bakery products will also contribute to the development of functional products. The microbial diversity of gluten-free bakery products made with sourdough is also quite large. In this context, the increased number of studies investigating appropriate starter cultures is critical in developing bakery products with stable sensory and microbial features. The production of sourdough requires a long fermentation period. The use of processes such as freeze-drying to facilitate the production of gluten-free sourdough foods on an industrial scale overcomes this disadvantage, allowing the sourdough to be brought into a ready-to-use form. Thus, it is essential to explore new technologies that will constitute an alternative to freeze-drying, such as spray drying, which will shorten the production process of gluten-free sourdough products and bring them into ready-to-use form.

References

Abedfar A, Hosseininezhad M, Rafe A (2020) Effect of microbial exopolysaccharide on wheat bran sourdough: rheological, thermal and microstructural characteristics. Int J Biol Macromol 154:371–379

Adepehin JO (2020) Microbial diversity and pasting properties of finger millet (*Eleusine coracana*), pearl millet (*Pennisetum glaucum*) and sorghum (*Sorghum bicolor*) sourdoughs. Food Biosci 37:100684

Aguilar N, Albanell E, Miñarro B, Capellas M (2016) Chestnut flour sourdough for gluten-free bread making. Eur Food Res Technol 242(10):1795–1802

Akinola SA, Osundahunsi OF (2017) Lactic acid bacteria and yeast diversities in spontaneously fermented millet sourdoughs. J Microbiol Biotechnol Food Sci 6(4):1030–1035

Alibašić H, Junuzović H, Selimović A, Selimović A, Brčina T (2020) Chemical composition and sensory properties of gluten-free crackers with buckwheat sourdough. Int J Res Appl Sci Biotechnol 7(4):108–113

Aponte M, Boscaino F, Sorrentino A, Coppola R, Masi P, Romano A (2013) Volatile compounds and bacterial community dynamics of chestnut-flour-based sourdoughs. Food Chem 141(3):2394–2404

Arendt EK, O'Brien CM, Schober TJ, Gallagher E, Gormley TR (2002) Development of gluten-free cereal products. Farm Food 12:21–27

Arendt EK, Morrissey A, Moore MM, Dal Bello F (2008) Gluten-free breads. In: Arendt EK, Dal Bello F (eds) Gluten-free cereal products and beverages. Academic/Elsevier, pp 289–319

Arendt EK, Moroni A, Zannini E (2011) Medical nutrition therapy: use of sourdough lactic acid bacteria as a cell factory for delivering functional biomolecules and food ingredients in gluten free bread. Microb Cell Factories 10(1):1–9

Axel C, Röcker B, Brosnan B, Zannini E, Furey A, Coffey A, Arendt EK (2015) Application of *Lactobacillus amylovorus* DSM19280 in gluten-free sourdough bread to improve the microbial shelf life. Food Microbiol 47:36–44

Axel C, Brosnan B, Zannini E, Furey A, Coffey A, Arendt EK (2016) Antifungal sourdough lactic acid bacteria as biopreservation tool in quinoa and rice bread. Int J Food Microbiol 239:86–94

Banwo K, Fasuyi TO, Olojede AO (2021) Potentials of *Lactobacillus plantarum* and *Pichia kudriavzevii* in co-fermentation of sourdough from millet. Int J Food Sci Technol 56(2):857–864

Barbaro MR, Cremon C, Stanghellini V, Barbara G (2018) Recent advances in understanding non-celiac gluten sensitivity [version 1; peer review: 2 approved]. F1000Research 7(F1000 Faculty Rev):1631

Bascuñán KA, Vespa MC, Araya M (2017) Celiac disease: understanding the gluten-free diet. Eur J Nutr 56(2):449–459

Bender D, Schönlechner R (2020) Innovative approaches towards improved gluten-free bread properties. J Cereal Sci 91:102904

Bender D, Fraberger V, Szepasvári P, D'Amico S, Tömösközi S, Cavazzi G et al (2018) Effects of selected lactobacilli on the functional properties and stability of gluten-free sourdough bread. Eur Food Res Technol 244(6):1037–1046

Biesiekierski JR (2017) What is gluten? J Gastroenterol Hepatol 32:78–81

Blanco CA, Ronda F, Pérez B, Pando V (2011) Improving gluten-free bread quality by enrichment with acidic food additives. Food Chem 127(3):1204–1209

Brandt MJ (2019) Industrial production of sourdoughs for the baking branch–an overview. Int J Food Microbiol 302:3–7

Caeiro C, Pragosa C, Cruz MC, Pereira CD, Pereira SG (2022) The role of pseudocereals in celiac disease: reducing nutritional deficiencies to improve well-being and health. J Nutrit Metab 2022:8502169

Calasso M, Vincentini O, Valitutti F, Felli C, Gobbetti M, Di Cagno R (2012) The sourdough fermentation may enhance the recovery from intestinal inflammation of coeliac patients at the early stage of the gluten-free diet. Eur J Nutr 51(4):507–512

Campo E, del Arco L, Urtasun L, Oria R, Ferrer-Mairal A (2016) Impact of sourdough on sensory properties and consumers' preference of gluten-free breads enriched with teff flour. J Cereal Sci 67:75–82

Caponio GR, Difonzo G, de Gennaro G, Calasso M, De Angelis M, Pasqualone A (2022) Nutritional improvement of gluten-free breadsticks by olive cake addition and sourdough fermentation: how texture, sensory, and aromatic profile were affected? Front Nutr 9:830932

Cappa C, Lucisano M, Raineri A, Fongaro L, Foschino R, Mariotti M (2016) Gluten-free bread: influence of sourdough and compressed yeast on proofing and baking properties. Foods 5(4):69

Cappelli A, Oliva N, Cini E (2020) A systematic review of gluten-free dough and bread: dough rheology, bread characteristics, and improvement strategies. Appl Sci 10(18):6559

Carbó R, Gordún E, Fernández A, Ginovart M (2020) Elaboration of a spontaneous gluten-free sourdough with a mixture of amaranth, buckwheat, and quinoa flours analyzing microbial load, acidity, and pH. Food Sci Technol Int 26(4):344–352

Casper JL, Atwell WA (2014) Gluten-free: baked products. AACC International, St. Paul

Chavan RS, Chavan SR (2011) Sourdough technology—a traditional way for wholesome foods: a review. Compr Rev Food Sci Food Saf 10(3):169–182

Chiş MS, Adriana P, Stan L, Muresan V, Vlaic RA, Man S et al (2018) *Lactobacillus plantarum* ATCC 8014 in quinoa sourdough adaptability and antioxidant potential. Rom Biotechnol Lett 23(3):13581–13591

Chiş MS, Păucean A, Stan L, Suharoschi R, Socaci SA, Man SM et al (2019) Impact of protein metabolic conversion and volatile derivatives on gluten-free muffins made with quinoa sourdough. CyTA-J Food 17(1):744–753

Chiş MS, Păucean A, Man SM, Bonta V, Pop A, Stan L et al (2020) Effect of rice flour fermentation with *Lactobacillus spicheri* DSM 15429 on the nutritional features of gluten-free muffins. Foods 9(6):822

Chochkov R, Savova-Stoyanova D, Papageorgiou M, Rocha JM, Gotcheva V, Angelov A (2022) Effects of teff-based sourdoughs on dough rheology and gluten-free bread quality. Foods 11(7):1012

Ciclitira PJ, Ellis HJ, Lundin KE (2005) Gluten-free diet—what is toxic? Best Pract Res Clin Gastroenterol 19(3):359–371

Coda R, Rizzello CG, Gobbetti M (2010) Use of sourdough fermentation and pseudo-cereals and leguminous flours for the making of a functional bread enriched of γ-aminobutyric acid (GABA). Int J Food Microbiol 137(2–3):236–245

Coda R, Di Cagno R, Gobbetti M, Rizzello CG (2014) Sourdough lactic acid bacteria: exploration of non-wheat cereal-based fermentation. Food Microbiol 37:51–58

Corsetti A, Settanni L (2007) Lactobacilli in sourdough fermentation. Food Res Int 40(5):539–558

Curiel JA, Coda R, Limitone A, Katina K, Raulio M, Giuliani G et al (2014) Manufacture and characterization of pasta made with wheat flour rendered gluten-free using fungal proteases and selected sourdough lactic acid bacteria. J Cereal Sci 59(1):79–87

Dar YL (2013) Advances and ongoing challenges in the development of gluten-free baked goods. Cereal Foods World 58(6):298–304

De Vuyst L, De Vin F, Vaningelgem F, Degeest B (2001) Recent developments in the biosynthesis and applications of heteropolysaccharides from lactic acid bacteria. Int Dairy J 11(9):687–707

De Vuyst L, Vrancken G, Ravyts F, Rimaux T, Weckx S (2009) Biodiversity, ecological determinants, and metabolic exploitation of sourdough microbiota. Food Microbiol 26(7):666–675

Deora NS, Deswal A, Mishra HN (2014) Alternative approaches towards gluten-free dough development: recent trends. Food Eng Rev 6(3):89–104

Di Cagno R, Rizzello CG, De Angelis M, Cassone A, Giuliani G, Benedusi A et al (2008) Use of selected sourdough strains of *Lactobacillus* for removing gluten and enhancing the nutritional properties of gluten-free bread. J Food Prot 71(7):1491–1495

Di Cagno R, Barbato M, Di Camillo C, Rizzello CG, De Angelis M, Giuliani G et al (2010) Gluten-free sourdough wheat baked goods appear safe for young celiac patients: a pilot study. J Pediatr Gastroenterol Nutr 51(6):777–783

Dingeo C, Difonzo G, Paradiso VM, Rizzello CG, Pontonio E (2020) Teff type-i sourdough to produce gluten-free muffin. Microorganisms 8(8):1149

Diowksz A, Kordialik-Bogacka E (2014) Stimulation of gluten-free sourdough fermentation. Acta Aliment 43(2):225–231

Drakula S, Novotni D, Mustač NČ, Voučko B, Krpan M, Hruškar M, Ćurić D (2021) Alteration of phenolics and antioxidant capacity of gluten-free bread by yellow pea flour addition and sourdough fermentation. Food Biosci 44:101424

Dubrovskaya N, Savkina O, Kuznetsova L, Parakhina O (2018) The development of gluten-free sourdough bread technology with rowan powder. Agron Res 16(S2):1360–1372

Edema MO, Sanni AI (2008) Functional properties of selected starter cultures for sour maize bread. Food Microbiol 25(4):616–625

Edema MO, Emmambux MN, Taylor JR (2013) Improvement of fonio dough properties through starch modification by sourdough fermentation. Starch-Stärke 65(9–10):730–737

El Khoury D, Balfour-Ducharme S, Joye IJ (2018) A review on the gluten-free diet: technological and nutritional challenges. Nutrients 10(10):1410

Falade AT, Emmambux MN, Buys EM, Taylor JR (2014) Improvement of maize bread quality through modification of dough rheological properties by lactic acid bacteria fermentation. J Cereal Sci 60(3):471–476

Foschia M, Horstmann S, Arendt EK, Zannini E (2016) Nutritional therapy–facing the gap between coeliac disease and gluten-free food. Int J Food Microbiol 239:113–124

Franco W, Pérez-Díaz IM, Connelly L, Diaz JT (2020) Isolation of exopolysaccharide-producing yeast and lactic acid bacteria from quinoa (*Chenopodium quinoa*) sourdough fermentation. Foods 9(3):337
Franco W, Evert K, Van Nieuwenhove C (2021) Quinoa flour, the germinated grain flour, and sourdough as alternative sources for gluten-free bread formulation: impact on chemical, textural and sensorial characteristics. Fermentation 7(3):115
Gallagher E, Gormley TR, Arendt EK (2004) Recent advances in the formulation of gluten-free cereal-based products. Trends Food Sci Technol 15(3–4):143–152
Galle S, Schwab C, Dal Bello F, Coffey A, Gänzle MG, Arendt EK (2012) Influence of in-situ synthesized exopolysaccharides on the quality of gluten-free sorghum sourdough bread. Int J Food Microbiol 155(3):105–112
Galli V, Venturi M, Coda R, Maina NH, Granchi L (2020) Isolation and characterization of indigenous *Weissella confusa* for in situ bacterial exopolysaccharides (EPS) production in chickpea sourdough. Food Res Int 138:109785
Gänzle MG (2014) Enzymatic and bacterial conversions during sourdough fermentation. Food Microbiol 37:2–10
Giuliani G, Benedusi A, Di Cagno R, Rizzello CG, De Angelis M, Gobbetti M, Cassone A (2010) Process of microbic biotechnology for completely degrading gluten in flours. W.O. Patent 2010073283-A2, 1 July 2010
Gobbetti M (1998) The sourdough microflora: interactions of lactic acid bacteria and yeasts. Trends Food Sci Technol 9(7):267–274
Greco L, Gobbetti M, Auricchio R, Di Mase R, Landolfo F, Paparo F et al (2011) Safety for patients with celiac disease of baked goods made of wheat flour hydrolyzed during food processing. Clin Gastroenterol Hepatol 9(1):24–29
Green PH, Lebwohl B, Greywoode R (2015) Celiac disease. J Allergy Clin Immunol 135(5):1099–1106
Hager AS, Wolter A, Czerny M, Bez J, Zannini E, Arendt EK, Czerny M (2012) Investigation of product quality, sensory profile and ultrastructure of breads made from a range of commercial gluten-free flours compared to their wheat counterparts. Eur Food Res Technol 235(2):333–344
Hamad SH, Dieng MC, Ehrmann MA, Vogel RF (1997) Characterization of the bacterial flora of Sudanese sorghum flour and sorghum sourdough. J Appl Microbiol 83(6):764–770
Hansen AS (2012) Sourdough bread. In: Hui YH, Evranuz EO, Arroyo-López FN, Fan L, Hansen AS, Jaramillo-Flores ME, Rakin M, Schwan RF, Zhou W (eds) Handbook of plant-based fermented food and beverage technology. CRC Press, Boca Raton, pp 493–515
Houben A, Höchstötter A, Becker T (2012) Possibilities to increase the quality in gluten-free bread production: an overview. Eur Food Res Technol 235(2):195–208
Hüttner EK, Dal Bello F, Arendt EK (2010) Identification of lactic acid bacteria isolated from oat sourdoughs and investigation into their potential for the improvement of oat bread quality. Eur Food Res Technol 230(6):849–857
Jagelaviciute J, Cizeikiene D (2021) The influence of non-traditional sourdough made with quinoa, hemp and chia flour on the characteristics of gluten-free maize/rice bread. LWT 137:110457
Jekle M, Houben A, Mitzscherling M, Becker T (2010) Effects of selected lactic acid bacteria on the characteristics of amaranth sourdough. J Sci Food Agric 90(13):2326–2332
Kaim U, Wilk M (2022) Pseudocereals – source of essential amino acids and bioactive compounds in gluten-free diet. In: Goluch Z (ed) Medycyna i zdrowie we współczesnym świecie. ArchaeGraph, Łódź, pp 97–111
Katina K, Arendt E, Liukkonen KH, Autio K, Flander L, Poutanen K (2005) Potential of sourdough for healthier cereal products. Trends Food Sci Technol 16(1–3):104–112
Lazaridou A, Biliaderis CG (2009) Gluten-free doughs: rheological properties, testing procedures–methods and potential problems. In: Gallagher E (ed) Gluten-free food science and technology. Wiley-Blackwell, London, pp 52–82
Lebwohl B, Ludvigsson JF, Green PH (2015) Celiac disease and non-celiac gluten sensitivity. BMJ 351

Lopez HW, Ouvry A, Bervas E, Guy C, Messager A, Demigne C, Remesy C (2000) Strains of lactic acid bacteria isolated from sour doughs degrade phytic acid and improve calcium and magnesium solubility from whole wheat flour. J Agric Food Chem 48(6):2281–2285

Lundin KE, Brottveit M, Skodje G (2022) Chapter 9- Noncoeliac gluten sensitivity. In: Schiepatti A, Sanders DS (eds) Coeliac disease and gluten-related disorders. Academic/Elsevier, pp 177–195

Lynch KM, Coffey A, Arendt EK (2018) Exopolysaccharide producing lactic acid bacteria: their techno-functional role and potential application in gluten-free bread products. Food Res Int 110:52–61

Maidana SD, Finch S, Garro M, Savoy G, Gänzle M, Vignolo G (2020a) Development of gluten-free breads started with chia and flaxseed sourdoughs fermented by selected lactic acid bacteria. LWT-Food Sci Technol 125:109189

Maidana SD, Ficoseco CA, Bassi D, Cocconcelli PS, Puglisi E, Savoy G, Vignolo G, Fontana C (2020b) Biodiversity and technological-functional potential of lactic acid bacteria isolated from spontaneously fermented chia sourdough. Int J Food Microbiol 316:108425

Mariotti M, Cappa C, Picozzi C, Tedesco B, Fongaro L, Lucisano M (2017) Compressed yeast and type I gluten-free sourdough in gluten-free breadmaking. Food Bioprocess Technol 10(5):962–972

Meroth CB, Hammes WP, Hertel C (2004) Characterisation of the microbiota of rice sourdoughs and description of *Lactobacillus spicheri* sp. nov. Syst Appl Microbiol 27(2):151–159

Mert ID, Campanella OH, Sumnu G, Sahin S (2014) Gluten-free sourdough bread prepared with chestnut and rice flour. In 9th Baltic conference on food science and technology "food for consumer well-being", 239

Milani J, Maleki G (2012) Hydrocolloids in food industry. In: Valdez B (ed) Food industrial processes-methods and equipment. IntechOpen, Croatia, Rijeka, pp 17–38

Montemurro M, Pontonio E, Rizzello CG (2021) Design of a "clean-label" gluten-free bread to meet consumers demand. Foods 10(2):462

Moore MM, Juga B, Schober TJ, Arendt EK (2007) Effect of lactic acid bacteria on properties of gluten-free sourdoughs, batters, and quality and ultrastructure of gluten-free bread. Cereal Chem 84(4):357–364

Moore MM, Dal Bello F, Arendt EK (2008) Sourdough fermented by *Lactobacillus plantarum* FST 1.7 improves the quality and shelf life of gluten-free bread. Eur Food Res Technol 226(6):1309–1316

Moroni AV, Dal Bello F, Arendt EK (2009) Sourdough in gluten-free bread-making: an ancient technology to solve a novel issue? Food Microbiol 26(7):676–684

Moroni, A. V., Arendt, E. K., Morrissey, J. P., Dal Bello, F. (2010). Development of buckwheat and teff sourdoughs with the use of commercial starters. Int J Food Microbiol, 142(1–2), 142–148, 142

Moroni AV, Arendt EK, Dal Bello F (2011) Biodiversity of lactic acid bacteria and yeasts in spontaneously-fermented buckwheat and teff sourdoughs. Food Microbiol 28(3):497–502

Mygdalia AS, Nouska C, Hatzikamari M, Biliaderis CG, Lazaridou A (2022) A sourdough process based on fermented chickpea extract as leavening and anti-staling agent for improving the quality of gluten-free breads. Food Res Int 159:111593

Nami Y, Gharekhani M, Aalami M, Hejazi MA (2019) *Lactobacillus*-fermented sourdoughs improve the quality of gluten-free bread made from pearl millet flour. J Food Sci Technol 56(9):4057–4067

Naqash F, Gani A, Gani A, Masoodi FA (2017) Gluten-free baking: combating the challenges-a review. Trends Food Sci Technol 66:98–107

Niewinski MM (2008) Advances in celiac disease and gluten-free diet. J Am Diet Assoc 108(4):661–672

Nionelli L, Rizzello CG (2016) Sourdough-based biotechnologies for the production of gluten-free foods. Foods 5(3):65

Nissen L, Bordoni A, Gianotti A (2020a) Shift of volatile organic compounds (VOCs) in gluten-free hemp-enriched sourdough bread: a metabolomic approach. Nutrients 12(4):1050

Nissen L, Samaei SP, Babini E, Gianotti A (2020b) Gluten free sourdough bread enriched with cricket flour for protein fortification: antioxidant improvement and Volatilome characterization. Food Chem 333:127410

Novotni D, Čukelj N, Smerdel B, Bituh M, Dujmić F, Ćurić D (2012) Glycemic index and firming kinetics of partially baked frozen gluten-free bread with sourdough. J Cereal Sci 55(2):120–125

Ogunsakin AO, Vanajakshi V, Anu-Appaiah KA, Vijayendra SVN, Walde SG, Banwo K et al (2017) Evaluation of functionally important lactic acid bacteria and yeasts from Nigerian sorghum as starter cultures for gluten-free sourdough preparation. LWT-Food Sci Technol 82:326–334

Olojede AO, Sanni AI, Banwo K (2020) Rheological, textural and nutritional properties of gluten-free sourdough made with functionally important lactic acid bacteria and yeast from Nigerian sorghum. LWT 120:108875

Păcularu-Burada B, Georgescu LA, Vasile MA, Rocha JM, Bahrim GE (2020) Selection of wild lactic acid bacteria strains as promoters of postbiotics in gluten-free sourdoughs. Microorganisms 8(5):643

Papadimitriou K, Zoumpopoulou G, Georgalaki M, Alexandraki V, Kazou M, Anastasiou R, Tsakalidou E (2019) Chapter 6-Sourdough bread. In: Galanakis CM (ed) Innovations in traditional foods. Elsevier-Woodhead Publishing, Cambridge, UK, pp 127–158

Păucean A, Man SM, Chiş MS, Mureşan V, Pop CR, Socaci SA et al (2019) Use of pseudocereals preferment made with aromatic yeast strains for enhancing wheat bread quality. Foods 8(10):443

Plessas S (2021) Innovations in sourdough bread making. Fermentation 7(1):29

Poutanen K, Flander L, Katina K (2009) Sourdough and cereal fermentation in a nutritional perspective. Food Microbiol 26(7):693–699

Rai S, Kaur A, Chopra CS (2018) Gluten-free products for celiac susceptible people. Front Nutr 5:116

Ramos L, Alonso-Hernando A, Martínez-Castro M, Morán-Pérez JA, Cabrero-Lobato P, Pascual-Maté A et al (2021) Sourdough biotechnology applied to gluten-free baked goods: rescuing the tradition. Foods 10(7):1498

Reale A, Konietzny U, Coppola R, Sorrentino E, Greiner R (2007) The importance of lactic acid bacteria for phytate degradation during cereal dough fermentation. J Agric Food Chem 55(8):2993–2997

Rinaldi M, Paciulli M, Caligiani A, Scazzina F, Chiavaro E (2017) Sourdough fermentation and chestnut flour in gluten-free bread: a shelf-life evaluation. Food Chem 224:144–152

Rizzello CG, De Angelis M, Di Cagno R, Camarca A, Silano M, Losito I et al (2007) Highly efficient gluten degradation by lactobacilli and fungal proteases during food processing: new perspectives for celiac disease. Appl Environ Microbiol 73(14):4499–4507

Rizzello CG, Montemurro M, Gobbetti M (2016) Characterization of the bread made with durum wheat semolina rendered gluten free by sourdough biotechnology in comparison with commercial gluten-free products. J Food Sci 81(9):H2263–H2272

Różyło R, Rudy S, Krzykowski A, Dziki D, Gawlik-Dziki U, Różyło K, Skonecki S (2015) Effect of adding fresh and freeze-dried buckwheat sourdough on gluten-free bread quality. Int J Food Sci Technol 50(2):313–322

Różyło R, Rudy S, Krzykowski A, Dziki D, Siastała M, Polak R (2016) Gluten-free bread prepared with fresh and freeze-dried rice sourdough-texture and sensory evaluation. J Texture Stud 47(5):443–453

Rühmkorf C, Jungkunz S, Wagner M, Vogel RF (2012) Optimization of homoexopolysaccharide formation by lactobacilli in gluten-free sourdoughs. Food Microbiol 32(2):286–294

Ruiz Rodríguez L, Vera Pingitore E, Rollan G, Cocconcelli PS, Fontana C, Saavedra L, Vignolo G, Hebert EM (2016a) Biodiversity and technological-functional potential of lactic acid bacteria isolated from spontaneously fermented quinoa sourdoughs. J Appl Microbiol 120(5):1289–1301

Ruiz Rodríguez L, Vera Pingitore E, Rollan G, Martos G, Saavedra L, Fontana C, Hebert EM, Vignolo G (2016b) Biodiversity and technological potential of lactic acid bacteria isolated from spontaneously fermented amaranth sourdough. Lett Appl Microbiol 63(2):147–154

Sáez GD, Saavedra L, Hebert EM, Zárate G (2018) Identification and biotechnological characterization of lactic acid bacteria isolated from chickpea sourdough in northwestern Argentina. LWT 93:249–256
Saturni L, Ferretti G, Bacchetti T (2010) The gluten-free diet: safety and nutritional quality. Nutrients 2(1):16–34
Scherf KA, Wieser H, Koehler P (2018) Novel approaches for enzymatic gluten degradation to create high-quality gluten-free products. Food Res Int 110:62–72
Schober TJ, Bean SR, Boyle DL (2007) Gluten-free sorghum bread improved by sourdough fermentation: biochemical, rheological, and microstructural background. J Agric Food Chem 55(13):5137–5146
Shewry P (1992) Exploring the structure and functionality of wheat gluten proteins. Biotechnol Biotechnol Equip 6(2):5–10
Shewry PR, Tatham AS (1997) Disulphide bonds in wheat gluten proteins. J Cereal Sci 25(3):207–227
Šmídová Z, Rysová J (2022) Gluten-free bread and bakery products technology. Foods 11(3):480
Sterr Y, Weiss A, Schmidt H (2009) Evaluation of lactic acid bacteria for sourdough fermentation of amaranth. Int J Food Microbiol 136(1):75–82
Tamani RJ, Goh KKT, Brennan CS (2013) Physico-chemical properties of sourdough bread production using selected lactobacilli starter cultures. J Food Qual 36(4):245–252
Tieking M, Korakli M, Ehrmann MA, Gänzle MG, Vogel RF (2003) In situ production of exopolysaccharides during sourdough fermentation by cereal and intestinal isolates of lactic acid bacteria. Appl Environ Microbiol 69(2):945–952
Ua-Arak T, Jakob F, Vogel RF (2016) Characterization of growth and exopolysaccharide production of selected acetic acid bacteria in buckwheat sourdoughs. Int J Food Microbiol 239:103–112
Valerio F, Bavaro AR, Di Biase M, Lonigro SL, Logrieco AF, Lavermicocca P (2020) Effect of amaranth and quinoa flours on exopolysaccharide production and protein profile of liquid sourdough fermented by *Weissella cibaria* and *Lactobacillus plantarum*. Front Microbiol 11:967
Vogelmann SA, Seitter M, Singer U, Brandt MJ, Hertel C (2009) Adaptability of lactic acid bacteria and yeasts to sourdoughs prepared from cereals, pseudocereals and cassava and use of competitive strains as starters. Int J Food Microbiol 130(3):205–212
Wieser H (2007) Chemistry of gluten proteins. Food Microbiol 24(2):115–119
Wolter A, Hager AS, Zannini E, Arendt EK (2014a) Influence of sourdough on in vitro starch digestibility and predicted glycemic indices of gluten-free breads. Food Funct 5(3):564–572
Wolter A, Hager AS, Zannini E, Czerny M, Arendt EK (2014b) Impact of sourdough fermented with *Lactobacillus plantarum* FST 1.7 on baking and sensory properties of gluten-free breads. Eur Food Res Technol 239(1):1–12
Wolter A, Hager AS, Zannini E, Galle S, Gänzle MG, Waters DM, Arendt EK (2014c) Evaluation of exopolysaccharide producing *Weissella cibaria* MG1 strain for the production of sourdough from various flours. Food Microbiol 37:44–50
Yang Q, Rutherfurd-Markwick K, Mutukumira AN (2021) Identification of dominant lactic acid bacteria and yeast in rice sourdough produced in New Zealand. Curr Res Food Sci 4:729–736
Zannini E, Pontonio E, Waters DM, Arendt EK (2012) Applications of microbial fermentations for production of gluten-free products and perspectives. Appl Microbiol Biotechnol 93(2):473–485
Zhou Y, She X, Zhu S, Zhou X (2022) The study of microbial diversity and volatile compounds in Tartary buckwheat sourdoughs. Food Chem X 14:100353

Index

A
Acidification, 2, 9–11, 49, 62, 64, 65, 95, 105, 111, 117, 132, 146, 149, 152, 262, 268, 282, 286, 287, 289, 290, 311, 322, 324, 325, 329, 345, 379, 380, 383, 444, 446, 462, 469, 470, 477, 478, 484
Agri-food industry, 286, 294, 299
Allergy, 293, 324, 387, 425, 443, 444, 450, 451, 464
Antioxidant, 12–14, 50, 82, 133, 146–148, 150, 237, 267, 268, 288, 291–293, 317, 333, 334, 357, 362, 363, 365, 385, 386, 388, 470, 479–482

B
Bacillus, 43, 45, 47, 48, 53, 72, 74, 85, 115, 149, 150, 171, 172, 174–177, 181, 182, 202, 215, 243, 326, 332, 333, 397, 409–416, 473, 474
Bacteria identification, 37, 410
Bioactive compounds, 36, 115, 224, 245, 247, 263, 282, 286, 288, 299, 310, 334, 344, 354, 358, 380, 461, 476, 479, 480
Biodiversity, 37, 41, 42, 51, 62–67, 73, 75, 76, 113, 180, 213, 216, 218
Biomolecular approach, 132
Biomolecular identification, 131–158
Biopreservative, 114, 296, 364, 403
Bread, 1, 23, 61, 94, 131, 177, 217, 245, 260, 282, 310, 343, 380, 395, 425, 462
Breadmaking, 1, 5, 27, 33–35, 46, 50, 52, 282, 379, 380, 383–387, 399, 403, 408, 438, 439, 463, 464, 468, 470, 480, 483, 485

C
Celiac disease, 10, 293, 387, 388, 425, 443, 444, 451, 464, 465, 482, 484
Central Europe, 61, 67–82, 405
Cereals, 4, 27, 61, 93, 152, 176, 211, 240, 264, 287, 321, 343, 385, 398, 428, 464
Clean-label, 62, 345, 347–367, 463
Contamination, 6, 113, 149, 154, 169, 170, 173, 175, 177–181, 184–202, 292, 332, 333, 364, 397–400, 402–405, 407–409, 411, 485

D
Dependent culture analyses, 3, 133, 134, 139, 154–158, 191, 212, 213, 216, 217, 219
Dried sourdough powders, 265–274

E
Encapsulation, 257, 258, 260–262, 264, 361, 362
Exopolysaccharides (EPSs), 2, 5, 10, 11, 81, 108, 116, 117, 145, 151, 232, 236, 268, 282, 284–287, 293, 295, 298, 299, 310, 316–321, 354, 355, 364–367, 380, 461, 463, 466, 467, 469, 475–478, 482, 483

F
Fermentation, 1, 22, 61, 95, 131, 175, 211, 223, 257, 282, 309, 343, 379, 408, 425, 462
Fermentation stages, 7, 8, 64, 67

E. B. Ceresino et al. (eds.), *Sourdough Microbiota and Starter Cultures for Industry*, https://doi.org/10.1007/978-3-031-48604-3

Flours, 1, 21, 61, 94, 132, 171–173, 268, 287, 326, 344, 382, 400, 426, 461
Food industry, 117, 132, 219, 223, 234, 238, 242, 244, 246, 281–299, 317, 331, 345, 358, 364, 366, 465, 466, 470, 476
Functional food, 107, 110, 122, 132, 274, 334, 364
Functional metabolites, 53, 389
Fungi, 3, 21, 44, 66, 83, 98, 113, 149, 180, 184, 185, 198, 199, 201, 217, 244, 284, 326, 399–402, 404, 405, 417, 483

G
Gamma-aminobutyric acid (GABA), 82, 291, 295, 298, 354, 355, 365, 381–383, 479, 481
Gluten-free (GF), 10, 36, 75, 151, 264, 320, 388, 398, 432, 461
Glycemic index (GI), 36, 267, 268, 273, 291, 296, 380, 463, 479–481

I
Immobilization, 257–274
Immunoreactivity, 426, 441, 443–446, 449–451, 484
Industrial applications, 67, 146, 224–247, 265, 287, 357, 360, 361, 367, 438, 470
Irritable bowel syndrome (IBS), 13, 120, 267, 425, 426, 430–434, 439, 440, 443, 451
Italy, 23–27, 36–41, 111, 136, 137, 141, 154, 171, 175, 182, 186, 192, 199, 200, 284, 293, 313, 315, 348, 407

L
Lactic acid bacteria (LAB), 2–12, 21, 22, 37, 38, 40–43, 45–53, 61–68, 72–76, 80, 81, 83–85, 95, 103, 105, 107–115, 117, 119, 120, 132, 134, 138, 139, 145–154, 157, 171, 173, 178, 180, 184, 201, 211, 213–217, 219, 223–244, 248, 263, 265–269, 271, 273, 282, 283, 285–298, 309–335, 343–347, 351, 352, 354, 355, 359, 364, 379, 380, 382–384, 386, 388, 389, 403, 408, 409, 414, 425, 426, 428, 434, 435, 439, 444, 446–451, 461, 462, 467, 469–471, 474–476, 478–485

M
Metabiotics, 345, 358–367
Metabolic engineering, 224–248, 359, 367
Metagenomics, 53, 213–219, 380, 388
Microbiota, 1–4, 6–14, 22, 23, 34, 35, 37, 41, 42, 44, 46, 48, 51–53, 61–85, 99–103, 118–121, 132, 134, 138, 139, 145, 146, 153, 169, 170, 175, 179–183, 185–189, 192–198, 212, 214, 216, 261, 267, 270, 282, 286, 287, 289, 293, 294, 296, 297, 310, 311, 313, 323, 343, 344, 346–347, 356, 362, 400, 414, 416, 431–433, 462, 469, 471, 472, 474, 478
Mineral bioavailability, 267, 268, 379, 479, 482
Mycotoxins, 149, 175, 176, 198–202, 292, 324, 361, 363, 365, 398–401, 404–409, 417, 483

N
Northern Europe and Baltic grains, 93–122

O
Omics, 213, 214

P
Portugal, 23–33, 38–40, 44, 48, 53, 171, 175, 186, 407
Probiotic, 3, 12, 14, 73, 115, 118, 122, 153–154, 223, 245, 263, 284, 292, 294, 295, 298, 359, 361–363, 365, 366, 411, 413, 432, 451

R
Rope, 172, 176, 177, 271, 326, 333, 398, 399, 409–414

S
Sourdough, 1, 21, 61, 94, 131, 169, 211, 239, 259, 281, 309, 343, 379, 397, 425, 461
Sourdough culturing, 132, 149
Sourdough diversity, 216, 217
Sourdough enzymes, 264
Sourdough fermentation, 2, 4–11, 13, 22, 23, 36, 37, 41, 44–50, 52, 53, 61, 64, 67, 73, 79, 81, 84, 96, 104–106, 114, 116, 120, 122, 131, 132, 151, 158, 175, 214, 217, 219, 262, 263, 267–271, 273, 284–287, 289–292, 296, 299, 309, 312, 313, 319–324, 330, 333, 334, 344, 347–351, 380, 382, 385, 386, 408, 426, 428, 434, 436–439, 444–446, 450, 451, 462, 463, 466–470, 474, 476–483, 485
Sourdough for novel products, 93–122
Sourdough microflora, 274

Sourdough products biofunctionalities and health properties, 118–122
Sourdough starter, 2, 4, 5, 52, 61, 63, 67, 105, 114, 217, 261, 262, 270, 281–299, 309, 311, 345
Sourdough starter cultures, 52, 105, 270, 294, 298, 299
Sourdough technology, 52, 80, 284, 335, 362, 461, 470, 475

T
Technological purposes, 310
Traditional bakery goods, 27
Traditional bread, 1, 27–33, 37, 107, 310, 483

U
Undesirable microorganisms, 84, 169–202

V
Vitamins, 36, 50, 62, 119–121, 139, 152, 169, 224, 227, 230, 237, 239–242, 247, 267, 334, 343, 347, 358, 380, 384, 395, 403, 425, 426, 465, 479

W
Wheat sensitivity, 425

Y
Yeast identification, 157
Yeasts, 1, 21, 61, 95, 132, 170, 211, 223, 261, 282, 309, 343, 379, 396, 425, 461
Yeast strains, 4, 5, 121, 146, 147, 224, 247, 344–347, 351, 352, 354–357, 359–367, 470

MIX
Papier aus verantwortungsvollen Quellen
Paper from responsible sources
FSC® C105338

FSC
www.fsc.org

If you have any concerns about our products,
you can contact us on
ProductSafety@springernature.com

In case Publisher is established outside the EU,
the EU authorized representative is:
Springer Nature Customer Service Center GmbH
Europaplatz 3, 69115 Heidelberg, Germany

Printed by Libri Plureos GmbH
in Hamburg, Germany